Drug Delivery Using Nanomaterials

Emerging Materials and Technologies

Series Editor: *Boris I. Kharissov*

Bioengineering and Biomaterials in Ventricular Assist Devices
Eduardo Guy Perpétuo Bock

Semiconducting Black Phosphorus: From 2D Nanomaterial to Emerging 3D Architecture
Han Zhang, Nasir Mahmood Abbasi, and Bing Wang

Biomass for Bioenergy and Biomaterials
Nidhi Adlakha, Rakesh Bhatnagar, and Syed Shams Yazdani

Energy Storage and Conversion Devices: Supercapacitors, Batteries, and Hydroelectric Cell
Anurag Gaur, A.L. Sharma, and Anil Arya

Nanomaterials for Water Treatment and Remediation
Srabanti Ghosh, Aziz Habibi-Yangjeh, Swati Sharma, and Ashok Kumar Nadda

2D Materials for Surface Plasmon Resonance-Based Sensors
Sanjeev Kumar Raghuwanshi, Santosh Kumar, and Yadvendra Singh

Functional Nanomaterials for Regenerative Tissue Medicines
Mariappan Rajan

Uncertainty Quantification of Stochastic Defects in Materials
Liu Chu

Recycling of Plastics, Metals, and Their Composites
R.A. Ilyas, S.M. Sapuan, and Emin Bayraktar

Viral and Antiviral Nanomaterials: Synthesis, Properties, Characterization, and Application
Devarajan Thangadurai, Saher Islam, and Charles Oluwaseun Adetunji

Drug Delivery Using Nanomaterials
Yasser Shahzad, Syed A.A. Rizvi, Abid Mehmood Yousaf, and Talib Hussain

Nanomaterials for Environmental Applications
Mohamed Abou El-Fetouh Barakat and Rajeev Kumar

For more information about this series, please visit: https://www.routledge.com/Emerging-Materials-and-Technologies/book-series/CRCEMT

Drug Delivery Using Nanomaterials

Edited by
Yasser Shahzad, Syed A.A. Rizvi,
Abid Mehmood Yousaf, and Talib Hussain

CRC Press is an imprint of the
Taylor & Francis Group, an **informa** business

First edition published 2022
by CRC Press
6000 Broken Sound Parkway NW, Suite 300, Boca Raton, FL 33487-2742

and by CRC Press
4 Park Square, Milton Park, Abingdon, Oxon, OX14 4RN

© 2022 selection and editorial matter, Yasser Shahzad, Syed A.A. Rizvi, Abid Mehmood Yousaf and Talib Hussain; individual chapters, the contributors

CRC Press is an imprint of Taylor & Francis Group, LLC

This book contains information obtained from authentic and highly regarded sources. While all reasonable efforts have been made to publish reliable data and information, neither the author[s] nor the publisher can accept any legal responsibility or liability for any errors or omissions that may be made. The publishers wish to make clear that any views or opinions expressed in this book by individual editors, authors or contributors are personal to them and do not necessarily reflect the views/opinions of the publishers. The information or guidance contained in this book is intended for use by medical, scientific or health-care professionals and is provided strictly as a supplement to the medical or other professional's own judgement, their knowledge of the patient's medical history, relevant manufacturer's instructions and the appropriate best practice guidelines. Because of the rapid advances in medical science, any information or advice on dosages, procedures or diagnoses should be independently verified. The reader is strongly urged to consult the relevant national drug formulary and the drug companies' and device or material manufacturers' printed instructions, and their websites, before administering or utilizing any of the drugs, devices or materials mentioned in this book. This book does not indicate whether a particular treatment is appropriate or suitable for a particular individual. Ultimately it is the sole responsibility of the medical professional to make his or her own professional judgements, so as to advise and treat patients appropriately. The authors and publishers have also attempted to trace the copyright holders of all material reproduced in this publication and apologize to copyright holders if permission to publish in this form has not been obtained. If any copyright material has not been acknowledged please write and let us know so we may rectify in any future reprint.

Except as permitted under U.S. Copyright Law, no part of this book may be reprinted, reproduced, transmitted, or utilized in any form by any electronic, mechanical, or other means, now known or hereafter invented, including photocopying, microfilming, and recording, or in any information storage or retrieval system, without written permission from the publishers.

For permission to photocopy or use material electronically from this work, access www.copyright.com or contact the Copyright Clearance Center, Inc. (CCC), 222 Rosewood Drive, Danvers, MA 01923, 978-750-8400. For works that are not available on CCC please contact mpkbookspermissions@tandf.co.uk

Trademark notice: Product or corporate names may be trademarks or registered trademarks and are used only for identification and explanation without intent to infringe.

Library of Congress Cataloging-in-Publication Data
Names: Shahzad, Yasser, editor. | Rizvi, Syed A. A., editor. |
Yousaf, Abid Mehmood, editor. | Hussain, Talib (Assistant professor of pharmacy) editor.
Title: Drug delivery using nanomaterials / edited by Yasser Shahzad, Syed
A.A. Rizvi, Abid Mehmood Yousaf and Talib Hussain.
Description: First edition. | Boca Raton : CRC Press, 2022. | Includes bibliographical references and index. Identifiers: LCCN 2021038007 (print) | LCCN 2021038008 (ebook) | ISBN 9780367767938 (hardback) | ISBN 9780367767945 (paperback) | ISBN 9781003168584 (ebook)
Subjects: MESH: Drug Delivery Systems--methods | Nanostructures--therapeutic use | Drug Carriers--therapeutic use
Classification: LCC RS201.N35 (print) | LCC RS201.N35 (ebook) | NLM QV 785 | DDC 615.1/9--dc23
LC record available at https://lccn.loc.gov/2021038007
LC ebook record available at https://lccn.loc.gov/2021038008

ISBN: 978-0-367-76793-8 (hbk)
ISBN: 978-0-367-76794-5 (pbk)
ISBN: 978-1-003-16858-4 (ebk)

DOI: 10.1201/9781003168584

Typeset in Times
by MPS Limited, Dehradun

Contents

Preface vii
Editors ix
Contributors xi

Chapter 1 Introduction to Nanomaterials for Drug Delivery 1

Akhlesh Kumar Jain and Keerti Mishra

Chapter 2 Manufacturing Techniques for Nanoparticles in Drug Delivery 23

Daniel Real, María Lina Formica, Matías L. Picchio, and Alejandro J. Paredes

Chapter 3 Pharmacokinetics of Nanomaterials/Nanomedicines 49

Mulham Alfatama and Zalilawati Mat Rashid

Chapter 4 Bioinspired Nano-Formulations: Materials, Approaches, and Applications in Drug Delivery 85

Jahanzeb Mudassir and Muhammad Sohail Arshad

Chapter 5 3D-Printed Nanocrystals for Oral Administration of the Drugs ... 109

Lucía Lopez-Vidal, Daniel Andrés Real, Alejandro J. Paredes, Juan Pablo Real, and Santiago Daniel Palma

Chapter 6 Functional Nanomaterials 135

Imran Saleem, Yousef Rasmi, Leyla Fath-Bayati, and Zohreh Arabpour

Chapter 7 Trigger-Sensitive Nanoparticle for Drug Delivery 155

Hadiqa Nazish Raja, Basalat Imran, and Fakhar Ud Din

Chapter 8 Metal Organic Frameworks for Drug Delivery 181

Saima Zulfiqar, Shahzad Sharif, and Muhammad Yar

Chapter 9 Graphene and Graphene-Based Nanomaterials: Current Applications and Future Perspectives 209

Abid Hussain, Yuhua Weng, and Yuanyu Huang

Chapter 10 Nanomaterials for Lungs Targeting 229

Keerti Mishra and Akhlesh Kumar Jain

Chapter 11 Nanostructured Drug Carriers for Nose-to-Brain Drug Delivery: A Novel Approach for Neurological Disorders 257

Talita Nascimento da Silva, Emanuelle Vasconcellos de Lima, Anna Lecticia Martinez Martinez Toledo, Julia H. Clarke, and Thaís Nogueira Barradas

Chapter 12 Carbon Nanotubes in Cancer Therapy 287

Renu Sankar, V.K. Ameena Shirin, Chinnu Sabu, and K. Pramod

Chapter 13 Applications of Silica-Based Nanomaterials for Combinatorial Drug Delivery in Breast Cancer Treatment 311

Mubin Tarannum and Juan L. Vivero-Escoto

Chapter 14 Cationic Gemini Surfactants as Genes and Drug Carriers 329

Mays Al-Dulaymi, Anas El-Aneed, and Ildiko Badea

Chapter 15 Stability of Nanomaterials 355

Mulham Alfatama, Ahmed R. Gardouh, and Abd Almonem Doolaane

Chapter 16 Nanotoxicology and Regulatory Aspects of Nanomaterials and Nanomedicines 381

Nashwa Osman and Imran Saleem

Index 403

Preface

After the drug discovery and envelopment process, safe delivery of optimum dose through suitable formulations has been a constant challenge, especially when drugs are very toxic, have poor solubility, and undesirable clearance profile. With recent advancement in synthetic technologies, a variety of customized nanomaterials can be synthesized to mimic the bioenvironment and can be equipped with various targeting and imaging moieties beside being laden with drug or desired drug combinations. A substantial number of books have provided information representing the snapshot of the advancement at a specific period of time in this everchanging field. This book not only covers the advancements previously made in the field of nanomaterial-based drug delivery systems, but it also includes current advancements, thus covering all the aspects that needed consideration for a successful and marketable nano-formulation.

Drug Delivery Using Nanomaterials is a unique comprehensive book that encompasses diverse facets, such as fabrication, characterization, administration, 3D printing, trigger sensitivity, toxicity, stability, and many more, of nanomedicine under its canopy. The striking feature of this book is that it provides information in a bottom-up approach manner, i.e., basic background information serves as a framework for understanding the advanced concepts later followed, thus making it a user-friendly resource for readers at large. A number of world-renowned connoisseurs from assorted backgrounds, including academia, research, and industry, have contributed to this book. The introductory chapters furnish a general overview of the whole process involved in synthesis and characterization of the nanomaterials for pharmaceutical formulations. Every chapter sequentially adds on and builds from the previous one, all of which represent a progressive pathway from basic to more advanced ones. Each chapter imparts knowledge and information from numerous areas to furnish the fundamentals of novel applications and outcomes of nanomedicine, as well as a landmark for the future.

This book will serve the students, academicians, and industrialists as it contains introductory level information as well advanced level current practices in nano-pharmaceutical development. The text is delivered in a concise, direct, and soft language to stimulate interest and enliven your readership.

Yasser Shahzad
Syed A.A. Rizvi
Abid Mehmood Yousaf
Talib Hussain

Editors

Yasser Shahzad is an Assistant Professor at Department of Pharmacy, COMSATS University Islamabad (CUI), Lahore Campus. Dr. Yasser won a prestigious Nicholson Scholarship to pursue PhD from University of Huddersfield, UK. He also remained a Postdoctoral Fellow at Pharmacen, North-West University South Africa. He has a particular interest in advanced drug delivery research at the interface of pharmaceutics, chemistry, and cosmetics, and publishes regularly in internationally recognized peer-reviewed journals. He has published over 65 peer-reviewed articles in the form of journal articles, editorials, book chapters, and oral/poster presentations. He also serves as Editorial Board member of various international journals. Dr. Yasser holds active research collaborations with various international universities in the UK, the USA, and South Africa.

Syed A.A. Rizvi is a Professor of Pharmaceutical Sciences and a broadly trained scientist, educator, and healthcare contributor with vast clinical and translational research experience, driving innovation within pharmaceutical product development platforms. He is highly motivated and committed to advancing medical innovation and education, discovering new therapies, improving patient outcomes, and playing an instrumental role in scientific and business strategies. Dr. Rizvi has extensive training and background spanning over 24 years of experience in industrial, academic, administration, and healthcare settings. He has published over 200 peer-reviewed articles, including journal publications, book chapters, books, US patents, and oral/poster presentations. His publications include highly rated articles and cover articles. He also serves on the editorial boards/panels of different journals in various capacities and reviews manuscripts and books for over 150 well-known scientific journals and books. Dr. Rizvi enjoys teaching, cooking, and practicing martial arts (Hakko-Ryu Jujutsu and black belt in Okinawan Goju Ryu Karate).

His research interests incorporate clinical and translational research including clinical trials, graduate medical education (GME) pharmaceutical formulations (direct compressed tablets, topicals and liquids); ionic liquid formation and cocrystallization of the pharmaceutical compounds; nano and micro-encapsulation of the drugs for targeting and controlled release; multistep synthesis of small organic molecules; pharmaceutical analysis (forensics and method development); design and synthesis of novel monomeric and polymeric surfactants for separation of chiral and achiral compounds in micellar electrokinetic chromatography (MEKC); method development and validation for separations of achiral and chiral compounds utilizing MEKC-UV/MS and HPLC-UV-MS; application of cryogenic high resolution scanning electron microscopy (Cryo-HRSEM) and analytical ultracentrifuge for characterization of polymeric surfactants; chemical transformation of pharmacologically active compounds using microbial whole cultures; isolation and pharmacology of compounds from fungal metabolites and medicinal plants; fragrance and aroma chemistry.

Abid Mehmood Yousaf is an Assistant Professor of Pharmacy in the Department of Pharmacy, COMSATS University Islamabad, Lahore campus, Pakistan. He has also served as Assistant Professor in the Faculty of Pharmacy, University of Central Punjab, Lahore, Pakistan (2016–2018). He earned a PharmD at the University of the Punjab, Lahore, Pakistan in 2009. In 2011, he was enrolled in the MS-PhD program at College of Pharmacy, Hanyang University, South Korea. He performed research in the Lab of Physical and Industrial Pharmacy under the supervision of Prof. Dr. Han-Gon Choi, a distinguished professor of Pharmacy in South Korea. He earned a PhD in pharmacy in 2015. Dr. Abid has numerous publications in peer-reviewed SCI journals of pharmacy. He has also contributed a number of chapters to the other books. The focus of his research has been optimization, development, in vitro and in vivo characterization of various nanoparticulated drug delivery systems, such as solid dispersions, nanoencapsulations, nanospheres, SNEDDS, nanofibers, binary and ternary type polymeric nanocomposites, and so on. He has supervised the research of several master's students in pharmaceutics. Many of his students have published his supervised work in well-reputed journals. Dr. Abid is a motivated researcher and academician.

Talib Hussain is an Assistant Professor in the Department of Pharmacy, COMSATS University Islamabad (CUI), Lahore campus. Prior to join CUI, he served in Primary and Secondary Health Department Punjab as Deputy Drug Controller/Provincial Drug Inspector. He obtained pharmacy (2006) and MPhil (2009) degrees at Bahauddin Zakariya University Multan. He earned a PhD in pharmacy at the University of Huddersfield, UK in 2014. His primary research was based on formulation development of poorly soluble drugs through bespoke microwave techniques. Moreover, isothermal titration calorimetry was employed to investigate the drug–excipient compatibility. He has developed active collaborations and expertise in the field of drug delivery, such as solid dispersions, nanoparticles, microemulsions, liposomes, and hydrogels for oral as well as for topical drug delivery systems. He has published more than 40 articles in original research, reviews, and book chapters.

Contributors

Mays Al-Dulaymi
Avro Life Science
and
School of Pharmacy
University of Waterloo
Kitchener, Ontario, Canada

Mulham Alfatama
Faculty of Pharmacy
Universiti Sultan Zainal Abidin, Besut Campus
Besut, Malaysia

Zohreh Arabpour
Department of Tissue Engineering and Applied Cell Sciences
School of Advanced Technologies in Medicine
and
Iranian Tissue Bank and Research Center, Gen
Cell and Tissue Research Institute, Tehran University of Medical Sciences
Tehran, Iran

Muhammad Sohail Arshad
Faculty of Pharmacy
Bahauddin Zakariya University
Multan, Pakistan

Ildiko Badea
College of Pharmacy and Nutrition
University of Saskatchewan
Saskatoon, Saskatchewan, Canada

Thaís Nogueira Barradas
Faculdade de Farmácia
Universidade Federal de Juiz de Fora
Juiz de Fora, Brazil

Julia H. Clarke
Programa de Pós-graduação em Ciências Farmacêuticas
Universidade Federal do Rio de Janeiro
Rio de Janeiro, Brazil

Fakhar Ud Din
Department of Pharmacy
Nanomedicine Research Group
Quaid-i-Azam University
Islamabad, Pakistan

Abd Almonem Doolaane
Department of Pharmaceutical Technology
Kulliyyah of Pharmacy
International Islamic University Malaysia (IIUM)
Selangor, Malaysia

Anas El-Aneed
College of Pharmacy and Nutrition
University of Saskatchewan
Saskatoon, Saskatchewan, Canada

Leyla Fath-Bayati
Department of Tissue Engineering and Regenerative Medicine
School of Medicine
Qom University of Medical Sciences
Qom, Iran
and
Department of Tissue Engineering and Applied Cell Sciences
School of Advanced Technologies in Medicine
Tehran University of Medical Sciences (TUMS)
Tehran, Iran

María Lina Formica
Unidad de Investigación y Desarrollo en
  Tecnología Farmacéutica (UNITEFA)
CONICET
and
Departamento de Farmacia
Facultad de Ciencias Químicas
Universidad Nacional de Córdoba
Córdoba, Argentina

Ahmed R. Gardouh
Department of Pharmaceutics and
  Industrial Pharmacy
Faculty of Pharmacy
Suez Canal University
Ismailia, Egypt
and
Department of Pharmaceutical
  Sciences
Faculty of Pharmacy
Jadara University
Irbid, Jordan

Yuanyu Huang
School of Life Science
Advanced Research Institute of
  Multidisciplinary Science
Institute of Engineering Medicine
Key Laboratory of Molecular Medicine
  and Biotherapy
Beijing Institute of Technology
Beijing, China
and
School of Materials and the Environment
Beijing Institute of Technology
Zhuhai, China

Abid Hussain
School of Life Science
Advanced Research Institute of
  Multidisciplinary Science
Institute of Engineering Medicine
Key Laboratory of Molecular Medicine
  and Biotherapy
Beijing Institute of Technology
Beijing, China

Basalat Imran
Department of Pharmacy
Nanomedicine Research Group
Quaid-i-Azam University
Islamabad, Pakistan

Akhlesh Kumar Jain
School of Pharmaceutical Sciences
Guru Ghasidas Central University
Bilaspur, India

Lucía Lopez-Vidal
Unidad de Investigación y
  Desarrollo en Tecnología
  Farmacéutica (UNITEFA)
CONICET
and
Departamento de Farmacia
Facultad de Ciencias Químicas
Universidad Nacional de
  Córdoba
Córdoba, Argentina

Keerti Mishra
School of Pharmaceutical
  Sciences
Guru Ghasidas Central University
Bilaspur, India

Jahanzeb Mudassir
Faculty of Pharmacy
Bahauddin Zakariya University
Multan, Pakistan

Talita Nascimento da Silva
Programa de Pós-graduação em
  Ciências Farmacêuticas
Universidade Federal do
  Rio de Janeiro
Rio de Janeiro, Brazil

Contributors

Nashwa Osman
Pharmacy and Biomolecular
  Sciences
Liverpool John Moores University
Liverpool, UK
and
Faculty of Medicine
Sohag University
Sohag, Egypt

Santiago Daniel Palma
Unidad de Investigación y Desarrollo
  en Tecnología Farmacéutica
  (UNITEFA)
CONICET
and
Departamento de Farmacia
Facultad de Ciencias Químicas
Universidad Nacional de
  Córdoba
Córdoba, Argentina

Alejandro J. Paredes
School of Pharmacy
Queen's University Belfast
Medical Biology Centre
Belfast, UK

Matías L. Picchio
Departamento de Química
  Orgánica
Facultad de Ciencias Químicas
Universidad Nacional de Córdoba
Córdoba, Argentina

K. Pramod
College of Pharmaceutical Sciences
Government Medical College
Kozhikode, India

Hadiqa Nazish Raja
Department of Pharmacy
Nanomedicine Research Group
Quaid-i-Azam University
Islamabad, Pakistan

Zalilawati Mat Rashid
Faculty of Bioresources and Food Industry
Universiti Sultan Zainal Abidin, Besut
  Campus
Besut, Malaysia

Yousef Rasmi
Department of Biochemistry
Faculty of Medicine
and
Cellular and Molecular Research Center
Urmia University of Medical Sciences
Urmia, Iran

Daniel Andrés Real
Departamento de Química
  Farmacológica y Toxicológica
Facultad de Ciencias Químicas y
  Farmacéuticas
Universidad de Chile
and
Advanced Center for Chronic Diseases
  ACCDiS
Santiago, Chile

Juan Pablo Real
Unidad de Investigación y Desarrollo
  en Tecnología Farmacéutica
  (UNITEFA) CONICET
and
Departamento de Farmacia
Facultad de Ciencias Químicas
Universidad Nacional de Córdoba
Córdoba, Argentina

Chinnu Sabu
College of Pharmaceutical Sciences
Government Medical College
Kozhikode, India

Imran Saleem
School of Pharmacy and Biomolecular
  Sciences
Liverpool John Moores University
Liverpool, UK

Renu Sankar
College of Pharmaceutical Sciences
Government Medical College
Kozhikode, India

Shahzad Sharif
Materials Chemistry Laboratory
Department of Chemistry
Government College University
  Lahore (GCUL)
Lahore, Pakistan

V.K. Ameena Shirin
College of Pharmaceutical Sciences
Government Medical College
Kozhikode, India

Mubin Tarannum
Department of Chemistry
The University of North Carolina at
  Charlotte
Charlotte, North Carolina, USA

**Anna Lecticia Martinez Martinez
  Toledo**
Instituto de Macromoléculas
  Eloisa Mano
Universidade Federal do Rio de Janeiro
Rio de Janeiro, Brazil

Emanuelle Vasconcellos de Lima
Programa de Pós-graduação em
  Ciências Farmacêuticas
Universidade Federal do Rio de Janeiro
Rio de Janeiro, Brazil

Juan L. Vivero-Escoto
Department of Chemistry
The University of North Carolina at
  Charlotte
Charlotte, North Carolina,
USA

Yuhua Weng
School of Life Science
Advanced Research Institute of
  Multidisciplinary Science
Institute of Engineering Medicine
Key Laboratory of Molecular Medicine
  and Biotherapy
Beijing Institute of Technology
Beijing, China

Muhammad Yar
Interdisciplinary Research Center in
  Biomedical Materials
COMSATS University Islamabad,
  Lahore Campus
Lahore, Pakistan

Saima Zulfiqar
Materials Chemistry Laboratory
Department of Chemistry
Government College University
  Lahore (GCUL)
and
Interdisciplinary Research Center in
  Biomedical Materials
COMSATS University Islamabad,
  Lahore Campus
Lahore, Pakistan

1 Introduction to Nanomaterials for Drug Delivery

Akhlesh Kumar Jain and Keerti Mishra
Guru Ghasidas Central University

CONTENTS

1.1 Nanomaterials 1
1.2 Significance of Nanomaterials in Drug Delivery 2
1.3 Types of Nanomaterials 3
    1.3.1 Nanoparticles 3
    1.3.2 Solid Lipid Nanoparticles (SLNs) 3
    1.3.3 Gold Nanoparticles 5
    1.3.4 Liposomes 8
    1.3.5 Dendrimers 11
    1.3.6 Nanoshells and Nanopores 12
    1.3.7 Nanotubes 13
    1.3.8 Quantum Dots (QDs) 13
1.4 Application of Nanomaterials in Drug Delivery 13
    1.4.1 Nanomaterials for the Treatment of Cancer 13
    1.4.2 Nanomaterials for Gene Delivery 14
    1.4.3 Nanomaterials as Ocular Drug Carriers 14
    1.4.4 Nanomaterials for Drug Delivery to the Brain 15
    1.4.5 Nanomaterials-Loaded Contact Lenses 15
    1.4.6 Accumulation of Nanomaterials in the Retina 16
1.5 Conclusions 16
References 17

1.1 NANOMATERIALS

The term "nanomaterial" may be defined as material having the surface structure, exterior dimension, or internal structure in the nano size range (1–1000 nm) (Cooke and Atkins, 2016; Jong and Borm, 2008). "Nano", being a Latin word, means dwarf, small in size. Size reduction of any material into nano range can cause alteration in various properties of material, including an increased surface area, increased surface-area-to-volume ratio, increased solubility, enhanced interaction with biomolecules that leads to increased drug uptake across cell membrane, and other beneficiary transformation in material's magnetic, thermal, and electrical

DOI: 10.1201/9781003168584-1

property (Ochekpe et al., 2009). Thus, nanomaterials are frequently recognized to demonstrate diverse physical, chemical, or biological effects when observed against large-scale corresponding dosages that could be intently valuable for drug delivery carriers (Wilczewska et al., 2012). Nanotechnology is a branch associated with science that deals with research and innovation of materials or devices, generally, on the nanoscale of atoms and molecules. In brief, nanotechnology is management of material in atomic, molecular, and supramolecular scale in plentiful sectors offering development of novel technology, chiefly in the area of nanomedicine (Silva, 2004). Richard Feynman, in 1959, was the first person who familiarized the idea of nanotechnology by illustrating it as a significant field for scientific research in the future. At present, application of nanotechnology has been seen by various novel nanodevices in drug delivery. Significant attention towards the growth of nanotechnology for novel drug delivery has been observed (Feynman, 1960).

1.2 SIGNIFICANCE OF NANOMATERIALS IN DRUG DELIVERY

Nanomaterials have been a center of attraction for decades due to their excellent properties as a drug delivery system. Increase in the quantity of surface atoms or molecules in the nanomaterials to the entire quantity of atoms or molecules in material, leads to increase in surface area of nanomaterials, which further leads to better attachment of drug, probe, protein, or other therapeutic material with nanomaterial (Hadjipanayis et al., 2010). Genes or proteins could be delivered orally through nanomaterials. Because of the small size of nanoparticles, they are gathered in high quantities in small, unapproachable areas like inflamed tissues or cancerous cells by enhanced permeability and retention effect (EPR), occasionally initiating lymphatic tissue impairment (Jong, 2008). Also, for the treatment of intracellular infections, reticuloendothelial cell targeting can be effortlessly accomplished through passive targeting through facilitating drug molecules towards the macrophages of the liver and spleen (Sahoo et al., 2007). Nanomaterials must be safe, soluble, biocompatible, and bioavailable. Moreover, they must be free flowing, i.e., should not block blood vessels, with least toxicity and invasiveness, so nanomaterials could be used in targeting of a specific area in a desirable safe concentration (Webster et al., 2013). Also, nanomaterials must be able to protect drug molecules from early drug degradation either by enzymatic and hydrolytic reactions; along with that, nanoparticles must be able to evade the drug from presystemic metabolism. Nanomaterials coated with hydrophilic polymers have longer circulation time, significantly enhancing the efficacy of drugs with shorter half-life; drug monitoring can be accomplished simply by sustained-release formulation and DNA delivery (Chakroborty et al., 2013). Moreover, nanomaterials also chiefly contribute in enhancing dissolution, therapeutic onset of action, and dose reduction, and in minimizing drug-induced toxicity. Nanomaterials also avoid the untimely loss of drugs caused by rapid clearance and metabolism (Sahoo et al., 2007). Nanomaterials also offer inordinate bio-adhesion, by which an increase in retention time of drugs in the body is observed. Nanomaterials are a nano-sized submicron colloidal system including nanoparticles, nanoliposomes, nanocapsules, nanospheres, nanopores, nanoshells, nanotubes, nanocrystals, dendrimers, fullerence, and quantum dots. These nanomaterials have potentials to revolutionize the conventional drug delivery carriers.

Afterwards, nano-robotics, chip-based nanoparticles, magnetic nanoparticles decorated with particular antibodies, and hollow virus capsids have recently been used as drug formulation. Consequently, nanocarriers are the obvious choice for use as smart drug carriers and to formulate the reported drugs in a novel carrier, which could enhance the safety and efficacy profile and patient compliance as well, resulting in an economical treatment (Attia et al., 2019).

1.3 TYPES OF NANOMATERIALS

1.3.1 NANOPARTICLES

Nanoparticles are a submicron-sized colloidal system made up of polymers with particle size ranging from 1 to 1000 nm having encapsulated drug or evenly distributed throughout the matrix or sometimes surface associated or bonded (Lee and Kim, 2005). Further, nanocarriers react as a single entity in terms of their properties and also demonstrate few unique properties compared to other materials (Nagavarma et al., 2012). Nanoparticles are either prepared by or used for the transportation of diverse materials like gold, silver, silica, or some specific heavy metals, drug molecules, quantum dots, and other similar materials (Praetorius and Mandal, 2007). Nanoparticles provide noteworthy benefits by active site selection and overcome multiple drug resistance. In this regard, nanoparticles could be conjugated with surface ligand or antibody in order to make them appropriate for active targeting. Polymeric nanoparticles adhere on the cell surface and releases the drug through diffusion. Although entire polymeric nanoparticle could also enter into the cell by endocytosis. Interaction of nanoparticles with the receptors present in the cell surface initiates the formation of endosome. Later, lysosomal enzymes lyse the formed endosome that leads to release of nanoparticles into the cytoplasm. Polymeric nanoparticles are broadly explored as drug nanocarriers (Singh and Lillard, 2009). For the composition of nanoparticles, polymers that are easily hydrolysed and are biodegradable are mostly preferred, such as poly (D, L-lactide-co-glycolide) (PLGA), polylactide (PLA), or poly ethylene glycol (PEG) (Panyam and Labhasetwar, 2003). Additionally, natural polysaccharides including chitosan, cyclodextrin, and dextran also have been extensively investigated. (Table 1.1) (Chan et al., 2009). Method of preparation of nanoparticles includes basic techniques such as emulsion-solvent evaporation, double emulsification (Figure 1.1), emulsion-diffusion, nanoprecipitation or solvent displacement, precipitation-chemical cross-linking, coacervation or ionic gelation, salting out, dialysis, and spray drying (Abhilash, 2010; Tamizhrasi et al., 2009; Tiruwa, 2015; Yoo et al., 1999). Characterization of nanoparticles is based on their particle size, morphology, entrapment efficiency, drug loading, in vitro drug release, in vivo studies, toxicity, and stability studies.

1.3.2 SOLID LIPID NANOPARTICLES (SLNS)

SLNs in simple words could be defined as the lipid nanoparticles. These could by lipospheres, solid lipid nanospheres, or particles that remain solid at body temperature (37 °C) and have a diameter smaller than 1000 nm (Pardeike et al., 2009).

TABLE 1.1
Advantages of Different Polymers Used for the Preparation of Nanoparticles

Polymer Type	Advantages
Natural Polysaccharides: Chitosan, Chitin, Alginate, Cyclodextrin, Dextran, Starch, Carrageenan	Biodegradable, stable, hydrophilic, nontoxic, low cost, and ease of manipulation
Polypeptides/proteins: Gelatin, Albumin, Collagen, Casein, Fibrinogen	Degradable, biocompatible, low antigenicity, nontoxic, and easily available
Synthetic homopolymers: Poly (lactide), Poly (lactide-co-glycolide), Poly (epsilon-caprolactone), Poly (isobutylcyanoacrylate), Poly(isohexylcyanoacrylate), Poly (n-butylcyanoacrylate), Poly (acrylate), Poly (mathacrylate)	Biocompatible and biodegradable materials
Copolymers: Poly (lactide)- poly (ethylene glycol), Poly (lactide-co-glycolide)- poly (ethylene glycol), Poly (epsilon-caprolactone)- poly (ethylene glycol), Poly (hexadecylcyanoacrylate-co-poly (ethylene glycol) cyanoacrylate)	Biocompatible and biodegradable materials
Colloid stabilisers: Dextran, Pluronic F68, Poly (vinyl alcohol),	Biodegradable, hydrophilic, and nontoxic

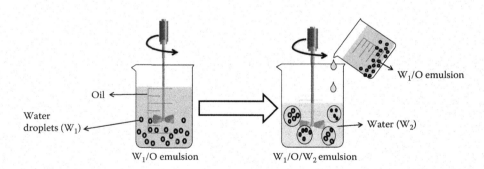

FIGURE 1.1 Preparation of nanoparticles through double emulsification method.

Lipid, emulsifier, and water or solvent are some chief compositions of solid lipid nanoparticles. Triglycerides, partial glycerides, steroids, fatty acids, and wax are some commonly used lipids in the preparation of SLNs. For stabilization of lipid dispersion of SLNs, innumerable emulsifiers and their compounds have been exploited. In order to avoid particle agglomeration more proficiently, mixture of two or more than two emulsifiers are preferred (Cavalli et al., 1993). A strong lipophilic matrix is formed in SLNs where drugs are loaded for subsequent discharge (Mukherjee et al., 2009; Pardeike et al., 2009). SLNs offer a variety of properties such as smaller size, larger surface area, higher loading of drug, and segment interaction in the interface, and are attractive for their capabilities to improve the effectiveness of therapeutic drugs (Cavalli et al., 1993). Lipid matrix in SLNs is made up of physiological lipids that reduce the risk of serious and chronic disease, which is the obvious advantage of SLN. Limited drug encapsulation and leakage while storage are some negative aspects of SLNs (Jenning et al., 2000), which can be overcome by certain strategies. SLNs are prepared by high shear homogenization, ultrasonication/high speed homogenization, solvent emulsification/evaporation (Figure 1.2), micro emulsion technique, spray drying, or double emulsion method (Mukherjee et al., 2009). Key parameters for which SLNs are characterized are size range and its distribution, surface charge, degree of crystallinity, polymorphism, simultaneous presence of associates subcolloidal structures such as micelles, liposome, or nanoparticles, release kinetics, and surface morphology (Mukherjee et al., 2009). A couple of SLN formulations reported by different researchers are illustrated in Table 1.2 (Duan et al., 2020).

1.3.3 GOLD NANOPARTICLES

Gold nanoparticles consist of a small gold particle with a diameter of 1 to 100 nm with or without surface decorated ligands (Figure 1.3) (Chen et al., 2014). It is very easy to synthesize gold nanoparticles of dissimilar morphology, such as into spheres, rods, cages, or stars, which enables the gold nanoparticles to attain anticipated characteristics like increased solubility in water, enhanced size dispersion, and other desired surface functionalities. Gold nanoparticles have excellent biocompatibility and are nontoxic. The size and shape of these nanoparticles are easy

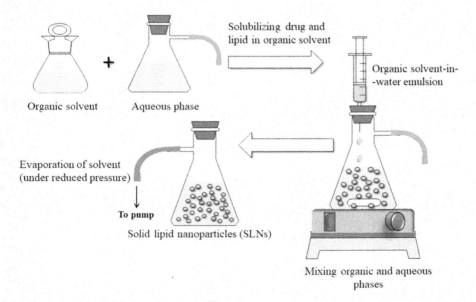

FIGURE 1.2 Method of preparation of SLNs by solvent evaporation technique.

TABLE 1.2
SLN Formulations Reported by Different Researchers

Drug	Lipid	Surfactant/Emulsifier	Co-Surfactant	Method for Preparation
Amphotericin B	Compritol® ATO 888, Precirol ATO 5 and stearic acid	Pluronic® F-68, Pluronic® F-127	—	Solvent diffusion method
	Compritol® ATO 888 (glycerylbehenate), glycerylpalmitostearate (Precirol® ATO 5), medium chain triglyceride	Tween 20, Pluronic® F-127, Cremophor RH40, polyoxyethylene (40) stearate (Myrj 52)		High-pressure homogenization
Baclofen	Stearic acid	Epikuron 200 (92% phosphatidylcholine)	Propionic acid, butyric acid, and sodium taurocholate	Multiple (w/o/w) warm, microemulsion
Buspirone HCl	Cetyl alcohol, Spermaceti	Pluronic® F-68, Tween 80	—	Emulsification-evaporation followed by ultrasonication
Camptothecin	Soybean lecithin, stearic acid	Pluronic® F-68, Tween 80	Glycerol, PEG 400, PPG	Hot HPH
Carvedilol	Stearic acid	Pluronic® F-68	Sodium taurocholate and ethanol	Microemulsion
Clozapine	Trimyristin, tripalmitin, tristearin, soy phosphatidylcholine	Pluronic® F-68	—	Ultrasonication method
Curcumin	Compritol 888 ATO	Soy lecithin, Tween 80		Microemulsion
	Tristearin	Polyoxyethylene (10) stearyl ether (Brij®S10), polyoxyethylene (100) stearyl ether (Brij® S100)		Oil-in-water emulsion technique
Cyclosporine A	Imwitor® 900	Tagat®S, sodium cholate	—	HPH, hot HPH
Diazepam	Compritol 888 ATO, Imwitor® 900	Pluronic® F-68, Tween 80	—	Ultrasound techniques modified high-shear homogenization

(*Continued*)

TABLE 1.2 (Continued)
SLN Formulations Reported by Different Researchers

Drug	Lipid	Surfactant/Emulsifier	Co-Surfactant	Method for Preparation
Doxorubicin hydrochloride	Glycerylcaprate	Polyethylene glycol 660 hydroxystearate (Solutol®HS15)	–	Ultrasonic homogenization
Ibuprofen	Trilaurin, tripalmitin, stearic acid	Pluronic®F127, sodium taurocholate	–	Solvent-free high-pressure homogenization (HPH)
Ketoprofen	Beeswax and carnauba wax	Tween 80, egg lecithin	–	Microemulsion technique
Lopinavir	Compritol 888 ATO (glycerylbehenate)	Pluronic®F127	–	Hot homogenization, ultrasonication
Lovastatin	Triglyceride and phosphatidylcholine 95%	Pluronic®F68	–	Hot homogenization, ultrasonication
Methotrexate	Stearic acid, monostearin, tristearin, and Compritol 888 ATO	l-α-Soya lecithin, and Sephadex G-50	–	Solvent diffusion method
Nevirapine	Steric acid, Compritol 888 ATO	Dimethyldioctadecyl ammonium bromide (DODAB), Tween 80, Lecithin	1-Butanol	Microemulsion
Nitrendipine	Triglyceride and phosphatidylcholine	Pluronic®F68	–	Hot homogenization, ultrasonication method
Praziquantel	Hydrogenated castor oil	Poly vinyl alcohol (PVA)	–	Hot homogenization, ultrasonication
Quercetin	Glyceryl monostearate, soy lecithin	Tween-80 and PEG 400	–	Emulsification-solidification
Rifampicin	Stearic acid	PVA	–	Emulsion-solvent diffusion
Tobramycin	Stearic acid	Epikuron 200	Sodium tauro-cholate	Microemulsion
Vinpocetine	Glyceryl monostearate, soy lecithin, polyoxyethylene hydrogenated castor oil	Tween 80	–	Ultrasonic-solvent emulsification

Source: Reproduced under Creative Commons License 4.0 from (Duan et al., 2020).

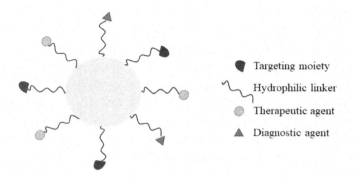

FIGURE 1.3 Illustration of a gold nanoparticle attached with surface ligands.

to modify; thus, it becomes easy to manage to their conjugating and optical properties, due to which they have received special attention. Further, it is well known that gold is oxidation resistant under physiological conditions that allow an unrestricted interaction of gold with the biological environment. In the gene therapy and imaging, DNA or drug delivery is achieved through gold NPs (Huang et al., 2007; Peng et al., 2009; Santos et al., 2018; Wei et al., 2007).

For the fabrication of gold nanoparticles, numerous approaches have been discovered. One of the approaches discovered by Turkevich was based on reduction of gold salts by reducing agent, which later starts nucleation of the gold ions leading to formation of nanoparticles. A stabilizing agent may be incorporated to avoid aggregation in the synthesis process (Turkevitch et al., 1951). Nanotechnology has undergone so much advancement that it offers several ways to synthesize stable and uniformly dispersed nanoparticles with size smaller than 6 nm. The Brust-Schiffrin scheme is a great advanced technique, by which huge amounts of higher quality monolayer-protected gold clusters can be synthesized (Brust et al., 1994). In the Brust-Schiffrin scheme, a phase-transfer agent (tetra octyl ammonium bromide) is used for relocating ions of gold from an aqueous solution of gold salt to toluene. Further, nonpolar organic thiol, usually a thiol-terminated long-chain alkane is incorporated, following the rapid addition of a reducing agent. The thiol-mediated covering of the gold cluster finally breaks the accumulation of the reduced gold atoms. Thiol ligand forms a stable organic monolayer, because of the strong gold-sulfur bonding. Size of the gold clusters can be modified effortlessly by altering the ratio of gold and ligand (Brust et al., 1994).

1.3.4 Liposomes

Liposomes, discovered by Alec Bangham in 1960, may be defined as the spherical bilayer vesicle generally in the size range between 25 nm and 2.5 μm, consisting of phospholipids and steroids (Landesman-Milo et al., 2013). Size range of the lipid vesicle in liposomes plays an essential role in determination of the distribution and in vivo retention time. On the basis of size and number of existing bilayers, liposomes are divided into two broad categories: multilamellar and unilamellar vesicles. Multilamellar vesicles, as name suggests, are more than one lipid bilayer. Unilamellar vesicles are single-layered lipid vesicles, further categorized as large and small unilamellar vesicles (Amarnath and Sharma, 1997). Liposomes have been an engrained formulation approach for the improvement in overall performance of drug delivery. Liposomes are considered to be the best vehicle for drug delivery because the structure of the membrane is similar to the cell membrane and because it facilitates the installation of drugs in them. Being a drug delivery carrier, liposome demonstrates a bunch of benefits (Table 1.3) (Yingchoncharoen et al., 2016). It has been advocated that liposomes excels in stabilizing therapeutic chemicals, improving their natural distribution and can be used with hydrophilic and hydrophobic drugs as well (Chapoy-Villanueva et al., 2015). For treatment of cancer, liposomes have been used through active and passive targeting (Figure 1.4). Several methods have been used for preparation of liposomes, such as mechanical dispersion method (sonication, French pressure cell: extrusion, freeze-thawed liposomes, thin film hydration method, microemulsification, membrane extrusion, and dried reconstituted vesicles method), solvent dispersion method (ether injection, ethanol injection, reverse phase evaporation method), detergent removal method (dialysis, detergent removal of mixed micelles, gel permeation chromatography), and dilution method (Akbarzadeh et al., 2013). Further, liposomes are evaluated for their average size and size distribution, shape, surface charge, encapsulation efficiency, drug loading, chemical composition, lamellarity, stability, etc. Liposomes, due to their wide application, have been used for the delivery of several drugs. Liposomes show precise potential for the intracellular of ribosomes, antisense molecules, DNA, or proteins/peptides. A specific affected site could be

TABLE 1.3
Advantages and Disadvantages of Liposomes

Advantages	Disadvantages
Liposomes increase efficacy and the therapeutic index of the drug.	Low solubility
Liposome increase stability via encapsulation.	Short half-life
Liposomes are nontoxic, flexible, biocompatible, completely biodegradable, and nonimmunogenic for systemic and nonsystemic administrations.	Sometimes phospholipid undergoes oxidation and hydrolysis-like reaction
Liposomes reduce the toxicity of the encapsulated agent.	Leakage and fusion of encapsulated drug/molecules
Liposomes help reduce the exposure of sensitive tissues to toxic drugs.	Production cost is high
Liposomes cause the site avoidance effect.	Fewer stables
Liposomes have the flexibility to couple with site-specific ligands to achieve active targeting.	–
Liposomes sustain the release of drugs, and prolong drug action time. They also improve the drug treatment index.	

Source: Reproduced under Creative Common License 2.0 from (Akbarzadeh et al., 2013).

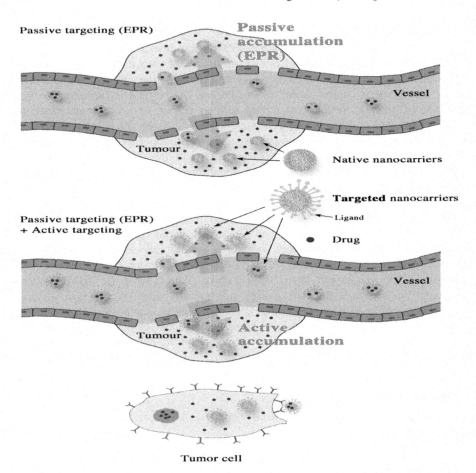

FIGURE 1.4 Active and passive targeting of liposome nanocarriers for the improvement in the liposomal efficiency to tumor sites. Reproduced under Creative Commons License 4.0 from (Attia et al., 2019).

treated by targeting specific cells using liposomes and result in improved effectiveness and reduced toxicity compared to naked drug/conventional drug carriers.

Liposomes in the nano-size range are termed nanoliposomes, which are capable of encapsulating drugs with small molecular weight, proteins, peptides, and imaging agents as well. In brief, these are the nano-sized bilayer phospholipid vesicles, generally unilamellar with an interior aqueous core (Patel et al., 2011; Yousefi et al., 2009). Nanoliposomes offer sustained release kinetics of cargo, hence capable of prolonged retention of drugs at the desired site. Water-soluble drugs are frequently incorporated in the aqueous core, and lipophilic drugs are loaded in the phospholipid layer (Patel et al., 2009). Unlike conventional vesicles, nanoliposome avoids faster recognition and elimination clearance via macrophages. Nanoliposome plays a significant role as a smart carrier for passive or active delivery of drugs (Kumar et al., 2010). Because of permeable vascular anatomy of the tumor structure, nanoliposomes are primarily deposited into the tumor tissue by means of passive targeting and release drugs locally for sustained period. Surface-decorated nanoliposome with antibody or specific ligands directly reach the predetermined site via active targeting without producing any deleterious effect on normal cells. Targeted nanoliposomal drug delivery is far more effectual than nontargeted drug delivery systems. For the treatment of blood cancer, C6-ceremide ligand-induced nanoliposome is used, which straight-forward picked up the highly expressed leukemic cells and diminished the over-expression of survival on the cell surface (Kumar et al., 2012). The long circulatory form of nanoliposomes is made up of surface coating of polyethylene glycol (PEG) that modify the absorption, distribution, and elimination of therapeutics in terms of greater half-life (Dadashzadeh et al., 2008).

1.3.5 Dendrimers

Dendrimers may be defined as the nano-sized three-dimensional structure composed of tree-like branches with diameter ranging from 2.5 to 10 nm (Figure 1.5). Dendrimers may be formed through synthetic as well as from natural monomers, such as monosaccharides, nucleotides, or amino acids. Drug molecules may be encapsulated, complexed, or conjugated with dendrimers, having shown therapeutic applications. Generally, polyamidoamines and polypropyleneimines are two derivates widely used for the synthesis of dendrimers (Santos et al., 2020; Souto et al. 2019). A central core having symmetrically developed branches around it, when enlarged enough, usually takes up a spheroidal three-dimensional structure in dispersion. Two similar chemical functional groups are added repeatedly on a central core; however, dissimilar chemical functional groups can be added after one junction of branching with similar functional groups (Albertazzi et al., 2013).

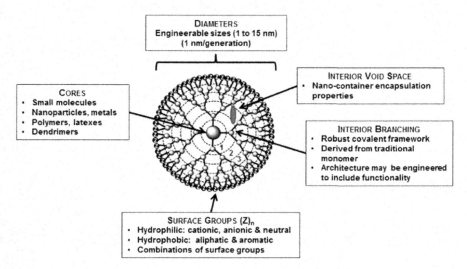

FIGURE 1.5 Basic structure of a dendrimer. Reproduced under Creative Commons License 4.0 from (Santos et al., 2020).

Due to their outstanding biological characteristics, dendritic structures have widespread pharmaceutical and biomedical application; however, the presence of positive surface charge limits its clinical uses. Altogether, morphologically dendrimers with almost spherical to compressed spheroids or ameba-like structures, particularly with electric charges, provide a shape similar to starfish. As per synthesis technique of dendrimers, their number and pattern of branches varies. Small molecules are kept in the interior cavity of dendrimer and are momentarily restrained there either with hydrogen, hydrophobic forces, or covalent bonds (Kannan et al., 2012). Dendrimers are widely used for transferring genetic material for the treatment of cancer or other viral diseases because of their excellent properties like high density of functional groups, multi-valency, mono-dispersity, and well-defined shape. Polyamidoamines (PAMAM) dendrimers are extensively used for gene delivery. Some other categories of dendrimers widely used are polyethyleneimine (PEI) dendrimers, glycodendrimers, peptide dendrimers, polypropilimine dendrimers, etc. (Santos et al., 2020). Furthermore, dendrimers are characterized for their structure, optical activity, molecular dynamics, purity, surface engineering, surface structure, size and shape, molecular mass and weight, chemical composition, and internal structure (Madaan et al., 2014).

1.3.6 Nanoshells and Nanopores

Nanoshells may be defined as a type of nanoparticles containing a dielectric core, such as silica coated with a thin metallic coating such as with gold (Fuchigami et al., 2012). Nanoshell includes a spherical core enclosed in a shell/thin coating of metallic layer of thickness near about 1 to 20 nm. Materials used for the preparation of nanoshell can be manufactured from metals, semiconductors, or insulators. Type of material used and core-to-shell ratio is very essential for determining overall property of nanoshells (Viswanathan et al., 2017). Nanoshells can be used as a drug carrier for both the purpose of imaging and therapy (Kherlopian et al., 2008). Tissue targeting by nanoshells can be effortlessly achieved by using appropriate immunological methods. Moreover, in hydrogel polymer nanoshells can be smoothly implanted (Lowery et al., 2006). Since it has been recognized that nanoshells have photothermal therapeutic capability to destroy bacteria and tumor cells, they are stimulated with infrared radiation so that they can absorb energy of the infrared radiation and accordingly generate heat. Through this method, risk of trauma to the adjacent tissues is minimized (Pattani and Tunnell, 2012). For enhanced treatment efficacy, nanoshells are incorporated with various antitumor drugs for their efficient delivery towards cancerous cells. For better binding between target and nanoshells, they are sometimes functionalized with active targeting ligands, such as antibodies, aptamers, and peptides (Viswanathan et al., 2017). Nanoshells are presently being explored for the treatment of diabetes and cancer by inhibition of tumor micrometastasis. Apart from this, nanoshells could be also used significantly as a diagnostic tool in the whole blood immunoassays (Hirsch et al., 2003).

Nanopores are structures having porous wafers; these pores are usually 20 nm in diameter and allow oxygen, glucose, and other small molecules to permeate. Nanopore devices can be utilized for the protection of relocated tissues against the

Nanomaterials for Drug Delivery

host immune system and also allow tissues to explore the advantage of transplantation (Freitas, 2005). Nanopore devices containing β-cells of the pancreas could be transplanted in the recipient's body who does not have enough β-cells. Another application of nanopores is in DNA sequencing. Also, for the differentiation of purines from pyrimidines, nanopores are being developed (Yadav et al., 2011).

1.3.7 NANOTUBES

Nanotubes are simple nano tube-like structures. According to the material used, these nanotubes can be subjected as carbon nanotube, DNA nanotube, inorganic nanotube, membrane nanotube, etc. (Reilly, 2007). Integration of ammonium or carboxylic groups in the structure of carbon nanotubes can increase their solubility. Nanotubes solve the problem of transportation of peptides, proteins, nucleic acid, or other drug molecules. By increasing the surface-to-volume ratio, efficient transportation of active drug components can be achieved. Nanotubes after oral administration are absorbed and are transported to the effect-relevant sites by blood or lymphatic circulation. While orally administered, physically shortened nanotubes are sometimes absorbed through the columnar cells of intestinal mucous membrane (Prato et al., 2008; Kumar et al., 2011). In gene therapy, DNA can be either joined or connected to the tip of nanotubes or can be integrated within the tubes; thus, the capability of nanotubes to transport DNA through cell membranes is very important (Mc-Devitt et al., 2007).

1.3.8 QUANTUM DOTS (QDs)

Quantum dots (QDs) may be defined as the minute semiconductor of nanocrystals with size less than 10 nm. On external stimulation of light, QDs could be designed to shine in various colors. By carefully commanding the biomolecule conjugation process, QDs may be exploited for targeting of various biomarkers. QDs can also be conjugated to different biomolecules, such as biotin, transferrin, immunoglobulins, nucleic acids, or peptides. By tagging QDs with biomolecules, they can be utilized as extremely delicate probes. QDs can also be used for imaging purposes, likewise, in case of tumor staging and therapy planning. They are used for imaging of the sentinel node in patients suffering from cancer. QDs successfully detect breast cancer at the initial stage, making treatment easy. They may provide new intuitions for better explaining the aspects of cancer, real-time imaging, and screening of tumors (Gangrade, 2011; Gao and Dave, 2007; Pisanic et al., 2014).

1.4 APPLICATION OF NANOMATERIALS IN DRUG DELIVERY

1.4.1 NANOMATERIALS FOR THE TREATMENT OF CANCER

Cancer-causing cells are more susceptible to anticancer drugs compared to normal cells, but unfortunately sometimes anticancer drugs cause damage to normal cells instead of cancerous cells. Thus, research is focused on efforts to inhibit the growth of cancerous cells rather than normal cells through specific tissue targeting with optimum dose and frequency (Surendiran et al., 2009). The penetrating capability of nanomaterials for therapeutic and diagnostic agents in cytotoxic cells is higher compared to conventional therapy (Malam et al., 2009). For taking maximum benefit from the morphology and growth process of cancerous cells, like rapid proliferation, leaky tumor vasculature, and antigen expression, nanomaterials are designed (Byrne et al., 2008). Nanomaterials offer a bunch of merits over conventional therapy like preventing drug degradation, preventing drug interaction with the biological environment, improvement in the intracellular penetration of therapeutic agents, increasing the drug absorption in a particular tissue, and in regulating the pharmacokinetic-based drug tissue profile as per requirement (Ruoslahti et al., 2010).

1.4.2 Nanomaterials for Gene Delivery

Many years ago, it was very difficult to treat genetic disorders such as diabetes mellitus, alpha-1 antitrypsin deficiency, or cystic fibrosis as these disorders were induced due to the absence of enzymatic factors, which was initially either because of absence of genes or defect in genes. In the treatment of genetic disorders, nanomaterials have emerged as a great approach for the transfer of genetic materials (Davis and Cooper, 2007).

Until the recent era, gene therapy was the only treatment considered for genetic disorders, but now, gene therapy has been used as a tool for the treatment of diseases including cancer, nervous or heart diseases, and other than genetic disorders (Alex and Sharma, 2013). PEGylated nanoliposomes with galactose are quickly taken up in liver Kupffer cells and target liver cells efficiently. These kinds of nanoliposomes are useful for the treatment of liver disorders, such as hemochromatosis or Wilson's disorder. Thus, cationic nanoliposomes are well-thought-out for the possible nonviral human gene delivery system (Pathak et al., 2008). One more approach for the administration of nanoliposomes is by the use of complex of ligand and receptor with the EGF-EGFR (epidermal growth factor receptor) system for site-specific targeting, as EGFs are small proteins having binding affinity towards EGFR. Instead of encapsulating negatively charged genetic material in the nanoliposomes, plasmid DNA is incorporated in cationic lipids, which is itself another great approach, leading to the establishment of lipoplexes where the progression is determined by the electrostatic interactions between plasmid DNA and cationic lipids (Bunuales et al., 2011). Plasmid-incorporated lipoplexes directly reach to the diseased cells by infusing either endosomal membrane or plasma. In nanoliposomes, unloading of the gene depends on the nature and type of lipid composed, which regulate the release pattern. Neutral lipids such as 1,2-dioleoyl-sn-glycero-3-phosphoethanolamine (DOPE) assist nanoliposomes in their fusion with the endosomal membrane or plasma by recognizing and weakening the phospholipids in a tumbler manner, causing release of nucleic acid into the cytoplasm (Alex and Sharma, 2013).

1.4.3 Nanomaterials as Ocular Drug Carriers

Nanomaterials loaded with drugs increase the duration of residence as well as drug penetration capability in deep layers of ocular structure. In addition, nanomaterials also reduce the loss of the drug through the precorneal area and toxicity (Kesavan et al., 2011). Nanomaterials can be easily used for targeting various targeted eye areas such as the cornea and the retina including the choroid either by superficial or intravitreal injection. As the retina bears no lymph system, nanomaterials could be used efficiently for delivering drugs in the retina, as the retinal and choroidal neovascularization have comparable situations for solid tumors; apart from this, drugs may be delivered to the targeted ocular system by EPR effect (Yasukawa, 2004). Thus, nanomaterials can be successfully used to deliver ocular drugs to treat various ocular diseases including diabetic retinopathy, corneal diseases, or glaucoma. Nanomaterials such as nanosuspensions, SLNs, or nanoliposomes demonstrate superior results in cases of ocular therapeutic efficacy, as these are very competent in crossing the membrane blood retinal barrier in the eye (Behar-Cohen, 2004).

1.4.4 NANOMATERIALS FOR DRUG DELIVERY TO THE BRAIN

Presence of complex structures such as endothelial cells, astroglia, astrocytes, pericytes, or perivascular mast cells in the brain makes the structure of the blood brain barrier (BBB) very complex and difficult to permeate. These cells collectively prevent the passageway of cells and drug molecules circulating towards the brain (Petty and Lo, 2002). Organization of these vascular layers of capillary endothelial cells present in the brain in such a way that they are very tightly interconnected with each other; hence, it is difficult to cross by normal drugs. Out of various available delivery systems, nano-sized (1–100 nm) drug carriers work efficiently as an entire unit in order for transportation to cross the BBB (Rabanel et al., 2012). Along with targeting of drugs in the brain, nanomaterials also increase the permeability of drugs through the BBB. It completely depends on the physicochemical and biomimetic properties of the drug, whether it will cross the BBB or not, and not its chemical structure (Youns et al., 2011). However, if certain nanomaterials are unable to cross the BBB, they may be transformed into stealth nanomaterials, so that the reticuloendothelial system can be avoided, which will increase the duration of the drug in circulation and the drug will remain in the blood for a prolonged period of time and conjugated with functional groups, which could translocate through the BBB (Gabathuler, 2010).

1.4.5 NANOMATERIALS-LOADED CONTACT LENSES

For topical delivery of drugs in the eye, contact lenses containing nanoparticles loaded with ophthalmic drugs are preferred nowadays, which slowly diffuse out from the lens matrix and offer constant drug release. Contact lenses soaked in sterile isotonic solution release drugs for a limited time duration (Gulsen and Chauhan, 2004). Duration for drug release can be easily increased significantly by entrapping the drug prior to nanomaterials, before dispersing in the contact lens. It will also prevent the interaction between the drug and polymerization mixture, leading to additional resistance to drug release, since initially the drug should diffuse out from the nanomaterial and afterwards should penetrate in the surface of nanomaterials to reach out at the contact lens matrix (Gulsen and Chauhan, 2005).

1.4.6 ACCUMULATION OF NANOMATERIALS IN THE RETINA

Accumulation of nanomaterials in the eye delivers better perceptions in respect to various parameters such as drug bioavailability, duration of action, and cellular uptake and drug toxicity. Accumulation of drugs in the retina and drainage of drugs from the ocular tissues depends on a number of factors like size of particles present in nanomaterials, composition of nanomaterials, surface charge, or mode of administration of nanomaterials (Amrite et al., 2008). It was observed that particles larger than 2 μm were not able to permeate through vitreous cavity, and were drained out within 6 days, whereas particles with size equivalent to or lesser than 200 nm were consistently dispersed in the cavity of vitreous humor and internal limiting membrane. Moreover, particles with size equivalent to 50 nm were able to penetrate the retinal barrier and were present there after 2 months of administration through injection (Kim et al., 2009). Also, the surface chemistry of nanomaterials has a very significant role in its distribution; nanomaterials with positive charge easily adhere to the anionic components of the vitreous cavity and accumulate there (Koo et al., 2012), whereas negatively charged nanomaterials diffuse out through the vitreous cavity and penetrate the retina to be occupied by muller cells (Peeters et al., 2015). As vigorous interactions occur between the nonviral vectors, which are positively charged, and vitreous humor having negative charge, vitreous humor acts as a rate-limiting barrier for ocular drug delivery of nonviral vectors (Koo et al., 2012).

Cationic polyethyleneimine (PEI) nanoparticles deposited within the core of vitreous humor do not reach to the retina due to the existence of various vitreal barrier; however, positively charged chitosan glycosylated nanoparticles and chitosan glycosylated-cationic polyethyleneimine nanoparticles, easily penetrate the vitreal barrier and reach to the inner rate-limiting membrane due to the presence of glycol groups (Kim et al., 2009).

1.5 CONCLUSIONS

Limitations associated with conventional drug delivery vehicles lead to investigation of smart alternative carriers for drug delivery. In this regard, nanocarriers have emerged as an outstanding choice. Due to the exponential increase in surface area at the nanoscale, nanocarriers have the capability to impact physiological interactions from the molecular level to the systemic level and have targeted in vivo potential. The scope of nanomaterials has increased many folds in the past two decades as nanomaterials are presently prepared by various organic and inorganic material with an unexpected regulation in their size, shape, surface charge, encapsulation efficiency, drug loading, and drug release. But unfortunately, the clinical transformation of nanomaterials is relatively slower, and only limited products like liposomes or micelles are commercially available. Regulatory guidelines for assuring the robustness of preparation techniques and characterization of nanocarriers for safety assurance are also essential. At initial steps, nanocarriers are characterized with respect to surface charge and ligand density; afterwards they are correlated with their behaviors in respective cell models. Nanocarriers in blood or biological fluids are covered with protein aura, which in the end indicates the in vivo fates and therapeutic response of the drug. For recognition of discrepancy among in vitro and in vivo behaviors of nanocarriers, various research groups have shifted towards models that involve early in vivo proof of concept studies. Furthermore, clinical extrapolative values of some animal models are freshly revisited, in terms of their significance to human diseases and the capability to summarize disease development. In a nutshell, it is vital for the researchers to understand the limitations and challenges of present methodologies and explore a novel pathway for nanocarriers, which can forecast the clinical findings at the initial stage of product development with superior dependability.

REFERENCES

Abhilash, M. "Potential Applications of Nanoparticles." *Indian Journal of Pharmaceutical and Biological Research* 1, no. 1(2010): 1–12.

Akbarzadeh, A., Rezaei-Sadabady, R., Davaran, S., Joo, S.W., Zarghami, N., Hanifehpour, Y., Samiei, M., Kouhi, M., and Nejati-Koshki, K. "Liposome: Classification, Preparation, and Applications." *Nanoscale Research Letters* 8, no. 1(2013): 102.

Albertazzi, L., Gherardini, L., Brondi, M., Sulis, S.S., Bifone, A., Pizzorusso, T., et al. "In vivo Distribution and Toxicity of PAMAM Dendrimers in the Central Nervous System Depend on Their Surface Chemistry." *Molecular Pharmacology* 10, no. 1(2013): 249–260.

Alex, S.M., and Sharma, C.P. "Nanomedicine for Gene Therapy." *Drug Delivery and Translational Research* 3 (2013): 437–445.

Amarnath, S., and Sharma, U.S. "Liposomes in Drug Delivery: Progress and Limitations." *International Journal of Pharmaceutics* 154 (1997): 123–140.

Amrite, A.C., Edelhauser, H.F., Singh, S.R., and Kompella, U.B. "Effect of Circulation on the and Ocular Tissue Distribution of 20 nm Nanoparticles After Periocular Administration." *Molecular Vision* 14 (2008): 150–160.

Attia, M.F., Anton, N., Wallyn, J., Omran, Z., and Vandamme, T.F. "An Overview of Active and Passive Targeting Strategies to Improve the Nanocarriers Efficiency to Tumour Sites." *Journal of Pharmacy and Pharmacology* 71, no. 8(2019): 11585–1198.

Behar-Cohen, F. "Drug Delivery to Target the Posterior Segment of the Eye." *Medical Sciences* 20, no. 6-7(2004): 701–706.

Brust, M., Walker, M., Bethell, D., Schiffrin, D.J., and Whyman, R.J. "Synthesis of Thiol-derivatised Gold Nanoparticles in a Two-phase Liquid-Liquid System." *Journal of the Chemical Society, Chemical Communications* (1994): 801–802.

Bunuales, M., Duzgunes, N., Zalba, S., Garrido, M.J., and Ilarduya, C.T. "Efficient Gene Delivery By EGF Lipoplexes *In Vitro* And *In Vivo.*" *Nanomedicine* 6, no. 1(2011): 89–98.

Byrne, J.D., Betancourt, T., and Brannon-Peppas, L. "Active Targeting Schemes for Nanoparticle Systems in Cancer Therapeutics." *Advanced Drug Delivery Reviews* 60 (2008): 1615–1626.

Cavalli, R., Caputo, O., and Gasco, M.R. "Solid Lipospheres of Doxorubicin and Idarubicin." *International Journal of Pharmaceutics* 89 (1993): R9–R12.

Chakroborty, G., Seth, N., and Sharma, V. "Nanoparticles and Nanotechnology: Clinical, Toxicological, Social, Regulatory and Other Aspects of Nanotechnology." *Journal of Drug Delivery and Therapeutics* 3, no. 4(2013): 138–141.

Chan, J.M., Zhang, L., Yuet, K.P., et al. "PLGA-Lecithin-PEG Core-shell Nanoparticles for Controlled Drug Delivery." *Biomaterials* 30, no. 8(2009): 1627–1634.

Chapoy-Villanueva, H., Martinez-Carlin, I., Lopez-Berestein, G., and Chavez-Reyes, A. "Therapeutic Silencing of HPV 16 E7 by Systemic Administration of siRNA-Neutral DOPC Nanoliposome in a Murine Cervical Cancer Model with Obesity." *Journal of Balkan Union of Oncology* 20 (2015): 1471–1479.

Chen, X., Li, Q.W., and Wang, X.M. "Gold Nanostructures for Bioimaging, Drug Delivery and Therapeutics." *Precious Metals for Biomedical Applications* (2014): 163–176.

Cooke, J.P., and Atkins, J. "Nanotherapeutic Solutions for Cardiovascular Disease." *Methodist Debakey Cardiovascular Journal* 12 (2016): 132–133.

Dadashzadeh, S., Vali, A.M., and Rezaie, M. "The Effect of Peg Coating on *In Vitro* Cytotoxicity and *In Vivo* Disposition of Topotecan Loaded Liposomes in Rats." *Pharmaceutical Nanotechnology* 353 (2008): 251–259.

Davis, P.B., and Cooper, M.J. "Vectors for Airway Gene Delivery." *AAPS Journal* 9 (2007): E11–E17.

Duan, Y., Dhar, A., Patel, C., Khimani, M., Neogi, S., Sharma, P., et al. "A Brief Review on Solid Lipid Nanoparticles: Part and Parcel of Contemporary Drug Delivery Systems." *RSC Advances* 10 (2020): 26777.

Feynman, R. *Engineering & Science Magazine.* USA: California Institute of Technology, 1960.

Freitas, R.A. "Current Status of Nanomedicine and Medical Nanorobotics." *Journal of Computational and Theoretical Nanoscience* 2 (2005): 1–25.

Fuchigami, T., Kitamoto, Y., and Namiki, Y. "Size-tunable Drug-delivery Capsules Composed of a Magnetic Nanoshell." *Biomatter* 2 (2012): 313–320.

Gabathuler, R. "Approaches to Transport Therapeutic Drugs Across the Blood-Brain Barrier to Treat Brain Diseases." *Neurobiology of Disease* 37, no. 1(2010): 48–57.

Gangrade, S.M. "Nanocrystals-a way for carrier free drug delivery." *Pharma Buzz* 6 (2011): 26–31.

Gao, X., and Dave, S.R. "Quantum Dots for Cancer Molecular Imaging." *Advances in Experimental Medicine and Biology* 620 (2007): 57–73.

Gulsen, D., and Chauhan, A. "Ophthalmic Drug Delivery Through Contact Lenses." *Investigative Ophthalmology & Visual Science* 45, no. 7(2004): 2342–2347.

Gulsen, D., and Chauhan, A. "Dispersion of Microemulsion Drops in Hema Hydrogel: A Potential Ophthalmic Drug Delivery Vehicle." *International Journal of Pharmaceutics* 292, no. 1-2(2005): 95–117.

Hadjipanayis, C.G., Machaidze, R., and Kaluzova, M., et al. EGFRvIII Antibody-conjugated Iron Oxide Nanoparticles for Magnetic Resonance Imaging-guided Convection-enhanced Delivery and Targeted Therapy of Glioblastoma." *Cancer Research* 70, no. 15(2010): 6303–6312.

Hirsch, L.R., Stafford, R.J., Bankson, J.A., Sershen, S.R., Rivera, B., and Price, R.E. "Nanosell-mediated Near-infrared Thermal Therapy of Tumors Under Magnetic Resonance Guidance." *The Proceedings of the National Academy of Sciences* 100 (2003): 13549–13554.

Huang, X., Jain, P.K., El-Sayed, I.H., and El-Sayed, M.A. "Gold Nanoparticles: Interesting Optical Properties and Recent Applications in Cancer Diagnostics and Therapy." *Nanomedicine* 2 (2007): 681–693.

Jenning, V., Gysler, A., Schafer-Korting, M., and Gohla, S.H. "Vitamin A Loaded Solid Lipid Nanoparticles for Topical Use: Occlusive Properties and Drug Targeting to the Upper Skin." *European Journal of Pharmaceutics and Biopharmaceutics* 49, no. 3(2000): 211–218.

Jong, W.H.D., and Borm, P.J.A. "Drug Delivery and Nanoparticle Applications and Hazards." *International Journal of Nanomedicine* 3, no. 2(2008): 133–149.

Jong, D.W.H., and Borm, P.J. "Drug Delivery and Nanoparticles: Applications and Hazards." *International Journal of Nanomedicine* 3, no. 2(2008): 133–149.

Kannan, S., Dai, H., Navath, R.S., Balakrishnan, B., Jyoti, A., Janisse, J., et al. "Dendrimer-based Postnatal Therapy for Neuroinflammation and Cerebral Palsy in a Rabbit Model." *Science Translational Medicine* 4, no. 130(2012): 130ra46.

Kesavan, K., Balasubramaniam, J., Kant, S., Singh, P.N., and Pandit, J.K. "Newer Approaches for Optimal Bioavailability of Ocularly Delivered Drugs: Review." *Current Drug Delivery* 8, no. 2(2011): 172–193.

Kherlopian, A.R., Song, T., Duan, Q., Neimark, M.A., Po, M.J., Gohagan, J.K., et al. "A Review of Imaging Techniques for Systems Biology." *BMC Systems Biology* 2 (2008): 74.

Kim, H., Robinson, S.B., and Csaky, K.G. "Investigating the Movement of Intravitreal Human Serum Albumin Nanoparticles in the Vitreous and Retina." *Pharmaceutical Research* 26, no. 2(2009): 329–337.

Kim J.H., Kim J.H., Kim K.W., Kim M.H., and Yu Y.S. "Intravenously Administered Gold Nanoparticles Passing Through the Blood Retinal Barrier Depending on the Particle Size, and Induce no Retinal Toxicity." *Nanotechnology* 20, no. 50(2009): 505101.

Koo, H., Moon, H., Han, H., Na, J.H., Huh, M.S., Park, J.H., et al. "The Movement of Self-Assembled Amphiphilic Polymeric Nanoparticles in the Vitreous and Retina After Intravitreal Injection." *Biomaterials* 33, no. 12(2012): 3485–3493.

Kumar, K.P.S., Bhowmik, D., and Deb, L. "Recent Trends in Liposomes Used as Novel Drug Delivery System." *Journal of Pharmaceutical Innovation* 1, no. 1(2012): 26–34.

Kumar, A., Badde, S., Kamble, R., and Pokharkar, V.B. "Development and Characterization of Liposomal Drug Delivery System for Nimesulide." *International Journal of Pharmacy and Pharmaceutical Sciences* 2, no. 4(2010): 87–89.

Kumar, P.S., Kumar, S., Savadi, R.C., and John, J. "Nanodentistry: A Paradigm Shift-from Fiction to Reality." *Journal of Indian Prosthodontic Society* 11 (2011): 1–6.

Landesman-Milo, D., Goldsmith, M., Leviatan, B.S., Witenberg, B., Brown, E., Leibovitch, S., et al. "Hyaluronan Grafted Lipid-based Nanoparticles as RNAi Carriers for Cancer Cells." *Cancer Letters* 334 (2013): 221–227.

Lee, M., and Kim, S.W. "Polyethylene Glycol Conjugated Copolymers for Plasmid DNA Delivery." *Pharmaceutical Research* 22 (2005): 1–10.

Lowery, A.R., Gobin, A.M., Day, E.S., Halas, N.J., West, J.L. "Immunonano Shells for Targeted Photothermal Ablation of Tumor Cells." *International Journal of Nanomedicine* 1, no. 2(2006): 149–154.

Madaan, K., Kumar, S., Poonia, N., Lather, V., and Pandita, D. "Dendrimers in Drug Delivery and Targeting: Drug-dendrimer Interactions and Toxicity Issues." *Journal of Pharmacy & Bioallied Sciences* 6, no. 3(2014): 139–150.

Malam, Y., Loizidou, M., and Seifalian, A.M. "Liposomes and Nanoparticles: Nanosized Vehicles for Drug Delivery in Cancer." *Trends in Pharmacological Sciences* 30, no. 11(2009): 592–599.

Mc-Devitt, M.R., Chattopadhyay, D., Kappel, B.J., Jaggi, J.S., Schiffman, S.R., Antczak, C., et al. "Tumor Targeting with Antibody Functionalized, Radiolabelled Carbon Nanotubes." *Journal of Nuclear Medicine* 48 (2007): 1180–1189.

Mukherjee, S., Ray, S., and Thakur, R.S. "Solid Lipid Nanoparticles: A Modern Formulation Approach in Drug Delivery System." *Indian Journal of Pharmaceutical Sciences* 71, no. 4(2009): 349–358.

Nagavarma, B.V., Ayuz, A., Hemant, K.S., et al. "Different Techniques For Preparation of Polymeric Nanoparticles-A Review." *Asian Journal of Pharmaceutical and Clinical Research* 5, no. 3(2012): 1–8, 16–23.

Ochekpe, N.A., Olorunfemi, P.O., and Ngwuluka, N.C. "Nanotechnology and Drug Delivery Part 1: Background and Applications." *Tropical Journal of Pharmaceutical Research* 8, no. 3(2009): 265–274.

Panyam, J., and Labhasetwar, V. "Biodegradable Nanoparticles for Drug and Gene Delivery to Cells and Tissue." *Advanced Drug Delivery Reviews* 55, no. 3(2003): 329–347.

Pardeike, J., Hommoss, A., and Muller, R.H. "Lipid Nanoparticles (SLN, NLC) in Cosmetic and Pharmaceutical Dermal Products." *International Journal of Pharmaceutics* 366, no. 1-2(2009): 170–184.

Patel, R.P., Patel, H., and Baria, A.H. "Formulation and Evaluation of Liposomes of Ketokonazole." *International Journal of Drug Delivery Technology* 1, no. 1(2009): 16–23.

Patel, S., Bhirde, A.A., Rusling, J.F., Chen, X., Gutkind, J.S., and Patel, V. "Nano Delivers Big: Designing Molecular Missiles for Cancer Therapeutics." *Pharmaceutics* 3, no. 1(2011): 34–52.

Pathak, A., Vyas, S.P., and Gupta, K.C. "Nano-vectors for Efficient Liver Specific Gene Transfer." *International Journal of Nanomedicine* 3 (2008): 31–49.

Pattani, V.P., and Tunnell, J.W. "Nanoparticle-mediated Photothermal Therapy: A Comparative Study of Heating for Different Particle Types." *Lasers in Surgery and Medicine* 44 (2012): 675–684.

Peeters, L., Sanders, N.N., Braeckmans, K., Boussery, K., Van-de Voorde, J., De-Smedt, S.C., et al. "Vitreous: A Barrier to Nonviral Ocular Gene Therapy." *Investigative Ophthalmology & Visual Science* 46, no. 10(2015): 3553–3561.

Peng, G., Tisch, U., and Adams, O. "Diagnosing Lung Cancer in Exhaled Breath Using Gold Nanoparticles." *Nature nanotechnology* 4 (2009): 669–673.

Petty, M.A., and Lo, E.H. "Junctional Complexes of the Blood Brain Barrier: Permeability Changes in Neuroinflammation." *Progress in Neurobiology* 68, no. 5(2002): 311–323.

Pisanic, T.R. 2nd, Zhang, Y., and Wang, T.H. "Quantum Dots in Diagnostics and Detection: Principles and Paradigms." *The Analyst* 139, no. 12(2014): 2968–2981.

Praetorius, N.P., and Mandal, T.K. "Engineered Nanoparticles in Cancer Therapy." *Recent Patents on Drug Delivery & Formulation* 1, no. 1(2007): 37–51.

Prato, M., Kostarelos, K., and Bianco, A. "Functionalized Carbon Nanotubes in Drug Design and Discovery." *Accounts of Chemical Research* 41, no. 1(2008): 60–68.

Rabanel, J.M., Aoun, V., Elkin, I., Mokhtar, M., and Hildgen, P. "Drug Loaded Nanocarriers: Passive Targeting and Crossing of Biological Barriers." *Current Medicinal Chemistry* 19, no. 19(2012): 3070–3102.

Reilly, R.M. "Carbon Nanotubes: Potential Benefits and Risks of Nanotechnology in Nuclear Nuclear Medicine." *Journal of Nuclear Medicine* 48, no. 7(2007): 1039–1042.

Ruoslahti, E., Bhatia, S.N., and Sailor, M.J. "Targeting of Drugs and Nanoparticles to Tumors." *Journal of Cell Biology* 188, no. 6(2010): 759–768.

Sahoo, S.K., Parveen, S., and Panda, J.J. "The Present and Future of Nanotechnology in Human Healthcare." *Nanomedicine: Nanotechnology, Biology and Medicine* 3, no. 1(2007): 20–31.

Santos, N.S., Emerson, S.B., and Ralph, S.O. "Nanoradiopharmaceuticals in Current Molecular Medicine." In *Fundamentals of Nanoparticles*, edited by A. Barhoum, and A.S.H. Makhlouf, 553–569, 2018.

Santos, A., Veiga, F., Figueiras, A. "Dendrimers as Pharmaceutical Excipients: Synthesis, Properties, Toxicity and Biomedical Applications." *Materials* 13 (2020): 65.

Silva, G.A. "Introduction to Nanotechnology and Its Applications to Medicine." *Surgical Neurology* 61, no. 3(2004): 216–220.

Singh, R., and Lillard, J.W. "Nanoparticle-Based Targeted Drug Delivery." *Experimental and Molecular Pathology* 86, no. 3(2009): 215–223.

Souto, E.B., Souto, S.B., Campos, J.R., Severino, P., Pashirova, T.N., Zakharova, L.Y., et al. "Nanoparticle Delivery Systems in the Treatment of Diabetes Complications." *Molecules* 24, no. 23(2019): 4209.

Surendiran, A., Sandhiya, S., Pradhan, S.C., and Adithan, C. "Novel Applications of Nanotechnology in Medicine." *Indian Journal of Medical Research* 130 (2009): 689–701.

Tamizhrasi, S., Shukla, A., Shivkumar, T., et al. "Formulation and Evaluation of Lamivudine Loaded Polymethacrylic Acid Nanoparticles." *International Journal of PharmTech Research* 1, no. 3(2009): 411–415.

Tiruwa, R. "A Review on Nanoparticles Preparation and Evaluation Parameter." *Indian Journal of Pharmaceutical and Biological Research* 4, no. 2(2015): 27–31.

Turkevitch, J., Stevenson, P.C., and Hillier, J. "A Study of the Nucleation and Growth Processes in the Synthesis of Colloidal Gold." *Discussions of the Faraday Society* 11 (1951): 55–75.

Viswanathan, P., Muralidaran, Y., and Ragavan, G. "Challenges in Oral Drug Delivery: A Nano-based Strategy to Overcome." *Nanostructures for Oral Medicine* (2017): 173–201.

Webster, D.M., Sundaram, P., and Byrne, M.E. "Injectable Nanomaterials for Drug Delivery: Carriers, Targeting Moieties, and Therapeutics." *European Journal of Pharmaceutics and Biopharmaceutics* 84, no. 1(2013): 1–20.

Wei, X.L., Mo, Z.H., Li, B., and Wei, J.M. "Disruption of Hep G2 Cell Adhesion by Gold Nanoparticle and Paclitaxel Disclosed by In Situ QCm Measurement." *Colloids and Surfaces B: Biointerfaces* 59 (2007): 100–104.

Wilczewska, A.Z., Niemirowicz, K., Markiewicz, K.H., and Car, H. "Nanoparticles as Drug Delivery Systems." *Pharmacological Reports* 64, no. 5(2012): 1020–1037.

Yadav, A., Ghune, M., and Jain, D.K. "Nanomedicine Based Drug Delivery System." *Journal of Advanced Pharmacy Education & Research* 1, no. 4(2011): 201–213.

Yasukawa, T. "Drug Delivery Systems for Vitreoretinal Diseases." *Progress in Retinal and Eye Research* 23, no. 3(2004): 253–281.

Yingchoncharoen, P., Kalinowski, D.S., and Richardson, D.R. "Lipid-based Drug Delivery Systems in Cancer Therapy: What Is Available and What Is Yet to Come." *Pharmacological Reviews* 68 (2016): 701–787.

Yoo, H.S., Oh, J.E., and Lee, K.H., et al. "Biodegradable Nanoparticles Containing PLGA Conjugate for Sustained Release." *Pharmaceutical Research* 16 (1999): 1114–1118.

Youns, M., Hoheisel, J.D., and Efferth, T. "Therapeutic and Diagnostic Applications of Nanoparticles." *Current Cancer Drug Targets* 12, no. 3(2011): 357–365.

Yousefi, A., Esmaeili, F., Rahimian, S., Atyabi, F., and Dinarvand R. "Preparation and In Vitro Evaluation of a Pegylated Nano-liposomal Formulation Containing Docetaxel." *Scientia Pharmaceutica* 77 (2009): 453–464.

2 Manufacturing Techniques for Nanoparticles in Drug Delivery

Daniel Real[1], María Lina Formica[2], Matías L. Picchio[2], and Alejandro J. Paredes[3]
[1]Universidad de Chile
[2]Universidad Nacional de Córdoba
[3]Queen's University Belfast

CONTENTS

2.1 Introduction..24
2.2 Polymeric Nanoparticles 24
    2.2.1 Polymerization..25
        2.2.1.1 Emulsion Polymerization 26
        2.2.1.2 Interfacial Polymerization 27
        2.2.1.3 Controlled Radical Polymerization 28
        2.2.1.4 Microfluidics-Assisted Nanofabrication..............................29
    2.2.2 Preformed Polymers 30
        2.2.2.1 Nanoprecipitation...31
        2.2.2.2 Dialysis 32
        2.2.2.3 Salting Out..33
        2.2.2.4 Supercritical Fluids Technology 33
2.3 Lipid-Based Nanocarriers..34
    2.3.1 Low-Energy Methods..36
        2.3.1.1 Microemulsion and Double Emulsion Methods................36
        2.3.1.2 Phase Inversion Methods 36
    2.3.2 High-Energy Methods 37
        2.3.2.1 High-Pressure Homogenization..37
        2.3.2.2 Ultrasonication or High-Speed Homogenization................39
    2.3.3 Solvent-Based Methods...39
2.4 Nanocrystals 40
    2.4.1 Drug Precipitation (Bottom-Up Techniques) 42
    2.4.2 Confinement Methods (Top-Down Techniques)................................42
        2.4.2.1 High-Pressure Homogenization..42
        2.4.2.2 Media Milling...43

2.4 Conclusion 44
Acknowledgments.. 44
References.. 44

2.1 INTRODUCTION

Nanotechnology has become an important tool for the development of novel products for prevention, treatment, and diagnosis of multiple acute and chronic disorders. Nanoparticle-based drug delivery systems are currently one of the most valuable strategies for addressing low solubility and tissue selectivity in drug delivery. Nanoparticles (NPs) can increase the solubility and permeation of drugs and produce their accumulation in target tissues, ultimately leading to improved therapeutic results (Pakzad et al., 2020). NPs present sizes between 1 and 1000 nm, and the active pharmaceutical ingredient can be adsorbed, encapsulated, trapped, or covalently bound to the NPs matrix or surface. Moreover, NPs may also be formed only by the solid drug, as in nanosuspensions, leading to a greater drug loading capacity.

When the active is loaded into nanocarriers, a sustained release is achievable, enabling the use of these NPs to develop long-acting formulations. Moreover, encapsulating the drug into nanocarriers can protect it from degradation in the body, i.e., by a low pH in the gastrointestinal tract or by the presence of plasmatic enzymes (Jiang et al., 2018; Lin et al., 2015; Weng et al., 2017). However, NPs formulation is a complex state-of-the-art that combines the selection of suitable excipients and techniques. The comprehension of this vast field is fundamental for pharmaceutical scientists and pharmacy students. Consequently, the present primary goal of this chapter is to discuss different manufacturing techniques for NPs with drug delivery purposes.

Drug NPs can be prepared using various methods depending on the intended application and the drug's physicochemical characteristics. An adequate manufacture approach should be selected to obtain optimum formulations regarding process yield, reproducibility, particle size distribution, drug loading, and surface properties (Real et al., 2020). It is commonly accepted that manufacturing techniques can be categorized into two main groups, the "bottom-up" and "top-down" approaches. Whereas the bottom-up techniques imply building up the particles from dissolved ions molecules or monomers, the top-down techniques involve the application of energy to break down large particles, leading to colloidal materials with particle sizes in the submicrometer scale.

Among the various nanoparticle-based drug delivery systems available, polymeric NPs (PNPs), lipid-based nanocarriers, and drug nanocrystals (NCs) have had the most significant impact on nanomedicine, becoming a critical component of a continuously increasing number of healthcare products. The following sections describe different techniques for the formulation of polymeric and lipid-based NPs and drug nanocrystals as summarized in Figure 2.1.

2.2 POLYMERIC NANOPARTICLES

Natural and synthetic polymers are two of the most studied materials to produce NPs in pharmaceutical sciences. PNPs are a diverse group of nanocarriers with

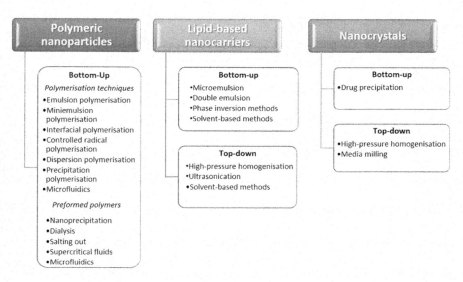

FIGURE 2.1 Main manufacturing techniques for the design of PNPs, lipid-based nanocarriers, and drug nanocrystals.

considerable advantages based on the biocompatibility and biodegradability of the materials used. Generally, biopolymers are biocompatible and biodegradable because of their elimination from the body by natural metabolic pathways (Real et al., 2013). However, they could be immunogenic and present batch-to-batch variations in their physicochemical properties (Kumari, 2010). On the other hand, synthetic polymers have a controlled chemical composition, but their elimination from the body strictly depends on their physicochemical properties (i.e., water solubility, presence of moieties susceptible to hydrolysis, or enzyme degradation). Generally, NPs can be classified into nanocapsules and nanospheres. Whereas nanocapsules act as reservoir-like systems, where a solid shell surrounds a liquid or semisolid core; nanospheres are solid matrix particles. In both cases, the active compound can be entrapped inside or adsorbed on the surface of PNPs.

The type of polymer, the drug's physicochemical properties, and the final desired characteristics of the PNPs should be analyzed before selecting a specific PNPs formulation method. The primary methodologies for PNPs formulation are based on monomers' polymerization or the self-assembly of preformed polymers, as described as follows.

2.2.1 POLYMERIZATION

The polymerization of monomers is a bottom-up strategy widely used in PNPs formulation. This technique can be used to design PNPs to obtain tailored properties for a particular application. The following section discusses different formulation methods for the production of PNPs based on the polymerization of monomers.

2.2.1.1 Emulsion Polymerization

Emulsion polymerization technique involves the reaction of free radicals with relatively hydrophobic monomer molecules within submicron polymer particles dispersed in a continuous medium, most often water (de La Cal et al., 2008). These polymeric dispersions are called latexes. Surfactants are generally required to stabilize the colloidal system; otherwise, latex particles nucleated during the early stage of polymerization may experience significant coagulation to reduce the interfacial free energy (Karsa and Tauer, 2020). The monomers are initially dispersed in water in the presence of the surfactants, which adsorb on the monomer droplets' surface (1–10 μm in size), stabilizing them. Commonly, the amount of surfactant exceeds that needed to completely cover the monomer droplets and forms micelles (5–100 nm in size) that are swollen with monomer. Thermal or redox initiators could be used depending on whether the process is carried out at elevated temperatures (75–90 °C) or whether a high initiation rate is required (Tauer and Müller, 2003). As shown in Figure 2.2, the polymerization process initiates in the aqueous phase where radicals react with the monomer dissolved, forming hydrophobic oligoradicals able to enter into the system's organic phases. As the micelles' total area is greater than that of the droplets (about three orders of magnitude), entry of

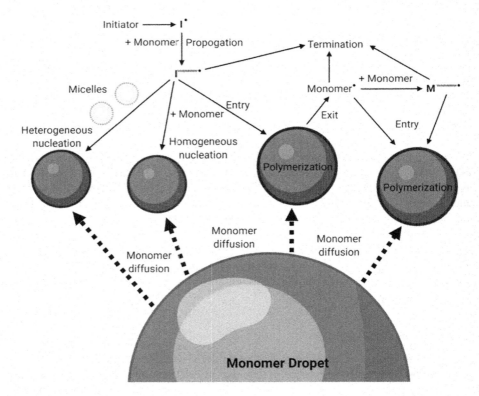

FIGURE 2.2 Schematic description of the mechanisms involved in the emulsion polymerization technique.

radicals into the micelles occurs preferentially. Then, oligoradicals grow rapidly in this monomer-rich environment forming a polymer particle. This process of formation of polymer particles by the entry of radicals into micelles is called heterogeneous nucleation (Harkins, 1947). Alternatively, polymer particles can also be formed when the oligoradicals grow in the aqueous phase beyond the length at which they are still soluble in water and precipitate. The precipitated polymer chains are stabilized by the emulsifier present in the aqueous phase, and monomer diffuses into the new organic phase, which allows fast growth of the polymer chain. This process is commonly called homogeneous nucleation (Thickett and Gilbert, 2007). Both homogeneous and heterogeneous nucleation may take place simultaneously in a given system. However, as a rule of thumb, homogeneous nucleation prevails for monomers of relatively high water-solubility whereas heterogeneous nucleation prevails for water-insoluble monomers.

2.2.1.2 Interfacial Polymerization

Interfacial polymerization (IP) is a class of step-growth polymerization taking place at the interface of two immiscible phases (Raaijmakers and Benes, 2016). IP has evolved as a robust and effective approach to synthesize a wide range of functional polymer materials. Figure 2.3 shows common polymerization systems for the preparation of microcapsules and NPs by IP including liquid/monomer–solid (Lm–S), liquid/monomer–liquid (Lm–L), liquid/monomer-in-liquid (Lm-in-L), and liquid/monomer-in-liquid/monomer (Lm-in-Lm), based on liquid-solid, liquid-liquid, and liquid-in-liquid emulsion interfaces (Zhang et al., 2020).

In Lm–S IP, the monomers are dissolved in the liquid solution, and then they polymerize at the L–S interface, resulting in hybrid materials. This type of IP can be an exciting alternative for polymer/noble metal hybrid NPs, widely used in nanomedicine for photothermal therapy/photoacoustic imaging (Kango et al., 2013), as it can significantly improve their dispersibility and stability. In Lm–L IP, the initiators

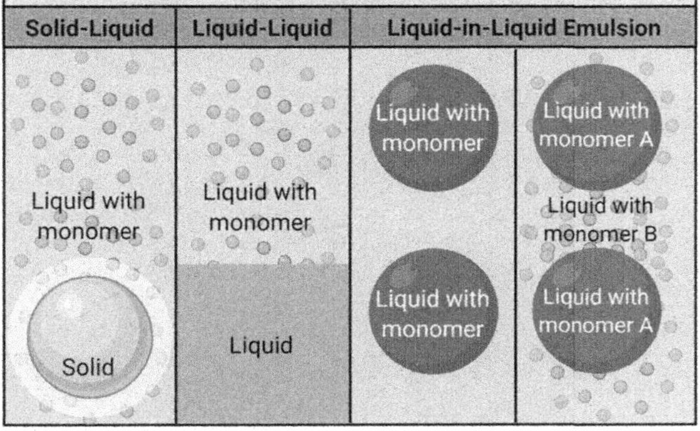

FIGURE 2.3 Different types of IP interfaces usually employed for nanoparticle preparation.

and monomers are dissolved in separate liquid phases. When initiators and monomers diffuse to the liquid-liquid interface, polymerization is activated. This interfacial system can produce homogeneous films or NPs. For Lm-in-L IP an interface is formed when the monomers are dissolved in emulsified liquid phase droplets, while the initiators are dissolved in the outer liquid phase. After free-radicals generation via UV light or thermal activation, they can diffuse into the L-in-L interface and initiate the polymerization, producing monodisperse capsules (Kang and Reichmanis, 2012).

Interestingly, the thickness of the shell layer could be controlled by the polymerization time. Although this approach is highly effective in producing uniform capsules, their size is usually in the range of micrometers and is challenging to tune, while nanocapsules are more attractive in drug delivery. Therefore, the preparation of monodisperse and size-tunable capsules remains a challenge for IP. Finally, in the Lm-in-Lm interface polymerization system, two types of monomers are simultaneously present in the emulsion's outer and inner liquid phases, while the initiator is dissolved in the outer liquid phase. After free radicals are generated, they diffuse into the Lm-in-L interface, initiating the copolymerization only at the interface. This type of IP is usually used to prepare composite polymer capsules and Janus NPs for drug delivery (Khoee and Nouri, 2018).

2.2.1.3 Controlled Radical Polymerization

In general terms, controlled radical polymerization (CRP) is based on a rapid dynamic equilibration between active growing radicals and an excess of dormant species to decrease termination events (Matyjaszewski and Spanswick, 2005). Different CRP methods have been developed, and the three most employed are: nitroxide mediated polymerization (NMP), transition-metal-catalyzed atom transfer radical polymerization (ATRP), and reversible addition-fragmentation chain transfer (RAFT) polymerization (Figure 2.4).

NMP technique is based on reversible recombination of macroradicals (i.e., growing polymer chains, $P^{\bullet}$) and nitroxide derivatives ($R_2NO^{\bullet}$, R = alkyl group) that form an alkoxyamine (R_2NOP, dormant chains), which leads to a low radical

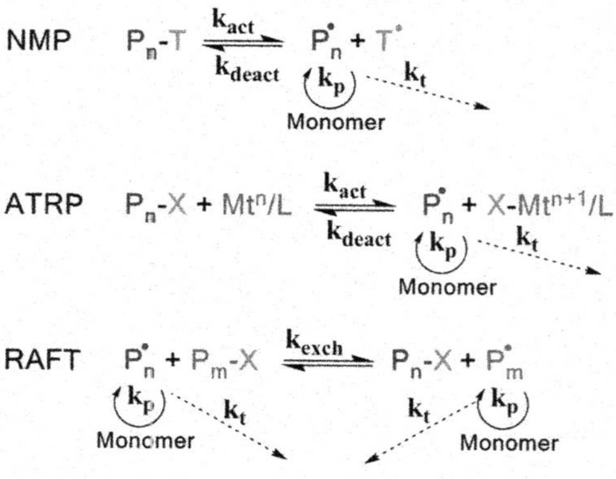

$$\text{NMP} \quad P_n\text{-}T \xrightleftharpoons[k_{deact}]{k_{act}} P_n^{\bullet} + T^{\bullet}$$

$$\text{ATRP} \quad P_n\text{-}X + Mt^n/L \xrightleftharpoons[k_{deact}]{k_{act}} P_n^{\bullet} + X\text{-}Mt^{n+1}/L$$

$$\text{RAFT} \quad P_n^{\bullet} + P_m\text{-}X \xrightleftharpoons{k_{exch}} P_n\text{-}X + P_m^{\bullet}$$

FIGURE 2.4 Main types of controlled radical polymerization.

concentration and decreases the irreversible termination reactions (Nicolas et al., 2013). As a result, most dormant living chains can grow until the monomer is consumed, producing a polymer with equal chain length and a re-activable chain end. It allows the preparation of polymers with controlled molecular weight and narrow distribution.

In the ATRP method, a transition metal catalyst (M_t^n–Y/L, where Y is a counterion and L is a complexing ligand) produces homolytic cleavage of an alkyl halogen (R–X, initiator) bond, which generates X–M_t^{n+1}–Y/L and an active radical (R$^\bullet$) (Matyjaszewski and Xia, 2001). Then, R$^\bullet$ can propagate, terminate, or be reversibly deactivated by X–M_t^{n+1}–Y/L to form a dormant (macro)alkyl halide (P_n-X). Transition metals such as copper, ruthenium, iron, or nickel and nitrogen-based ligands are classically employed as catalyst systems in ATRP. This method enables accurately controlling the molecular architecture, molecular weight, and polydispersity of a great variety of polymers encompassing styrenes, (meth)acrylates, (meth)acrylamides, acrylonitrile, and so on.

Like ATRP and NMP, RAFT is another reversible-deactivation radical polymerization technique but is more versatile and allows for better control over molecular weights and polydispersity (Keddie, 2014). This polymerization method relies on adding a chain transfer agent (RAFT agent) to a conventional free radical polymerization medium. RAFT polymerization is characterized by four different steps: initiation, addition–fragmentation, reinitiation, and equilibration. In the first step, free radicals are generated from the initiator, and the subsequent addition of monomer creates active polymer chains ($P_n^\bullet$). In the addition-fragmentation step, the polymer chains combine with the RAFT agent, giving an active intermediate and releasing a homolytic leaving group (R$^\bullet$). This step is reversible, and the active intermediate can lose either the cleavable group (R$^\bullet$) or the polymeric chain ($P_n^\bullet$). Re-initiation can start with R$^\bullet$ by addition of a monomer and forming a new active polymer ($P_m^\bullet$). This active chain goes through the addition–fragmentation or equilibration steps. Active polymer chains ($P_n^\bullet$ and $P_m^\bullet$) are in equilibrium between the active and dormant (bound to the thiocarbonyl compound) stages. Thus, when one polymer chain is in the dormant stage, the other chain is active in polymerization.

2.2.1.4 Microfluidics-Assisted Nanofabrication

Microfluidics refers to the science and technology of systems that process small amounts of fluids (in the order of nanoliters (nL) to picoliters (pL)), using micron size channels (Shrimal et al., 2020). Microfluidics offers reduced batch-to-batch variation, low particle size dispersity, low reagent consumption, and high reproducibility due to miniaturization of reaction environment and continuous mode of operation. Nanoparticle synthesis in microfluidic systems is performed in tubular micro-reactors and more commonly in lab-on-a-chip devices. Based on the reacting phase involved, microfluidic systems can be divided into two general categories, single-phase continuous flow system (SFCFS) (Jahn et al., 2008) and multiphase flow system (MFS) (Ding et al., 2020), as illustrated in Figure 2.5.

In SFCFS, two or more miscible solvents are mixed in a continuous flow reactor (in a laminar flow regimen) where nucleation and growth occur. The factors that affect the nanoprecipitation process in microfluidics are reaction time, temperature,

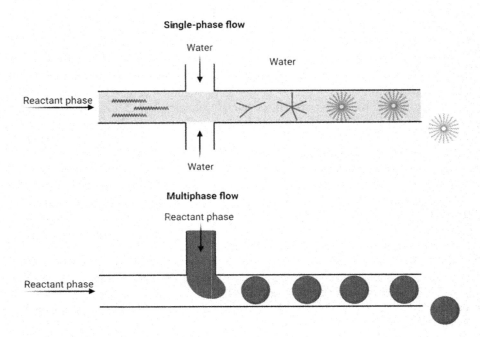

FIGURE 2.5 Scheme of the two central microfluidics systems.

concentration of solvent, flow rate ratio of solvent to antisolvent, and mixing patterns. By rapid and tunable mixing, strict control over precipitation can be achieved, and the homogenous environment in this microreactor yields small NPs with a narrow particle size distribution. Therefore, continuous flow microfluidic reactor provides an excellent platform for drug nanoprecipitation with enhanced controllability, reproducibility, homogeneity of the size characteristics, and scalability.

On the other hand, MFS involves two or more immiscible fluids in segmented phases. The flow phases can be liquid-liquid flows (often water-in-oil or oil-in-water) or gas-liquid flows. This system is used to produce emulsions, microdroplets, microparticles (Shang et al., 2017), and droplets formed by shear forces generated at the immiscible fluids' interface. The tiny volume of the droplet, typically on the subnanoliter scale, favors efficient thermal exchange and high reaction speed. Besides, droplets generated in the microfluidic channels can be manipulated independently and therefore serve as individual units for polymerization reactions, resulting in a highly reproducible nanoparticle production with a small size distribution. The parameters that play an essential role in MFS are flow rate, viscosity, and surfactant concentration.

2.2.2 Preformed Polymers

Residuals of the polymerization process (monomer, oligomer, surfactant, etc.) could be harmful if they are present in NPs developed for medical purposes, demanding careful purification of the colloidal material after the manufacture process (Vauthier

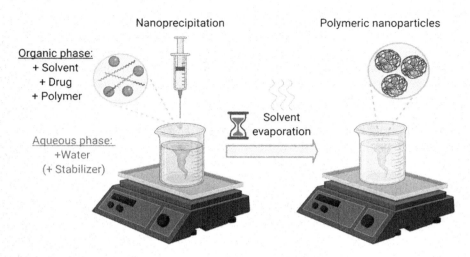

FIGURE 2.6 Schematic representation of nanoprecipitation process.

and Bouchemal, 2009). One strategy that can be used to overcome this issue involves employing preformed polymers as described in the following subsections.

2.2.2.1 Nanoprecipitation

The nanoprecipitation technique is based on the interfacial deposition of a polymer after displacement of the organic solvent from a lipophilic solution to an aqueous phase as schematized in Figure 2.6. A water-miscible solvent of intermediate polarity is used to dissolve the polymer, and this solution is added dropwise, stepwise, in one shot, or by controlled addition rate into an aqueous phase under agitation. NPs are formed immediately in an attempt to evade the water molecules because of the fast diffusion of the organic phase into the aqueous solution. The Marangoni effect seems to govern this process since a decrease in the interfacial tension between the two phases increases the surface area, due to the rapid diffusion, and leads to the formation of small droplets of the organic solvent (Pinto Reis et al., 2006). For ensuring the colloidal suspension's stability, surfactants could be incorporated in the process, though their presence is not required to guarantee nanoparticle formation. Different solvents such as acetone, ethanol, or methanol could be used. The miscibility and the ease with which they are evaporated are critical points to be analyzed. Generally, the organic phase is poured over the aqueous phase, yet this could be reversed without compromising the nanoparticle formation. It is also possible to use more than one organic or aqueous phase as long as miscibility, solubility, and insolubility requirements are met.

The experimental design variables are the key factors that define the final nanoparticle properties (Crucho and Barros, 2017). An increase in the polymer molecular weight or the polymer concentration usually increases particle size due to higher viscosity of the organic phase, which obstructs solvent diffusion, resulting in larger nanodroplets. It is possible to control the PNP physicochemical properties by judiciously adjusting the nature and concentration of the components, injection rate of the organic phase, mixing speed, aqueous phase/organic phase proportion, and fluid dynamics (Nguyen et al., 2018). A relevant drawback of this technique is the low association efficacy of hydrophilic drugs since they can diffuse to the aqueous phase during polymer precipitation. Varying the solvent composition or changing the pH could be practical strategies to modify the drug's solubility for improving its encapsulation efficiency.

This method is extensively used due to its reproducibility, simplicity, and rapidity. However, one of the main challenges to be considered for optimizing a nanoprecipitation process is finding a proper polymer/drug/solvent/nonsolvent system, which permits fruitful nanoparticle formation and drug encapsulation.

2.2.2.2 Dialysis

The dialysis technique is governed by a similar mechanism to that previously described for the nanoprecipitation method, but with a different experimental setup (Crucho and Barros, 2017). In this procedure, semipermeable membranes or dialysis tubes, with a suitable molecular weight cut-off, are used as a physical barrier for the polymer. As seen in Figure 2.7, the drug and polymer are dissolved in an organic solvent, placed inside the dialysis bag, and dialysed against a miscible nonsolvent. The progressive diffusion of the nonsolvent inside the tube turns the solvent less able to dissolve the polymer. Additionally, polymer aggregation due to an increase in the interfacial tension leads to the formation of a colloidal suspension of NPs. Several experimental parameters can be optimized to modulate the particle size, morphology, and distribution of the obtained PNPs, such as dialysis molecular weight cut-off, solvent/nonsolvent pair, polymer concentration, temperature, and solvent mixing speed. This method finds a significant drawback in the large volume of counter dialysing medium that could produce a premature release of the nanoparticle payload due to the long duration of the process.

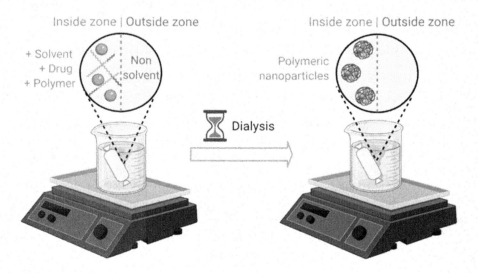

FIGURE 2.7 Formulation of PNPs by dialysis method.

2.2.2.3 Salting Out

The salting-out technique consists of separating a water-miscible solvent from aqueous solution via the salting-out effect (Galindo-Rodriguez et al., 2004). The solvent used to prepare a polymer and drug solution should be totally miscible with water (usually acetone). Emulsification is conducted with an aqueous phase containing a high concentration of salting-out agents (electrolytes, such as calcium chloride, magnesium acetate, magnesium chloride, or nonelectrolytes like sucrose) and a colloidal stabilizer. The saturated aqueous phase prevents the solvent from mixing with water. The emulsified droplets are then diluted in water. A sudden drop of salt concentration in the continuous phase causes the extraction of organic solvent and precipitation of polymer-drug NPs, as illustrated in Figure 2.8. Afterwards, both the salting-out agents and the solvent should be eliminated by cross-flow filtration. The main drawbacks are the extensive washing steps and the limited application to lipophilic drugs (Mendoza-Munoz et al., 2012). The principal advantages of salting-out are the minimum stress produced in proteins and heat-sensitive substances encapsulation.

2.2.2.4 Supercritical Fluids Technology

The preceding described methods require the use of organic solvents and surfactants that could be hazardous for the environment and physiological systems. In this sense, supercritical fluids have emerged as an interesting alternative due to the use of an ecofriendly methodology, and homogenous and easy scale-up (Adschiri and Yoko, 2018). This procedure is based on using a fluid heated and compressed above its critical temperature and critical pressure. In such conditions, the fluid behaves like a gas but with the solvating properties and density of a liquid. Supercritical carbon dioxide is the most broadly used supercritical fluid because it has mild critical conditions and is nontoxic, inexpensive, abundant, nonflammable, and environmentally benign.

Two strategies have been used to formulate NPs using supercritical fluids: rapid expansion of supercritical solution (RESS) and rapid expansion of supercritical solution into a liquid solvent (RESOLV) (Byrappa et al., 2008). In conventional

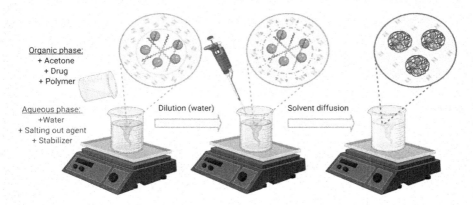

FIGURE 2.8 Flow diagram of the salting technique for the manufacture of PNPs.

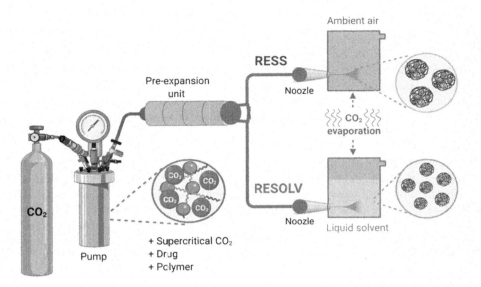

FIGURE 2.9 Supercritical fluid methods. RESS (Rapid expansion of supercritical solution) and RESOLV (Rapid expansion of supercritical solution into liquid solvent).

RESS, the solutes are dissolved in an SCF to form a solution, followed by the fast expansion of the solution across a capillary nozzle into the ambient air. In RESOLV, the supercritical solution's expansion befalls into a liquid solvent instead of ambient air (Figure 2.9). Although no organic solvents are used in the RESS method, the particles obtained using this procedure are microsized rather than nanosized. In RESOLV the presence of liquid solvent in the expansion jet seems to suppress the particle growth, making it possible to obtain nanosized particles.

The main drawback of this technology is that most polymers exhibit non-solubility or poor solubility in supercritical fluids. Additionally, it requires a high investment for high-pressure equipment.

2.3 LIPID-BASED NANOCARRIERS

Lipid-based nanocarriers are drug delivery systems formed from a lipid phase and an aqueous phase, stabilized by surfactants. These systems are composed of biocompatible and biodegradable components and offer advantages related to controlled release, stability, drug loading, and surface versatility. Moreover, lipid-based nanocarriers can be used for targeted delivery, administered by different routes and even co-loading both lipophilic and hydrophilic drugs (Formica et al., 2021; Valetti et al., 2013). In this way, a broad range of lipid-based nanocarriers has been developed with drug delivery purposes, including liposomes, micelles, nanoemulsions, lipid nanocapsules (LNC), solid lipid NPs (SLN), and nanostructured lipid carriers (NLC), the structures of which are illustrated in Figure 2.10. Briefly, liposomes are small (30–1000 nm) artificial vesicles of spherical shape consisting of one or more lipid bilayers mainly composed of amphipathic phospholipids, bearing

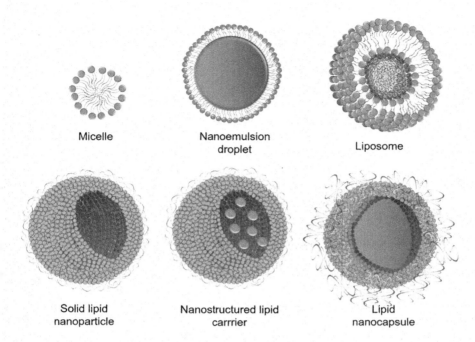

FIGURE 2.10 Schematic representation of lipid-based nanocarriers.

a nonpolar tail and polar head, and enclosing an interior aqueous space (Puri et al., 2009). Depending on the number of bilayers, they can be multilamellar vesicles or unilamellar vesicles. Lipid micelles are colloidal dispersions with a hydrophobic core, and a hydrophilic shell formed spontaneously from amphiphilic molecules or surfactant agents at specific concentrations and temperatures (Valetti et al., 2013). On the other hand, nanoemulsions are translucent oil-in-water or water-in-oil systems (50–200 nm), characterized by excellent stability in suspension due to its tiny droplet size and significant steric stabilization between droplets.

In the case of LNC (20–100 nm), they are composed of an oily liquid triglyceride core surrounded by a polymer layer made from a mixture of lecithin and polyethylene glycol-based polymers, representing a hybrid structure between polymeric nanocapsules and liposomes (Heurtault et al., 2002). The SLN and NLC are colloidal systems (50–1000 nm) that offer similar advantages that liposomes and microemulsions, but more useful for drug control release and drug protection against chemical degradation (Puri et al., 2009). While SLN consists of solid lipid core with a monolayer coating of surfactants (Anderluzzi et al., 2019), NLC presents a less organized solid lipid matrix that results from a blend of solid lipids and liquid lipids (oils) (Zhang et al., 2013), which are developed as the second generation of the former in order to increase drug loading and prevent drug leakage.

Although the various lipid-based nanocarriers described above present different structures and components, all their manufacturing methods generally involve several steps and require the input of different levels of energy. Among the manufacturing processes for nonliposomal lipid nanocarriers (namely, nanoemulsions,

LNC, SLN, and NLC), different approaches have been explored, which can be divided into low-energy, high-energy, and organic solvent-based methods, as described below.

2.3.1 Low-Energy Methods

Low-energy methods imply the formation of the NPs with non or very low-energy inputs and simple experimental setups using, e.g., magnetic stirring (Gordillo-Galeano and Mora-Huertas, 2018). Other methods require surfactants that respond with changes in the hydrophilic/lipophilic balance as a function of temperature (Heurtault et al., 2002). The microemulsion, double emulsion, and phase inversion temperature are some of the most common methods used within this group of techniques.

2.3.1.1 Microemulsion and Double Emulsion Methods

The microemulsion method is based on the spontaneous formation of the microemulsions by a high ratio of surfactant/lipid, in which a blend composed of preheated oil and solid melted lipid containing the drug is added to an aqueous surfactant solution at the same temperature under mechanical stirring. Then, the hot microemulsion is added to cold water (2–4 °C) under stirring, leading to the precipitation of lipids as small particles (Salvi and Pawar, 2019).

In the double emulsion method, an aqueous solvent containing a hydrophilic drug is added to a melted mixture of lipids to form a first clear water-in-oil microemulsion. Then, the solvent is removed in a rotary evaporator and this microemulsion is dispersed in an aqueous solution that contains the hydrophilic stabilizing agent to form a double emulsion water-in-oil-in-water. In some cases, this double emulsion is diluted in cold water to obtain NPs, but in others, nanometric drops are separated by evaporation or filtration (Amoabediny et al., 2018).

2.3.1.2 Phase Inversion Methods

The phase inversion methods are based on the insignificant interfacial tension achieved when the surfactant curvature is inverted to drive emulsification easily. These are classified into phase inversion temperature and phase inversion composition methods, whether a change in temperature or composition triggers emulsification, respectively, as illustrated in Figure 2.11 (Aparicio-Blanco et al., 2019).

The phase inversion temperature method is an organic solvent-free process used to prepare LNC, in which all components (lipids, surfactants, water, electrolytes) of the formulation are mixed under stirring and subjected to repeated heating and cooling cycles, leading to a successive reversal of phases from an oil-in-water emulsion to a water-in-oil emulsion. Each phase inversion zone crossing promotes a decrease in the size of the drops. Finally, a fast cooling-dilution is induced to achieve an oil-in-water emulsion, by adding cold water (2 °C) in the phase inversion zone. In this method, the hydrophilic/lipophilic balance of ethoxylated surfactants commonly used in the formulation, may change with temperature and become less hydrophilic on heating (Heurtault et al. 2002, 2003).

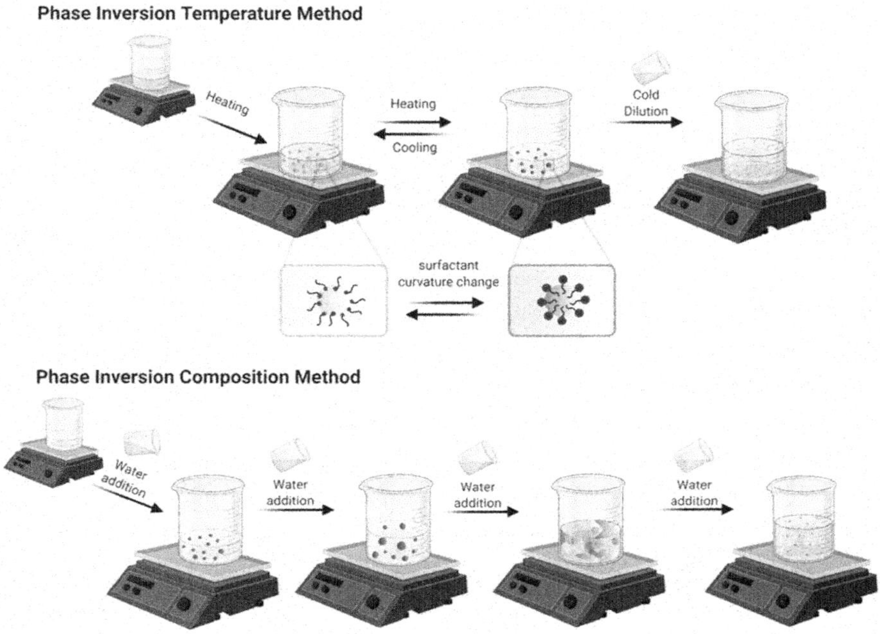

FIGURE 2.11 Schematic representations of phase inversion methods.

In the phase inversion composition method, water is progressively added over a mixture of the lipids and surfactant, to form an oil-in-water emulsion (Figure 2.11). The water addition leads to the hydration of surfactant chains and spontaneous change in their curvature from negative to zero. Therefore, the surfactant hydrophilic-lipophilic properties are balanced, and bicontinuous or lamellar structures are formed (Solans and Solé, 2012).

2.3.2 High-Energy Methods

The methods based on high-energy require a considerable energy application to achieve NPs with suitable size distribution. In these methods, the input of energy is provided by a mechanical device that applies high-shear and distortion forces, as well as high pressures. As a rule of thumb, smaller particles or droplet sizes are obtained with increasing energy inputs. However, the degree of energy applied must be evaluated due to the high shear stress involved during these processes may lead to degradation and loss of activity of the active molecules. Typically high-energy procedures involve high-pressure homogenization or ultrasonication (Aparicio-Blanco et al., 2019).

2.3.2.1 High-Pressure Homogenization

High-pressure homogenization (HPH) is a commonly used technique for preparing lipid-based nanocarriers due to its reproducibility and ease of scaling up. In this method, the lipid is melted at 5–10 °C above its melting point, and the drug is

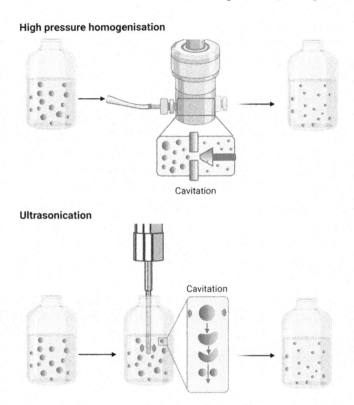

FIGURE 2.12 Schematic representation of the decrease in the droplet size of nanoemulsions by high-pressure homogenization and ultrasonication.

dissolved in the molten lipid. Then, the drug-lipid dispersion is subjected to high pressure (100–2000 bar) through a narrow gap for several subsequent cycles using a high-pressure homogenizer (Figure 2.12). The particle size is decreased to the submicron range by the impact of shear stress, cavitation force, and turbulence.

To obtain SLN or NLC, HPH can be carried out by either hot or cold homogenization (Mehnert and Mäder, 2001). In the hot HPH, liquid and solid lipids are mixed and heated above the solid lipid's melting temperature, and then, the drug is dispersed in the lipid melt. Meanwhile, the surfactant is dissolved in distilled water (aqueous phase), and heated at the same temperature as the lipid melt. The drug-lipid mixture is combined with the aqueous phase by high-speed stirring and/or high-shear homogenization (pre-emulsion) and then, passed through a high-pressure homogenizer at varied pressures for 3–5 cycles. In general, a greater number of HPH cycles leads to a decrease in the nanoemulsion's droplet size. Finally, the emulsion is cooled at room temperature under stirring, and solid lipid recrystallization takes place. Importantly, this formulation approach is not suitable for heat-sensitive drugs.

In the cold HPH method, lipids are melted above the solid lipid's melting temperature, followed by the dispersion of the drug in the hot lipid molten. Then, the drug-lipid mixture is subjected to HPH. The resultant emulsion is rapidly cooled with liquid nitrogen or dry ice. After solidification, dried lipids are ground by milling to microparticles which are dispersed in a cold surfactant solution. This pre-emulsion is subjected to high shear homogenization, ultrasonication, or homogenized using a high-pressure homogenizer at room temperature (or below) to obtain SLN or NLC (Amoabediny et al., 2018).

2.3.2.2 Ultrasonication or High-Speed Homogenization

In this technique, the liquid and solid lipids are mixed and melted at 5–10 °C temperature above the solid lipid's melting point, followed by the dispersion of the drug in this lipid melted. The surfactant is dissolved in distilled water and heated at the same temperature as the melted lipid. Then, the drug-lipid mixture is combined with the aqueous phase, and the resultant pre-emulsion is mixed in a high-speed homogenizer. Subsequently, ultrasound is applied with a sonotrode in direct contact with this emulsion for a predetermined time to reduce particle size. Afterward, the sample is cooled at room temperature for solid lipid recrystallization. The reduction of droplet size in the emulsion is mainly due to the acoustic cavitation process (Figure 2.12). This technique is rapid, and the required equipment is commonly available, making it an attractive approach for the preparation of lipid-based nanocarriers. However, probe sonication could cause contamination of formulation with metal particles (Gordillo-Galeano and Mora-Huertas, 2018; Salvi and Pawar, 2019).

2.3.3 SOLVENT-BASED METHODS

The solvent-based methods are a group of techniques in which the addition of organic solvents to the system is the basis for drop size reduction, such as emulsification-solvent evaporation and emulsification-solvent diffusion techniques. Likewise, these methods can be also considered low-energy or high-energy depending on the level of energy input.

The emulsification-solvent evaporation method involves using an organic solvent immiscible with water, such as chloroform or cyclohexane, which dissolves the drug and lipids. The mixture of lipids is dissolved in the water-saturated organic solvent followed by the addition of the drug and its subsequent emulsification in an organic solvent-saturated aqueous solution of the stabilizing agent under mechanical stirring or ultrasonication. Then, the solvent is evaporated at room temperature and reduced pressure, resulting in NPs formed in the aqueous phase by precipitation of the lipid (Figure 2.13). Although this technique can be easily scaled up, and it is one of the most frequently used in the preparation of SLN or NLC, the residual organic solvent may remain in the formulation (Amoabediny et al., 2018; Gordillo-Galeano and Mora-Huertas, 2018).

On the other hand, the emulsification-solvent diffusion method involves using a partially water-miscible solvent that dissolves the lipids, such as methanol, ethanol, or acetone. Drug and lipids dissolved in the saturated solvent are sonicated and maintained in a water bath at high temperature to obtain a clear lipid phase. Then, a saturated aqueous phase containing the surfactant is added to the lipid phase under

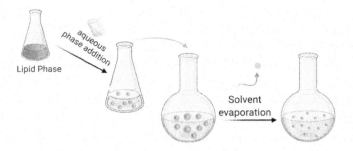

FIGURE 2.13 Schematic representations of solvent-based methods.

stirring to form an oil-water emulsion, followed by water addition to initiate solvent diffusion and nanoparticle formation. This dispersion is stirred at room temperature for cooling, and the solvent is removed by vacuum evaporation or lyophilization (Amoabediny et al., 2018).

2.4 NANOCRYSTALS

NCs are solid particles of a drug with crystalline characteristics and sizes in the nanometric range, typically between 200 and 500 nm (Müller and Keck, 2012). NCs are surrounded by a layer of stabilizer to prevent aggregation and can be prepared either as suspensions, commonly named as nanosuspensions, or as NCs, a term that frequently refers to the solid material obtained after solvent removal. There are three main features related to the reduction in the drug particle size of the drug to the submicron range: 1) Increased dissolution rate, which, according to the Noyes-Whitney equation, is linked to an augmented surface of the drug being exposed to the solvent for dissolution, as described in Figure 2.14a (Noyes and Whitney, 1897). 2) Increased saturation solubility, which is related to the augmented curvature and dissolution pressure of the drug from the NCs, leading to a more significant amount of dissolved molecules of the active compound available for absorption through physiological barriers, and posterior distribution in the body (Figure 2.14b) (Mauludin et al., 2009). 3) Increased adhesiveness, since drug NPs can interact with physiological barriers in a greater magnitude due to their enlarged surface and amount of contact points (Figure 2.14c) (Müller et al., 2011). These factors, all together lead to enhanced in vivo performances of poorly soluble drugs when nanocrystallized, allowing to increase its oral absorption and therapeutic response (Paredes, Bruni, et al., 2018; Paredes, Litterio et al., 2018; Pensel et al., 2018).

In comparison with other nanoparticle-based drug delivery systems, NCs have a series of distinctive advantages. On the one hand, since they are composed of 100% drug, their drug loading capacity is significantly higher than that observed for other nanocarriers, in which the drug payload is conditioned by the need for carrier materials and is frequently less than 10% (Shen et al., 2017). NCs can also be produced using easily scalable methods, such as bead milling and HPH. Moreover, as NCs are solid particles, their long-term stability is higher than that observed for soft NPs formed by assembled polymers, which can be sensitive to pH, temperature,

Nanoparticles in Drug Delivery

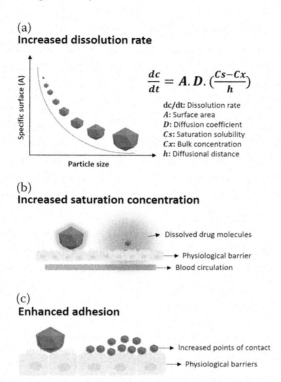

FIGURE 2.14 Main advantages of NPs in drug delivery due to the reduction of its particle size. (a) Exponential increase of the surface factor (A in the Noyes-Whitney equation) with the reduction of particle size to the submicron range. (b) A larger number of dissolved molecules are available for absorption with NPs in comparison to drug microparticles. (c) Increased adhesiveness with physiological barriers.

salinity, etc. Another critical factor is that NCs can be obtained at both the laboratory and industrial scales using techniques that are free of organic solvents. This fact, together with the possibility of producing NCs at neutral pH, makes this platform an attractive alternative for the delivery of drugs with low aqueous solubility, from both the toxicological and the industrial perspective.

NCs can be prepared using two main groups of techniques, namely the bottom-up or top-down techniques. While the bottom-up approaches involve the precipitation of the NCs from solutions of the drug, the top-down approaches are based on applying large amounts of energy to fragment drug microparticles, which are generally previously suspended in aqueous solutions of different steric or ionic stabilizers (Fontana, 2018; Rabinow, 2004; Xia et al., 2014). Among the first group of techniques, drug precipitation is the most frequently used method, followed by spray-based techniques such as nanospray drying, the aerosol flow reactor method, spraying of low-boiling point solvent under ambient conditions, and electrospraying of low-electrical conducting solutions.

2.4.1 DRUG PRECIPITATION (BOTTOM-UP TECHNIQUES)

Drug precipitation is a bottom-up approach that has been widely used to prepare NCs. This inexpensive technique allows the preparation of NCs easily, especially at the laboratory scale. Here, the active is dissolved in a suitable organic solvent and precipitated by mixing the organic phase with an aqueous solution of stabilizers under controlled conditions of agitation and temperature. Despite this approach's practicality, there are a series of considerations that must be kept in mind. On the one hand, the drug must be in solution before precipitation, and this is not always a possible scenario given that many actives are insoluble even in organic solvents, thus leading to low process yields. The particle sizes and distributions (polydispersity index) obtained with bottom-up approaches are higher than those of top-down approaches, which can be explained by the co-occurrence of rapid crystal growth and the need for stabilization of these novel surfaces being created. Finally, the organic solvents can be entrapped within the formed NCs. Altogether, these drawbacks make bottom-up approaches a less attractive alternative for pharmaceutical industries, which also is reflected in the fact that most NCs-based products commercially available are obtained by top-down techniques such as HPH or media milling (Möschwitzer, 2013; Tuomela et al., 2016).

2.4.2 CONFINEMENT METHODS (TOP-DOWN TECHNIQUES)

2.4.2.1 High-Pressure Homogenization

In high-pressure or "piston-gap" homogenization, a suspension containing the drug and stabilizers is repeatedly forced to pass through a thin gap at extremely high pressures (500–1500 bar) (Keck and Müller, 2006). For this purpose, the sample is placed in a cylindrical container and pumped by pneumatic or electric motors towards a small gap created between a forcer and the equipment's walls. When the sample passes through this gap, the sudden reduction in the diameter leads to a sharp increase in the dynamic pressure, with simultaneous reduction of the static pressure, producing the cavitation phenomenon responsible for particle size reduction (Keck and Müller, 2006). During the homogenization process, the drug particles break at their weakest points or imperfections, and therefore, the reduction of the particle mean diameter is frequently concentrated in the first homogenization cycle at a constant pressure. When these imperfections reduce less and less, the crystals become perfect and higher pressures are required to achieve smaller particle sizes (Keck and Müller, 2006; Paredes, 2016). In HPH, three main factors affect the performance of the process and final particle size of the formulation, namely the applied pressure, the number of homogenization cycles, and the temperature. Due to the significant amount of energy applied to the drug suspension, NCs obtained by this technique tend to be uniform in particle diameter, decreasing the chances of Ostwald ripening and thus enhancing the stability of the sample. Another advantage of HPH is the availability of industrial equipment for scaling up NCs production, making this process an accepted technique by the pharmaceutical industry. The materials' transference from the homogenizer to the suspension, due to the wearing produced by particle collision and cavitation, is a risk that should be considered. However, previous research works demonstrated that heavy metals in a nanosuspension of a model compound produced by HPH using 20 cycles at 1500 bar were negligible (Krause et al., 2000).

2.4.2.2 Media Milling

Media milling, also known as bead or wet bead milling, is a widely used top-down technique for the production of drug NCs, and has been applied to a wide variety of actives (Camiletti et al., 2020; Malamatari et al., 2018; Melian et al., 2020). Media mills consist of a milling chamber, a rotor coupled to a motor, and a jacket for coolant circulation to prevent sample overheating. Some models permit the sample's recirculation by separating the beads from the drug suspension using rotating gaps or sieves with proper mesh sizes. In media mills, the particle size reduction principle is related to shear forces, pressure, collisions, and mechanical attrition produced when the drug suspension is agitated in the presence of micrometric beads of hard and dense materials such as yttrium-stabilized zirconium oxide, alumina, stainless steel, glass, and highly cross-linked polystyrene and methacrylate (Malamatari et al., 2018). As depicted in Figure 2.15, when the system is agitated, drug particles and beads collide with each other and with the milling chamber walls, increasing the specific energy of the system and producing millions of stress events that lead to particle confinement (Afolabi et al., 2014). As the process is carried out in a liquid medium, the aggregation and compaction of the milled particles is prevented, which is an important advantage over the dry milling techniques. Moreover, the liquid may serve as a lubricant in the process and, with the help of the stabilizers, coat the surface of the newly formed NCs, therefore improving their stability (Loh et al., 2014; Möschwitzer and Müller, 2006). The main variables at play during media milling are the percentage of drug and stabilizer in the initial slurry, the motor's rotational speed (agitation), the amount of milling media, the processing time, and the intrinsic hardness of the drug crystal lattice (Paredes et al., 2020). This versatile technique permits the production of NCs with high process yields and very narrow particle size distributions. Furthermore, media milling is hugely flexible in terms of batch size, which may range from a few milliliters of suspension using simple laboratory setups, to cubic meters in industrial mills (Romero et al., 2016).

On the drawback side is the erosion and transference of material from milling media to drug suspension due to the intense agitation forces in the milling chamber (Juhnke et al., 2012). These residual materials have been reported to contaminate drug suspension, affecting the chemical and physical stability of newly exposed

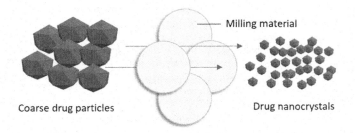

FIGURE 2.15 Schematic representation of drug particle confinement by media milling.

areas and formed particles (Loh et al., 2014). Having this in mind, continuous monitoring of wearing in the milling parts, as well as the quantification of impurities in the final product, is a necessity when using media milling.

2.4 CONCLUSION

The potential of NPs in diagnosis, therapy, theranostics, and drug delivery is substantial. During the past decades, a variety of NPs have been developed, and a considerable number of nanoparticle-based products are already available for users. However, the design of drug delivery nanocarriers is a complex process that implies the selection of nanoparticle type and an optimum manufacturing method.

Future perspectives in NPs' manufacturing are related to process optimization to improve yields and simplify synthesis procedures to allow easy scale-up. In this sense, both the reduction of the energetic input required to prepare NPs and the avoidance of organic solvents are critical for nanomedicines to reach the population at large. Moreover, post-processing techniques and the incorporation of drug NPs in final pharmaceutical forms, such as tablets, capsules, microarray patches, etc. are still big challenges to be solved for expanding drug delivery nanomaterials from lab to bedside.

ACKNOWLEDGMENTS

Daniel Real received funding from Agencia Nacional de Investigación y Desarrollo (ANID), Chile: Fondecyt Postdoctoral 3200384, Fondap 15130011. María Lina Formica and Matías L. Picchio gratefully acknowledge the Universidad Nacional de Córdoba (Argentina) and CONICET (Argentina) for financial support.

REFERENCES

Adschiri, T., and Yoko, A. "Supercritical fluids for nanotechnology." *J. Supercrit. Fluids* 134 (2018): 167–175.

Afolabi, A., Akinlabi, O., and Bilgili, E. "Impact of process parameters on the breakage kinetics of poorly water-soluble drugs during wet stirred media milling: A microhydrodynamic view." *Eur. J. Pharm. Sci.* 51 (2014): 75–86.

Amoabediny, G., Haghiralsadat, F., Naderinezhad, S., Helder, M. N., Akhoundi Kharanaghi, E., Mohammadnejad Arough, J., and Zandieh-Doulabi, B. "Overview of preparation methods of polymeric and lipid-based (niosome, solid lipid, liposome) nanoparticles: A comprehensive review." *Int. J. Polym. Mater. Polym. Biomater.* 67 (2018): 383–400.

Anderluzzi, G., Lou, G., Su, Y., and Perrie, Y. "Scalable manufacturing processes for solid lipid nanoparticles." *Pharm. Nanotechnol.* 7 (2019): 444–459.

Aparicio-Blanco, J., Sebastián, V., Rodríguez-Amaro, M., García-Díaz, H. C., and Torres-Suárez, A. I. "Size-tailored design of highly monodisperse lipid nanocapsules for drug delivery." *J. Biomed. Nanotechnol.* 15 (2019): 1149–1161.

Byrappa, K., Ohara, S., and Adschiri, T. "Nanoparticles synthesis using supercritical fluid technology–towards biomedical applications." *Adv. Drug Deliv. Rev.* 60 (2008): 299–327.

Camiletti, B. X., Camacho, N. M., Paredes, A. J., Allemandi, D. A., Palma, S. D., and Grosso, N. "Self-dispersible nanocrystals of azoxystrobin and cyproconazole with increased efficacy against soilborne fungal pathogens isolated from peanut crops." *Powder Technol.* 372 (2020): 455–465.

Crucho, C. I. C., and Barros, M. T. "Polymeric nanoparticles: A study on the preparation variables and characterization methods." *Mater. Sci. Eng. C* 80 (2017): 771–784.

de La Cal, J. C., Leiza, J. R., Asua, J. M., Buttè, A., Storti, G., and Morbidelli M. "Emulsion polymerization." In *Handbook of Polymer Reaction Engineering*, edited by T. Meyer and J. Keurentjes, 249–322. Weinheim: Wiley-VCH, 2008.

Ding, Y., Howes, P. D., and Demello, A. J. "Recent advances in droplet microfluidics." *Anal. Chem.* 92 (2020): 132–149.

Fontana, F., Figueiredo, P., Zhang, P., Hirvonen, J. T., Liu, D., and Santos, H. A. "Production of pure drug nanocrystals and nano co-crystals by confinement methods." *Adv. Drug Deliv. Rev.* 131 (2018): 3–21.

Formica, M. L., Legeay, S., Bejaud, J., Montich, G. G., Ullio Gamboa, G. V., Benoit, J.-P., and Palma, S. D. "Novel hybrid lipid nanocapsules loaded with a therapeutic monoclonal antibody - Bevacizumab - and Triamcinolone acetonide for combined therapy in neovascular ocular pathologies." *Mater. Sci. Eng. C* 119 (2021): 111398.

Galindo-Rodriguez, S., Allémann, E., Fessi H., and Doelker, E. "Physicochemical parameters associated with nanoparticle formation in the salting-out, emulsification-diffusion, and nanoprecipitation methods." *Pharm. Res.* 21 (2004): 1428–1439.

Gordillo-Galeano, A., and Mora-Huertas, C. E. "Solid lipid nanoparticles and nanostructured lipid carriers: A review emphasizing on particle structure and drug release." *Eur. J. Pharm. Biopharm.* 133 (2018): 285–308.

Harkins, W. D. "A general theory of the mechanism of emulsion polymerization." *J. Am. Chem. Soc.* 69 (1947): 1428–1444.

Heurtault, B., Saulnier, P., Pech, B., Proust, J.-E., and Benoit, J.-P. "A novel phase inversion-based process for the preparation of lipid nanocarriers." *Pharm. Res.* 19 (2002): 875–880.

Heurtault, B., Saulnier, P., Pech, B., Venier-Julienne, M.-C., Proust, J. E., Phan-Tan-Luu, R., and Benoit, J. P. "The influence of lipid nanocapsule composition on their size distribution." *Eur. J. Pharm. Sci.* 18 (2003): 55–61.

Jahn, A., Reiner, J. E., Vreeland, W. N., DeVoe, D. L., Locascio, L. E., and Gaitan, M. "Preparation of nanoparticles by continuous-flow microfluidics." *J. Nanoparticle Res.* 10 (2008): 925–934.

Jiang, S., Franco, Y. L., Zhou, Y., and Chen, J. "Nanotechnology in retinal drug delivery." *Int. J. Ophthalmol.* 11 (2018): 1038–1044.

Juhnke, M., Märtin, D., and John, E. "Generation of wear during the production of drug nanosuspensions by wet media milling." *Eur. J. Pharm. Biopharm.* 81, (2012): 214–222.

Kang, J. H., and Reichmanis, E. "Low-threshold photon upconversion capsules obtained by photoinduced interfacial polymerization." *Angew. Chemie - Int. Ed.* 51 (2012): 11841–11844.

Kango, S., Kalia, S., Celli, A., Njuguna, J., Habibi, Y., and Kumar, R. "Surface modification of inorganic nanoparticles for development of organic-inorganic nanocomposites—a review." *Prog. Polym. Sci.* 38 (2013): 1232–1261.

Karsa, D. R., and Tauer, K. "The role of emulsifiers in the kinetics and mechanisms of emulsion polymerization." In *Surfactants in Polymers, Coatings, Inks and Adhesives*, edited by D. R. Karsa, 32–70. Oxford: Blackwell Publishing, 2020.

Keck, C. M., and Müller, R. H. "Drug nanocrystals of poorly soluble drugs produced by high pressure homoginisation." *Eur. J. Pharm. Biopharm.* 62 (2006): 3–16.

Keddie, D. J. "A guide to the synthesis of block copolymers using reversible-addition fragmentation chain transfer (RAFT) polymerization." *Chem. Soc. Rev.* 43 (2014): 496–505.

Khoee, S., and Nouri, A. "Preparation of Janus nanoparticles and its application in drug delivery." In *Design and Development of New Nanocarriers*. 2018, pp. 145–180. https://doi.org/10.1016/B978-0-12-813627-0.00004-1

Krause, K. P., Kayser, O., Mäder, K., Gust, R., and Müller, R. H. "Heavy metal contamination of nanosuspensions produced by high-pressure homogenisation." *Int. J. Pharm.* 196 (2000): 169–172.

Kumari, A., Yadav, S. K., and Yadav, S. C. "Biodegradable polymeric nanoparticles based drug delivery systems." *Colloids Surfaces B Biointerfaces* 75 (2010): 1–18.

Lin, T. C., Hung, K. H., Peng, C. H., Liu, J. H., Woung, L. C., Tsai, C. Y., Chen, S. J., Chen, Y. T., and Hsu, C. C. "Nanotechnology-based drug delivery treatments and specific targeting therapy for age-related macular degeneration." *J. Chinese Med. Assoc.* 78 (2015): 635–641.

Loh, Z. H., Samanta, A. K., and Sia Heng, P. W. "Overview of milling techniques for improving the solubility of poorly water-soluble drugs." *Asian J. Pharm. Sci.* 10 (2014): 255–274.

Malamatari, M., Taylor, K. M. G., Malamataris, S., Douroumis, D., and Kachrimanis, K. "Pharmaceutical nanocrystals: production by wet milling and applications." *Drug Discov. Today* 23 (2018): 534–547.

Matyjaszewski, K., and Xia, J. "Atom Transfer Radical Polymerization." *Chem. Rev.* 101 (2001): 2921–2990.

Matyjaszewski, K., and Spanswick, J. "Controlled/living radical polymerization." *Mater. Today* 8 (2005): 26–33.

Mauludin, R., Müller, R. H., and Keck, C. M. "Development of an oral rutin nanocrystal formulation." *Int. J. Pharm.* 370 (2009): 202–209.

Mehnert, W., and Mäder, K. "Solid lipid nanoparticles production, characterization and applications." *Adv. Drug Deliv. Rev.* 47 (2001): 165–196.

Melian, M. E., Paredes, A., Munguía, B., Colobbio, M., Ramos, J. C., Teixeira, R., Manta, E., Palma, S., Faccio, R., and Domínguez, L. "Nanocrystals of novel valerolactam-fenbendazole hybrid with improved in vitro dissolution performance." *AAPS PharmSciTech* 21 (2020): 1–15.

Mendoza-Munoz, N., Quintanar-Guerrero, D., and Allemann, E. "The impact of the salting-out technique on the preparation of colloidal particulate systems for pharmaceutical applications." *Recent Pat. Drug Deliv. Formul.* 6 (2012): 236–249.

Möschwitzer, J. P. "Drug nanocrystals in the commercial pharmaceutical development process." *Int. J. Pharm.* 453 (2013): 142–156.

Möschwitzer, J., and Müller, R. H. "New method for the effective production of ultrafine drug nanocrystals." *J. Nanosci. Nanotechnol.* 6 (2006): 3145–3153.

Müller, R. H., and Keck, C. M. "Twenty years of drug nanocrystals: Where are we, and where do we go?" *Eur. J. Pharm. Biopharm.* 80 (2012): 1–3.

Müller, R. H., Gohla, S., and Keck, C. M. "State of the art of nanocrystals--special features, production, nanotoxicology aspects and intracellular delivery." *Eur. J. Pharm. Biopharm.* 78 (2011): 1–9.

Nguyen, T. N. G., Tran, V.-T., Duan, W., Tran, P. H. L., and Tran, T. T. D. "Nanoprecipitation for poorly water-soluble drugs." *Curr. Drug Metab.* 18 (2018): 1000–1015.

Nicolas, J., Guillaneuf, Y., Lefay, C., Bertin, D., Gigmes, D., and Charleux, B. "Nitroxide-Mediated Polymerization." *Prog. Polym. Sci.* 38 (2013): 63–235.

Noyes, A. A., and Whitney, W. R. "The rate of solution of solid substances in their own solutions." *J. Am. Chem. Soc.* 19 (1897): 930–934.

Pakzad, Y., Fathi, M., Omidi, Y., Zamanian, A., and Mozafari, M. "Nanotechnology for ocular and optic drug delivery and targeting." In *Nanoengineered Biomaterials for Advanced Drug Delivery*, 499–523, Elsevier, 2020. https://doi.org/10.1016/B978-0-08-102985-5.00021-8

Paredes, A. J., Llabot, J. M., Sánchez Bruni, S., Allemandi, D., and Palma, S. D. "Self-dispersible nanocrystals of albendazole produced by high pressure homogenization and spray-drying." *Drug Dev. Ind. Pharm.* 42 (2016): 1564–1570.

Paredes, A. J., Bruni, S. S., Allemandi, D., Lanusse, C., and Palma, S. D. "Albendazole nanocrystals with improved pharmacokinetic performance in mice." *Ther. Deliv.* 9 (2018): 89–97.

Paredes, A. J., Litterio, N., Dib, A., Allemandi, D. A., Lanusse, C., Bruni, S. S., and Palma, S. D. "A nanocrystal-based formulation improves the pharmacokinetic performance and therapeutic response of albendazole in dogs." *J. Pharm. Pharmacol.* 70 (2018): 51–58.

Paredes, A. J., Camacho, N. M., Schofs, L., Dib, A., del, M. Zarazaga, P., Litterio, N., Allemandi, D. A., Sánchez Bruni, S., Lanusse, C., and Palma, S. D. "Ricobendazole nanocrystals obtained by media milling and spray drying: Pharmacokinetic comparison with the micronized form of the drug." *Int. J. Pharm.* 585, (2020): 119501.

Pensel, P., Paredes, A., Albani, C. M., Allemandi D., Sanchez Bruni S., Palma S. D., and Elissondo M. C. "Albendazole nanocrystals in experimental alveolar echinococcosis: Enhanced chemoprophylactic and clinical efficacy in infected mice." *Parasitol.* 251 (2018): 78–84.

Pinto Reis, C., Neufeld, R. J., Ribeiro, A. J., and Veiga, F. "Nanoencapsulation I. methods for preparation of drug-loaded polymeric nanoparticles." *Nanomedicine Nanotechnology, Biol. Med.* 2 (2006): 8–21.

Puri, A., Loomis, K., Smith, B., Lee, J.-H., Yavlovich, A., Heldman, E., and Blumenthal, R. "Lipid-based nanoparticles as pharmaceutical drug carriers: From concepts to clinic." *Crit. Rev. Ther. Drug Carrier Syst.* 26 (2009): 523–580.

Raaijmakers, M. J. T., and Benes, N. E. "Current trends in interfacial polymerization chemistry." *Prog. Polym. Sci.* 63 (2016): 86–142.

Rabinow, B. E. "Nanosuspensions in drug delivery." *Nat. Rev. Drug Discov.* 3 (2004): 785–796.

Real, D. A., Hoffmann, S., Leonardi, D., Goycoolea, F. M., and Salomon, C. J. "A quality by design approach for optimization of Lecithin/Span 80 based nanoemulsions loaded with hydrophobic drugs." *J. Mol. Liq.* (2020): 114743.

Real, D. A., Martinez, M. V., Frattini, A., Soazo, M., Luque, A. G., Biasoli, M. S., Salomon, C. J., Olivieri, A. C., and Leonardi, D. "Design, characterization, and in vitro evaluation of antifungal polymeric films." *AAPS PharmSciTech* 14 (2013): 64–73.

Romero, G. B., Keck, C. M., and Müller, R. H. "Simple low-cost miniaturization approach for pharmaceutical nanocrystals production." *Int. J. Pharm.* 501, (2016): 236–244.

Salvi, V. R., and Pawar, P. "Nanostructured lipid carriers (NLC) system: A novel drug targeting carrier." *J. Drug Deliv. Sci. Technol.* 54 (2019): 255–267.

Shang, L., Cheng, Y., and Zhao, Y. "Emerging droplet microfluidics." *Chem. Rev.* 117 (2017): 7964–8040.

Shen, S., Wu, Y., Liu, Y., and Wu, D. "High drug-loading nanomedicines: Progress, current status, and prospects." *Int. J. Nanomedicine* 12 (2017): 4085–4109.

Shrimal, P., Jadeja, G., and Patel, S. "A review on novel methodologies for drug nanoparticle preparation: Microfluidic approach." *Chem. Eng. Res. Des.* 153 (2020): 728–756.

Solans, C., and Solé, I. "Nano-emulsions: Formation by low-energy methods." *Curr. Opin. Colloid Interface Sci.* 17 (2012): 246–254.

Tauer, K., and Müller, H. "On the role of initiator in emulsion polymerization." *Colloid Polym. Sci.* 281 (2003): 52–65.

Thickett, S. C., and Gilbert, R. G. "Emulsion polymerization: State of the art in kinetics and mechanisms." *Polymer (Guildf).* 48 (2007): 6965–6991.

Tuomela, A., Saarinen, J., Strachan, C. J., Hirvonen, J., and Peltonen, L. "Production, applications and in vivo fate of drug nanocrystals." *J. Drug Deliv. Sci. Technol.* 34 (2016): 1–11.

Valetti, S., Mura, S., Stella, B., and Couvreur, P. "Rational design for multifunctional non-liposomal lipid-based nanocarriers for cancer management: Theory to practice." *J. Nanobiotechnology* 11, Suppl 1 (2013): S6.

Vauthier, C., and Bouchemal, K. "Methods for the preparation and manufacture of polymeric nanoparticles." *Pharm. Res.* 26 (2009): 1025–1058.

Weng, Y., Liu, J., Jin, S., Guo, W., Liang, X., and Hu, Z. "Nanotechnology-based strategies for treatment of ocular disease." *Acta Pharm. Sin. B* 7 (2017): 281–291.

Xia, D., Gan, Y., and Cui, F. "Application of precipitation methods for the production of water-insoluble drug nanocrystals: Production techniques and stability of nanocrystals." *Curr. Pharm. Des.* 20 (2014): 408–435.

Zhang, F., Fan, J-b., and Wang, S. "Interfacial polymerization: From chemistry to functional materials." *Angew. Chemie - Int. Ed.* 59 (2020): 21840–21856.

Zhang, K., Lv, S., Li, X., Feng, Y., Li, X., Liu, L., Li, S., and Li, Y. "Preparation, characterization, and in vivo pharmacokinetics of nanostructured lipid carriers loaded with oleanolic acid and gentiopicrin." *Int. J. Nanomedicine* 8 (2013): 3227–3239.

3 Pharmacokinetics of Nanomaterials/ Nanomedicines

Mulham Alfatama and Zalilawati Mat Rashid
Universiti Sultan Zainal Abidin, Besut Campus

CONTENTS

3.1 Introduction 49
3.2 Classification of Nanosystems Based on Pharmacokinetic Properties 50
    3.2.1 Lipid Nanosystems 50
    3.2.2 Polymeric Nanoparticles 52
    3.2.3 Dendrimers 52
    3.2.4 Micelles 53
    3.2.5 Engineered Nanoparticles 53
3.3 Different Administration Routes of Nanoparticles 54
    3.3.1 Transdermal Delivery and Role of Nanoparticles 54
        3.3.1.1 Challenges to Transdermal Delivery and Nanotechnology 54
        3.3.1.2 Pharmacokinetics of Transdermal Nanoparticles 55
    3.3.2 Brain Drug Delivery and the Role of Nanoparticles 56
        3.3.2.1 Challenges to Brain Drug Delivery and Nanotechnology . 56
        3.3.2.2 Pharmacokinetics of Brain Drug Delivery of Nanoparticles 63
    3.3.3 Oral Delivery and the Role of Nanoparticles 64
        3.3.3.1 Challenges to Oral Delivery and Nanotechnology 64
        3.3.3.2 Pharmacokinetics of Oral Nanoparticles 70
    3.3.4 Inhalation Delivery of Nanodrugs 72
3.4 Conclusion 78
References 79

3.1 INTRODUCTION

Nanocarriers have emerged in medicine through many shapes and compositions, such as liposomes, nanoemulsion, dendrimers, solid lipid nanoparticles, polymeric nanoparticles, and micelles. The applications of nanomedicine are superior to the conventional ones and widely highlighted with regard to safety, improved efficacy, enhanced physiochemical properties, and better pharmacodynamic/pharmacokinetics

DOI: 10.1201/9781003168584-3

profiles of pharmaceutical encapsulants (Choi and Han 2018). The key control of the kinetic characteristics of nanomedicine is the formulation compositions.

Various formulations-based nanomedicines have been evolved and applied commercially for clinical and nonclinical uses. Nanoparticles have demonstrated unique attributes, including efficient absorption and transport across different tissues in the body, prolonged circulation in the blood, reduced immune response and inflammation, increased accumulation in the target site, and higher affinity to the receptors at the target tissue. These properties over conventional medicine are mediated by the physiochemical characteristics (e.g., size, chemical composition, and surface) of the nanoparticles (Onoue, Yamada, and Chan 2014). The endeavors to design nanomedicine with predetermined characteristics are to confer them ability treating uncurbable diseases by conventional therapeutics, as nanodrug enables improved specificity, targetability, safety, biocompatibility, broad therapeutic ranges of rapid development of new therapeutics, and/or remodeling the pharmacokinetics profiles in vivo (Choi and Han 2018). Many drugs have been incorporated into nanocarriers with the aim of expanding the efficacy and minimizing the side effects and toxicities via modulating the physiochemical characteristics and pharmacokinetic profiles of the payload (Dawidczyk et al. 2014). Particularly, prolonged circulation half-life, improved bioavailability, reduced drug concentration, and dose frequency can be achieved with oral administration of nanodrugs (Dawidczyk, Russell, and Searson 2014). Therefore, the advancement applications of nanomedicine can be highly ameliorated by controlling the pharmacokinetic properties. This chapter highlights the recent understanding of nontechnology-based drug delivery with regard to the pharmacokinetic profiles.

3.2 CLASSIFICATION OF NANOSYSTEMS BASED ON PHARMACOKINETIC PROPERTIES

Nanosystems demonstrate various pharmacokinetic properties depending on their formulations. Accordingly, the advantages and drawbacks of the commonly used nanosystem formulations are summarized in Table 3.1 (Choi and Han 2018).

3.2.1 LIPID NANOSYSTEMS

Lipid-based nanosystems including emulsions, liposomes, solid lipid nanoparticles (SLNs), nanostructure lipid carriers (NLCs), and lectin-modified solid lipid nanoparticles could be employed to control the formulation deterioration, metabolism, and potentially expand the systemic exposure. SLNs are composed of lipids with covered surfactant shell that could alleviate their distribution (Souto et al. 2020). SLNs are solid at room temperature and their advantages over liposomes and emulsions include the modulation of the bioactive substance release capacity and the improvement of skin hydration. SLNs have been reported to enhance both the integrated substances and particulates physicochemical stability. Furthermore, SLNs and NLCs protect the encapsulated bioactive substances from enzymatic degradation as well as the inhibition of trans-epidermal water loss. However, SLNs have disadvantages such as uncontrolled release from the carrier and finite load-

Pharmacokinetics of Nanomedicines

TABLE 3.1
Specific Pharmacokinetic Properties of Nanosystems

Nanosystems	Formulations	Pharmacokinetic Properties*		Other Properties*
		Advantages	Disadvantages	
Lipid nanosystems	Solid lipid nanoparticles (SLCs) Nanolipid carriers (NLCs) Liposomes Niosomes Transferosomes Ethosomes Emulsion Lectin-modified solid lipid	• Formulated substances degradation or metabolism • Enhanced systemic exposure • Delivery of specific drug	• Quick removal by reticulo-endothelial system (RES) uptake • Limitation of possible routes of administration	• Low antigenicity and toxicity • Surfactant cytotoxicity effect
Polymeric nanoparticles	Ethyl cellulose/casein Albumin Chitosan PLGA PLGA alginate PLA-PEG Hydrogel	• Stability of drug release in vivo • Drug retention time could potentially be increased	• Requirement of early burst protection • Limitation of possible routes of administration	• Low immunogenicity • Requirement for expulsion of nondegradable polymer
Dendrimers	PEGylated polylysine Polylysine Poly(amidoamine) Lactoferrin-conjugated	• Good solubility • Enhanced permeation Release control • Delivery of specific drug	• Limitation of possible routes of administration	• Blood toxicity • Low immunogenicity
Micelles		• Enhanced permeation • Good solubility • Enhanced systemic exposure	• Inadequate consistent emission	• Low immunogenicity • Surfactant cytotoxicity effect
Engineered/Inorganic-based nanoparticles	Nanocrystal Magnetic and/or metallic NPs Solumatrix fine particle Nanosized amorphous	• Enhanced systemic exposure • Longer bioavailability in mucus • Several administration routes	• Inadequate consistent emission	• NSAIDs gastric mucosal irritation relief • Toxicity due to increase C_{max}

* (Choi and Han 2018)

bearing capacity which brought upon the development of NLCs. The NLCs have a superior loading capacity and reduced active substance leaching upon storage compared to SLNs (Khosa, Reddi, and Saha 2018). NLCs consist of a mixture of both solid and liquid lipids (oils). The liquid part is either surrounded by the solid counterpart or localized at the surface of solid platelets which has a surfactant layer. Liposomes are closed colloidal lipid vesicles mainly used to encapsulate hydrophilic substances inside the core and lipophilic substances between the lipid bilayers. They have diverse morphological in terms of high hydration, temperature, and composition, while their deformability decreases with the increase of the number of cholesterols in their structure (Briuglia et al. 2015). Other lipid vesicles such as niosomes, transferosomes, and ethosomes have been developed by alteration of the liposome's composition (Chenthamara et al. 2019).

3.2.2 Polymeric Nanoparticles

The polymeric substances that are commonly used in nanoparticle formulations originated from natural sources are chitosan, alginate, cellulose, casein, and albumin, while the synthetic materials are polyalkylcyanoacylates, poly-lactic acid (PLA), poly-ε-caprolactone (PCL), poly-glycolic acid (PGA), or their co-polymers as polylactic-co-glycolic acid (PLGA) (Shakeri et al. 2020). These nanosystems commonly have characteristics of relatively stable drug release and prolonged duration of action (Choi and Han 2018). Chitosan is extensively studied natural polymer for effective delivery of macromolecules across biological surfaces compared to PLGA (Chenthamara et al. 2019). Chitosan is a cationic polysaccharide derived from partial *N*-deacetylation of chitin which could be extracted from crustacean shells. The reaction of chitin in hot alkaline media forms a copolymer that contains *N*-acetyl-glucosamine and *N*-glucosamine sequences linked by β-(1,4)-glycosidic linkages. It possesses attractive characteristics, such as biocompatible, biodegradable, nonallergenic, and nontoxic that render it a potential candidate for nanoparticulate-based delivery systems (Shakeri et al. 2020).

3.2.3 Dendrimers

Dendrimers are hyperbranched star-shaped macromolecules with nanosized dimensions, consisting of three elements: a symmetric central core (the interior dendritic branches), and an outer surface reaching the outside periphery (end-groups) that can be physicochemically or functionally modified (Abbasi et al. 2014; Wijagkanalan, Kawakami, and Hashida 2011). Contrary to linear polymer and polymeric micelles, dendrimers can be developed by controllable size, charge, and chemical characteristics of the dendrimers part and varied end-groups, endowing an excellent platform for biodistribution and pharmacokinetic targeting. They have shown good membrane permeation, controlled release, selective delivery of bioactive compounds, and solubility enhancement. Decades of studies have shown advanced generations of dendrimers with potential enhanced release and accumulation times in brain drug delivery. The ability of temozolomide (TMZ)-Lactoferrin (Lf) NPs in tumor regression investigated in glioma mice model revealed significantly improved TMZ accumulation in the brain which is can be attributed to an efficient crossing of NPs and accumulation of Lf in the brain endothelial tissue (Kumari et al. 2017). However, some potential disadvantages of these carriers have been reported including a complex synthesis process as well as limitations of the administration route, cases of immunogenicity, and blood toxicity. The pharmacokinetics could be enhanced for efficient therapy using ligand-modified dendrimers recognizable by the matching receptors of the particular tissue for better accumulation (Wijagkanalan, Kawakami, and Hashida 2011).

3.2.4 MICELLES

Micelles are dynamic colloidal amphiphilic polymers that assemble into nanoscopic spherical core-shell structures. The polymeric micelles are synthesized from amphiphilic di- or triblock copolymers, in which the hydrophobic blocks constitute the core while the hydrophilic micelles are the outer structure (Allen, Maysinger, and Eisenberg 1999). The hydrophobic core functions as a microenvironment for the embodied lipophilic drugs, while the other layer stabilizes the interface between the core and the exterior environment (Torchilin 2007). Their advantages include the enhanced permeability, solubility, bioavailability, and retention effect of the active ingredients, though some disadvantages involving the inadequate consistent release and the surfactant cytotoxicity are also recorded. Differing the sizes and composition of the hydrophilic and hydrophobic parts would influence the size, loading capacity, blood bioavailability, and properties of the polymeric micelles (Torchilin 2007). Micelles have been studied as nanocarriers for hydrophobic drugs, such as camptothecin, doxorubicin, and paclitaxel. The micelle blocks specific targeting to the required areas and allow direct administration of the drugs at improved systemic exposure and reduce clearance time (Choi and Han 2018).

3.2.5 ENGINEERED NANOPARTICLES

Engineered nanoparticles include nanocrystals, magnetic and/or metallic nanoparticles, inorganic-based nanoparticles, nanosized amorphous particles, and solumatrix fine particles. Their advantages involve improved systemic exposure and decreased retention in the mucosal layer, as well as various routes of administration. The magnetic and/or metallic nanoparticles composed of iron derivatives such as magnetic, paramagnetic, and superparamagnetic can be useful in particular for skin cell labeling or targeting, and early diagnosis of skin diseases applications (Wahajuddin and Arora 2012). Meanwhile, nanomaterials like titanium dioxide and zinc oxide have become the key ingredients in many cosmetics and sunscreen formulations due to their ability in reflecting and scattering UV light (Chenthamara et al. 2019). On the other hand, quantum dot (QD) nanoparticles have unique optical characteristics, including strong and photostable fluorescence emission, which made them suitable candidates as luminescent nanoprobes and nanocarriers for drugs delivery applications. A QD nanocarrier systems potentially enable timely and specific detection, observation, and focus treatments of disease areas as well as improvement of the drugs stability, bioavailability, targeted absorption, distribution, and metabolism (Zhao and Zhu 2016).

3.3 DIFFERENT ADMINISTRATION ROUTES OF NANOPARTICLES

3.3.1 TRANSDERMAL DELIVERY AND ROLE OF NANOPARTICLES

Human skin functions as a barrier and interface between humans and their surroundings. It is the human's largest organ, with a surface area of 1.8 to 2.0 m^2 (Prow et al. 2011). Its main regions are the epidermis, dermis, and hypodermis. The outermost epithelium of the body, the epidermis, is a protective multilayered region that acts multifunctionally as the physical barrier that restricts pathogens penetration, the chemical barrier as well as a barrier against water and electrolyte loss (Hänel et al. 2013).

The epidermis is made up of closely packed cells consisting of more than 85% of stratified keratinocytes (flattened keratins layers). These cells are continuously regenerated from the formation of new keratinocytes at stratum basale, to the transport upward across stratum spinosum, stratum granulosum, stratum lucidum, and stratum corneum (SC, the outermost layer). SC comprises thick matrix densely packed corneocytes (dead keratinocytes) (10–15 μm) that are embedded within lipid lamellae (Elias and Menon 1991).

In recent decades, transdermal drug delivery (TDD) researches involve the development of nanocarriers and nanoparticles for the purposes of overcoming the SC barrier, thus enhancing the delivery of therapeutic substances into and across the skin. Several advantages of nanocarriers over penetration by using chemical enhancers include sustainable drug release over an extended period of time and protection of the encapsulated substances from biochemical influences.

3.3.1.1 Challenges to Transdermal Delivery and Nanotechnology

TDD system utilizes intact skin as the route of administration of a topical drug into the systemic circulation. The foremost challenge in TDD is drug permeability across stratum corneum and the keratinocytes multilayered "bricks and mortar" arrangement, which is mainly responsible for the epidermal barrier against external materials as well as for water diffusion and holding capacity of the skin. Early researches on TDD covered the kinetics of the small molecule's delivery from transdermal patches with further development by the incorporation of chemical enhancers. The nicotine patches became the first most successful TDD in medicine especially in public health (Prausnitz and Langer 2009).

TDD has become a good alternative to oral delivery and hypodermic injection due to various advantages such as avoiding the first-pass liver metabolism that could prematurely metabolize drugs. TDD enables an administration of lower drug dosage with reduced toxicity (Wiedersberg and Guy 2014) as well as prolongs the duration of drug release (Prausnitz and Langer 2009). However, TDD systems still possess several disadvantages including the limited number of drugs amenable for permeation, inconsistency of drug release, toxicity-related problems, and incapable of burst release formulation (Alexander et al. 2012). Recent successful TDD only

Pharmacokinetics of Nanomedicines

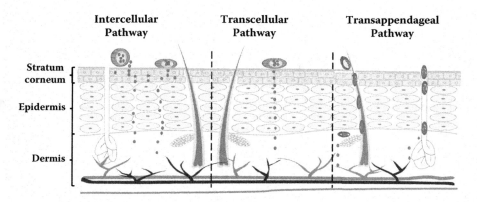

FIGURE 3.1 Schematic representation of different pathways of transdermal drug delivery systems of nanoparticles.

involved drugs having low molecular masses of up to a few hundred Daltons and displayed octanol-water partition coefficients. Delivery of hydrophilic drugs including peptides and molecules with higher molecular weight (>500 Da) via transdermal system has posed particular challenges (Chenthamara et al. 2019; Prausnitz and Langer 2009).

The ability to release the encapsulated drug before absorption through the skin or translocate intact into skin layers without being degraded, followed by the release of the drug are the crucial characteristics of a good nanosystem in TDD. Liposomes are the first nanoparticulate systems with effective delivery in TDD (Bouwstra and Honeywell-Nguyen 2002). The first liposome-encapsulated drug, Doxil (PEGylated liposome-encapsulated doxorubicin) was officially approved by the US FDA in 1995 for the treatment of AIDS related Kaposi's sarcoma (Chenthamara et al. 2019).

3.3.1.2 Pharmacokinetics of Transdermal Nanoparticles

The permeation of a particle in TDD system is likely to occur through three possible pathways which are the intercellular, transcellular, and transappendageal pathways (Figure 3.1). A particle possibly permeates via a combination of these routes, depending on the physiochemical properties of the molecule (Alexander et al. 2012; Desai, Patlolla, and Singh 2010). The intercellular pathway involves diffusion through the continuous lipid domains occupying the keratinocytes intercellular spaces. This pathway presents a significant challenge as a drug must successively permeate into and diffuse through the due to the "brick and mortar" arrangement and the stratified bilayers nature of the intercellular domain structure. This is generally acknowledged as the common permeation route of small uncharged molecules (Gandhi et al. 2012). The transcellular pathway involves diffusion through dead keratinocytes, the corneocytes. The corneocytes highly hydrate keratins provide an aqueous condition for the hydrophilic drug to pass through the skin after several partitioning and diffusion steps which eventually affect the drug bioavailability. The transappendageal pathway (shunt route) involves penetration across the skin hair follicles, sebaceous glands, and sweat glands. Despite providing a continuous channel directly across the stratum corneum, the appendageal pathway is constricted by the appendages small surface area (commonly 0.1% of surface area of the skin), thus limiting the spaces available for the drug application (Alexander et al. 2012).

The liposomes and transferosomes retain the drug in stratum corneum and only deformed in the lowest layer. The deformable liposomes also potentially followed the transappendageal route for skin penetration (Bouwstra and Honeywell-Nguyen 2002). For nanolipid systems, the penetration is suggested to be via the narrow intercellular space of the SC (Chauhan and Sharma 2019). The current researches revealed the translocation ability attributed to only minimal peptide sequence (amino acids) of CPPs such as trans-acting activator of transcription (TAT) (Chaulagain et al. 2018). The frequently used nanocarriers for TDD are liposomes, polymeric nanoparticles, and nanoemulsions. Some recently developed representative nanoplatforms and the pharmacokinetic summary for TDD are highlighted in Table 3.2.

3.3.2 BRAIN DRUG DELIVERY AND THE ROLE OF NANOPARTICLES

Blood brain barrier (BBB) is considered the most significant barrier which inhibits many drugs from entering the central nervous system (CNS), thus constricting the delivery of drugs to the brain via the blood circulatory system (Pathan et al. 2009). BBB is constituted of central nervous system (CNS) microvasculature that composed of highly specialized endothelial cells with astrocytes and pericytes as the primary components, together with other components, such as microglial cells, basement membrane, smooth muscles, microglia, and adjacent neurons (Obermeier, Daneman, and Ransohoff 2013). The neighboring endothelial cells are held tightly together by three distinctive interendothelial cell junctions; the tight, adherent, and gap junctions which act as the physical barriers (Caprifico et al. 2020).

Nanosystems have been used to transport therapeutics across the BBB without disrupting the brain physiological function. Nanosystems resolve the difficulty of crossing the BBB by shrinking of the endothelial cells and opening endothelium tight junctions for the delivery of chemotherapeutics. In anticancer therapy, amphiphilic-formed liposomes and polymeric nanosystems are the most notable and extensively exploited for brain drug delivery (Chen and Liu 2012). These nanocarriers are varied in particle sizes (range from 10 to 1000 nm), can carry a range of drug payload as well as able to control drug release (Tawfik et al. 2020). These noninvasive brain drug delivery methods are suitable alternatives for treatments of neurodegenerative diseases, while could also mitigate the high cost and risk factors of conservative surgery, radiotherapy, and chemotherapy treatments.

3.3.2.1 Challenges to Brain Drug Delivery and Nanotechnology

Methods for drug delivery across BBB can generally be categorized into direct injection and implantation, permeability enhancers, chemical modifications, and nanoplatforms via the intravenous (IV) and intranasal pathways (Chenthamara et al. 2019). Efficient entry into CNS systemic circulation via nasal administration is limited to lipophilic low molecular weight drugs. The application of nanosystems in BBB drug delivery have been reported to overcome the nasal administration

TABLE 3.2
Pharmacokinetics Summary for Selected Transdermal Drug Delivery Nanosystems

Nanosystem	Drug	Formulation	Dosage Form	Pharmacokinetic Profiles	Ref.
Nanostructured lipid carrier (NLC)	Flurbiprofen (Arthritis treatment)	Flurbiprofen (FP), coconut oil, stearic acid, soya lecithin	Dispersion PS (nm): 214; ZP (mV): −30.70; EE (%): 92.58	• Animal model: Male albino Wistar rats (230 ± 10 g) Dose: 4 mg/kg Groups: 6 rats each (FP-NLC, commercial gel) • Results for commercial gel: $C_{max} = 38.42 \pm 1.37$ µg/mL; $t_{max} = 6.00 \pm 0.70$ h; $AUC = 195.73 \pm 44$ µg·h/mL • Results for FP-NLC gel nanosystem: $C_{max} = 34.18 \pm 1.28$ µg/mL; $t_{max} = 8.00 \pm 0.45$ h; $AUC = 335.52 \pm 56$ µg·h/mL	(Kawadkar et al. 2013)
Solid lipid nanoparticles (SLN)	Aconitine (Intractable pain treatment)	Aconitine (AN), oil (Compritol® 888 ATO), surfactant (Cremophor® EL), co-surfactant (Transcotol P)	Microemulsion EE (%): 87.81	• Animal model: Male Sprague-Dawley (300 ± 20 g) Groups: 5 rats each (AN-SLN, AN tincture) • Results for AN-SLN nanosystem: $C_{max} = 1.21 \pm 0.11$ µg/mL; $t_{max} = 300$ min; $AUC_{0-t} = 346.31 \pm 25.21$ µg·min/mL • Results for AN tincture: $C_{max} = 0.34 \pm 0.14$ µg/mL; $t_{max} = 60$ min; $AUC_{0-t} = 152.94 \pm 28.95$ µg·min/mL	(Zhang et al. 2015)
					(Alam et al. 2016)

(Continued)

TABLE 3.2 (Continued)
Pharmacokinetics Summary for Selected Transdermal Drug Delivery Nanosystems

Nanosystem	Drug	Formulation	Dosage Form	Pharmacokinetic Profiles	Ref.
Nanostructured lipid carriers (NLC)	Pioglitazone (Type 2 diabetes treatment)	Pioglitazone (PZ), apifil, labrasol, Carbopol®, Tween-80, triethanolamine	DispersionPS (nm): 166.05:1; ZP (mV): -27.5EE (%): 84.56%	• Animal model: Albino Wistar rats (100-150 g) Dose: 1 mg/kg Groups: 6 rats each (PZ tablet, PZ-PNLG-TTS) • Results for PZ oral tablet: $C_{max} = 65.67 \pm 61.41$ ng/ml; $t_{max} = 2.14 \pm 0.21$ h; $AUC_{0-\infty}$: 578.21 ± 18.45 ng·h/mL • Results for PZ-PNLG–TTS patch nanosystem: $C_{max} = 54.19 \pm 14.67$ ng/ml; $t_{max} = 8.57 \pm 1.98$ h; $AUC_{0-\infty}$: 1461.54 ± 76.34 ng·h/mL	(Joshi et al. 2017)
Solid lipid nanoparticles (SLN)	Colchicine (Acute gout treatment)	Colchicine (C), kolliwax glyceryl monostearate, Tween-20, sodium lauryl sulfate	Pellets PS (nm): 107 nm; ZP (mV): -17.4 mV; EE (%): 37.25%	• Animal model: Wistar rats (230-250 g) Dose: 5 mg/kg Groups: 6 rats each (colchicine patch, C-SLNs) • Results for free colchicine patch: $C_{max} = 39.74 \pm 1.38$ ng/ml; $t_{max} = 4$ h; $AUC_{0-\infty}$: below 10 ng·h/mL • Results for C-SLNs patch: $C_{max} = 37.75 \pm 3.38$ ng/ml; $t_{max} = 6$ h; $AUC_{0-\infty}$: 30.27 ± 2.80 ng·h/mL	

(Continued)

TABLE 3.2 (Continued)
Pharmacokinetics Summary for Selected Transdermal Drug Delivery Nanosystems

Nanosystem	Drug	Formulation	Dosage Form	Pharmacokinetic Profiles	Ref.
Nanostructured lipid carrier (NLC)	Rivastigmine (Dementia treatment)	Rivastigmine (RG), oil, Tween-80, span-80, eudragit E-100 (EE-100), poly-butyl methacrylate-co-methyl methacrylate (PBMACMA)	NLC in patchNLC: PS (nm): 134.60 ± 15.10;ZP (mV): −11.80 ± 2.24;EE (%): 70.56 ± 1.20	• Animal model: Albino Wistar rats (200-250 g) Dose: 30 mg RG/0.25 cm^2 Groups: 6 rats each (RG-NLCs-PBMACMA, RG-NLCs-EE-100, commercial RG-Exelon® patches) • Results for commercial RG-Exelon® patch: C_{max} = 16.92 ± 0.31 µg/mL; t_{max} = 8.03 ± 0.11 h; AUC_{0-72} = 552.17 ± 3.65 µg·h/mL • Results for NLCs-PBMACMA patch nanosystem: C_{max} = 16.94 ± 0.58 µg/mL; t_{max} = 36.10 ± 0.25 h; AUC_{0-72} = 865.70 ± 5.88 µg·h/mL • Results for RG-NLCs-EE-100 patch nanosystem: C_{max} = 16.87 ± 0.62 µg/mL; t_{max} = 24.14 ± 0.18 h; AUC_{0-72} = 823.87 ± 5.52 µg·h/mL	(Chauhan and Sharma 2019)
Nanolipid carrier (NLC), Nanoemulsion (NE)	Triptolide (TPL) (Autoimmune and inflammatory diseases treatment)	TPL, Compritol 888 ATO (solid lipid), Capryol 90 (liquid lipid), mixed surfactants	NLCPS (nm): 139.6.0 ± 2.53; ZP (mV): −36.03 ± 2.41 mV	• Animal model: Male Sprague-Dawley rats (200 ± 10 g) Dose: TPL-NLCs 400 µg/kg, 0.5 g TPL-NE Groups: 3 rats each (TPL-NLCs, TPL-NE)	(Gu et al. 2019)

(Continued)

TABLE 3.2 (Continued)
Pharmacokinetics Summary for Selected Transdermal Drug Delivery Nanosystems

Nanosystem	Drug	Formulation	Dosage Form	Pharmacokinetic Profiles	Ref.
		(Tween 80, Transcutol HP, soybean oil)		• Results for TDD TPL-NLCs nanosystem: C_{max} = 1378.81 ± 37.68 ng/ml; t_{max} = 2.5 ± 0.46 h; $AUC_{0-\infty}$; 7954.93 ± 61.76 ng·h/mL; AUC_{0-12}; 6180.20 ± 81.50 ng·h/mL • Results for TDD TPL-NE nanosystem: C_{max} = 976.21 ± 3.34 ng/ml; t_{max} = 2.00 ± 0.43 h; $AUC_{0-\infty}$; 7208.27 ± 40.65 ng·h/mL; AUC_{0-12}; 6523.17 ± 30.13 ng·h/mL	(Sita and Vavia 2020)
Nanoemulsion (NE)	Bromocriptine mesylate (Parkinson's disease treatment)	Bromocriptine mesylate (BCM), oil, surfactant, co-surfactant, and co-solvent, carbopol 974P	GelPS (nm): 160; ZP (mV): −20.4; Drug content: 99.45 ± 1.9%	• Animal model: Adult male Wistar rats • Dose: 30 mg/kg • Groups: 6 rats each (TDD BCM-NE, ODD aqueous BCM) • Results for TDD BCM-NE gel nanosystem: C_{max} = 37.2 ± 4.22 ng/mL; t_{max} = 8 h; AUC_{0-48} = 562.54 ± 77.55 ng·h/mL • Results for ODD BCM suspension: C_{max} = 84.04 ± .04 ng/mL; t_{max} = 1 h; AUC_{0-48} = 204.96 ± 51.93 ng·h/mL	
Polymeric nanoparticles	Rasagiline mesylate (Parkinson's disease treatment)	Rasagiline mesylate (RM), poly (lactic-co-glycolic acid) (PLGA)	Film	• Animal model: Adult male Wistar rats (250 - 300 g) • Dose: 0.088 mg/kg	(Bali and Salve 2020)

(Continued)

TABLE 3.2 (Continued)
Pharmacokinetics Summary for Selected Transdermal Drug Delivery Nanosystems

Nanosystem	Drug	Formulation	Dosage Form	Pharmacokinetic Profiles	Ref.
Solid lipid nanoparticles (SLN), Nanolipid carrier (NLC)	Ropinirole (Parkinson's disease treatment)	Ropinirole (RP)dynasan-114,caproyl 90,soylecithin,carbopol 934 poloxamer 188, tween 20, water	HydrogelPS (nm):210.6 - 193.2;Drug content (%): 98.2 – 99.6	Groups: 5 rats each (TDD RM-NPs, ODD RM-NPs, IV RM-NPs) • Results for TDD RM-NPs patch nanosystem: $C_{max} = 89.36 \pm 4.92$ ng/mL; $t_{max} = 24 \pm 4.58$ h; $AUC_{0-t} = 4505.40 \pm 34.99$ ng·h/mL • Results for ODD RM-NPs nanosystem: $C_{max} = 94.29 \pm 5.20$ ng/mL; $t_{max} = 2 \pm 0.31$ h; $AUC_{0-t} = 675.06 \pm 21.67$ ng·h/mL • Results for IV RM-NPs nanosystem: $C_{max} = 97.34 \pm 5.10$ ng/mL; $t_{max} = 0.5 \pm 0.02$ h; $AUC_{0-t} = 388.41 \pm 19.75$ ng·h/mL • Animal model: Male Wistar rats (180–210 g) • Dose: 1.1 mg/kg • Groups: 6 rats each (topical administered, hydrogels RP-C, RP-SLN-C and RP-NLC-C) • Results for free RP-C: $C_{max} = 6.1 \pm 1.3$ μg/mL; $t_{max} = 3$ h; $AUC_{tot} = 34.8 \pm 2.9$ μg·g/mL • Results for RP-SLN-C nanosystem:	(Dudhipala and Gorre 2020)

(Continued)

TABLE 3.2 (Continued)
Pharmacokinetics Summary for Selected Transdermal Drug Delivery Nanosystems

Nanosystem	Drug	Formulation	Dosage Form	Pharmacokinetic Profiles	Ref.
				• $C_{max} = 8.1 \pm 1.7$ μg/mL; $t_{max} = 6$ h; $AUC_{tot} = 69.8 \pm 5.6$ μg·h/mL • Results for RP-NLC-C nanosystem: $C_{max} = 8.9 \pm 1.4$ μg/mL; $t_{max} = 6$ h; $AUC_{tot} = 76.8 \pm 4.8$ μg·h/mL;	
Invasomes	Agomelatine (AGM), (Antidepressant)	Agomelatine (AGM), phosphatidylcholine (1:10), limonene (1.5% w/v)]	GelAGM-I2-LFU-C4: PS (nm): 313; ZP (mV): −64; EE (%): 78.6	• Animal model: Adult male albino rabbits (2.0–2.5 kg) Dose: 1 mgGroups: 3 rabbits each (TDD AGM-I2-LFU-C4; ODD AGM aqueous dispersion) • Results for TDD AGM-I2-LFU-C4 gel nanosystem: $C_{max} = 178.93 \pm 20.20$ ng/mL; $t_{max} = 0.5$ (0.5–1) h; $AUC_{0-\infty} = 364.42 \pm 46.87$ ng·h/mL; $AUC_{0-12h} = 354.47 \pm 44.73$ ng·h/mL • Results for ODD AGM aqueous dispersion: $C_{max} = 27.94 \pm 5.50$ ng/mL; $t_{max} = 0.5$ (0.5–0.5) h; $AUC_{0-\infty} = 49.94 \pm 7.40$ ng·h/mL; $AUC_{0-12h} = 48.85 \pm 7.09$ ng·h/mL	(Tawfik et al. 2020)

*PS = particle size; ZP = zeta potential; EE = entrapment efficiency; C_{max} = peak plasma concentration; T_{max} = time to reach C_{max}; AUC = area under the curve; TDD = transdermal drug delivery; ODD = oral drug delivery; IV = intravenous delivery.

constraint and improves drug retention time in the CNS (Ding et al. 2020). The ideal characteristics of nanosystems for BBB drug delivery should possess particle size with diameter less than 100 nm, BBB-targeted moiety (receptor-mediated endocytosis or monocytes or macrophages uptake), ability to evade the reticuloendothelial systems, prolong the duration of circulation, carry small molecules, peptides, proteins, or nucleic acids, as well as exhibiting nontoxicity and noninflammatory effects, stability in blood (without aggregation and dissociation), controlled or modulated profiles of drug release, biodegradable, biocompatible, and manufacturing cost-effectiveness (Bhaskar et al. 2010).

The biocompatibility of the natural vesicles such as liposomes and niosomes is due to the presence of the phospholipid bilayer. They have been reported to easily recognize and target the ligand, remain circulated in the blood, while evading the body immune system (Chen et al. 2016). Meanwhile, polymeric nanosystems offer certain advantages for application in BDD which include sustainable release, low toxicity, biodegradability, biocompatibility, and mucoadhesive. They could improve drugs permeation and bioavailability by reducing enzymatic and hydrolytic degradation (Jena, Mcerlean, and Mccarthy 2020).

3.3.2.2 Pharmacokinetics of Brain Drug Delivery of Nanoparticles

The mechanisms for drug transport across BBB involve paracellular pathway, transcellular pathway, carrier-mediated transport, receptor-mediated transcytosis, adsorptive-mediated transcytosis, and transport proteins (Figure 3.2). Paracellular pathway is the diffusion of small hydrophilic substances (molecular weight <500 Da) through a space between neighboring endothelial cells. Transcellular pathway is the distribution of solute particle through the endothelial cells. Receptor-mediated transcytosis (RMT) mechanism uses the cell surface receptors to explicitly bind with the NPs ligand, forming an intracellular vesicle through membrane enfolding. Adsorptive mediated transcytosis (AMT) is a technique of transporting charged nanosystems or macromolecules across BBB. In general, various drugs in their free

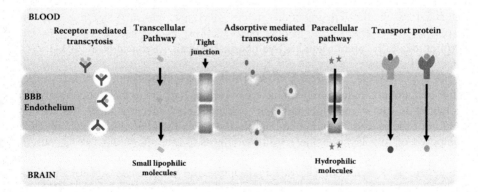

FIGURE 3.2 Schematic representation of different pathways of blood brain drug delivery systems of nanoparticles.

form enter the brain via the transcellular pathway, while most nanosystems apply adsorptive mediated transcytosis and receptor-mediated transcytosis as the two major pathways in delivering neurotherapeutics (Ding et al. 2020).

The size and zeta potential of NPs have been found to play important role in NPs endocytosis by brain endothelial cells. For example, previous studies indicated the effectiveness of NPs exhibiting dimensions smaller than 200 nm and a positive ZP value in crossing the BBB (Table 3.3). In addition, polymeric NPs functionalized with several ligands such as lactoferrin, transferrin, and lipoprotein have exhibited a better transcytosis across BBB (Kumari et al. 2017; Shakeri et al. 2020). Besides, multifunctional or self-assembled protein nanovesicles could also potentially carry drugs that normally exhibit poor ability to cross the BBB. For example, the SSTR2 peptide tagged nanoencapsulated DIM (SSTR2-pep-DIM-NP) could control drug release at a longer duration compared to nanoencapsulated DIM (DIM-NP) (Bhowmik et al. 2017). Some studies reported the advantages of inorganic NPs such as better stability and distinct substance- and size-dependent physicochemical properties compared to polymeric and biomimetic NPs. However, inorganic NPs could also disrupt BBB properties and function (Ding et al. 2020). The pharmacokinetic summary of some of the recently developed representative nanoplatforms for transdermal drug delivery are highlighted in Table 3.3.

3.3.3 ORAL DELIVERY AND THE ROLE OF NANOPARTICLES

Oral delivery is among the most convenient route of drug administration owing to fundamental characteristics including ease and self-administration, painless, enhanced patient compliance, and relatively cheaper compared to injections. The selection of route of administration is influenced by various factors including characteristic of the drug, location, and type of the disease. However, oral route is also associated with disadvantages and limitations that hinder the research and development of poorly soluble, stable, and biocompatible drugs as well as protein delivery. Also, oral drugs usually exhibit low intestinal membrane permeability and immediate release that may cause toxicity. In addition, drugs given orally are subjected to chemical degradation, mucus clearance, gastric emptying, and intestinal motility. This results in poor bioavailability as a consequence of low absorption of orally administered active pharmaceuticals (Agrawal et al. 2014).

3.3.3.1 Challenges to Oral Delivery and Nanotechnology

After oral administration, drug should cross various barriers to the blood circulation including the low stomach pH and high acid, intestinal lumen, the layer of mucus on the top of the intestinal epithelium, and lastly the epithelium cells. The stomach is composed of four layers namely from outermost to innermost: the serosa, muscularis externa, submucosa, and mucosa. It is coated by mucus membrane with glands for gastric acid secretion. The intestinal epithelium is the main site for drug absorption assisted by villi to increase the surface area. The physiology of the GIT can hinder the drug absorption and bioavailability as a result of poor mucosal permeability and premature drug degradation prior reaching to sites of absorption (Lin et al. 2017). The presence of P-glycoproteins mediated efflux in the epithelial cell

TABLE 3.3
Pharmacokinetics Summary for Selected Blood Brain Drug Delivery Nanosystems

Nanosystem	Drug	Formulation	Dosage Form	Pharmacokinetic Profiles	Ref
Solid lipid nanoparticles (SLNs)	Haloperidol (Antipsychotic medication)	Haloperidol (HP), tween-80	Formulation PS (nm): 115.1; ZP (mV): −16.7; EE (%): 71.56	• Animal model: Albino Wistar rats (200–250 g) • Dose: 0.893 mg/kg I.N • Groups: 6 rats each (HP-SLNs, HP solution). • Results for free: $C_{max} = 90.13$ ng/mL; $t_{max} = 2$ h; $AUC_{0-\infty} = 683.15$ ng·h/mL • Results for HP-SLNs nanosystem: $C_{max} = 329.177$ ng/mL; $t_{max} = 2$ h; $AUC_{0-\infty} = 2389.17$ ng·h/mL	(Yasir and Sara 2014)
Polymeric nanoparticles	3,30-Diindolylmethane (Glioblastoma treatment)	3,30-Diindolylmethane (DIM), Poly (lactic-co-glycolic acid), SSTR2 peptide	Suspension PS (nm): 28–98; EE (%): 70%	• Animal model: Adult Sprague-Dawley rats (180–220 g) • Dose: 2.5 mg/kg • Groups: 6 rats each (3,3´-diindolylmethane (DIM), nanoencapsulated DIM (DIM-NP) and SSTR2 peptide tagged nanoencapsulated DIM (SSTR2-pep-DIM-NP)). • Results for free DIM: $C_{max} = 1.6$ μg/mL; $t_{max} = 2$ h; $AUC_{0-\infty} = 19$ μg·h/mL • Results for DIM-NP nanosystem:	(Bhowmik et al. 2017)

(Continued)

TABLE 3.3 (Continued)
Pharmacokinetics Summary for Selected Blood Brain Drug Delivery Nanosystems

Nanosystem	Drug	Formulation	Dosage Form	Pharmacokinetic Profiles	Ref

Solid lipid nanoparticles (SLNs)	Agomelatine (AGM) (Melatonin receptor agonist)	Agomelatine (AGM), Gelucire 43/01, glyceryl tripalmitate, glyceryl tristearate	DispersionPS (nm): 167.7; ZP (mV): −17.9; EE (%): 91.25	• C_{max} = 45 μg/mL; t_{max} = 8 h; $AUC_{0-\infty}$ = 236.21 μg·h/mL
				• Results for SSTR2 pep-DIM-NP nanosystem:
				  C_{max} = 85 μg/mL; t_{max} = 8 h; $AUC_{0-\infty}$ = 288.34 μg·h/mL
				• Animal model: Adult rats (70–250 g)
				• Dose: 2.14 μg/g
				• Groups: rats each (AGM solution I.V, AGM-SLN-14).
				• Results for free AGM:
				  C_{max} = 227.00 ng/mL; t_{max} = 5 min; AUC_{0-t} = 10407.65 ng·min/mL
				• Results for AGM-SLN-14 nanosystem:
				  C_{max} = 165.62 ng/mL; t_{max} = 5 min; AUC_{0-t} = 12499.60 ng·min/mL
					(Fatouh, Elshafeey, and Abdelbary 2017)

Polymer-lipid nanoparticle (NP)	Docetaxel (DTX) (anti-mitotic drug)	Docetaxel (DTX), amphiphilic polymer, dodecylamine, ethyl arachidate, Pluronic® F-68	EmulsionPS (nm): 100.1; ZP (mV): −48; EE (%): 98.2	• Animal model: Immune deficiency (SCID) mice • Dose: 20 mg/kg DTX • Groups: 3 mice each (DTX-NP, Taxotere® commercial drug). • Results for free drug DTX-NP: C_{max} = 0.78 × 10^3 ng/g; t_{max} = 0.64 h; AUC_{0-24h} = 4.83 × 10^3 ng·h/g	(He et al. 2017)

(Continued)

TABLE 3.3 (Continued)
Pharmacokinetics Summary for Selected Blood Brain Drug Delivery Nanosystems

Nanosystem	Drug	Formulation	Dosage Form	Pharmacokinetic Profiles	Ref
Nanoparticle	Temozolomide (Glioma treatment)	Temozolomide (TMZ), DMSO, Lactoferrin (Lf), PBS	Dispersion PS (nm): 70 ± 10; EE (%): 42 ± 4.9	• Results for free Taxotere®: $C_{max} = 0.25 \times 10^3$ ng/g; $t_{max} = 0.43$ h; $AUC_{0-24h} = 0.95 \times 10^3$ ng·h/g • Animal model: Adult C57BL/6 mice (22–24 g) • Dose: 10 mg/kg • Groups: 3 mice each (TMZ, TMZ-loaded lactoferrin nanoparticles (TMZ-LfNPs)) • Results for free TMZ: $t_{1/2} = 68 \pm 2.13$ min; AUC = 4639.66 ± 109.24 µg·min/mL; Mean residence time (MRT) = 88 ± 7.4 min • Results for TMZ-LfNPs nanosystem: $t_{1/2} = 360 \pm 70.86$ min; AUC = 10998 ± 754 µg·min/mL; Mean residence time (MRT) = 367 ± 18.33 min	(Kumari et al. 2017)
Solid lipid nanoparticles (SLN), Nanolipid carrier (NLC)	Curcumin	SLN: Curcumin, acetyl palmitate, tween 80; NLC: Curcumin, acetyl palmitate, tween 80, liquid lipid	SLN: PS (nm): 204.76 ± 0.36 EE (%): 82 ± 0.49; NLC: PS (nm): 117.36 ± 1.36; EE (%): 94 ± 0.74	• Animal model: Male Sprague-Dawley rats (200–250 g) • Dose: 4 mg/kg • Groups: 3 rats each (curcumin solution, Cur-SLN, Cur-NLC) • Results for free curcumin solution: $C_{max} = 0$ ng/g; $t_{max} = -$ h; $AUC_{0-t} = 0$ ng·h/g	(Malvajerd et al. 2019)

(Continued)

TABLE 3.3 (Continued)
Pharmacokinetics Summary for Selected Blood Brain Drug Delivery Nanosystems

Nanosystem	Drug	Formulation	Dosage Form	Pharmacokinetic Profiles	Ref
		(oleic acid), cholesterol		• Results for Cur-SLN nanosystem: $C_{max} = 114.22 \pm 58.21$ ng/g; $t_{max} = 0.5 \pm 0.28$ h; $AUC_{0-t} = 116.31 \pm 31.45$ ng·h/g • Results for Cur-NLC nanosystem: $C_{max} = 390.30 \pm 35.93$ ng/g; $t_{max} = 1 \pm 0.28$ h; $AUC_{0-t} = 505.76 \pm 38.47$ ng·h/g	
Polymeric nanoparticles	Selegiline (Parkinson's disease treatment)	Selegiline (SG) hydrochloride, sodium tripolyphosphate (STPP), and chitosan oligosaccharide (>95% deacylation)	Formulation PS (nm): 341.6; ZP (mV): 3.4; EE (%): 92.20	• Animal model: male Sprague-Dawley rats (300–400 g) • Dose: 1 mg/kg I.N • Groups: 3 rats each (SG solution, SG termosensitive gel, SGNP) • Results for free SG solution: $C_{max} = 1.67 \pm 0.04$ ng/g; $t_{max} = 5$ min; $AUC_{0-24h} = 5.84 \pm 0.19$ ng·h/g • Results for SG gel: $C_{max} = 0.93 \pm 0.03$ ng/g; $t_{max} = 5$ min; $AUC_{0-24h} = 4.25 \pm 0.13$ ng·h/g • Results for SGNP nanosystem: $C_{max} = 3.93 \pm 0.04$ ng/g; $t_{max} = 5$ min; $AUC_{0-24h} = 6.42 \pm 0.019$ ng·h/g	(Sridhar et al. 2018)
Solid lipid nanoparticles (SLN)	Acyclovir (Herpes simplex virus (HSV) treatment)	Acyclovir (CV), Compritol, Pluronic® F68, lecithin	Suspension PS (nm): 320; ZP (mV): 34.9; EE (%): 86.41	• Animal model: Male New Zealand white rabbits (1.8–2.2 kg) • Dose: 62 mg/kg • Groups: 6 rabbits each (ACV, ACV-SLN)	(El-Gizawy, El-Maghraby, and Hedaya 2019)

(Continued)

TABLE 3.3 (Continued)
Pharmacokinetics Summary for Selected Blood Brain Drug Delivery Nanosystems

Nanosystem	Drug	Formulation	Dosage Form	Pharmacokinetic Profiles	Ref
				• Results for free ACV: Terminal volume of distribution (V) = 0.17 (mg)µg/mL; $t_{1/2}$ = 1.83 h; $AUC_{0-\infty}$ = 124.30 µg·h/mL • Results for ACV-SLN nanosystem: Terminal volume of distribution (V) = 0.97 (mg)µg/mL; $t_{1/2}$ = 3.66 h; $AUC_{0-\infty}$ = 233.26 µg·h/mL		
Nanoparticle, Microemulsion	Quetiapine (Antipsychotic)	Quetiapine fumarate, capmule MCM-EP, tween-80, chitosan, labrasol, transcutol-P, sodium tripolyphosphate	NP:PS (nm): 131.08 ZP (mV): 34.4 ME: PS (nm): 29.75; ZP (mV): 20.99	• Animal model: Adult Sprague-Dawley rats (250-300 g) • Dose: 2.3 mg/kg (IV) • Groups: 4 rats each (Quetiapine-loaded nanoparticles (QNP), Quetiapine-loaded microemulsion (QME)) • Results for QNP nanosystem: C_{max} = 83.21 ng/g; t_{max} = 15 min; AUC_{0-240} = 8152.12 ng·min/g • Results for QME nanosystem: C_{max} = 59.73 ng/g; t_{max} = 30 min; AUC_{0-240} = 8816.71 ng·min/g	(Shah, Khunt, and Misra 2021)

*PS = particle size; ZP = zeta potential; EE = entrapment efficiency; C_{max}, peak plasma concentration; T_{max} = time to reach C_{max}; AUC = area under the curve; TDD = transdermal drug delivery; IN = internasal delivery; IV = intravenous delivery.

membrane is another barrier against drug absorption. The physiochemical properties of the drug on the other hand may alter the pharmacokinetics attributes. Poor drug solubility impacts absorption negatively as the drug must be soluble at the absorption site. Moreover, inadequate partition coefficient negates drug permeation through lipid membrane. In addition, drug affinity for the first-pass metabolism determines the proportions in the blood and hence the bioavailability (Sharma, Sharma, and Jain 2016).

Conventional approaches to formulate poorly water-soluble drugs are associated with delayed onset of action, low bioavailability, poor dosage consistency, inability to attain steady state plasma concentration, and side effects. Thus, poor patient compliance may arise due to under- or over dose effects. These limitations can be resolved through a novel and effective drug delivery systems that enable reduction of the doses frequency and size, site-targeting specificity, improved permeability, and enhanced oral bioavailability. Nanotechnology-based drug delivery is a favorable approach for potent drugs with limited clinically values due to their low solubility, poor absorption, insufficient bioavailability, and other inappropriate biopharmaceutical profiles (Delmar and Bianco-Peled 2016; Gupta, Bhandari, and Sharma 2009). The most recent and commonly used types of nanodrug-based systems are polymeric nanoparticles, nanoemulsion, liposomes, solid lipid nanoparticles, dendrimers, micelles, and carbon nanotubes that offer targeted, sustained, and controlled drug release properties.

3.3.3.2 Pharmacokinetics of Oral Nanoparticles

Oral delivery of nanoparticles has attracted a considerable attention to deliver wide range of active pharmaceutical ingredients to treat various diseases. Proteins and peptides delivered orally via nanotechnology have been reported excessively in recent decades with main aim of protection from pH and enzymatic degradation in the stomach. Shi et al. have prepared liraglutide nanoparticles for oral delivery to enhance the half-life and avoid the daily subcutaneous injection to maintain the blood glucose levels in the normal range in type 2 diabetics. Despite the numerous studies of GLP-1 analogs as orally administered dosage form, the structural chemistry is decomposed easily by gastric acidic fluids, digestive enzymes, and intestinal secretions. On the other hand, peptide absorption is also restrained by gastrointestinal barriers. Foreign materials are trapped and removed by the action of negatively and naturally charged mucin and mucus barrier, respectively. This poses a big obstacle to the peptide to cross the epithelium cells into the blood circulation. Epithelial cells and the tight junctions between them are forming the epithelial structure that selectively permits the molecules passage. Peptide can be transported through the intercellular spaces after the opening of the tight junctions or via transcellular pathway, crossing the two main structural membranes of epithelial cells, namely: the apical side and the basal side, while combating the destruction effects by lysosomes, endosomes, and Golgi apparatus (Figure 3.3) (Shi et al. 2020). In another study by Purvin et al., the pharmacokinetics of the solid lipid nanoparticles loaded zidovudine were studied in rats after oral administration. Compared to free- zidovudine, nanodrug showed 31.25% higher and 1.83-fold increase of area under the curve and mean resident time, respectively, in plasma and

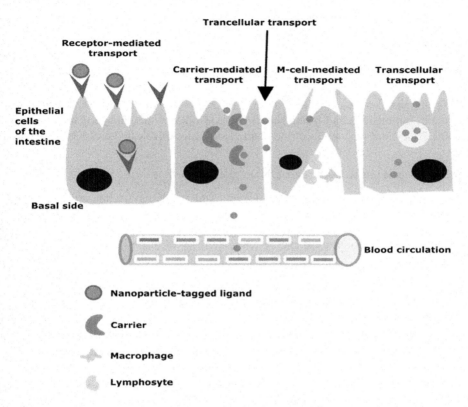

FIGURE 3.3 Schematic representation of different mechanisms of nanoparticles uptake in the intestinal epithelium.

tissue using reverse phase high performance liquid chromatography. Moreover, the elimination half-life of the nanodrug was prolonged and the drug distribution pattern was altered (Purvin et al. 2014).

Zinc oxide (ZnO) nanoparticles were prepared at a range size of 20 nm to 70 nm and different surface charges to investigate their pharmacokinetics attributes after a single dose orally administered to the rats (Paek et al. 2013). The effect of the particle size on biodistribution was negated, whereas the negatively charged nanoparticles exhibited enhanced kinetic profiles in the blood circulation compared to the positive ones. The dissolution of ZnO nanoparticles was reduced at the gastric medium than the zinc ions. Despite the increased of excretion kinetics for the smaller particle size, this study concluded that surface charge of the nanoparticles was the key factor to alter the pharmacokinetic profiles of ZnO nanoparticles.

The physiochemical attributes of nanoparticles including surface area, surface charge, particle size, and shape, significantly alter the pharmacokinetic properties. Despite the general consensus that the surface of the nanoparticles affects the intestinal absorption, there is a lack of standardizations. However, there is a crucial point that scientists generally agree that crossing the mucosal layer is mediated by the propensities of the nanoparticles surface (Cao et al. 2019). Many approaches have been applied to facilitate the penetration ability of the nanoparticles via manipulating the physiochemical properties of the nanoparticles and introducing a functional group on their surface to confer targeting property (Rao, Yajurvedi, and Shao 2008). It is commonly recognized that nanoparticles are easily eternalized into the cells and tissues compared to large molecules, which bring about increased tissue accumulations and prolonged circulation time of the nanoparticles in vivo. Table 3.4 displays examples of the pharmacokinetics of selected oral nanosystems.

3.3.4 Inhalation Delivery of Nanodrugs

Nanotherapeutics-based pulmonary delivery via inhalation techniques has been attracting considerable attention over the recent years and has been an investigational research area for several decades. The unique physiological characteristics of the lungs allowing both local and systemic delivery of therapeutics including high blood flow exposure, large absorptive area, slow surface clearance, and thin alveolar epithelial layer. The appealing noninvasive attribute of inhaled drugs enables localized administration to the lungs with improved bioavailability and efficiency of various formulations to a specific site of action followed by systemic distribution. Pulmonary local delivery is highly preferred over conventional invasive techniques in overcoming toxicities, degradations in the gastrointestinal tract, and hepatic first-pass metabolism (Shen and Minko 2020). Conversely, systemic administration is usually associated with increased adverse effects due to drug distribution to both diseased and healthy tissues. Nonetheless, most of the drugs, nucleic acids, and proteins are unable to be administrated into the lungs in their free form, thus, demanding a special formulation or nanoformulation for inhalation delivery.

Noninvasive lung delivery of drugs is a great choice for local effect oriented therapy, such as treatment of airways diseases, lung cancer, chronic obstructive pulmonary disease, and cystic fibrosis (Karathanasis et al. 2005). Lung exhibits reduced drug metabolism driven by enzyme activities intracellularly and extracellularly, hence drug bioavailability could be enhanced, compared to other delivery approaches like injections or oral (Loira-Pastoriza, Todoroff, and Vanbever 2014). Moreover, lung delivery permits rapid onset of action, higher absorption rate, and reduced drug doses. However, the inhaled drug localization, adsorption, and penetration is faced by physiological barriers including macrophages, mucus, and ciliated cells (Ruge, Kirch, and Lehr 2013). Drug deposition determines the mechanism of clearance, where upper airways deposition of drug is removed by ciliated cells, whereas alveolar macrophages take place for drug removal at lower airways localization (Sung, Pulliam, and Edwards 2007). Upon approaching to the lower part of the airway, drugs are engulfed and lysosomal digested by macrophages after labeling it as a foreign substance. The bioavailability of chemotherapeutics at the cancer cells of the lungs is the key determinant of successful therapy. This is mediated by controlling several drug-related factors, such as drug efflux, aqueous solubility, stability, dissolution rate, and alveolar macrophages clearance (Bohr et al. 2020) (Figure 3.4).

TABLE 3.4
Pharmacokinetics Summary for Selected Oral Nanosystmes

Nanosystem	Drug	Formulation	Dosage Form	Pharmacokinetic Profiles	Ref
Nanosuspension	Darunavir/ritonavir	Alginate + chitosan	Microparticle	• Animal model: rats (Sprague-Dawley) • Dose: ritovanir + daruvanir at 25 mg/kg of. • Treatment: free drug, nanodrug and nanodrug-loaded microparticle (NDM). • Results for free drug: $C_{max} = 0.14$ μg/mL; $t_{max} = 2.63$ h; $AUC_\infty = 1.17$ μg·h/mL; • Results for nanodrug: $C_{max} = 0.11$ μg/mL; $t_{max} = 1.75$ h; $AUC_\infty = 1.35$ μg·h/mL; • Results for NDM: $C_{max} = 0.38$ μg/mL; $t_{max} = 2.75$ h; $AUC_\infty = 2.67$ μg·h/mL.	(Augustine et al. 2018)
Nanoemulsion	Vitamin K1	Glycocholic acid (surfactant), soybean lecithin, Transcutol HP (co-surfactant)	Tablets	• Animal model: beagle dogs. • Dose: single 10 mg dose. • Groups: 6 healthy beagle dogs (n=2) (nanosystem tablets and commercial vitaminK1 tablets). • Results for nanosystem tablet: $C_{max} = 575.46$ ng/mL; $t_{max} = 1.67$ h; $AUC_\infty = 1716.33$ ng·h/mL.	(Tong et al. 2018)

(Continued)

TABLE 3.4 (Continued)
Pharmacokinetics Summary for Selected Oral Nanosystmes

Nanosystem	Drug	Formulation	Dosage Form	Pharmacokinetic Profiles	Ref
Nanoemulsion	Finasteride (FSD)	Anise oil, methanol, ethanol, propanol, butanol, Tween 80	Lyophilized tablet (LT)	• Results for commercial tablet: C_{max} = 249.23 ng/mL; t_{max} = 2.0 h; AUC_∞ = 866.14 ng·h/mL. • Human model: healthy male volunteers. • Dose: 5 mg of FSD. • Groups: nanosystem tablets; FSD-LTs; marketed tablets (Proscar®). • Results for nanosystem tablets: C_{max} = 44.635 ng/mL; t_{max} = 1.5 h; AUC_∞ = 721.66 ng·h/mL. • Results for FSD-LTs: C_{max} = 37.794 ng/mL; t_{max} = 2 h; AUC_∞ = 444.08 ng·h/mL. • Results for marketed tablets: C_{max} = 29.150 ng/mL; t_{max} = 3 h; AUC_∞ = 487.64 ng·h/mL.	(Ahmed et al. 2018)
Nanoemulsion	Rosuvastatin	Tween 80, Cremophore RH 40 (surfactants), labrafac, oleic acid, labrafil (oils), propylene glycol (co-surfactant)	Tablet	• Human model: healthy male volunteers. • Dose: 10 mg. • Groups: nanosystem tablet and commercial tablets (Crestor®). • Data for nanosystem tablet:	(Salem et al. 2018)

(Continued)

TABLE 3.4 (Continued)
Pharmacokinetics Summary for Selected Oral Nanosystems

Nanosystem	Drug	Formulation	Dosage Form	Pharmacokinetic Profiles	Ref
Nanocrystals	Rebamipide (REB)	HPMC K4M, HPMC E5, Pluronic F68, PVP K30	Tablet	• C_{max} = 66 521 (ng/mL); t_{max} = 2 h; AUC_∞ = 648 219 (ng·h/mL). • Results for Crestor® tablets: C_{max} = 23 885 (ng/mL); t_{max} = 3 h; AUC_∞ = 264 210 (ng·h/mL). • Animal model: male Sprague-Dawley rats. • Dose: 10 mg/kg. • Groups: rebamipide nanocrystal tablets and a reference formulation of Mucosta® tablets. • Results for nanocrystal tablets: C_{max} = 543.40 ng/mL; t_{max} = 1.67 h; AUC_∞ = 2622.30 ng·min/mL. • Results for Mucosta® tablets: C_{max} = 281.50 ng/mL; t_{max} = 1.08 h; AUC_∞ = 1187.40 ng·min/mL.	(Guo, Wang, and Xu 2015)
Polymeric nanoparticles	Daunorubicin	PLGA	Macromolecule of chitosan as a coat	• Animal model: Wister rats (n=6) • Dose: 10 mg/kg • Groups: free drug, nanodrug • Results for free drug: C_{max} = 44.65; t_{max} = 2.00; $t_{1/2}$ 54.55; AUC_∞ = 690.15. • Results for uncoated nanodrug:	(Ahmad et al. 2019)

(Continued)

TABLE 3.4 (Continued)
Pharmacokinetics Summary for Selected Oral Nanosystmes

Nanosystem	Drug	Formulation	Dosage Form	Pharmacokinetic Profiles	Ref
Nanoparticle	Zidovudine	Lactoferrin	–	• $C_{max} = 318.55$; $t_{max} = 4.00$; $t_{1/2} = 119.39$; $AUC_\infty = 26932.44$ • Results for coated nanodrug: $C_{max} = 591.33$; $t_{max} = 4.00$; $t_{1/2} = 152.70$; $AUC_\infty = 61992.38$ • Animal model: Male and female Wistar rats. • Dose: 10mg/kg • Groups: free drug, nanodrug • Results for free drug for male rats: $C_{max} = 37.67$; $t_{max} = 1$; $t_{1/2} = 1.759$; $AUC_\infty = 139.73$ • Results for free drug for female rats: $C_{max} = 35.45$; $t_{max} = 1$; $t_{1/2} = 1.92$; $AUC_\infty = 126.74$ • Results for nanodrug for male rats: $C_{max} = 49.12$; $t_{max} = 2$; $t_{1/2} = 3.07$; $AUC_\infty = 1270.57$ • Results for nanodrug for male rats: $C_{max} = 46.17$; $t_{max} = 2$; $t_{1/2} = 3.27$; $AUC_\infty = 1485.24$	(Kumar et al. 2015)
Polymeric nanoparticles	Atorvastatin	Ethylcellulose	Capsules	Animal model: new-Zeeland rabbits Dose: 10 mg /kgGroups: Results	(Shaker et al. 2020)

(Continued)

TABLE 3.4 (Continued)
Pharmacokinetics Summary for Selected Oral Nanosystmes

Nanosystem	Drug	Formulation	Dosage Form	Pharmacokinetic Profiles	Ref
Nanoemulsoin	Raloxifene hydrochloride	Soy lecithin-chitosan	-	Animal model: Female Wistar rats Dose: 15 mg/kg Groups: Results for free drug: C_{max} (ng/ml)= 515; t_{max} = 1.79; $t_{1/2}$ = 1.26; AUC_{0-12} (ng./h) = 2517 Results for Lipitor®: C_{max} (ng/ml)= 635; t_{max} = 2.07; $t_{1/2}$ = 2.90; AUC_{0-12} = 4367 Results for nanodrug: C_{max} (ng/ml)= 940; t_{max} = 3.15; $t_{1/2}$ = 3.38; AUC_{0-12} = 8759	(Murthy et al. 2020)
Solid lipid nanoparticles	Cilnidipine	Compritol 888-poloxamer 188	Suspension	Animal model: Female Wistar rats Dose: 15 mg/kg Groups: Results for free drug: C_{max} (ng/ml)= 186; t_{max} = 2.1; $t_{1/2}$ = 8.7; AUC_{0-t} (µg./h) = 1.53 Results for nanodrug: C_{max} (ng/ml)= 794.5; t_{max} = 1.2; $t_{1/2}$ = 5.06; AUC_{0-t} (µg./h) = 6.46 Animal model: Male Wistar rats Dose: 10 mg/kg Groups: Results for free drug: C_{max} (ng/ml)= 363.6; t_{max} = 1.5; AUC_{0-t} (ng./h) = 2316.1 Results for nanodrug: C_{max} (ng/ml)= 572.4; t_{max} = 2; AUC_{0-t} (ng./h) = 5588.6	(Diwan et al. 2020)

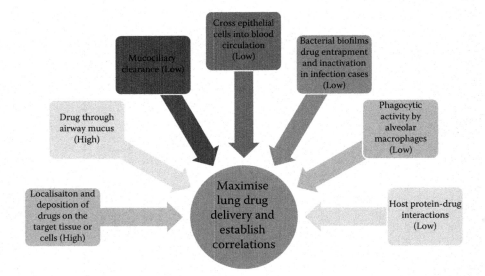

FIGURE 3.4 Ideal parameters for efficient nanodrug for lung delivery.

The lungs are in direct contact with the environment and are the first port of entry for inhaled nanomaterials into the body. They consist of two segments, the airways that made of trachea, bronchi, and bronchioles and the alveoli, the gas exchange areas. Different segment structures allow nanoparticles depositions with three principal mechanisms, namely: inertial impaction, gravitational sedimentation, and Brownian diffusion, based on their size (Praphawatvet, Peters, and Williams III 2020). Particles possess a mass median aerodymanic diameter (MMAD) greater than 5 μm will exhibit inertial impaction deposition while transiting through the oropharynx and large airways. Moreover, particles with a size range of 1 to 5 μm will be subjected to sedimentation mediated by gravitational force in smaller airways and bronchioles. In addition, particles of less than 0.5 μm MMAD are significantly deposited by diffusion-based on the Brownian motion (Courrier, Butz, and Vandamme 2002).

3.4 CONCLUSION

Nanoparticles exhibit potential for enhancing the efficacy of drugs with low bioavailability or a narrow therapeutic window, such as nucleic acid-based drugs and anticancer therapeutics. The pharmacokinetics (PK) and tissue distribution of the nanocarriers largely determine their therapeutic effect and associated toxicity. Physiochemical attributes of the nanoparticles, including surface chemistry, size, and surface charge, are essential factors that define their PK and biodistribution. Moreover, the intracellular pathway of the nanoparticles after cellular internalization that affects the drug bioavailability were discussed. In addition, strategies for overcoming barriers for intracellular delivery and drug release were presented.

REFERENCES

Abbasi, Elham, Sedigheh Fekri Aval, Abolfazl Akbarzadeh, Morteza Milani, and Hamid Tayefi Nasrabadi. "Dendrimers: Synthesis, Applications, and Properties." *Nanoscale Research Letters* (2014): 1–10. https://doi.org/10.1186/1556-276X-9-247.

Agrawal, U., R. Sharma, M. Gupta, and S.P. Vyas. "Is Nanotechnology a Boon for Oral Drug Delivery?" *Drug Discovery Today* 19, no. 10(2014): 1530–1546. https://doi.org/10.1016/j.drudis.2014.04.011.

Ahmad, Niyaz, Rizwan Ahmad, Md Aftab Alam, Farhan Jalees Ahmad, Mohd Amir, Faheem Hyder Pottoo, Md Sarafroz, Mohammed Jafar, and Khalid Umar. "Daunorubicin Oral Bioavailability Enhancement by Surface Coated Natural Biodegradable Macromolecule Chitosan Based Polymeric Nanoparticles." *International Journal of Biological Macromolecules* 128 (2019): 825–838.

Ahmed, Tarek A, Khalid M El-Say, Khaled M Hosny, and Bader M Aljaeid. "Development of Optimized Self-Nanoemulsifying Lyophilized Tablets (SNELTs) to Improve Finasteride Clinical Pharmacokinetic Behavior." *Drug Development and Industrial Pharmacy* 44, no. 4(2018): 652–661.

Alam, Sohrab, Mohammed Aslam, Anam Khan, Syed Sarim Imam, Mohammed Aqil, Yasmin Sultana, and Asgar Ali. "Nanostructured Lipid Carriers of Pioglitazone for Transdermal Application: From Experimental Design to Bioactivity Detail." *Drug Delivery* 23, no. 2(2016): 601–609. https://doi.org/10.3109/10717544.2014.923958.

Alexander, Amit, Shubhangi Dwivedi, Ajazuddin, Tapan K Giri, Swarnlata Saraf, Shailendra Saraf, and Dulal Krishna Tripathi. "Approaches for Breaking the Barriers of Drug Permeation through Transdermal Drug Delivery." *Journal of Controlled Release: Official Journal of the Controlled Release Society* 164, no. 1(2012): 26–40. https://doi.org/10.1016/j.jconrel.2012.09.017.

Allen, Christine, Dusica Maysinger, and Adi Eisenberg. "Nano-Engineering Block Copolymer Aggregates for Drug Delivery." *Colloids and Surfaces B: Biointerfaces* 16 (1999): 3–27.

Augustine, Robin, Dana Levin Ashkenazi, Roni Sverdlov Arzi, Vita Zlobin, Rona Shofti, and Alejandro Sosnik. "Nanoparticle-in-Microparticle Oral Drug Delivery System of a Clinically Relevant Darunavir/Ritonavir Antiretroviral Combination." *Acta Biomaterialia* 74 (2018): 344–359.

Bali, Nikhil R., and Pramod S. Salve. "Impact of Rasagiline Nanoparticles on Brain Targeting Efficiency via Gellan Gum Based Transdermal Patch: A Nanotheranostic Perspective for Parkinsonism." *International Journal of Biological Macromolecules* 164 (2020): 1006–1024. https://doi.org/10.1016/j.ijbiomac.2020.06.261.

Bhaskar, Sonu, Furong Tian, Tobias Stoeger, Wolfgang Kreyling, Jesús M. de la Fuente, Valeria Grazú, Paul Borm, Giovani Estrada, Vasilis Ntziachristos, and Daniel Razansky. "Multifunctional Nanocarriers for Diagnostics, Drug Delivery and Targeted Treatment across Blood-Brain Barrier: Perspectives on Tracking and Neuroimaging." *Particle and Fibre Toxicology*. BioMed Central. (2010). https://doi.org/10.1186/1743-8977-7-3.

Bhowmik, Arijit, Sayak Chakravarti, Aparajita Ghosh, Rajni Shaw, Suman Bhandary, Satyaranjan Bhattacharyya, Parimal C. Sen, and Mrinal K. Ghosh. "Anti-SSTR2 Peptide Based Targeted Delivery of Potent PLGA Encapsulated 3,3'-Diindolylmethane Nanoparticles through Blood Brain Barrier Prevents Glioma Progression." *Oncotarget* 8, no. 39(2017): 65339–65358. https://doi.org/10.18632/oncotarget.18689.

Bohr, Adam, Nicolas Tsapis, Camilla Foged, Ilaria Andreana, Mingshi Yang, and Elias Fattal. "Treatment of Acute Lung Inflammation by Pulmonary Delivery of Anti-TNF-α SiRNA with PAMAM Dendrimers in a Murine Model." *European Journal of Pharmaceutics and Biopharmaceutics* 156 (2020): 114–120.

Bouwstra, J.A., and P.L. Honeywell-Nguyen. "Skin Structure and Mode of Action of Vesicles." *Advanced Drug Delivery Reviews* 54, (2002). https://doi.org/10.1016/S01 69-409X(02)00114-X.

Briuglia, Maria Lucia, Chiara Rotella, Amber McFarlane, and Dimitrios A. Lamprou. "Influence of Cholesterol on Liposome Stability and on in Vitro Drug Release." *Drug Delivery and Translational Research* 5, no. 3(2015): 231–242. https://doi.org/10.1007/s13346-015-0220-8.

Cao, Shu-jun, Shuo Xu, Hui-ming Wang, Yong Ling, Jiahua Dong, Rui-dong Xia, and Xiang-Hong Sun. "Nanoparticles: Oral Delivery for Protein and Peptide Drugs." *AAPS PharmSciTech* 20, no. 5(2019): 190.

Caprifico, Anna E., Peter J.S. Foot, Elena Polycarpou, and Gianpiero Calabrese. "Overcoming the Blood-Brain Barrier: Functionalised Chitosan Nanocarriers." *Pharmaceutics* 12, no. 11(2020): 1–20. https://doi.org/10.3390/pharmaceutics12111013.

Chauhan, Meenakshi Kanwar, and Pankaj Kumar Sharma. *Optimization and Characterization of Rivastigmine Nanolipid Carrier Loaded Transdermal Patches for the Treatment of Dementia. Chemistry and Physics of Lipids*. Vol. 224. Elsevier Ireland Ltd, 2019. https://doi.org/10.1016/j.chemphyslip.2019.104794.

Chaulagain, Bivek, Ankit Jain, Ankita Tiwari, Amit Verma, and Sanjay K. Jain. "Passive Delivery of Protein Drugs through Transdermal Route." *Artificial Cells, Nanomedicine and Biotechnology* 46, sup1(2018): 472–487. https://doi.org/10.1080/21691401.2018.1430695.

Chen, Yan, and Lihong Liu. "Modern Methods for Delivery of Drugs across the Blood-Brain Barrier." *Advanced Drug Delivery Reviews* 64, no. 7(2012): 640–665. https://doi.org/10.1016/j.addr.2011.11.010.

Chen, Ze, Pengfei Zhao, Zhenyu Luo, Mingbin Zheng, Hao Tian, Ping Gong, Guanhui Gao, et al. "Cancer Cell Membrane-Biomimetic Nanoparticles for Homologous-Targeting Dual-Modal Imaging and Photothermal Therapy." *ACS Nano* 10, no. 11(2016): 10049–10057. https://doi.org/10.1021/acsnano.6b04695.

Chenthamara, Dhrisya, Sadhasivam Subramaniam, Sankar Ganesh Ramakrishnan, Swaminathan Krishnaswamy, Musthafa Mohamed Essa, Feng-huei Lin, and M Walid Qoronfleh. "Therapeutic Efficacy of Nanoparticles and Routes of Administration." *Biomaterials Research* (2019): 1–29.

Choi, Young Hee, and Hyo Kyung Han. "Nanomedicines: Current Status and Future Perspectives in Aspect of Drug Delivery and Pharmacokinetics." *Journal of Pharmaceutical Investigation* 48, no. 1(2018): 43–60. https://doi.org/10.1007/s40005-017-0370-4.

Courrier, H.M., N. Butz, and Th F Vandamme. "Pulmonary Drug Delivery Systems: Recent Developments and Prospects." *Critical Reviews™ in Therapeutic Drug Carrier Systems* 19 (2002): 4–5.

Dawidczyk, Charlene M, Chloe Kim, Jea Ho Park, Luisa M Russell, Kwan Hyi Lee, Martin G Pomper, and Peter C Searson. "State-of-the-Art in Design Rules for Drug Delivery Platforms: Lessons Learned from FDA-Approved Nanomedicines." *Journal of Controlled Release* 187 (2014): 133–144.

Dawidczyk, Charlene M, Luisa M Russell, and Peter C Searson. "Nanomedicines for Cancer Therapy: State-of-the-Art and Limitations to Pre-Clinical Studies That Hinder Future Developments." *Frontiers in Chemistry* 2 (2014): 69.

Delmar, Keren, and Havazelet Bianco-Peled. "Composite Chitosan Hydrogels for Extended Release of Hydrophobic Drugs." *Carbohydrate Polymers* 136 (2016): 570–580.

Desai, Pinaki, Ram R. Patlolla, and Mandip Singh. "Interaction of Nanoparticles and Cell-Penetrating Peptides with Skin for Transdermal Drug Delivery." *Molecular Membrane Biology* (2010). https://doi.org/10.3109/09687688.2010.522203.

Ding, Shichao, Aminul Islam Khan, Xiaoli Cai, Yang Song, Zhaoyuan Lyu, Dan Du, Prashanta Dutta, and Yuehe Lin. "Overcoming Blood–Brain Barrier Transport:

Advances in Nanoparticle-Based Drug Delivery Strategies." *Materials Today* 37 (2020 August): 112–125. https://doi.org/10.1016/j.mattod.2020.02.001.

Diwan, Rimpy, Punna Rao Ravi, Nikita Shantaram Pathare, and Vidushi Aggarwal. "Pharmacodynamic, Pharmacokinetic and Physical Characterization of Cilnidipine Loaded Solid Lipid Nanoparticles for Oral Delivery Optimized Using the Principles of Design of Experiments." *Colloids and Surfaces B: Biointerfaces* (2020): 111073.

Dudhipala, Narendar, and Thirupathi Gorre. "Neuroprotective Effect of Ropinirole Lipid Nanoparticles Enriched Hydrogel for Parkinson's Disease: In Vitro, Ex Vivo, Pharmacokinetic and Pharmacodynamic Evaluation." *Pharmaceutics* 12, no. 5(2020): 1–24. https://doi.org/10.3390/pharmaceutics12050448.

El-Gizawy, Sanaa A., Gamal M. El-Maghraby, and Asmaa A. Hedaya. "Formulation of Acyclovir-Loaded Solid Lipid Nanoparticles: 2. Brain Targeting and Pharmacokinetic Study." *Pharmaceutical Development and Technology* 24, no. 10(2019): 1299–1307. https://doi.org/10.1080/10837450.2019.1667386.

Elias, P.M., and G.K. Menon. "Structural and Lipid Biochemical Correlates of the Epidermal Permeability Barrier." *Advances in Lipid Research*. Elsevier, 1991. https://doi.org/10.1 016/b978-0-12-024924-4.50005-5.

Fatouh, Ahmed M., Ahmed H. Elshafeey, and Ahmed Abdelbary. "Intranasal Agomelatine Solid Lipid Nanoparticles to Enhance Brain Delivery: Formulation, Optimization and in Vivo Pharmacokinetics." *Drug Design, Development and Therapy* 11, (2017 June): 1815–1825. https://doi.org/10.2147/DDDT.S102500.

Gandhi, Kamal, Anu Dahiya, Monika Taruna Kalra, and Khushboo Sigh. "Transdermal Drug Delivery System: A Review." *International Journal of Research in Pharmaceutical Sciences* 3, no. 3(2012): 379–388.

Gu, Yongwei, Xiaomeng Tang, Meng Yang, Dishun Yang, and Jiyong Liu. *Transdermal Drug Delivery of Triptolide-Loaded Nanostructured Lipid Carriers: Preparation, Pharmacokinetic, and Evaluation for Rheumatoid Arthritis. International Journal of Pharmaceutics*. Vol. 554. Elsevier B.V., 2019. https://doi.org/10.1016/j.ijpharm.201 8.11.024.

Guo, Yu, Yongjun Wang, and Lu Xu. "Enhanced Bioavailability of Rebamipide Nanocrystal Tablets: Formulation and in Vitro/in Vivo Evaluation." *Asian Journal of Pharmaceutical Sciences* 10, no. 3(2015): 223–229.

Gupta, Himanshu, Dinesh Bhandari, and Aarti Sharma. "Recent Trends in Oral Drug Delivery: A Review." *Recent Patents on Drug Delivery & Formulation* 3, no. 2(2009): 162–173.

Hänel, Kai H, Christian Cornelissen, Bernhard Lüscher, and Jens Malte Baron. "Cytokines and the Skin Barrier." *International Journal of Molecular Sciences* 14 (2013): 14. https://doi.org/10.3390/ijms14046720.

He, Chunsheng, Ping Cai, Jason Li, Tian Zhang, Lucy Lin, Azhar Z. Abbasi, Jeffrey T. Henderson, Andrew Michael Rauth, and Xiao Yu Wu. "Blood-Brain Barrier-Penetrating Amphiphilic Polymer Nanoparticles Deliver Docetaxel for the Treatment of Brain Metastases of Triple Negative Breast Cancer." *Journal of Controlled Release* 246 (2017): 98–109. https://doi.org/10.1016/j.jconrel.2016.12.019.

Jena, Lynn, Emma Mcerlean, and Helen Mccarthy. "Delivery across the Blood-Brain Barrier: Nanomedicine for Glioblastoma Multiforme Lipid Nanoparticle MGMT O6-Methylguanine Methyltransferase MI Mechanical Index MMPs Metalloproteinases MPS Mononuclear Phagocytic System MTX Methotrexate NPs Nanoparticles PBCA Poly(n-Butyl-2-Cyanoacrylate) PEG Poly(Ethylene Glycol)." *Drug Delivery and Translational Research* 10 (2020): 304–318. https://doi.org/10.1007/s13346-019-00679-2.

Joshi, Sumit Ashok, Sunil Satyappa Jalalpure, Amolkumar Ashok Kempwade, and Malleswara Rao Peram. "Fabrication and In-Vivo Evaluation of Lipid Nanocarriers

Based Transdermal Patch of Colchicine." *Journal of Drug Delivery Science and Technology* 41 (2017): 444–453. https://doi.org/10.1016/j.jddst.2017.08.013.

Karathanasis, Efstathios, Ananta Laxmi Ayyagari, Rohan Bhavane, Ravi V Bellamkonda, and Ananth V Annapragada. "Preparation of in Vivo Cleavable Agglomerated Liposomes Suitable for Modulated Pulmonary Drug Delivery." *Journal of Controlled Release* 103, no. 1(2005): 159–175.

Kawadkar, Jitendra, Ashwinkumar Pathak, Raj Kishore, and Meenakshi Kanwar Chauhan. "Formulation, Characterization and in Vitro-in Vivo Evaluation of Flurbiprofen-Loaded Nanostructured Lipid Carriers for Transdermal Delivery." *Drug Development and Industrial Pharmacy* 39, no. 4(2013): 569–578. https://doi.org/10.3109/03639045.2012.686509.

Khosa, Archana, Satish Reddi, and Ranendra N. Saha. "Nanostructured Lipid Carriers for Site-Specific Drug Delivery." *Biomedicine and Pharmacotherapy* 103 (2018 April): 598–613. https://doi.org/10.1016/j.biopha.2018.04.055.

Kumar, Prashant, Yeruva Samrajya Lakshmi, C. Bhaskar, Kishore Golla, and Anand K Kondapi. "Improved Safety, Bioavailability and Pharmacokinetics of Zidovudine through Lactoferrin Nanoparticles during Oral Administration in Rats." *PloS One* 10, no. 10(2015): e0140399.

Kumari, Sonali, Saad M. Ahsan, Jerald M. Kumar, Anand K. Kondapi, and Nalam M. Rao. "Overcoming Blood Brain Barrier with a Dual Purpose Temozolomide Loaded Lactoferrin Nanoparticles for Combating Glioma (SERP-17-12433)." *Scientific Reports* 7, no. 1(2017): 1–13. https://doi.org/10.1038/s41598-017-06888-4.

Lin, Chih-Hung, Chun-Han Chen, Zih-Chan Lin, and Jia-You Fang. "Recent Advances in Oral Delivery of Drugs and Bioactive Natural Products Using Solid Lipid Nanoparticles as the Carriers." *Journal of Food and Drug Analysis* 25, no. 2(2017): 219–234.

Loira-Pastoriza, Cristina, Julie Todoroff, and Rita Vanbever. "Delivery Strategies for Sustained Drug Release in the Lungs." *Advanced Drug Delivery Reviews* 75 (2014): 81–91.

Malvajerd, Soroor Sadegh, Amir Azadi, Zhila Izadi, Masoumeh Kurd, Tahereh Dara, Maryam Dibaei, Sharif Mohammad Zadeh, Hamid Akbari-Javar, and Mehrdad Hamidi. "Brain Delivery of Curcumin Using Solid Lipid Nanoparticles and Nanostructured Lipid Carriers: Preparation, Optimization, and Pharmacokinetic Evaluation." *ACS Chemical Neuroscience* 10, no. 1(2019): 728–739. https://doi.org/10.1021/acschemneuro.8b00510.

Murthy, Aditya, Punna Rao Ravi, Himanshu Kathuria, and Rahul Vats. "Self-Assembled Lecithin-Chitosan Nanoparticles Improve the Oral Bioavailability and Alter the Pharmacokinetics of Raloxifene." *International Journal of Pharmaceutics* 588 (2020): 119731.

Obermeier, Birgit, Richard Daneman, and Richard M. Ransohoff. "Development, Maintenance and Disruption of the Blood-Brain Barrier." *Nature Medicine*. (2013). https://doi.org/10.1038/nm.3407.

Onoue, Satomi, Shizuo Yamada, and Hak-Kim Chan. "Nanodrugs: Pharmacokinetics and Safety." *International Journal of Nanomedicine* 9 (2014): 1025.

Paek, Hee-Jeong, Youn-Joung Lee, Hea-Eun Chung, Nan-Hui Yoo, Jeong-A Lee, Mi-Kyung Kim, Jong Kwon Lee, Jayoung Jeong, and Soo-Jin Choi. "Modulation of the Pharmacokinetics of Zinc Oxide Nanoparticles and Their Fates in Vivo." *Nanoscale* 5, no. 23(2013): 11416–11427.

Pathan, Shadab, Zeenat Iqbal, Syed Zaidi, Sushma Talegaonkar, Divya Vohra, Gaurav Jain, Adnan Azeem, et al. "CNS Drug Delivery Systems: Novel Approaches." *Recent Patents on Drug Delivery & Formulation* 3, no. 1(2009): 71–89. https://doi.org/10.2174/187221109787158355.

Praphawatvet, Tuangrat, Jay I Peters, and Robert O Williams III. "Inhaled Nanoparticles—an Updated Review." *International Journal of Pharmaceutics* 587 (2020): 119671. https://doi.org/10.1016/j.ijpharm.2020.119671.

Prausnitz, Mark R, and Robert Langer. "Transdermal Drug Delivery." *Nature Biotechnology* 26, no. 11(2009): 1261–1268. https://doi.org/10.1038/nbt.1504.Transdermal.

Prow, Tarl W., Jeffrey E. Grice, Lynlee L. Lin, Rokhaya Faye, Margaret Butler, Wolfgang Becker, Elisabeth M.T. Wurm, et al. "Nanoparticles and Microparticles for Skin Drug Delivery." *Advanced Drug Delivery Reviews* 63, no. 6(2011): 470–491. https://doi.org/10.1016/j.addr.2011.01.012.

Purvin, Shah, Parameswara Rao Vuddanda, Sanjay Kumar Singh, Achint Jain, and Sanjay Singh. "Pharmacokinetic and Tissue Distribution Study of Solid Lipid Nanoparticles of Zidovudine in Rats." *Journal of Nanotechnology* 2014 (2014).

Rao, Sripriya Venkata Ramana, Kavya Yajurvedi, and Jun Shao. "Self-Nanoemulsifying Drug Delivery System (SNEDDS) for Oral Delivery of Protein Drugs: III. In Vivo Oral Absorption Study." *International Journal of Pharmaceutics* 362, no. 1–2(2008): 16–19.

Ruge, Christian A, Julian Kirch, and Claus-Michael Lehr. "Pulmonary Drug Delivery: From Generating Aerosols to Overcoming Biological Barriers—Therapeutic Possibilities and Technological Challenges." *The Lancet Respiratory Medicine* 1, no. 5(2013): 402–413.

Salem, Heba F, Rasha M Kharshoum, Abdel Khalek A Halawa, and Demiana M Naguib. "Preparation and Optimization of Tablets Containing a Self-Nano-Emulsifying Drug Delivery System Loaded with Rosuvastatin." *Journal of Liposome Research* 28, no. 2(2018): 149–160.

Shah, Brijesh, Dignesh Khunt, and Manju Misra. "Comparative Evaluation of Intranasally Delivered Quetiapine Loaded Mucoadhesive Microemulsion and Polymeric Nanoparticles for Brain Targeting: Pharmacokinetic and Gamma Scintigraphy Studies." *Future Journal of Pharmaceutical Sciences* 7, no. 1(2021): 1–12. https://doi.org/10.1186/s43094-020-00156-5.

Shaker, Mohamed A, Hossein M Elbadawy, Sultan S Al Thagfan, and Mahmoud A Shaker. "Enhancement of Atorvastatin Oral Bioavailability via Encapsulation in Polymeric Nanoparticles." *International Journal of Pharmaceutics* 592, (2020): 120077.

Shakeri, Shahryar, Milad Ashrafizadeh, Ali Zarrabi, Rasoul Roghanian, Elham Ghasemipour Afshar, Abbas Pardakhty, Reza Mohammadinejad, Anuj Kumar, and Vijay Kumar Thakur. "Multifunctional Polymeric Nanoplatforms for Brain Diseases Diagnosis, Therapy and Theranostics." *Biomedicines* 8, no. 13(2020): 1–29. https://doi.org/10.3390/biomedicines8010013.

Sharma, Mayank, Rajesh Sharma, and Dinesh Kumar Jain. "Nanotechnology Based Approaches for Enhancing Oral Bioavailability of Poorly Water Soluble Antihypertensive Drugs." *Scientifica* 2016, (2016).

Shen, Andrew M, and Tamara Minko. "Pharmacokinetics of Inhaled Nanotherapeutics for Pulmonary Delivery." *Journal of Controlled Release* 326, (2020): 222–244.

Shi, Yanan, Miaomiao Yin, Yina Song, Tengteng Wang, Shiqi Guo, Xuemei Zhang, Kaoxiang Sun, and Youxin Li. "Oral Delivery of Liraglutide-Loaded Poly-N-(2-Hydroxypropyl) Methacrylamide/Chitosan Nanoparticles: Preparation, Characterization, and Pharmacokinetics." *Journal of Biomaterials Applications* 35 (2020). https://doi.org/10.1177/0885328220947889

Sita, V.G., and Pradeep Vavia. "Bromocriptine Nanoemulsion-Loaded Transdermal Gel: Optimization Using Factorial Design, In Vitro and In Vivo Evaluation." *AAPS PharmSciTech* 21, no. 3(2020). https://doi.org/10.1208/s12249-020-1620-8.

Souto, Eliana B, Iara Baldim, Wanderley P. Oliveira, Rekha Rao, Nitesh Yadav, Francisco M. Gama, and Sheefali Mahant. "SLN and NLC for Topical, Dermal, and Transdermal Drug Delivery." *Expert Opinion on Drug Delivery* 17, no. 3(2020): 357–377. https://doi.org/10.1080/17425247.2020.1727883.

Sridhar, Vinay, Ram Gaud, Amrita Bajaj, and Sarika Wairkar. "Pharmacokinetics and Pharmacodynamics of Intranasally Administered Selegiline Nanoparticles with Improved Brain Delivery in Parkinson's Disease." *Nanomedicine: Nanotechnology, Biology, and Medicine* 14, no. 8(2018): 2609–2618. https://doi.org/10.1016/j.nano.2018.08.004.

Sung, Jean C, Brian L Pulliam, and David A Edwards. "Nanoparticles for Drug Delivery to the Lungs." *Trends in Biotechnology* 25, no. 12(2007): 563–570.

Tawfik, Mai Ahmed, Mina Ibrahim Tadros, Magdy Ibrahim Mohamed, and Sara Nageeb El-Helaly. "Low-Frequency versus High-Frequency Ultrasound-Mediated Transdermal Delivery of Agomelatine-Loaded Invasomes: Development, Optimization and in-Vivo Pharmacokinetic Assessment." *International Journal of Nanomedicine* 15, (2020): 8893–8910. https://doi.org/10.2147/IJN.S283911.

Tong, Yongtao, Yuli Wang, Meiyan Yang, Jiahui Yang, Lu Chen, Xiaoyang Chu, Chunhong Gao, Qian Jin, Wei Gong, and Chunsheng Gao. "Systematic Development of Self-Nanoemulsifying Liquisolid Tablets to Improve the Dissolution and Oral Bioavailability of an Oily Drug, Vitamin K1." *Pharmaceutics* 10, no. 3(2018): 96.

Torchilin, V.P. "Micellar Nanocarriers: Pharmaceutical Perspectives." *Pharmaceutical Research* 24, no. 1(2007): 1–16. https://doi.org/10.1007/s11095-006-9132-0.

Wahajuddin, and Sumit Arora. "Superparamagnetic Iron Oxide Nanoparticles: Magnetic Nanoplatforms as Drug Carriers." *International Journal of Nanomedicine* 7, (2012): 3445–3471. https://doi.org/10.2147/IJN.S30320.

Wiedersberg, Sandra, and Richard H. Guy. "Transdermal Drug Delivery: 30 + Years of War and Still Fighting!" *Journal of Controlled Release*. Elsevier B.V., 2014. https://doi.org/10.1016/j.jconrel.2014.05.022.

Wijagkanalan, Wassana, Shigeru Kawakami, and Mitsuru Hashida. "Designing Dendrimers for Drug Delivery and Imaging: Pharmacokinetic Considerations." *Pharmaceutical Research* 28, no. 7(2011): 1500–1519. https://doi.org/10.1007/s11095-010-0339-8.

Yasir, Mohd, and Udai Vir Singh Sara. "Solid Lipid Nanoparticles for Nose to Brain Delivery of Haloperidol: In Vitro Drug Release and Pharmacokinetics Evaluation." *Acta Pharmaceutica Sinica B* 4, no. 6(2014): 454–463. https://doi.org/10.1016/j.apsb.2014.10.005.

Zhang, Yong Tai, Meng Qing Han, Li Na Shen, Ji Hui Zhao, and Nian Ping Feng. "Solid Lipid Nanoparticles Formulated for Transdermal Aconitine Administration and Evaluated In Vitro and In Vivo." *Journal of Biomedical Nanotechnology* 11, no. 2(2015): 351–361. https://doi.org/10.1166/jbn.2015.1902.

Zhao, Mei Xia, and Bing Jie Zhu. "The Research and Applications of Quantum Dots as Nano-Carriers for Targeted Drug Delivery and Cancer Therapy." *Nanoscale Research Letters* 11, no. 1(2016). https://doi.org/10.1186/s11671-016-1394-9.

4 Bioinspired Nano-Formulations

Materials, Approaches, and Applications in Drug Delivery

Jahanzeb Mudassir and
Muhammad Sohail Arshad
Bahauddin Zakariya University

CONTENTS

4.1 Bioinspired Nano-Formulations: Introduction and Background 86
4.2 Approaches of Bioinspiration in Drug Delivery 87
    4.2.1 Bioinspired Stimulus-Responsive Release 88
        4.2.1.1 Endogenous Stimuli Responsive (*Host Environment-Responsive*) 88
        4.2.1.2 Exogenous Stimuli Responsive (*External Stimuli-Responsive*) 90
        4.2.1.3 Bioinspired Shielding Strategies for Nano-Formulations .. 91
        4.2.1.4 Biomimicking of Inner Architecture 92
        4.2.1.5 Biomimetic Movements of Bioinspired Nano-Formulations 92
4.3 Bioinspired Materials 93
    4.3.1 Naturally from Plants and Animals 93
    4.3.2 Synthetic Polymeric Materials 93
    4.3.3 Biotemplates and Biomimetics 94
4.4 Bioinspired Nanomaterials of Microbial Origin 94
    4.4.1 Bioinspired Materials from Bacterial Origin 94
    4.4.2 Bioinspired Materials from Viral Origin 95
4.5 Types of Bioinspired Nano-Formulations 96
    4.5.1 Based on Polymeric Hydrogels 96
    4.5.2 Based on Metallic Nanoparticles 96
    4.5.3 Based on Bioactive Silicate 96
    4.5.4 Based on Hydroxyapatite 97
4.6 Fabrication Methods for Bioinspired Nano-Formulations 97

DOI: 10.1201/9781003168584-4

    4.6.1   Top-Down Approach..97
    4.6.2   Bottom-Up Approach..98
4.7 Development of Bioinspired Nano-Formulations for Drug Delivery...........99
    4.7.1   Polymer-Based Bioinspired Nano-Formulations.............................99
    4.7.2   Lipid-Based Bioinspired Nano-Formulations.................................99
    4.7.3   Protein-Based Bioinspired Nano-Formulations............................100
4.8 Bioinspired-Based Nano-Formulations for Transdermal Delivery............101
    4.8.1   Skin Patches..101
    4.8.2   Skin Adhesives...101
4.9 Conclusions...102
References...103

4.1 BIOINSPIRED NANO-FORMULATIONS: INTRODUCTION AND BACKGROUND

Since the past decade, the field of pharmaceutical technology has incurred considerable efforts in evolution of novel strategies to overcome conventional limitations (such as solubility/stability) of active pharmaceutical ingredients (APIs) (Hussain, Mahmood, Arshad, Abbas, Ijaz, et al. 2020; Khan et al. 2013). The safety/efficiency profiles of drug substance have been improved by targeting to desired tissues and cells (Peppas 2013). The bioinspired drug delivery system with advanced functionalities have gained interest in formulation sciences because of versatile intrinsic characters like biocompatible, biodegradable, nonimmunogenic to the cells or tissues (Xu, Wu, and Zhou 2018). The development of bioinspired delivery system involves mimicking structural and functional aspects of formulations with disease microenvironment and promoting responsive drug release. For example, translation of physiologic features like shape, texture, surface, and mobility with cells and microorganisms would improve functions of membrane channels.

The advancement in the development of nano-formulations in biomedical and healthcare systems has led to novel applications in diagnosis and bio-sensing (Khan et al. 2019; Mishra et al. 2019; Mudassir, Darwis, and Yusof 2017; Singh, Sirbaiya, and Mishra 2019). The bio-sensing approach has been employed for the treatment of numerous diseases, such as cardiovascular, inflammation, infection, and cancer. Nano-formulations work at molecular level involving atomic and molecular interactions by exploiting responsive characterizes of materials (Mishra et al. 2019).

Numerous classes of drugs including carbohydrates, protein, and peptides, synthetic low and high molecular weight compounds have been successfully incorporated in these carrier systems (Mahmood et al. 2020). Bioinspired nano-formulations can be developed as stimuli responsive and or multi-layered having some inherent characteristics of self-regulation or remote system (Karimi et al. 2016). These delivery systems have shown efficiency in the management and diagnosis of numerous diseases. In this context, an understanding of the evolution process of these carriers could improve the design of bioinspired nano-formulations. Mimicking the mechanism by which nutrients and cell regulators travels within in vivo environment interact with biological system would revolutionize the drug delivery domain of therapeutics. The modifications of surfaces of bioinspired nano-formulations would also enable them to

Bioinspired Nano-Formulations

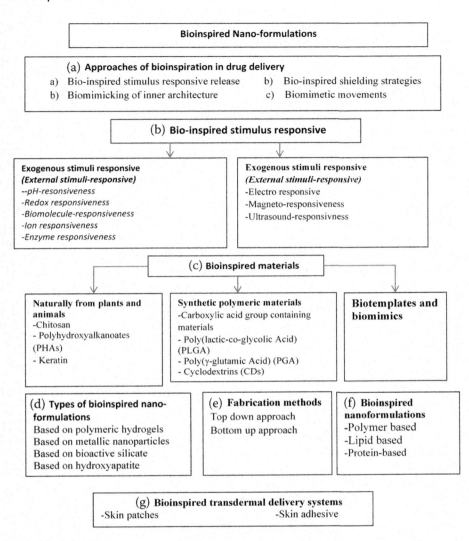

SCHEME 4.1 Flow diagram of various bioinspired approaches for drug delivery.

overcome physiological barriers and accurately target diseased site (Alvarez-Lorenzo and Concheiro 2013). The flow diagram is presented in Scheme 4.1.

4.2 APPROACHES OF BIOINSPIRATION IN DRUG DELIVERY

Since the past decade, bioinspired nano-formulations which respond to endogenous stimuli (i.e., host environment responsive) and exogenous stimuli (i.e., external stimuli responsive) have been studied for controlled and site specific delivery of loaded drugs such as genes, interferon, metal ions, protein and peptide drugs, and hormones (Mudassir, Darwis, and Khiang 2015; Raza, Rasheed, et al. 2019).

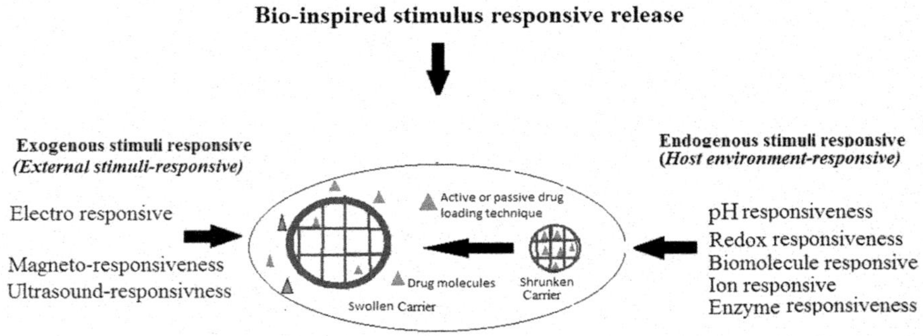

Bio-inspired stimulus responsive release

Exogenous stimuli responsive
(External stimuli-responsive)

Electro responsive
Magneto-responsiveness
Ultrasound-responsivness

Endogenous stimuli responsive
(Host environment-responsive)

pH responsiveness
Redox responsiveness
Biomolecule responsive
Ion responsive
Enzyme responsiveness

FIGURE 4.1 Bioinspired stimulus responsive release (exogenous and endogenous).

4.2.1 BIOINSPIRED STIMULUS-RESPONSIVE RELEASE

4.2.1.1 Endogenous Stimuli Responsive (*Host Environment-Responsive*)

pH-Responsiveness as Endogenous Stimuli:

The existence of pH difference with in biological environment is the basis for pH-responsive delivery systems. The significant difference in the pH value exists throughout the gastrointestinal tract, i.e., pH 2 and 7 in the stomach and colon, respectively. Figure 4.1 represents the mechanism of bioinspired stimulus-responsive release (exogenous and endogeneous). Furthermore, there is much difference in the pH of cancer cells and pH in the surroundings microenvironment of the normal tissue. The pH of tumor tissue is usually 0.5–1 unit lower that the normal tissue. It is believed that this reduction in the pH is ascribed to the metabolic reactions, such as glycolysis, resulting in accumulation of lactic acid. The cellular component also show some difference in the pH, e.g., pH at lysosomes, endosomes, and cytosol is 4.5–5, 5.5–6.0, and 7.4, respectively. Furthermore, wounds microenvironment can either be acidic or basic depending upon release of enzymes within biological environment (Colson and Grinstaff 2012, Mudassir et al. 2019).

These differences in the pH enable bioinspired nano-formulations to deliver the loaded drug to specific area resulting in improved therapeutic effect with minimum side effects. However, it is important to note that the activity of certain anticancer drugs is reduced at low pH of cancerous cells. Bioinspired nano-formulations offer an opportunity to overcome this problem by utilization of acidic environment of the tissues, to trigger the drug release. For this purpose, several polymers, including methacrylic acid (MAA), acrylic acid (AA), acrylonitrile (AN), polycarbonates, poly(e-caprolactone) (PCL), poly(lactic acid) (PLA), polyketals, and polyanhydrides have been investigated for pH-responsive bioinspired nano-formulations (Zhang, Luo, and Li 2007).

Redox Responsiveness as Endogenous Stimuli:

The existence of reductive environment within the living system (which is primarily due to the presences of glutathione at intracellular compartment) is the basis to develop reduction potential responsive bioinspired nano-formulations. For this purpose, the redox responsive character can be developed into the nanocarrier by introducing link, the cleavage of which produces free thiols in the reductive environment. Since this reduction environment is much extensive in tumor cells thus this phenomena is more efficient during cancer and gene therapy. The thiol links can be incorporated either between polymers blocks or within the backbone of polymer. The redox responsive delivery systems could be used for DNA, siRNA, antisense oligonucleotides, and anticancer agents (Schafer and Buettner 2001, Mintzer and Simanek 2009).

Biomolecule Responsiveness as Endogenous Stimuli:
The responsiveness to biological macromolecules, which generally exist in or overexpressed in certain pathological sites, has been widely studied. These bioinspired nano-formulations are designed to recognize and interact with biological macromolecules resulting in subsequent drug release. For instance, the glucose-responsive bioinspired nano-formulations that recognize the presence of glucose in biological fluids have been developed for delivery of insulin. Furthermore, biomolecule-responsive hydrogels have been formulated to sense the presence of glucose.

Ion Responsiveness as Endogenous Stimuli:
The ion-responsive polymeric bioinspired formulations are engineered by incorporating pendant basic or acidic functionalities on the side chain or back bone which are strongly influenced by the concentration of physiological ions. The drug release is triggered by the biological environment containing different ions. The ionization of greater number of pendant acidic groups consequent electrostatic repulsion among the negatively charged carboxyl groups. These phenomena manifest higher degree of swelling ratios at basic pH and vice versa (Zhang et al. 2005). The drug release is governed either by donating or accepting the ions from the biological media. Therefore, to develop ion-responsive carriers, the polymers with lower critical solution temperature (LCST) transitions are widely used. For example, poly (vinyl ether), poly (N-vinyl caprolactam), and poly (N-isopropylacrylamide) (PNIPAM), cellulose derivatives have been used to prepare pH-sensitive hydrogel for drug delivery and tissue engineering (Furyk et al. 2006). Furthermore, the ions exchange resins are also used for development of ion-sensitive polymers and are used for taste masking and sustained drug release applications.

Enzyme Responsiveness as Endogenous Stimuli:
The biochemical and metabolic processes involve large number of enzymes to carry out fundamental activities inside the living organism. These activities include structural transformations and transitions which are characterized by high selectivity and substrate specificity (Rasheed et al. 2018). The enzyme functioning has been utilized by formulation scientists to develop enzyme-sensitive bioinspired nano-formulations. Several enzyme-sensitive moieties have been recognized and incorporated in the main chain or side group linkages of carriers. The applications of enzyme responsiveness as an endogenous stimulus include imaging agents and biomarkers for the prognosis and diagnosis of certain diseases. The examples of enzymes investigated to develop enzyme-responsive bioinspired nano-formulations

include hydrolases, oxidoreductases, and azoreductase (Basel et al. 2011). It is pertinent to mention that proteases are usually over expressed during infectious diseases, especially in cancer and inflammation. In this context, nanocarriers sensitive to proteases are considered as promising candidate for the development of an effective drug delivery system.

4.2.1.2 Exogenous Stimuli Responsive *(External Stimuli-Responsive)*

The exogenous stimuli-responsive bioinspired formulations offer potential benefits to overcome inter-patient variability. Since the release of encapsulated active therapeutic agent is controlled by external factor, therefore, these systems present precise control on drug release. Different external stimuli reported includes electrical field, magnetic field, and ultrasound (Raza, Hayat, et al. 2019).

Electro Responsive as Exogenous Stimuli:

Electro-responsive bioinspired nano-formulations allow the sustained or on-demand release of loaded drug substances on application of weak electric field. The intensity of electric field around one volt is sufficient to achieve controlled drug release. Electro-responsive nanocarriers consist of polyelectrolytes, which are responsive to applied electric field. The carrier shows swelling and shrinking behavior, which subsequently results in the release of encapsulated drug moiety. Other mechanisms involved in controlling the release of model drug include disruption structure of carriers (Servant et al. 2013), oxidation reduction reaction (Jeon et al. 2011), and stimulation of thermo-response of carrier due to the heat produced from electric current (Ge et al. 2012). Recently, Neumann et al. (Hosseini-Nassab et al. 2017) synthesized electro-responsive drug loaded nanofilms composed of pH-responsive polymer, i.e., poly(methyl methacrylate-co-methacrylic acid). These systems release drug from carrier due to electrochemical reaction in the formulation resulting in change in local pH. The pH change induced by application of electrical field returned back shortly after the removal of stimuli. This phenomena offer advantages as they prevent off state drug release.

Magneto-Responsiveness as Exogenous Stimuli:

Magneto-responsive bioinspired nano-formulations is another choice that allow controlled release of incorporated drug on application of magnetism as exogenous stimuli. It can also provide on-demand controlled and sustained release of incorporated drugs. Magnetic field is traditionally being employed to the body for imaging in MRI as it can penetrate deep into body tissue (Wang and Kohane 2017). In addition to the imaging, the application of magnetic field includes controlled release of model drug. The mechanisms involved in the drug release include magnetic field-induced hyperthermia, which induces physicochemical change in the carrier and magnetic field mediated drug targeting. The hyperthermia induced by the application of magnetic field causes tumor inhibition and also provides opportunity of tumor imaging (Zhou et al. 2018). Thirunavukkarasu et al. (Thirunavukkarasu et al. 2018) prepared PLGA nanoparticles containing iron oxide (Fe_3O_4) loaded with doxorubicin (DOX). The heat generated by application of magnetic field triggered the release of loaded DOX. The authors reported that ~39% and ~57% drug was released at temperatures 37 °C and 45 °C, respectively, over 24 hour at pH 7.4.

Ultrasound-Responsiveness as Exogenous Stimuli:
Ultrasound (US) has been used as effective exogenous stimuli to obtain diagnostic imaging and to trigger the release of drug from bioinspired nano-formulations. US-mediated heat and mechanical oscillations results in cavitation within the carrier. This discontinuation in the carrier results in drug release. The advantages that US-based bioinspired delivery system offer noninvasive therapy of various illness using nonionizing radiation.

4.2.1.3 Bioinspired Shielding Strategies for Nano-Formulations

Among the numerous challenges faced by nano-formulations, one of the serious issues is the recognition and subsequent clearance by the immune system. The nano-formulations interact with serum proteins resulting in the formation of protein corona. The immune protein involved in this phenomenon includes immunoglobulin. The complex of nanoparticles and immune proteins are taken up by reticulo-endothelial system (RES) and cleared from general circulation. This results in failure to achieve desired objective of nano-formulations. This provides the basis for passive targeting of drug to these organs (Gulati, Stewart, and Steinmetz 2018).

In order to avoid the present challenge, PEGylation of nanoparticles has been used as effective strategy to overcome immune recognition. PEG being flexible hydrophilic polymer was extensively investigated as shielding agent (Abuchowski et al. 1977). The mechanism to avoid immune recognition includes decreased immune protein adsorption and antibody binding. PEG forms hydrophilic layer over the exposed surface of nanoparticles and block the protein interactions through steric hindrance. These changes then improve the circulation time of PEGylated nano-formulations by many folds. Doxil a PEGylated doxorubicin is clinically approved long circulating nano-formulation (Hatakeyama, Akita, and Harashima 2013). The half life of doxorubicin loaded liposomes was reported to increase from 10 minutes to over 40 hours (Jokerst et al. 2011). The effectiveness of shielding agent depends upon the physical characteristics of selected materials as well on the optimization of shielding process parameters. Other factors effecting the shielding include the molecular weight of the selected material and its concentration. In spite of huge success of using PEG to reduce immune clearance the recent challenge reported is the production of PEG-specific antibodies in the general population. PEG antibodies have been found to be detrimental for nano-formulations requiring repeat administration of formulation (Jokerst et al. 2011).

The prevention of immune response could also be achieved through mimicking the physiological mechanisms which inhibit adsorption of immune protein while promotes selective binding to target cell or tissue surface (Harwansh et al. 2019). For example, the hydrophilic coating material, i.e., glycocalyx (an oligosaccharides) suppresses nonspecific binding between tissues and organs by providing stealth layer around nanocarriers (Alvarez-Lorenzo and Concheiro 2013). To achieve these goals, the biomimicking polymer, i.e., glyco-saminoglycan dermatan sulfate, has been used to prepare bioinspired nano-formulations (Alvarez-Lorenzo and Concheiro 2013).

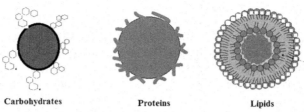

FIGURE 4.2 Representative examples of bioinspired shielding strategies.

The representative examples of bioinspired shielding strategies are presented in Figure 4.2.

4.2.1.4 Biomimicking of Inner Architecture

The modification of inner architecture of bioinspired nano-formulation is another promising approach to improve therapeutic efficiency. In this approach, the nanocarrier which is generally rigid and round shape structure is modified by mimicking the softness as spheroids and cylindrical shape as certain cells. The suitable examples of material to be used as bioinspired nano-formulations include genetically modified stem and dendritic cells and erythrocytes bacterial ghosts. In one of the approach, the surface architecture of nanocarrier is modified through conjugating anti-phagocytosis CD47 layer of erythrocytes which acts as an effective transporter for bioinspired nano-formulations. Erythrocyte CD47 layer is considered as novel biomimetic material which facilitates fabricating as disc like flexible nanocarriers (Alvarez-Lorenzo and Concheiro 2013; Pérez-Mitta et al. 2017; Pitchaimani, Nguyen, and Aryal 2018).

The internal biological microenvironment of target site imparts a significant role in triggering and release of loaded drug from nano-formulations. These environmental factors include pH, temperature, and concentration of various substances surrounding the drug target site. However, certain external factors such as ultrasound, light, and magnetic field could also trigger the release of therapeutics from nano-formulations (Pérez-Mitta et al. 2017). The greater detail of the materials influenced by the physiological environment is described in further session

4.2.1.5 Biomimetic Movements of Bioinspired Nano-Formulations

The effectiveness of bioinspired nano-formulations is based on the mobility of drug-loaded nanocarriers across the biological and physiological fluids. Thus regulating such movements is a great challenge. The researcher focuses to regulate the mobility of living cells and certain cells through wireless swimmers technology. The technology offered significant success in travelling up of nanocarriers to tissues in predictable manner. Most promising example includes mini robots made up of bio-mimicking polymers, such as polyethylene glycol or silicon dioxide-based formulations. It was reported that these bioinspired formulations showed simulating mobility as bacterial flagella, octopuses, myriapods, and spermatozoids (Alvarez-Lorenzo and Concheiro 2013).

4.3 BIOINSPIRED MATERIALS

The polymeric materials play a prime role in developing bioinspired nano-formulations. They are traditionally being used for controlled and targeted delivery. Moreover, these are intended to maximize loading capability and modulate biodistribution of drug in human body. The inherent characteristics of nanocarriers, such as tiny structure, surface morphology/texture, nanoscale surface mechanics, and the scaffold's surface chemistry, are mainly responsible to confer success to developed bioinspired nano-formulations. There are diverse kind of materials that could be used to develop bioinspired nano-formulations which are discussed in the following sections (Bangde et al. 2017; Merkle 2015).

4.3.1 NATURALLY FROM PLANTS AND ANIMALS

Polymers obtained through bioengineering or extraction from biological systems, like plants, animals, or microorganisms, are termed as natural polymers. For example, carbohydrates such as starch, cellulose; proteins, e.g., enzymes, keratin, and polyhydroxyalkanoates (PHAs) like poly-(3-hydroxybutyrate) [P(3HB)] (Iqbal et al. 2015a, 2015b). The functional characteristics of biopolymers depend upon its structure and other materials attributes such as crystalline and amorphous nature. Certain modifications through biological or physiochemical treatment can alter their structure as well as potential applications. For instance, (i) the properties of cellulose or poly(β-d-glucose) are due to crystalline nature; (ii) ethyl cellulose and cellulose acetate which are chemically modified cellulose, applied to achieve pH-dependent drug release (Stevens 2008). The examples of naturally occurring polymer include polyhydroxyalkanoates (PHAs) and keratin. PHAs is a family of compound produced abundantly by fermentation of carbon-rich substrate using bacteria. In the presence of stress conditions, such as unbalanced nutrition supply, the bacteria tend to accumulate PHAs as internal energy source in the form of granules (Stevens 2008). The most abundant source of keratin is chicken feather. Keratin was traditionally extracted from animal hooves with the help of lime. Later on, oxidative and reductive approaches were developed to extract keratin (Rouse and Van Dyke 2010).

4.3.2 SYNTHETIC POLYMERIC MATERIALS

The synthetic polymeric materials used to prepare bioinspired nano-formulations generally contribute in improving stability and regulate the drug release at cellular, tissue, and organ levels. Other advantages include enhancing the drug permeability, improving solubility, as well as bioavailability. Among the carboxylic acid group containing materials, which are used for the preparation of bioinspired nano-formulations, the poloxamers and their derivatives are widely used as inert biomimetic polymers (Ranjha et al. 2010; Ranjha, Mudassir, and Majeed 2011; Ranjha, Mudassir, and Zubair 2011). These are block copolymer composed of poly(ethylene oxide)-b-poly(propylene oxide)-b-poly(ethylene oxide) (PEO-PPO-PEO). Poloxamers show favorable characteristics, e.g., preparing self-assembled nano-formulations like polymeric micelles of water in soluble drug (Bangde et al. 2017, Harwansh et al. 2019; Merkle

2015). PLGA is biodegradable and biocompatible polymeric material and is approved for incorporation with high molecular weight proteins and peptide drugs for parenteral administration. PLGA act by triggering biological response and is widely used for drug delivery and site specific targeting. Poly(γ-glutamic Acid) (PGA) is water soluble and biodegradable polymeric material and is widely used to develop advanced pharmaceutical bioinspired nano-formulations. PGA could be conjugated with variety of materials. Few of PGA-based conjugated drugs are in clinical trial phase (Harwansh et al. 2019; Merkle 2015).

4.3.3 Biotemplates and Biomimetics

The materials used to develop bioinspired nano-formulations can be broadly classified into two subclasses, namely biotemplates and biomimics. The human biological system is made up of nanoscale self-assembly of large biomacromolecules. Scientists have focused to adopt the design of structural features of various biological system to develop drug carrier (Alvarez-Lorenzo and Concheiro 2013; Mishra et al. 2019).

Biological templates are defined as structural architecture that can complement the shape features of target molecules. These target molecules are used as molecular containers, catalysts for DNA assays and protein assays. An example of biotemplate is "protein cage". It comprises interior, exterior, and interface between subunits. Protein cages have also shown applications in designing protein-based nanomedicines or drug delivery systems. These could be developed either by self-assembling the protein subunits of same protein or combination of various proteins, thus, resulting in the formation of complete system. These systems have advantages of being biocompatible, biodegradable, and nontoxic. Ferritins is another example of biotemplate which store and sequester iron. This phenomena is well-established within biological setting. Some of the examples include i) nucleic acid storage, ii) bio-mineralization, sequestration, and iii) the transport and delivery of nucleic acids between diverse chemical environments (Mishra et al. 2019).

4.4 BIOINSPIRED NANOMATERIALS OF MICROBIAL ORIGIN

The nano-formulations based on bioinspired materials obtained from microbial sources offer excellent biomimetic properties. These features are used as biosensors and have been investigated for use as diagnostics and for targeted or personalized treatment. The applications of bacteria on different microbial entities, such as bacterium, virus, etc., are frequently engineered to obtain different bioinspired carriers. In pharmaceutical prospective, the majority of the microbial materials such as lipid and amphiphilic polymer is intended for cellular drug delivery because of their endogenous origin, transportation, and disposition. The efficiency of these carriers largely resides on biogenesis, isolation, and method of purification utilized.

4.4.1 Bioinspired Materials from Bacterial Origin

Micro vesicles (MVs) are present at the exterior surface of bacterial membrane frequently comprising lipid bilayer along with protein moieties. The primary features of

Bioinspired Nano-Formulations

these vesicles are interbacterial communication, virulence factor into host bacteria, and transport of secreted proteins (Caruana and Walper 2020). Moreover, MVs also contains species-specific functional components including outer membrane proteins, lipopolysaccharides, peptidoglycans, bacteriocins, enzymes, nucleic acid, lipoproteins, and MV-associated adhesion (Guerrero-Mandujano et al. 2017).

MVs can be extracted commonly using ultracentrifugation and ultrafiltration and subsequently subjected to purification using density gradient medium. The MVs have been used in formulating nanoparticle-based drug delivery systems. Various antibiotics such as ciprofloxacin, ceftriaxone, azithromycin, amikacin, ampicillin have been successfully loaded into MV isolated from different bacterial species as they provide basis for active targeting of antibiotics to the bacterial cells (Gao et al. 2018; Huang et al. 2020). Menina et al. modified the surface of liposome using extracellular adherence protein. The results showed decrease intracellular load of *Salmonella enterica* (Menina et al. 2019). The bacterial MV have also shown application in targeted delivery of anticancer drugs. In this context recently the doxorubicin (DOX)-loaded *K. pneumoniae* MVs and PTX-loaded cholera toxin subunit B-modified PLGA nanoparticles were developed and showed targeted delivery to monosialoganglioside GM1 (glycosphingolipid) (Kuerban et al. 2020).

Exopolysaccharides (EPSs) and some proteins such as lactin are involved in the formation of biofilm, which provide additional protection to the bacterial cells from host defence system. These have prompt number of activities such as antidiabetic, antioxidant, cholesterol lowering, antimicrobial, anticancer, and antibiofilm effect. Hung et al. prepared surfactin (SUR) lipopeptide (secreted by *B. subtilis*)-based nanocarriers containing doxorubicin by using solvent evaporation method. These nanoparticles reduced the tumor mass in MCF7/ADR-xenografted mice (Huang et al. 2018). In another study, nanoparticles were prepared from rhamnolipids derived from *P. aeruginosa* loaded with photosensitizer pheophorbide. The carrier showed tumor suppression effect in murine model of head and neck carcinoma (SCC7) following photodynamically irradiation (Yi et al. 2019).

4.4.2 Bioinspired Materials from Viral Origin

Virus are submicron size viral particles usually comprises of protein cover and nucleic acid in the core and possess unique evolving attribute which allows them to bind and penetrate host cells. Generally viruses enters the host cells via endocytic or fusogenic pathways and evade the immune surveillance of the host (Cagno et al. 2018). Bacteriophage are the viruses that generally infect and replicate in bacterial cells. It has been reported that interplay between phages and human gut microbiome generally regulate gut microbiota composition, and thus inspire phage associated therapies (Shkoporov et al. 2018).

The therapeutic applications of virus-like particles (VLPs) are well-recognized. VLPs are envelope proteins that assemble to develop structures that mimic the native protein virus. Notably, these do not possess viral genome. The major applications of VLPs include gene delivery, packaging of anionic nucleic acids, and transduction. The examples of VLPs include brome mosaic virus (BMV)-like particles and cowpea chlorotic mottle virus (CCMV)-like particles (Lam and

Steinmetz 2019). VLPs based on plant viruses have also been used for loading and delivery of anticancer drugs. More recently, VLPs have been engineered to obtain desired solubility, membrane stability, and cell internalization. It was also suggested that nanoscale structure of L particles can be modified for functional attributes, such as gene fusion, chemical conjugation, electroporation, as well as in terms of polymer and liposome fusion (Yamada et al. 2003).

Virosomes have also shown application as bioinspired nanocarriers. These are virus-derived proteins embedded within phospholipid bilayer membranes. For example, pre-S1/2-48 lipopeptide incorporated in poly(ethylene glycol)-(PEG)ylated liposomes presented enhanced targeting to hepatocytes in animal models (Zhang et al. 2015).

4.5 TYPES OF BIOINSPIRED NANO-FORMULATIONS

4.5.1 BASED ON POLYMERIC HYDROGELS

Recent trend had showed that bioinspired materials were widely used to fabricate hydrogels to achieve success in developing bioinspired formulations. Hydrogels sensitive to specific external stimuli, such as temperature, pH, electric field, light, magnetic field, and US has been designed using bioinspired materials for variety of biomedical applications (Fisher et al. 2010). For example, pH-sensitive wound dressing was obtained by crosslinking the materials by following polymerization techniques, such as a chain growth by UV polymerization, chain growth polymerization, mixed mode of step growth, and a step growth thiol-end photo click reaction. The obtained hydrogels exhibited acceptable mechanical properties, pH sensitivities swelling behavior without affecting their cyto-compatibility with NIH/3T fibroblasts suggesting that these hydrogels are effective in stage-responsive wound dressing (Kloxin et al. 2010).

4.5.2 BASED ON METALLIC NANOPARTICLES

The application of metallic nanoparticle in bioinspired nano-formulation involves the incorporation of these nanoparticles within certain protein cage structures. One of this approach involves a sequential sequestration of ions in the synthesis of ZnSe nanoparticles. In another approach, metal ions are incubated within the cages and are synthesized within the cages through the process of chemical reduction with $NaBH_4$ (Alam et al. 2017). For example, in one study ferritin-encapsulated silver nanoparticles were synthesized by genetically introducing silver-binding peptides within the ferritin cage and subsequently the silver ions bound to the peptides were chemically reduced to silver nanoparticles in situ (Iwahori et al. 2005). The developed metallic nanoparticles will then show application in the field of catalysis, sensing, and biomedical.

4.5.3 BASED ON BIOACTIVE SILICATE

Bioactive silicates are another type of nanomaterial that show higher degree of anisotropy and functionality. The interaction of these nanomaterials with biological macromolecules is substantially different from their respective macro, micro, and nano size ranges. It is because they possess a high surface-to-volume ratio. The biomedical and biological applications of bioactive silicates include imaging, disease-related diagnostics, therapeutics, as well as for variety of musculoskeletal tissue engineering applications (Mishra et al. 2019; Zreiqat et al. 2002).

4.5.4 Based on Hydroxyapatite

The nano hydroxyapatite is referred as bioinspired nanofiller which is a natural mineral found in hard tissues. The majority of applications of nHAp are linked with polymeric hydrogels network. Hydrogels and nanogels play a significant role in external and internal stimuli-responsive drug delivery vehicle with different drug release pharmacokinetics. Several hydrogels based on hydroxyapatite were developed. For example, temperature sensitive hydrogels based on chitosan-4-thiobutylamidine (CS-TBA), hydroxyapatite (HAp) and beta-glycerophosphate disodium (β-GP) (CS-TBA/HAp/β-GP gels) were synthesized (Mishra et al. 2019; Yang et al. 2014). These hydrogels showed high storage modulus, low toxicity and better degradation rate. Similarly pH-responsive controlled-release system based on meso-porous bioglass materials capped with mineralized HAp were synthesized and used as drug carriers. HAp was used to cap the pores and it restricted the drug release. The developed system showed pH sensitivity on degradation of HAp at acidic pH (Liu et al. 2017).

4.6 FABRICATION METHODS FOR BIOINSPIRED NANO-FORMULATIONS

Among the numerous approaches used to fabricate bioinspired nano-formulations the top-down and bottom-up approach have been extensively used. Through these techniques bioinspired architecture can be developed at macro, micro, and nanoscale level with desired theoretical, structural, and material characteristics (Mirkhalaf and Zreiqat 2020). The objective of these methods is to control geometric features including biphasic property of tips, size, as well as curvatures. These morphologic characteristics mimic the adhesive behavior on natural surface. The developed bioinspired material having adhesive property (with biologically inspired architecture) is desired to exhibit the following characteristics, such as biocompatible, adaptable, air permeability, and ability to simulate with nonflat surface like human skin. Different techniques used for development of bioinspired nanomaterials include multistep template molding, dip transfer method, particle-assisted replication, and partial wetting technique (Mirkhalaf and Zreiqat 2020). Typical examples to prepare bioinspired nano-formulations have been presented in Figure 4.3.

4.6.1 Top-Down Approach

Top-down approach involved the reduction of particle size within a biological system. The desired geometry is obtained through techniques such as laser engraving, etching,

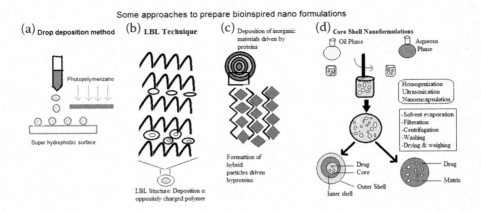

FIGURE 4.3 Typical examples to prepare bioinspired nano-formulations.

or deconstructive process. Furthermore, the top-down approaches are used to synthesize mono-disperse nanoparticles with controllable size. This approach is classified into physical, chemical, and biological methods.

Laser engraving has been used to create 3D and 2D architecture within transparent and on opaque materials, respectively. This technique involves focusing the nano decent laser beam within glasses or on the ceramic surfaces. The heat generated at focal points results in expansion of materials which elevates stress and consequent microcracks. These microcracks form weak interfaces of variety of shapes and arrangements of building blocks. Laser engraving has shown the advantage of easy tailoring the shape and size of glass ceramic building blocks. The size of these building blocks ranges from 100 μm to 100 mm (Gattass and Mazur 2008).

4.6.2 Bottom-Up Approach

In the bottom-up approach, the building blocks are designed at first stage, which are assembled into a specified architecture. Few examples of bottom-up techniques include layer-by-layer mineralization, 3D printing, freeze casting, and mixing or coating assembling approaches.

The mineralized carrier within biopolymer scaffold involves growth of mineral particles. This is accompanied by polymeric growth. This process results in the formation of materials with structure similar to nacre. Several techniques have been studied for the preparation biominerals. Some of these includes layer-by-layer assembly or deposition technique. LBL technique involved two stages, during first stage the substances is immersed in PEC while in second stage a suspension of nanoparticles is adsorbed (Meyers et al. 2008). In order to prepared multilayered materials the process is repeated several times. The thickness of the layer depend on assembly and number of layers of oppositely charge polyelectrolytes.

Freeze casting technique has been used since long to produce porous ceramics. More recently this technique has been used to design and develop various architectural materials having application in drug delivery. The first successful application of freeze casting includes the development of layered materials similar to osteons in cortical bone.

3D printing is one of the famous examples of bottom-up approach. 3D printing also follows the layer by layer methodology. In 3D printing, the 3D geometry is first sliced into layers, the adjacent layers are then added through extrusion, laser sintering, extrusion of a binder on a powder bed, or photo crosslinking (Hussain, Mahmood, Arshad, Abbas, Qamar, et al. 2020; Qamar et al. 2019). The approaches of 3D printing to make materials with dense architecture can be divided into following three groups; i) 3D printing through extrusion based techniques (both hard and soft phases); (ii) 3D printing using materials that contain micro- and nanometer-sized ceramic or glass inclusions; (iii) 3D printing of building blocks followed by manual assembly.

4.7 DEVELOPMENT OF BIOINSPIRED NANO-FORMULATIONS FOR DRUG DELIVERY

The advanced bioinspired drug delivery systems behave like supercomputer. They comprise separate components which control and regulate the drug release by sensing the biological microenvironment.

The unique characteristics of these biological macromolecules make them favorable for in vivo investigations. For example, erythrocytes may act as drug carrier by incorporating drug in the core of RBCs or attached onto their surface. Fimbriae and flagella filaments present on bacterial surface which could attach itself to human mucosa may be exploited to develop mucoadhesive delivery system.

Other most remarkable advantages of drug delivered through bioinspired nano-formulations include prolong stay of drug in human body and unique opportunity to overcome the host immune response. Certain microorganisms, e.g., bacteria, used to prepare bioinspired nano-formulations are generally nonpathogenic and safe.

4.7.1 POLYMER-BASED BIOINSPIRED NANO-FORMULATIONS

The polymer-based bioinspired nano-formulation generally comprise two major parts, i.e., a polymeric matrix and therapeutic drug. The polymer part of formulation results in drug encapsulation and shield drug from physiological environments and thus provides better stability. However, the nano-formulations are also designed to improves the solubility, enhance permeation, flexibility, surface character, and bioavailability and also provide a prolonged release of the drug. The ideal polymeric bioinspired nano-formulation are required to present nontoxic metabolic product.

4.7.2 LIPID-BASED BIOINSPIRED NANO-FORMULATIONS

Lipid-based bioinspired nano-formulations is one of the widely explored therapeutic strategies in biomedical and drug delivery applications. These formulations frequently encountered the challenges including safety and biocompatibility of building materials. Lipid-based nano-formulations are considered as effective choice to be used in cancer therapy (Rigon et al. 2016). There exist several different types of lipid-based formulations depending upon method of preparations and materials used (Čerpnjak et al. 2013).

Among the materials used for the preparation of solid lipid nanoparticles (SLN) the most commonly used materials includes glycerides, mixture of glycerides, or waxes. SLN can be loaded with drug which could be either water soluble or water in soluble. The amount of drug entrapped depends upon several factors, such as type of solid lipid used, lipid solubility of drug, and processing techniques (Mehnert and Mäder 2012).

As compared to conventional chemotherapeutics that show low accumulation at target site and presents significant off-target effects, the lipid-based formulations could maximize drug accumulation at target and prevents complications associated with conventional treatment choices. This function produces enhanced permeability and retention (EPR) effect (Maeda, Nakamura, and Fang 2013). The lipid-based nano-formulations loaded with transresveratrol (RES) were investigated for cytotoxicity in HaCat keratinocyted cell and transdermal delivery. The skin permeation study was performed by evaluating in vitro tyrosinase inhibitory activity. It was observed that after 24 hours, up to 45% of the RES nanostructure permeated through the skin. This study suggests that drug encapsulated lipid nanoparticles have therapeutic applications in skin pathology (Raja, Katas, and Wen 2015; Rigon et al. 2016).

The advantages of these formulations include: (i) capability for loading multiple therapeutic agents with diverse physicochemical characteristics, (ii) capable of incorporation simple modification for multifunctional biomedical applications, (iii) flexibility to control the size of nanocarriers. Other remarkable characters of these formulations are their resemblance with cell membrane.

The widely investigated lipid-based formulations include micelles, nanoemulsions, liposomes, core-shell-type lipid-polymer hybrids SLNs, biomimetic vesicles, and even blood cells.

4.7.3 Protein-Based Bioinspired Nano-Formulations

Since the past few decades, protein-based bioinspired nano-formulations have gained considerable attention in the field of cancer treatment. The modern approaches utilizing protein-based nano-formulations involve its application as theranostics. Theranostics are carriers which deliver therapeutic moiety along with diagnostic agent. These carriers incorporates suitable contrast agents and drugs which serves as image-guided disease-specific drug delivery for the treatment of cancer. Certain proteins such as HAS transferrin and lactoferrin have been intended for the fabrication of theranostics (Chen et al. 2015).

The albumin-based bioinspired nano-formulations have shown widespread applications in drug delivery and biomedical field. Different sources of albumins includes human serum albumin (HAS), bovine serum albumin (BSA), rat serum (RSA), and egg white (ovalbumin). HAS been also used as natural carrier for the fabrication of theranostic agents. The examples include HSA-dye complexes loaded with anticancer drug, HSA-based complexes, and HSA-coated complexes. Similarly, BSA has been used for cancer theranostics. These nano-formulations are fabricated to employ multiple functions in one platform (Sabu et al. 2018).

4.8 BIOINSPIRED-BASED NANO-FORMULATIONS FOR TRANSDERMAL DELIVERY

Among the several applications of bioinspired-based transdermal delivery systems, multiscale architectures such as skin patches, skin adhesive, and miniaturized suckers have been of high interest. The conventional approaches involving cyanoacrylate derivatives, catechol components, and suturing to skin provides string adhesion but poses some disadvantages including cytotoxicity, skin contamination, risks of infection, damages, and loss of wet adhesion make them less effective. Overall, bioinspired adhesive architecture have shown high adhesion, conformity, and repeatability to the human skin with minimum risk of contamination (Baik et al. 2019).

4.8.1 SKIN PATCHES

Skin patches having bioinspired architectures offer applications to the interfacial layers of integrated bioelectronics. To achieve diagnostics and therapeutics capabilities the flexible biointegrated devices have been widely investigated for application on the highly soft yet horny surfaces of human skin. In this context, the bioinspired architecture have been incorporated to bioelectronics which proved effective for conformal attachment to the engaged surface (Xu et al. 2014). It has shown maximum interfacial contact area to improve functionalities of devices. Such devices were used for detecting physical signals (like electrical stimulation, stain, and temperature), electrophysiological signals (like electroencephalogram (EEG), electrocardiogram (ECG), electrocorticogram (ECoG), and electromyographic (EMG) as well as collection of biochemical information like pH, glucose, hypoxia, enzymes, and other biomolecules (Schwartz et al. 2013; Xu et al. 2014). The developed bioinspired therapeutic systems have shown high sensitivity to detect biochemical, physical, and electrophysiological stimuli to control drug delivery.

4.8.2 SKIN ADHESIVES

Bioinspired skin adhesives are suitable for use on dry and wet skin. As compared to conventional skin adhesives the adhesion strengths of bioinspired architectures are highly effective. This is true especially for adhesiveness composed of acrylates and catechol components. The bioinspired skin adhesive nano-formulations may present firm tissue attachment; however, their chemical bonding may be get weak upon exposure to water. Other disadvantages include absence of multiple applications of adhesive surface, and inability to provide residue-free adhesion. In addition, they may cause skin damage or induce allergic response (Kudur et al. 2009; Watanabe et al. 2008).

The structural and material features from biologically inspired architecture for internal organs and skin may be classified as (i) mushroom-shaped beetle-inspired architecture for reversible skin patches (Heepe and Gorb 2014), (ii) endoparasite like microneedle-based skin adhesives (Yang et al. 2013), (iii) octopus-inspired suction cups for dry/wet adhesion (Chen and Yang 2017), and (iv) slug-like adhesive with energy dissipation layer (Bae et al. 2013).

The adhesive structure found in the footpads of lizards is considered as the most effective adhesive architecture in nature. The capability of these structures to enable lizards to walk freely on ceilings and vertical walls is due to the microscopic seta (foot hairs) which splits into hundreds of tiny nanoscale ends known as spatulas. Kwak et al. (Kwak, Jeong, and Suh 2011) investigated and developed the structural features of gecko foot hair over skin adhesive patch. In this study, micropillars and bulged tips were made from poly(dimethylsiloxane) (PDMS) for maximized adhesive shear over rough skin surface. In the absence of chemical adhesives, these micropillars offer an effective contamination-free adhesion to the skin. Other advantages includes biocompatibility, permeability of water, and ventilation of air.

The formation of skin adhesive with artificial microneedle (MN) was first investigated on the proboscences structures of endoparasitic worms. Yang et al. (Yang et al. 2013) developed the microneedle-based skin adhesives in which mechanical interlocking of needle-like proboscences were made up of water-responsive outer layer (polystyrene-*block*-poly(acrylic acid) (PS-*b*-PAA)) and supporting inner layer. These microneedles, upon skin penetration undergo swelling by adsorbing biological fluids of the body (maximum 40% swelling within 10 minutes). The swell-able microneedles present strong attachments to the skin surfaces (maximum ≈ 1.2 N cm^{-2}) and to wet mucous intestinal surface (maximum ≈ 4.5 N cm^{-2}) due to the mechanical interlocking between swollen microneedles and skin. As compared to conventional sutures and staple fixation, the microneedle-based skin adhesives offer minimal invasive to biological barriers and lower risks of surgical site infection. Microstructure-based polymeric architecture often requires skin invasion in order to provide interlocking (Yang et al. 2013).

The supramolecular biomaterials mimicking the marine mussels have been investigated for strong adhesion to wet surface. The suction cups of octopi like architecture have shown significant adhesive strengths on both we and dry surface. The octopus sucker is divided into two parts, i.e, upper portion is referred as infundibulum while distal portion resembles dome-like protuberance (acetabulum). The skin patches comprising structural architecture of suction cups provide stable adhesion on both dry and wet skin (Baik et al. 2017, 2018; Chen and Yang 2017).

Furthermore, the slug-like adhesives follows the analogy of viscous mucus secreted by slugs. These skin adhesive have applications in wound dressing and tissue repairing. The two major components of slug-like adhesives includes: (i) the interpenetrating positively charged polymer covalently binds with the surface of cells, (ii) dissipative matrix, which stabilize following the application of interfacial stress (Baik et al. 2019)

4.9 CONCLUSIONS

Bioinspired nano-formulations (BioIns-NFs) are designed to sense different physicochemical features of the host cells and in so doing they undergo structural changes resulting in an efficient drug release. In the modern age of pharmaceutical development where the formulations are desired to center gene delivery, biopharmaceuticals, it is inevitable to build a strong understanding about nature-based mechanisms that can minimize drug losses due to widespread tissue distribution and metabolism. Different approaches have been listed in this chapter for the fabrication of bioinspired nanocarriers that can deliver various drug substances to specific tissues or body organs. Improved therapeutic efficiency and safety of medication use remained salient features of this field.

REFERENCES

Abuchowski, Abraham, John R McCoy, Nicholas C Palczuk, Theo van Es, and Frank F Davis. "Effect of covalent attachment of polyethylene glycol on immunogenicity and circulating life of bovine liver catalase." *Journal of Biological Chemistry* 252, no. 11(1977): 3582–3586.

Alam, Md Sabir, Arun Garg, Faheem Hyder Pottoo, Mohammad Khalid Saifullah, Abu Izneid Tareq, Ovais Manzoor, Mohd Mohsin, and Md Noushad Javed. "Gum ghatti mediated, one pot green synthesis of optimized gold nanoparticles: Investigation of process-variables impact using Box-Behnken based statistical design." *International Journal of Biological Macromolecules* 104 (2017): 758–767.

Alvarez-Lorenzo, Carmen, and Angel Concheiro. "Bioinspired drug delivery systems." *Current Opinion in Biotechnology* 24, no. 6(2013): 1167–1173.

Bae, Won Gyu, Doogon Kim, Moon Kyu Kwak, Laura Ha, Seong Min Kang, and Kahp Y Suh. "Skin Patches: Enhanced Skin Adhesive Patch with Modulus-Tunable Composite Micropillars (Adv. Healthcare Mater. 1/2013)." *Advanced Healthcare Materials* 2, no. 1(2013): 1-1.

Baik, Sangyul, Jiwon Kim, Heon Joon Lee, Tae Hoon Lee, and Changhyun Pang. "Highly adaptable and biocompatible octopus-like adhesive patches with meniscus-controlled unfoldable 3D microtips for underwater surface and hairy skin." *Advanced Science* 5, no. 8(2018): 1800100.

Baik, Sangyul, Heon Joon Lee, Da Wan Kim, Ji Won Kim, Youngkwan Lee, and Changhyun Pang. "Bioinspired adhesive architectures: From skin patch to integrated bioelectronics." *Advanced Materials* 31, no. 34(2019): 1803309.

Baik, Sangyul, Youngjin Park, Tae-Jin Lee, Suk Ho Bhang, and Changhyun Pang. "A wet-tolerant adhesive patch inspired by protuberances in suction cups of octopi." *Nature* 546, no. 7658(2017): 396–400.

Bangde, Prachi, Sonal Atale, Anomitra Dey, Ashish Pandit, Prajakta Dandekar, and Ratnesh Jain. "Potential Gene Therapy Towards Treating Neurodegenerative Diseases Employing Polymeric Nanosystems." *Current Gene Therapy* 17, no. 2(2017): 170–183.

Basel, Matthew T, Tej B Shrestha, Deryl L Troyer, and Stefan H Bossmann. "Protease-sensitive, polymer-caged liposomes: A method for making highly targeted liposomes using triggered release." *ACS Nano* 5, no. 3(2011): 2162–2175.

Cagno, Valeria, Patrizia Andreozzi, Marco D'Alicarnasso, Paulo Jacob Silva, Marie Mueller, Marie Galloux, Ronan Le Goffic, Samuel T Jones, Marta Vallino, and Jan Hodek. "Broad-spectrum non-toxic antiviral nanoparticles with a virucidal inhibition mechanism." *Nature Materials* 17, no. 2(2018): 195–203.

Caruana, Julie C, and Scott A Walper. "Bacterial membrane vesicles as mediators of microbe–microbe and microbe–host community interactions." *Frontiers in Microbiology* 11 (2020): 432.

Čerpnjak, Katja, Alenka Zvonar, Mirjana Gašperlin, and Franc Vrečer. "Lipid-based systems as a promising approach for enhancing the bioavailability of poorly water-soluble drugs." *Acta Pharmaceutica* 63, no. 4(2013): 427–445.

Chen, Qian, Chao Liang, Chao Wang, and Zhuang Liu. "An imagable and photothermal "Abraxane-like" nanodrug for combination cancer therapy to treat subcutaneous and metastatic breast tumors." *Advanced Materials* 27, no. 5(2015): 903–910.

Chen, Ying-Chu, and Hongta Yang. "Octopus-inspired assembly of nanosucker arrays for dry/wet adhesion." *ACS Nano* 11, no. 6(2017): 5332–5338.

Colson, Yolonda L, and Mark W Grinstaff. "Biologically responsive polymeric nanoparticles for drug delivery." *Advanced Materials* 24, no. 28(2012): 3878–3886.

Fisher, Omar Z, Ali Khademhosseini, Robert Langer, and Nicholas A Peppas. "Bioinspired materials for controlling stem cell fate." *Accounts of Chemical Research* 43, no. 3(2010): 419–428.

Furyk, Steven, Yanjie Zhang, Denisse Ortiz-Acosta, Paul S Cremer, and David E Bergbreiter. "Effects of end group polarity and molecular weight on the lower critical solution temperature of poly (N-isopropylacrylamide)." *Journal of Polymer Science Part A: Polymer Chemistry* 44, no. 4(2006): 1492–1501.

Gao, Feng, Lulu Xu, Binqian Yang, Feng Fan, and Lihua Yang. "Kill the real with the fake: eliminate intracellular Staphylococcus aureus using nanoparticle coated with its extracellular vesicle membrane as active-targeting drug carrier." *ACS Infectious Diseases* 5, no. 2(2018): 218–227.

Gattass, Rafael R, and Eric Mazur. "Femtosecond laser micromachining in transparent materials." *Nature Photonics* 2, no. 4(2008): 219–225.

Ge, Jun, Evgenios Neofytou, Thomas J Cahill III, Ramin E Beygui, and Richard N Zare. "Drug release from electric-field-responsive nanoparticles." *ACS Nano* 6, no. 1(2012): 227–233.

Guerrero-Mandujano, Andrea, Cecilia Hernández-Cortez, Jose Antonio Ibarra, and Graciela Castro-Escarpulli. "The outer membrane vesicles: secretion system type zero." *Traffic* 18, no. 7(2017): 425–432.

Gulati, Neetu M, Phoebe L Stewart, and Nicole F Steinmetz. "Bioinspired shielding strategies for nanoparticle drug delivery applications." *Molecular Pharmaceutics* 15, no. 8(2018): 2900–2909.

Harwansh, Ranjit K, Rohitas Deshmukh, Md A Barkat, and Md Rahman. "Bioinspired polymeric-based core-shell smart nano-systems." *Pharmaceutical Nanotechnology* 7, no. 3(2019): 181–205.

Hatakeyama, Hiroto, Hidetaka Akita, and Hideyoshi Harashima. "The polyethyleneglycol dilemma: Advantage and disadvantage of PEGylation of liposomes for systemic genes and nucleic acids delivery to tumors." *Biological and Pharmaceutical Bulletin* 36, no. 6(2013): 892–899.

Heepe, Lars, and Stanislav N Gorb. "Biologically inspired mushroom-shaped adhesive microstructures." *Annual Review of Materials Research* 44(2014): 173–203.

Hosseini-Nassab, Niloufar, Devleena Samanta, Yassan Abdolazimi, Justin P Annes, and Richard N Zare. "Electrically controlled release of insulin using polypyrrole nanoparticles." *Nanoscale* 9, no. 1(2017): 143–149.

Huang, Weiwei, Qishu Zhang, Weiran Li, Mingcui Yuan, Jingxian Zhou, Liangqun Hua, Yongjun Chen, Chao Ye, and Yanbing Ma. "Development of novel nanoantibiotics using an outer membrane vesicle-based drug efflux mechanism." *Journal of Controlled Release* 317 (2020): 1–22.

Huang, Wenjing, Yan Lang, Abdul Hakeem, Yan Lei, Lu Gan, and Xiangliang Yang. "Surfactin-based nanoparticles loaded with doxorubicin to overcome multidrug resistance in cancers." *International Journal of Nanomedicine* 13 (2018): 1723.

Hussain, Amjad, Faisal Mahmood, Muhammad Sohail Arshad, Nasir Abbas, Qazi Amir Ijaz, Nadia Qamar, and Fahad Hussain. "Drug loading and printability of two different grades of prefabricated polyvinyl alcohol filaments for fused deposition modeling-based 3D printing." *Journal of 3D Printing in Medicine* 4, no. 2(2020): 105–112.

Hussain, Amjad, Faisal Mahmood, Muhammad Sohail Arshad, Nasir Abbas, Nadia Qamar, Jahanzeb Mudassir, Samia Farhaj, Jorabar Singh Nirwan, and Muhammad Usman Ghori. "Personalised 3D Printed Fast-Dissolving Tablets for Managing Hypertensive Crisis: In-Vitro/In-Vivo Studies." *Polymers* 12, no. 12(2020): 3057.

Iqbal, Hafiz MN, Godfrey Kyazze, Ian C Locke, Thierry Tron, and Tajalli Keshavarz. "Development of bio-composites with novel characteristics: Evaluation of phenol-induced antibacterial, biocompatible and biodegradable behaviours." *Carbohydrate Polymers* 131 (2015a): 197–207.

Iqbal, Hafiz MN, Godfrey Kyazze, Ian Charles Locke, Thierry Tron, and Tajalli Keshavarz. "Poly (3-hydroxybutyrate)-ethyl cellulose based bio-composites with novel characteristics for infection free wound healing application." *International Journal of Biological Macromolecules* 81 (2015b): 552–559.

Iwahori, Kenji, Keiko Yoshizawa, Masahiro Muraoka, and Ichiro Yamashita. "Fabrication of ZnSe nanoparticles in the apoferritin cavity by designing a slow chemical reaction system." *Inorganic Chemistry* 44, no. 18(2005): 6393–6400.

Jeon, Gumhye, Seung Yun Yang, Jinseok Byun, and Jin Kon Kim. "Electrically actuatable smart nanoporous membrane for pulsatile drug release." *Nano Letters* 11, no. 3(2011): 1284–1288.

Jokerst, Jesse V, Tatsiana Lobovkina, Richard N Zare, and Sanjiv S Gambhir. "Nanoparticle PEGylation for imaging and therapy." *Nanomedicine* 6, no. 4(2011): 715–728.

Karimi, Mahdi, Masoud Eslami, Parham Sahandi-Zangabad, Fereshteh Mirab, Negar Farajisafiloo, Zahra Shafaei, Deepanjan Ghosh, Mahnaz Bozorgomid, Fariba Dashkhaneh, and Michael R Hamblin. "pH-Sensitive stimulus-responsive nanocarriers for targeted delivery of therapeutic agents." *Wiley Interdisciplinary Reviews: Nanomedicine and Nanobiotechnology* 8, no. 5(2016): 696–716.

Khan, Arshad Ali, Jahanzeb Mudassir, Safia Akhtar, Vikneswaran Murugaiyah, and Yusrida Darwis. "Freeze-dried lopinavir-loaded nanostructured lipid carriers for enhanced cellular uptake and bioavailability: statistical optimization, in vitro and in vivo evaluations." *Pharmaceutics* 11, no. 2(2019): 97.

Khan, Arshad Ali, Jahanzeb Mudassir, Noratiqah Mohtar, and Yusrida Darwis. "Advanced drug delivery to the lymphatic system: lipid-based nanoformulations." *International Journal of Nanomedicine* 8 (2013): 2733.

Kloxin, April M, Christopher J Kloxin, Christopher N Bowman, and Kristi S Anseth. "Mechanical properties of cellularly responsive hydrogels and their experimental determination." *Advanced Materials* 22, no. 31(2010): 3484–3494.

Kudur, Mohan H, Sathish B Pai, H. Sripathi, and Smitha Prabhu. "Sutures and suturing techniques in skin closure." *Indian Journal of Dermatology, Venereology, and Leprology* 75, no. 4(2009): 425.

Kuerban, Kudelaidi, Xiwen Gao, Hui Zhang, Jiayang Liu, Mengxue Dong, Lina Wu, Ruihong Ye, Meiqing Feng, and Li Ye. "Doxorubicin-loaded bacterial outer-membrane vesicles exert enhanced anti-tumor efficacy in non-small-cell lung cancer." *Acta Pharmaceutica Sinica B* 10, no. 8(2020): 1534–1548.

Kwak, Moon Kyu, Hoon-Eui Jeong, and Kahp Y Suh. "Rational design and enhanced biocompatibility of a dry adhesive medical skin patch." *Advanced Materials* 23, no. 34(2011): 3949–3953.

Lam, Patricia, and Nicole F Steinmetz. "Delivery of siRNA therapeutics using cowpea chlorotic mottle virus-like particles." *Biomaterials Science* 7, no. 8(2019): 3138–3142.

Liu, Mengrui, Hongliang Du, Wenjia Zhang, and Guangxi Zhai. "Internal stimuli-responsive nanocarriers for drug delivery: Design strategies and applications." *Materials Science and Engineering: C* 71 (2017): 1267–1280.

Maeda, Hiroshi, Hideaki Nakamura, and Jun Fang. "The EPR effect for macromolecular drug delivery to solid tumors: Improvement of tumor uptake, lowering of systemic toxicity, and distinct tumor imaging in vivo." *Advanced Drug Delivery Reviews* 65, no. 1(2013): 71–79.

Mahmood, Faisal, Amjad Hussain, Muhammad Sohail Arshad, Nasir Abbas, Muhammad Irfan, Nadia Qamar, Fahad Hussain, and Muhammad Usman Ghori. "Effect of Solublising Aids on the Entrapment of Loratidine in Pre-Fabricated PVA Filaments used for FDM Based 3D-Printing." *Acta Poloniae Pharmaceutica-Drug Research* 77, no. 1(2020): 175–182.

Mehnert, Wolfgang, and Karsten Mäder. "Solid lipid nanoparticles: production, characterization and applications." *Advanced Drug Delivery Reviews* 64 (2012): 83–101.

Menina, Sara, Janina Eisenbeis, Mohamed Ashraf, M. Kamal, Marcus Koch, Markus Bischoff, Sarah Gordon, Brigitta Loretz, and Claus-Michael Lehr. "Bioinspired liposomes for oral delivery of colistin to combat intracellular infections by Salmonella enterica." *Advanced Healthcare Materials* 8, no. 17(2019): 1900564.

Merkle, Hans P. "Drug delivery's quest for polymers: Where are the frontiers?" *European Journal of Pharmaceutics and Biopharmaceutics* 97 (2015): 293–303.

Meyers, Marc André, Po-Yu Chen, Albert Yu-Min Lin, and Yasuaki Seki. "Biological materials: structure and mechanical properties." *Progress in Materials Science* 53, no. 1(2008): 1–206.

Mintzer, Meredith A, and Eric E Simanek. "Nonviral vectors for gene delivery." *Chemical Reviews* 109, no. 2(2009): 259–302.

Mirkhalaf, Mohammad, and Hala Zreiqat. "Fabrication and Mechanics of Bioinspired Materials with Dense Architectures: Current Status and Future Perspectives." *JOM* 72, no. 4(2020): 1458–1476.

Mishra, Supriya, Shrestha Sharma, Md N Javed, Faheem Hyder Pottoo, Md Abul Barkat, Md Sabir Alam, Md Amir, and Md Sarafroz. "Bioinspired nanocomposites: applications in disease diagnosis and treatment." *Pharmaceutical Nanotechnology* 7, no. 3(2019): 206–219.

Mudassir, Jahanzeb, Yusrida Darwis, and Peh Kok Khiang. "Prerequisite characteristics of nanocarriers favoring oral insulin delivery: Nanogels as an opportunity." *International Journal of Polymeric Materials and Polymeric Biomaterials* 64, no. 3(2015): 155–167.

Mudassir, Jahanzeb, Yusrida Darwis, Suriani Muhamad, and Arshad Ali Khan. "Self-assembled insulin and nanogels polyelectrolyte complex (Ins/NGs-PEC) for oral insulin delivery: characterization, lyophilization and in-vivo evaluation." *International Journal of Nanomedicine* 14 (2019): 4895.

Mudassir, Jahanzeb, Yusrida Darwis, and Siti Rafidah Yusof. "Synthesis, characterization and toxicological evaluation of pH-sensitive polyelectrolyte Nanogels." *Journal of Polymer Research* 24, no. 10(2017): 1–20.

Peppas, Nicholas A. "Historical perspective on advanced drug delivery: How engineering design and mathematical modeling helped the field mature." *Advanced Drug Delivery Reviews* 65, no. 1(2013): 5–9.

Pérez-Mitta, Gonzalo, Alberto G Albesa, Christina Trautmann, María Eugenia Toimil-Molares, and Omar Azzaroni. "Bioinspired integrated nanosystems based on solid-state nanopores: 'iontronic' transduction of biological, chemical and physical stimuli." *Chemical Science* 8, no. 2(2017): 890–913.

Pitchaimani, Arunkumar, Tuyen Duong Thanh Nguyen, and Santosh Aryal. "Natural killer cell membrane infused biomimetic liposomes for targeted tumor therapy." *Biomaterials* 160 (2018): 124–137.

Qamar, Nadia, Nasir Abbas, Muhammad Irfan, Amjad Hussain, Muhammad Sohail Arshad, Sumera Latif, Faisal Mehmood, and Muhammad Usman Ghori. "Personalized 3D printed ciprofloxacin impregnated meshes for the management of hernia." *Journal of Drug Delivery Science and Technology* 53 (2019): 101164.

Raja, Maria Abdul Ghafoor, Haliza Katas, and Thum Jing Wen. "Stability, intracellular delivery, and release of siRNA from chitosan nanoparticles using different cross-linkers." *PloS one* 10, no. 6(2015): e0128963.

Ranjha, Nazar M, Jahanzeb Mudassir, Tanveer Abbas, Mohammad AF Siam, and Abid Hussain. "In vitro evaluation of commercially available theophylline sustained release tablets in Pakistan." *Latin American Journal of Pharmacy* 29, no. 6(2010): 869–875.

Ranjha, Nazar M, Jahanzeb Mudassir, and Sheikh Zuhair Zubair. "Synthesis and characterization of pH-sensitive pectin/acrylic acid hydrogels for verapamil release study." *Iranian Polymer Journal* 20, no. 2(2011): 147–159.

Ranjha, Nazar Mohammad, Jahanzeb Mudassir, and Sajid Majeed. "Synthesis and characterization of polycaprolactone/acrylic acid (PCL/AA) hydrogel for controlled drug delivery." *Bulletin of Materials Science* 34, no. 7(2011): 1537–1547.

Rasheed, Tahir, Muhammad Bilal, Nedal Y Abu-Thabit, and Hafiz MN Iqbal. "The smart chemistry of stimuli-responsive polymeric carriers for target drug delivery applications." In *Stimuli Responsive Polymeric Nanocarriers for Drug Delivery Applications, Volume 1*, 61–99. Elsevier, 2018. https://doi.org/10.1016/B978-0-08-101997-9.00003-5

Raza, Ali, Uzma Hayat, Tahir Rasheed, Muhammad Bilal, and Hafiz MN Iqbal. ""Smart" materials-based near-infrared light-responsive drug delivery systems for cancer treatment: A review." *Journal of Materials Research and Technology* 8, no. 1(2019): 1497–1509.

Raza, Ali, Tahir Rasheed, Faran Nabeel, Uzma Hayat, Muhammad Bilal, and Hafiz Iqbal. "Endogenous and exogenous stimuli-responsive drug delivery systems for programmed site-specific release." *Molecules* 24, no. 6(2019): 1117.

Rigon, Roberta B, Naiara Fachinetti, Patrícia Severino, Maria HA Santana, and Marlus Chorilli. "Skin delivery and in vitro biological evaluation of trans-resveratrol-loaded solid lipid nanoparticles for skin disorder therapies." *Molecules* 21, no. 1(2016): 116.

Rouse, Jillian G, and Mark E Van Dyke. "A review of keratin-based biomaterials for biomedical applications." *Materials* 3, no. 2(2010): 999–1014.

Sabu, Chinnu, Christine Rejo, Sabna Kotta, and K. Pramod. "Bioinspired and biomimetic systems for advanced drug and gene delivery." *Journal of Controlled Release* 287 (2018): 142–155.

Schafer, Freya Q, and Garry R Buettner. "Redox environment of the cell as viewed through the redox state of the glutathione disulfide/glutathione couple." *Free Radical Biology and Medicine* 30, no. 11(2001): 1191–1212.

Schwartz, Gregor, Benjamin C-K Tee, Jianguo Mei, Anthony L Appleton, Do Hwan Kim, Huiliang Wang, and Zhenan Bao. "Flexible polymer transistors with high pressure sensitivity for application in electronic skin and health monitoring." *Nature Communications* 4 (2013): 1859.

Servant, Ania, Cyrill Bussy, Khuloud Al-Jamal, and Kostas Kostarelos. "Design, engineering and structural integrity of electro-responsive carbon nanotube-based hydrogels for pulsatile drug release." *Journal of Materials Chemistry B* 1, no. 36(2013): 4593–4600.

Shkoporov, Andrey N, Ekaterina V Khokhlova, C Brian Fitzgerald, Stephen R Stockdale, Lorraine A Draper, R Paul Ross, and Colin Hill. "ΦCrAss001 represents the most abundant bacteriophage family in the human gut and infects Bacteroides intestinalis." *Nature Communications* 9, no. 1(2018): 1–8.

Singh, Satya P, Anup Kumar Sirbaiya, and Anuradha Mishra. "Bioinspired Smart Nanosystems in Advanced Therapeutic Applications." *Pharmaceutical Nanotechnology* 7, no. 3(2019): 246–256.

Stevens, Molly M. "Biomaterials for bone tissue engineering." *Materials Today* 11, no. 5(2008): 18–25.

Thirunavukkarasu, Guru Karthikeyan, Kondareddy Cherukula, Hwangjae Lee, Yong Yeon Jeong, In-Kyu Park, and Jae Young Lee. "Magnetic field-inducible drug-eluting nanoparticles for image-guided thermo-chemotherapy." *Biomaterials* 180 (2018): 240–252.

Wang, Yanfei, and Daniel S Kohane. "External triggering and triggered targeting strategies for drug delivery." *Nature Reviews Materials* 2, no. 6(2017): 1–14.

Watanabe, Akihiro, Shunji Kohnoe, Rinshun Shimabukuro, Takeharu Yamanaka, Yasunori Iso, Hideo Baba, Hidefumi Higashi, Hiroyuki Orita, Yasunori Emi, and Ikuo Takahashi. "Risk factors associated with surgical site infection in upper and lower gastrointestinal surgery." *Surgery Today* 38, no. 5(2008): 404–412.

Xu, Lei, Shuo Wu, and Xiaoqiu Zhou. "Bioinspired nanocarriers for an effective chemotherapy of hepatocellular carcinoma." *Journal of Biomaterials Applications* 33, no. 1(2018): 72–81.

Xu, Sheng, Yihui Zhang, Lin Jia, Kyle E Mathewson, Kyung-In Jang, Jeonghyun Kim, Haoran Fu, Xian Huang, Pranav Chava, and Renhan Wang. "Soft microfluidic assemblies of sensors, circuits, and radios for the skin." *Science* 344, no. 6179(2014): 70–74.

Yamada, Tadanori, Yasushi Iwasaki, Hiroko Tada, Hidehiko Iwabuki, Marinee KL Chuah, Thierry VandenDriessche, Hideki Fukuda, Akihiko Kondo, Masakazu Ueda, and Masaharu Seno. "Nanoparticles for the delivery of genes and drugs to human hepatocytes." *Nature Biotechnology* 21, no. 8(2003): 885–890.

Yang, Chunyu, Wei Guo, Liru Cui, Di Xiang, Kun Cai, Huiming Lin, and Fengyu Qu. "pH-responsive controlled-release system based on mesoporous bioglass materials capped with mineralized hydroxyapatite." *Materials Science and Engineering: C* 36 (2014): 237–243.

Yang, Seung Yun, Eoin D O'Cearbhaill, Geoffroy C Sisk, Kyeng Min Park, Woo Kyung Cho, Martin Villiger, Brett E Bouma, Bohdan Pomahac, and Jeffrey M Karp. "A bio-inspired swellable microneedle adhesive for mechanical interlocking with tissue." *Nature Communications* 4, no. 1(2013): 1–10.

Yi, Gawon, Jihwan Son, Jihye Yoo, Changhee Park, and Heebeom Koo. "Rhamnolipid nanoparticles for in vivo drug delivery and photodynamic therapy." *Nanomedicine: Nanotechnology, Biology and Medicine* 19 (2019): 12–21.

Zhang, Kaipu, Yanling Luo, and Zhanqing Li. "Synthesis and characterization of a pH-and ionic strength-responsive hydrogel." *Soft Materials* 5, no. 4(2007): 183–195.

Zhang, Quan, Xuanmiao Zhang, Tijia Chen, Xinyi Wang, Yao Fu, Yun Jin, Xun Sun, Tao Gong, and Zhirong Zhang. "A safe and efficient hepatocyte-selective carrier system based on myristoylated preS1/21-47 domain of hepatitis B virus." *Nanoscale* 7, no. 20(2015): 9298–9310.

Zhang, Rhongsheng, Mingguo Tang, Adrian Bowyer, Robert Eisenthal, and John Hubble. "A novel pH-and ionic-strength-sensitive carboxy methyl dextran hydrogel." *Biomaterials* 26, no. 22(2005): 4677–4683.

Zhou, Xiaohan, Longchen Wang, Yanjun Xu, Wenxian Du, Xiaojun Cai, Fengjuan Wang, Yi Ling, Hangrong Chen, Zhigang Wang, and Bing Hu. "A pH and magnetic dual-response hydrogel for synergistic chemo-magnetic hyperthermia tumor therapy." *RSC Advances* 8, no. 18(2018): 9812–9821.

Zreiqat, H., C.R. Howlett, A. Zannettino, P. Evans, G. Schulze-Tanzil, C. Knabe, and M. Shakibaei. "Mechanisms of magnesium-stimulated adhesion of osteoblastic cells to commonly used orthopaedic implants." *Journal of Biomedical Materials Research: An Official Journal of The Society for Biomaterials, The Japanese Society for Biomaterials, and The Australian Society for Biomaterials and the Korean Society for Biomaterials* 62, no. 2(2002): 175–184.

5 3D-Printed Nanocrystals for Oral Administration of the Drugs

Lucía Lopez-Vidal[1], Daniel Andrés Real[2], Alejandro J. Paredes[3], Juan Pablo Real[1], and Santiago Daniel Palma[1]
[1]CONICET and Universidad Nacional de Córdoba
[2]Universidad de Chile and Advanced Center for Chronic Diseases ACCDiS
[3]Queen's University Belfast

CONTENTS

5.1 Introduction	110
5.1.1 The Challenges of Oral Drug Delivery	110
5.2 Pharmaceutical Nanocrystals	111
5.3 Techniques for Obtaining and Post-Processing Nanocrystals	113
5.3.1 Incorporation of Nanocrystals in Oral Solid Dosage Forms	116
5.4 3D Printing in the Pharmaceutical Industry	117
5.4.1 Advantages of 3DP Usage	117
5.4.2 Stages of 3DP	118
5.4.3 Types of 3DP	120
5.5 Oral Solid Formulations of Nanocrystals by 3DP	121
5.5.1 Inkjet-Based Printing Systems	121
5.5.1.1 Drop-on-Solid Deposition	122
5.5.1.2 Drop-on-Drop Deposition	122
5.5.2 Nozzle-Based Printing Systems	123
5.5.2.1 Pressure-Assisted Microsyringe Method	124
5.5.2.2 Fused Deposition Modeling	124
5.5.2.3 Melting Solidification Printing Process	126
5.6 Electromagnetic Radiation-Based Printing Systems	128
5.6.1 Selective Laser Sintering	128
5.6.2 Stereolithography	129
5.7 Conclusions and Future Prospects	131
References	131

5.1 INTRODUCTION

Currently, nanotechnology represents one of the most valuable strategies to face solubility and selectivity challenges in drug delivery. The development of nanosystems has proven to be effective in improving both formulation-related problems (such as low solubility or low dissolution rate) and in increasing effectiveness and reducing adverse reactions caused by different drugs (Vinaud et al., 2020). In this sense, the formulation of nanocrystals (NCs) has become the strategy of choice to improve the performance of poorly soluble drugs (Müller et al., 2011; Paredes et al., 2021). However, the conversion of these systems into solid dosage forms for oral administration is an issue, as their physical stability could be compromised if conventional processes such as compaction are applied. Therefore, it is necessary to explore alternative technologies that allow the development of solid formulations based on NCs without affecting their physicochemical properties. One of the strategies with the greatest potential is additive manufacturing method, e.g., 3D printing (3DP), which has been shown to be an innovative and versatile tool for the development of solid dosage forms with the potential for spatiotemporal control of the release of the active ingredients (Real et al., 2020). Employing these techniques, it is possible to adapt the formulations to different formulation strategies and – additionally, given the adaptability of these techniques – it is possible to customize these formulations by adjusting, e.g., the size and dosage of the formulations according to the characteristics of each patient (Konta et al., 2017).

Based on the above, this chapter discusses the challenges of oral drug delivery, how nanocrystals can contribute to overcoming these challenges, and the opportunities generated by 3DP in the production of pharmaceutical forms of nanocrystals for oral delivery.

5.1.1 THE CHALLENGES OF ORAL DRUG DELIVERY

The oral route of administration is considered the first choice when establishing pharmacotherapy. Due to the convenience, safety, and price, it results in a high degree of patient compliance. In the USA and Europe, 85% of the most widely sold drugs are administered by this route (Savjani et al., 2012). However, the development of pharmaceutical formulations for the oral administration of drugs with certain physicochemical characteristics, such as low solubility or low permeability, remains a major challenge for pharmaceutical sciences.

The bioavailability of a drug is defined as the "amount and speed with which it reaches the systemic circulation, becoming available to reach its site of action" (Marianne, 2007). Aqueous solubility plays a fundamental role in absorption after oral administration, so it is essential to achieve the desired concentration of the drug in the systemic circulation to generate the desired pharmacological response. In 1995, pharmacist Gordon L. Amidon proposed the Biopharmaceutical Classification System (BCS), which allows predicting the absorption of drugs at the intestinal level by grouping them into four categories according to their aqueous solubility and membrane permeability (table below) (Amidon 1995). Class II and IV drugs are those with low water solubility. The formulation of this type of drugs presents numerous problems since their oral bioavailability is incomplete, variable, and dependent on factors such as concomitant administration with food.

Solubility Permeability	High	Low
High	Class I	**Class II**
Low	Class III	**Class IV**

Over the past 10 years, the number of poorly soluble drugs has increased steadily, with an estimated 40% of drugs on the market and 90% of drugs in the discovery pipeline facing solubility problems (Paredes et al., 2021). In turn, 40% of all potential drug candidates were shelved as a result of intrinsic solubility problems that could not be resolved. In this context, the development of strategies to overcome the limitations encountered when formulating these drugs is of particular interest to the pharmaceutical industry.

Multiple techniques have been studied to improve drug solubility, including physical and chemical modifications (Junyaprasert and Morakul, 2015), the use of cosolvents (Santos Souza et al., 2017) the formation of salts (Flores-Ramos et al., 2017), solid dispersions (Real et al.), the formation of inclusion complexes with cyclodextrins (Real et al., 2018), etc. The application of nanotechnology concepts to the development of drug forms has proven to be one of the most valuable strategies to solve solubility and selectivity problems in drug release (Real et al., 2018). By reducing particle size to the nanometric scale, it is possible to improve the apparent saturation solubility, dissolution rate, and oral bioavailability of hydrophobic molecules, achieving an increase in their therapeutic response or improvements in their safety profile (Real et al., 2020).

Among the various nanoparticle-based drug delivery systems, it is necessary to highlight nanocrystals. The following section describes their general characteristics and discusses their advantages and disadvantages.

5.2 PHARMACEUTICAL NANOCRYSTALS

NCs are by definition solid particles composed of 100% of the active ingredient with crystalline characteristics and sizes in the nanometer range, typically between 250 and 750 nm. NCs are surrounded by a stabilizer layer and can be in suspension (nanosuspensions) or in solid state, the latter are obtained after solvent removal. These systems have unique characteristics when compared to the micronized forms of drugs as described below.

$$log \frac{Cs}{C\alpha} = \frac{2\sigma V}{2.303 RT \rho r}$$

- **Increased drug saturation concentration** (Mauludin et al., 2009): The saturation concentration (Cs) indicates the amount of substance that can be dissolved to saturation point in a given medium and at a given temperature. Although this concentration is defined as a constant for each drug, this is limited to particles with a size in the micrometer range or larger. When we refer to systems below 1000 nm, such as nanocrystals, this property can be modified by varying the particle radius (r) as shown by the Ostwald–Freundlich equation (Eslami and Elliott, 2014) (Equation 1). As we can see, this variable is inversely related to Cs so that, as r decreases, Cs increases.

Equation 1. Ostwald–Freundlich equation, where Cs is the saturation solubility, $C\alpha$ is the solubility of the solid consisting of large particles, σ is the interfacial tension of substance, V is the molar volume of the particle material, R is the gas constant, T is the absolute temperature, ρ is the density of the solid, and r is the radius.

$$\frac{C}{dt} = A \cdot D \frac{(Cs - C)}{h}$$

- **Increased dissolution rate:** As it can be seen in the Noyes–Withney equation (Equation 2) (Noyes and Whitney, 1897), the dissolution rate of a drug will be conditioned, among other factors, by the surface area available for dissolution and the saturation concentration seen above.

Equation 2. Noyes–Withney equation. Where dC/dt is the dissolution rate, A is the surface area available for dissolution, D is the diffusion coefficient of the compound, Cs is the saturation concentration of the drug in the medium, and h is the thickness of the diffusion layer.

The process of nanometrization of the particles leads to an increase in the exposed surface area, as illustrated in Figure 5.1, so that the surface area available for contact with the dissolution medium increases, making the dissolution process faster.

- **Enhanced adhesion to cell-surface membranes:** Drug delivery systems must cross several barriers to reach their site of action. One such barrier is the dynamic semi-permeable barrier of mucus. Mucus is a complex aqueous mixture of glycoproteins, lipids, and salts that coats the epithelial barriers of various organs (gastrointestinal, ocular, respiratory, etc.). These polymers form a negatively charged hydrophilic network that prevents or limits the absorption of mainly hydrophobic substances.

Permeation through this layer depends, among other things, on particle size. Although it is not a constant rule, several studies show that at sizes below 500 nm particles passively pass through the mucin network (Lai et al., 2009).

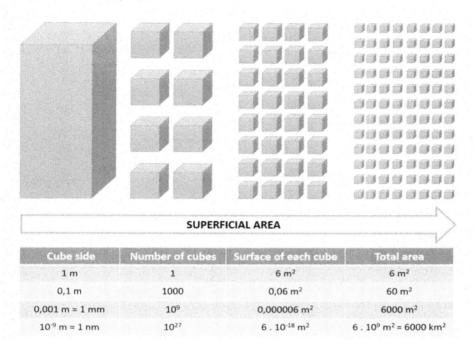

FIGURE 5.1 Increase in surface area due to the decrease in particle size.

In addition, the characteristics of mucus can be exploited to increase system retention at specific sites. The charge and conformation of the substances will impact adhesion to the mucus. The mucus layer is negatively charged, so it is possible to adhere positively charged systems to this barrier. The aforementioned increase in the exposed surface area of the nanosystems also increases the contact area, enhancing adhesion to mucus or other biological or cellular barriers. This mucoadhesive effect has been shown to contribute to increased drug absorption (Kaur et al., 2018).

To summarize, it can therefore be said that the increase in the specific surface area presented by NCs generates an increase in dissolution rate, saturation concentration, and mucoadhesiveness. It is feasible that these modifications could translate into improvements in the absorption and bioavailability of the drug, leading to an increase in the therapeutic response of poorly soluble drugs. Since it is possible to improve the biopharmaceutical characteristics of drugs through the development of nanocrystals, these systems have been extensively studied and various methodologies have been developed to produce them, as shown below.

5.3 Techniques for Obtaining and Post-Processing Nanocrystals

There are two main branches of methodologies for obtaining NCs (Figure 5.2). On the one hand, the "bottom-up" or ascending methods, in which a drug that is dissolved in controlled conditions is precipitated with the addition of stabilizers, thus

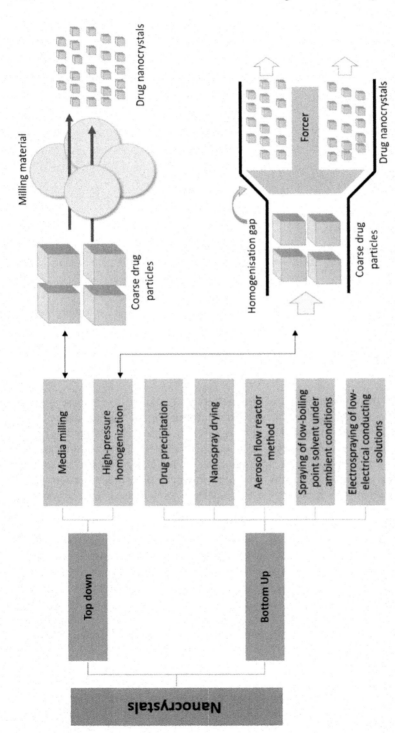

FIGURE 5.2 Top-down and bottom-up methods of nanocrystal formulation. Detail of the media milling and high-pressure homogenization mechanisms.

obtaining smaller and more uniform particles in size. This approach has some disadvantages that diminish the interest of the pharmaceutical industry. Among the disadvantages is the need to use organic solvents and the low transferability to large-scale manufacture (Van Eerdenbrugh et al., 2008). On the other hand, there are the "top-down" methods, in which the starting point is an aqueous suspension of a drug to which large amounts of energy are applied to achieve the breakup of the drug particles and the consequent nanometrization of the system. These methods are advantageous compared to bottom-up methods, as they are highly scalable, reproducible, and accepted by the pharmaceutical industry. Within top-down methodologies, two main techniques can be distinguished, high-pressure homogenization (HPH) and wet bead milling (WBM).

HPH is one of the most widely used techniques, both at the laboratory and industrial scale. It is performed by forcing a suspension through a duct with a decreasing diameter – until it reaches an opening of approximately 25 μm – at high pressure and velocity. This passage of the fluid through the tiny space in the homogenization gap results in cavitation. This phenomenon subjects the suspension to high turbulence and shear which, ultimately, causes the particle size reduction (Keck and Müller, 2006). This technique requires the application of high pressures that can lead to wear of the constituent materials of the equipment and consequent contamination of the sample with metallic components.

WBM or media milling is a flexible, highly reproducible, and scalable technique in which a grinding agent – usually consisting of zirconium, glass, steel, or ceramic beads or microspheres – is added to the drug-stabilizers suspension. The agitation applied to the system produces collisions between the drug particles, the grinding agent, and the walls of the equipment, which finally lead to decrease in particle size. The equipment used for this process consists of a chamber where the grinding material and the drug suspension are housed, inside which there is an agitator coupled to a motor.

Both HPH and WBM are high-energy techniques that allow obtaining NCs in the form of generally aqueous suspensions (nanosuspensions). In order to produce a solid form, it is then necessary to carry out a solvent removal process. Among the most commonly used techniques for this post-processing are spray drying and freeze-drying.

- Spray drying is a technology that makes it possible to obtain, in a simple and scalable way, dry powders from a liquid sample by drying with a hot gas. The fundamental principle is the rapid evaporation of the solvent by atomizing the liquid under vacuum and heat conditions. The equipment basically consists of three parts: an atomizer, a drying chamber where the volatile phase (e.g., water from a suspension) evaporates to form dry particles under controlled temperature and airflow conditions, and a cyclone that separates the formed particles. It may be necessary to add drying aids or carriers that increase the glass transition temperature of the material to prevent the generation of a gummy solid that adheres to the walls of the equipment, thus decreasing performance.

- Freeze-drying is based on the elimination of water or other solvents by freezing and subsequent sublimation (passage from the solid to the gaseous state without passing through the liquid state) at reduced pressure (primary drying) and the elimination of bound molecules by desorption (secondary drying). The whole process is carried out at low temperature and pressure and is therefore useful for drying thermolabile compounds. Sublimation or primary drying can be simplistically described in three steps: freezing, the sample is completely frozen; vacuum, the sample is subjected to a deep vacuum, below the triple point of water; and drying, thermal energy is added to the product, which causes the ice to sublime. The primary drying temperature must be kept as high as possible below the critical process temperature (collapse temperature for the amorphous substance).

When choosing a drying method, it is important to consider that although freeze-drying has numerous advantages, it is costly and time-consuming, unlike spray drying which is faster and cheaper.

5.3.1 Incorporation of Nanocrystals in Oral Solid Dosage Forms

Oral Solid Dosage Forms (OSDFs), especially tablets, are the first choice when establishing pharmacotherapy, as they allow for better patient compliance, and formulation, as they provide better storage stability (Tan et al., 2017). Despite this and the enormous advantages of NCs in improving the oral bioavailability of many drugs, few products currently use this technology. This is essentially because NCs are difficult to incorporate in OSDFs (tablets, capsules, pellets) due to two fundamental problems:

- The unfavorable flow properties of these powders, which make them unwieldy for processing, especially from an industrial point of view (Mauludin et al., 2009).
- The irreversible aggregation of nanoparticles by external forces (granulation, compression, mixing), which is related to the proportion of nanoparticles in the tablet: the higher the proportion of NCs in the powder mixture to be compressed, the higher the probability that nanocrystals will meet each other and undesired aggregation will occur (Müller and Junghanns, 2008). This sets a limit for loading tablets with NCs above which formulation by this methodology is counterproductive.

In other words, NCs are difficult to formulate into tablets and, at the same time, tablets can only carry a low proportion of NCs. An example is Rapamune, a medicine currently on the market consisting of tablets made from Sirolimus nanocrystals. The drug to tablet ratio is close to 1% (total tablet weight: 365 mg, the weight of active pharmaceutical ingredient (API) per tablet: 1 mg).

Other techniques for the formulation of NCs have been explored: Möschwitzer and Müller (Möschwitzer and Müller, 2013) succeeded in loading hydrocortisone acetate

NCs by fluidized bed granulation. The drug release they obtained was fast and complete; however, the loading capacity of the system obtained was low (close to 1%).

The incorporation of NCs in OSDFs requires the implementation of techniques that do not require specific flow control or the use of high pressures and, at the same time, that allow the redispersion of the nanocrystals, without significant size increases. In this sense, in recent years, 3DP techniques are a new strategy to consider for the development of pharmaceutical dosage forms. In the following section, the main 3DP techniques are described and the strategies that could be used to deliver NCs using these technologies are discussed.

5.4 3D PRINTING IN THE PHARMACEUTICAL INDUSTRY

Additive manufacturing, commonly known as 3DP, makes it possible to create solid objects from predesigned digital models. The printers add the material layer by layer until the three-dimensional digitized shape is complete.

3DP technology has been around for more than 30 years. Its inventor, Charles Hull, applied for a patent for the first additive manufacturing device in 1984. Hull designed his first commercially available, stereolithography-based 3D printer in 1988. From then to the present day, various 3DP processes have emerged. They have allowed to customize and fabricate complex structures that drive their use in many healthcare applications.

The introduction of 3DP in the pharmaceutical industry has the potential to generate a paradigm shift in the way drugs are obtained. Currently, large-scale drug production methods do not allow for the inter-individual variability of patients, which is an increasingly important factor in the therapeutic approach. In this context, 3DP does not come to compete with traditional drug manufacturing, as it can't do so. Its production times are 60 times longer and manufacturing costs are much higher. 3DP offers the possibility of bridging the existing therapeutic gap between traditional drug manufacturing and the need for individualized pharmacotherapy. At the same time, 3DP makes it possible to obtain solid pharmaceutical forms with unique and differential potentialities, which are impossible to achieve with industrial manufacturing.

The scientific and industrial interest in the subject of 3DP of drugs has increased from 2000 to the present day. The attention to 3DP of drugs grew mainly from 2015 when the FDA approved Spritam, the first 3D-printed drug, in August 2015. This fact is reflected in the number of publications, which has been increasing year after year (Figure 5.3).

The large number of publications observed is a reflection of the potential of this technology. The advantages that make 3DP one of the most promising techniques in the field of pharmaceutical sciences will be described below.

5.4.1 Advantages of 3DP Usage

In the pharmaceutical sector, 3DP technology represents a potentially versatile tool of simple and precise design for the production of OSDFs with unique and differential capabilities, such as:

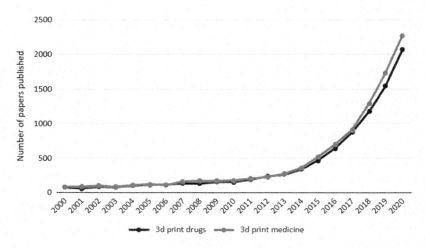

FIGURE 5.3 The number of published papers accumulated year by year using the terms "3d print drugs" and "3d print medicine" in the search engine ScienceDirect.

- Creating, with the same equipment and without changing assemblies, solid structures of different shapes and sizes without losing precision.
- Producing different solid dosage forms with the same equipment: printed tablets, liquid-filled capsules, films, vaginal ovules, etc.
- Creating innovative geometries that are difficult to obtain with traditional manufacturing, such as hollow or porous structures that allow the systems to float.
- Designing personalized OSDFs for the patient's pharmacotherapy, adapting the dose to the body mass and metabolic needs specific to the individualized treatment.
- Adjusting the kinetics of drug release by creating OSDFs with an immediate or modulated release, as required. Changing the shape and composition of the materials or even their arrangement in the OSDFs are some of the techniques that have been explored to modify drug release.
- Grouping numerous API, even incompatible ones, in the same OSDF. This combination, called "polypill", differs from the fixed-dose pharmacological combinations produced at the industrial level. The main difference is its ability to be personalized in terms of dosage and composition, varying and adjusting the dosage from patient to patient.

5.4.2 Stages of 3DP

As is described in this chapter, different methods of 3DP have been explored to produce solid dosage forms. Each technique requires the use of different raw materials and equipment; however, the basic procedure is the same and it can be divided into a series of steps (see Figure 5.4):

3D-Printed Nanocrystals

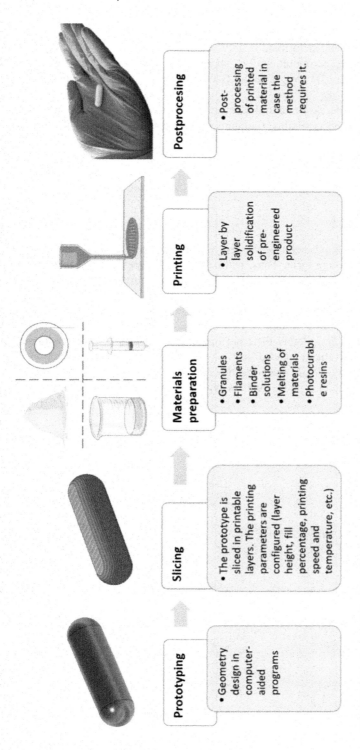

FIGURE 5.4 Stages of the 3D printing process.

1. Designing of the desired pharmaceutical shape by computer-aided design (CAD) programs in.stl format.
2. The.stl format is computationally "sliced" into printable layers and then transferred to the 3D programs in G-Code format so that it can execute the 3DP.
3. Preparation of the raw material needed for printing according to the 3DP method to be used (granules, filaments, binder solutions, fusion of the materials, etc.).
4. Additive manufacturing: the printer is configured and the raw materials are added and solidified layer by layer to produce the predesigned product.
5. Post-processing of the printed material if the method requires it.

5.4.3 Types of 3DP

Several 3DP methods have been developed, described, and patented for the production of the solid dosage form. Thus, within 3DP techniques, we can distinguish three main groups: printing systems based on inkjet, extrusion methods from nozzles, and systems based on the use of electromagnetic radiation (Figure 5.5).

The rationale behind 3DP is always the same: the material is added layer by layer until the desired 3D shape is obtained. However, each of the different techniques mentioned above uses completely different methodologies and presents particular characteristics with advantages and disadvantages when used in the production of pharmaceutical systems. While selecting the 3DP method, it is important to consider not only formulation aspects, such as type of materials, the melting point of

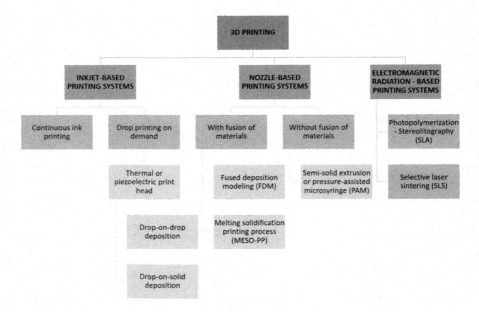

FIGURE 5.5 Classification of 3D printing methods.

3D-Printed Nanocrystals

the materials, mechanical properties, nature of the API, and desired release profile (immediate or delayed), but also the capacity for customization and decentralized production.

5.5 ORAL SOLID FORMULATIONS OF NANOCRYSTALS by 3DP

In this section, the different 3DP technological processes schematized in Figure 5.5 are described, analyzing in detail how these methodologies have been used or can be used for the process of obtaining OSDFs with nanocrystals.

5.5.1 INKJET-BASED PRINTING SYSTEMS

Inkjet printing encompasses two different types of techniques: continuous droplet printing (CDP) and droplet-on-demand (DOD) printing. Both are based on the theory of Lord Rayleigh who studied the stability of a column of nonviscous fluid.

In CDP, a high-pressure pump is used to generate a continuous flow of ink and a piezoelectric crystal causes the flow to break into droplets. The droplets are then directed by electrostatic plates to the substrate or waste to be recirculated (Figure 5.6) (Prasad and Smyth, 2016). They are called continuous because the droplets are propelled uninterruptedly, unlike DOD techniques in which the droplets are ejected when necessary.

DOD printing is considered to be more accurate and efficient in its use of ink than CDP. DOD printing can produce very small droplet volumes at very high speeds. The two most common types of actuators in print-on-demand are thermal and piezoelectric.

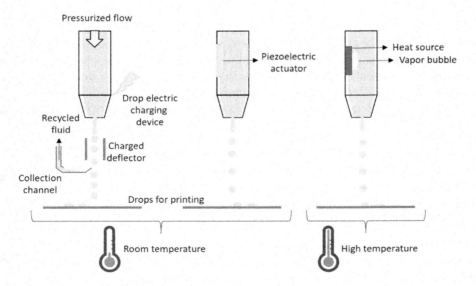

FIGURE 5.6 Fundamental principles of inkjet techniques.

Thermal printing uses a resistor that heats up rapidly, generating a vapor bubble in the ink and forcing it out through the print head (Prasad and Smyth, 2016). In this case, high temperatures are generated near the heating element, so ink degradation due to heat is a factor to be considered. It also requires the use of high vapor pressure solvents (volatile), which may limit its use in pharmaceutical technology. The piezoelectric one, on the other hand, does not operate by heat generation, does not require volatile solvents and has better control of droplet formation. These characteristics make it more suitable for application on the pharmaceutical field. Both heads can produce droplets characterized by a diameter ranging from 10 to 50 μm, which corresponds to a volume between 1 and 70 picoliters.

Usually, these methods use Newtonian liquid solutions as ink. However, due to the growing interest in 3DP in biomedical and pharmaceutical manufacturing, studies have been carried out to evaluate the "printability" of polymeric solutions and other liquids with complex rheological behavior (Derby, 2010). When a suspension is used as an ink, particle size, suspension stability, and their effect on fluid rheology must be considered. In general, it is recommended that the mean particle diameter of the suspension be two orders of magnitude smaller than the diameter of the jetting head orifice, to avoid clogging of the jetting head (orifices have typical diameters of 20–50 μm).

Within inkjet-based 3DP processes, there are two mechanisms of solidification of the printed material. When the printer head shoots the formulated droplets at each other to produce a solid layer of the build material it is called drop-on-drop deposition; when it shoots droplets onto a solid material, it is known as drop-on-solid deposition.

5.5.1.1 Drop-on-Solid Deposition

Drop-on-solid deposition (DSD), also referred to as binder deposition, creates a thin, uniform layer of powder on which the printhead precisely deposits a liquid binder. Once the layer formed solidifies, a new layer of powder is created on top of it and the process is repeated layer by layer until the three-dimensional shape created by the computer is obtained. Through this technology, it has been possible to obtain drugs with different release patterns and there is even a drug currently approved by the FDA that is obtained through this 3DP technique (Spritam). By using this technology, nanocrystals could be incorporated as a suspension in the inks or could also be part of the powder bed (Figure 5.7). Despite this, the process requires high control of the powder flow level and moisture content. Slow solidification times and shrinkage of bonded material are two other factors that can condition this process. Powder handling, which can lead to cross-contamination, and the work before (flow control, binder selection/preparation) and after (drying) printing would conspire against its use for decentralized (e.g., in a pharmacy office) production of customized OSDFs.

5.5.1.2 Drop-on-Drop Deposition

This technique is based on printing free forms that solidify drop by drop. Commonly injected materials include polymers and waxes, UV curable resins, solutions, suspensions, and complex multi-component fluids. Material injection

3D-Printed Nanocrystals

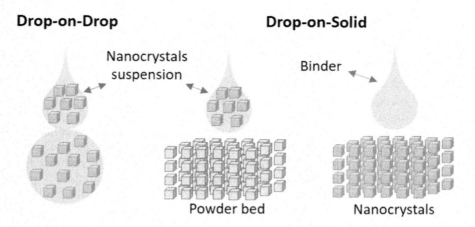

FIGURE 5.7 Printing of nanocrystals using inkjet printing.

differs substantially from binder deposition, resulting in an operation that is often more difficult to implement. The entire formulation needs to be designed for injection, followed by rapid solidification, and the product geometry becomes highly dependent on the droplet flight path, droplet impact, and surface wetting. One advantage that material injection has over binder injection and other methods is resolution; ink droplets are around 100 μm in diameter, and layer thicknesses for material injection are smaller than droplet sizes.

5.5.2 Nozzle-Based Printing Systems

These 3DP methods propose mixing the drug and excipients before the printing process. Then, the material is extruded through a nozzle and the 3D object previously designed in the digital file is formed layer by layer. As shown in Figure 5.8, this technology is further divided into those that do not melt the materials (Pressure-assisted

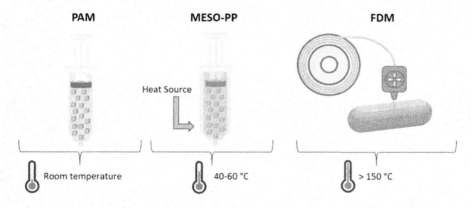

FIGURE 5.8 Nozzle-based printing methods.

Microsyringe (PAM)) and those that use temperature to melt the materials (Fused Deposition Modeling (FDM), and Solidification-Fusion 3D Printing Process (MESO-PP)).

5.5.2.1 Pressure-Assisted Microsyringe Method

The 3DP method known under the name of semisolid extrusion or PAM, is a printing process that uses as starting material (or ink) a semisolid formulation or paste that must have the property to form layers of filaments without collapsing during printing (Khaled et al., 2018). Since the impression material does not have to be melted but only plastically deformed, the printing process can be performed at room temperature, being a suitable method for thermally unstable drugs.

In general, the preparation of the inks used in this method is usually a simple process using fine powders combined with a suitable binder liquid. The choice of excipients used to form the paste or gel is decisive. The printability depends mainly on the viscosity of the formulation: if the viscosity is too high, the nozzle becomes clogged and printing is not possible, while if the viscosity is too low, the material will not take on the consistency of a 3D structure (it will collapse). The viscosity and rheological properties of the final formulation affect the extrudability, the consistency of the filament diameter, the bonding of the different layers, and the weight supported by the deposited layers. These are, therefore, two determining factors that must be studied if it is decided to use this technique. For example, the use of water-soluble excipients in water-based pastes would result in a low-viscosity paste and a non-consistent filament diameter leading to difficulties in retaining the shape of the printed objects. Therefore, to enable 3DP of water-soluble active ingredients, the inclusion of water-insoluble hydrophilic excipients may be necessary.

This printing technique requires post-processing treatment for solvent evaporation, solidification, and drying of the OSDFs. This process can require considerable time (24–48 h) depending on the solvent chosen (Khaled et al., 2015). This is one of its main disadvantages. On the other hand, the choice of solvents can be complicated because water requires a longer drying time and induces stability problems (chemical and microbial), while the use of organic solvents can limit their use in special groups of patients (e.g., for the pediatric population) and, in some cases, a determination of the residual solvent will be necessary according to certain pharmacopeias.

Although there are no antecedents of the use of this technique to obtain OSDFs containing nanocrystals, it could be promising in that it does not use high temperature or high pressure to obtain the desired shape.

5.5.2.2 Fused Deposition Modeling

FDM printers are used domestically, in research laboratories and industry, and are currently the most commercialized (Cunha-Filho et al., 2017). Due to the versatility and low cost of this technology, its use has spread widely around the world. These printers use thermoplastic polymer filaments as "ink", and these are molded at high temperature (generally > 150 °C) through the nozzle of the printer, delivering consecutive layers on a platform, which allows achieving the desired geometry as established by a computer program.

3D-Printed Nanocrystals

The FDM printing process involves several variables that can be configured to achieve optimal printing, such as nozzle diameter, resolution, plate and nozzle temperatures, travel and extrusion speed, fill percentage, and layer height. Additionally, the correct selection of the polymer to be used is a critical factor. Among the most commonly used thermoplastic polymers in FDM printing are polylactic acid (PLA), polyvinyl alcohol, and acrylonitrile butadiene styrene (ABS). However, PLA and ABS are insoluble in water allowing only the formulation of inert matrices that remain intact during the entire GI tract transit and can cause intestinal obstructions.

The biggest challenge faced by FDM for pharmaceutical application is that APIs must first be incorporated into the polymeric filaments. The most commonly used methods for incorporating APIs into filaments are soaking, spraying, and hot-melt extrusion (Figure 5.9). In the immersion (soaking) process, the prints or filaments are incubated in concentrated solutions of the drug. At the moment, this method allows only limited drug loading, usually below 10%. Beck et al. (2017) used this strategy for the development of nanocapsule-loaded tablets. These tablets were prepared by FDM from poly(e-caprolactone) and Eudragit1 RL100 filaments. The influence of the addition of mannitol as a channeling agent was studied. Tablets were soaked in deflazacort-loaded nanocapsules (particle size: 138 nm) to produce 3D-printed tablets (printlets) loaded with them. Effective tablet loading was demonstrated by SEM. The drug release profiles were influenced by the polymeric

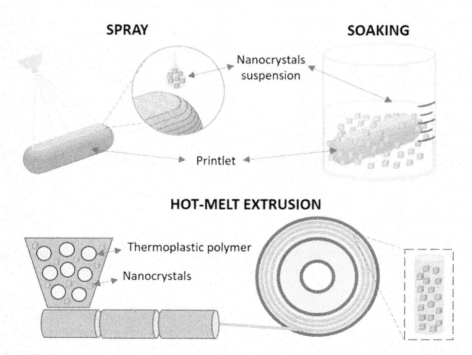

FIGURE 5.9 Methods of incorporating nanocrystals into FDM-printed solid dosage forms.

material used and by the presence of mannitol. However, the highest drug loading obtained was 0.27% w/w.

In a similar strategy, Shariatnia et al. (2019) developed a method to improve the mechanical performance of FDM-printed parts by incorporating cellulose nanocrystals (CNC). The incorporation of these nanocrystals between adjacent printed polymer layers was developed by spraying aqueous suspensions of CNCs during the printing process. The CNCs improved the interlayer adhesion and final strength of the printed parts. By incorporating aqueous suspensions of CNCs with concentrations of 0.5–1 wt%, it was possible to increase the interlayer strength of ABS parts by 44%, tensile strength by 33%, tensile modulus by 20%, and toughness by 33% in both longitudinal and transverse directions. However, when the CNC concentration was increased above 1%, agglomeration of nanocrystals occurred between the layers, which negatively affected the process.

In such a context, the use of hot-melt extrusion (HME) for the incorporation of nanosystems into filaments before the printing process may be a solution to overcome the loading limitations presented in the immersion and spray methods. HME is considered a promising technique in the pharmaceutical industry because it is a continuous process without solvents and without time-consuming steps. These characteristics make HME a suitable candidate for nanocrystal vehiculization. This methodology has been used to manufacture a variety of solid dosage forms, including tablets (Agrawal et al., 2016), granules, transdermal films (Repka et al., 2008), and lipid nanoparticles (Patil et al., 2014). In terms of nanocrystal incorporation, Baumgartner et al. (2014) used a milling technique to prepare a nanosuspension, which was then processed through the hot melt extruder into nanocrystals. This process was relatively slow, taking 24 hours to finish. Ye et al. (2016) used a similar strategy, combining the formulation of nanocrystals by HPH and HME. In the first instance, a nanosuspension was prepared using HPH, and then the same was extruded by HME with the help of Soluplus to obtain a solid dispersion of nanocrystals. This overall process took less than 1 hour to complete. By using this technique, an increase in dissolution rate was achieved as a result of the decreased particle size, increased surface area, and also due to improved wettability. The good stability was attributed to the maintained crystalline state of the drug. However, it is noteworthy that no publications combining HME and FDM printing for nanocrystal formulation have been developed so far.

5.5.2.3 Melting Solidification Printing Process

The melting solidification printing process (MESO-PP or FICIS in Spanish) is a method developed and patented at the National University of Cordoba (Argentina) (Real et al., 2020). The method uses a modified PAM printer, with an added thermostated jacket to control the temperature of the syringes. The MESO-PP procedure uses materials that have a melting temperature in the range of 40–60 °C. Among these excipients, we can mention certain lipids such as Gelucire or hydrophilic polymers such as polyethylene glycol, which when mixed with other components or by themselves form what we call "inks". These inks are mixed in a solid-state with one or more active ingredients (drugs) to make a premixed solid product (PSM), which can be used immediately or be stored in a stable form. At the time of printing, the whole solid mixture is heated to a temperature above the melting point of the ink, with continuous agitation greater than 150 RPM. As a result of the heating, the ink melts and the active ingredient is incorporated in suspended or dissolved form. Once the drug is incorporated, the mixture, loaded into a cartridge inside the printer's alloy tube, is ready to be heated (melted), extruded, and deposited in a controlled manner on a printing surface (which can be cooled) following the pattern designed by computer to create a 3D image from the layer to layer material aggregate.

This technique has a series of advantages, such as the use of safe materials, widely used in pharmaceutical technology (e.g., polymers and lipids) with the regulatory advantage that this brings. MESO-PP allows obtaining OSDFs without the need to use high temperatures, with the consequent improvement in stability (allowing the technique to be used in thermosensitive drugs) and costs due to the reduction in the application of heat. It does not use water or any other solvent and, thus, it does not require any type of post-processing (such as drying), which reduces the time to obtain a solid object. As a result, we obtain a simple, flexible, economical, and pharmaceutically "friendly" method, which allows to adapt the drug to different groups of patients and to obtain special geometries.

This technique presents a series of advantages, such as the use of safe materials, widely used in pharmaceutical technology (e.g., polymers and lipids) with the regulatory advantage that this brings. MESO-PP allows obtaining OSDFs without the need to use high temperatures, with the consequent improvement in stability (allowing the technique to be used in thermosensitive drugs) and in costs due to the reduction in the application of heat. It does not use water or any other solvent and, thus, it does not require any type of post-processing (such as drying), which reduces the time to obtain a solid object. As a result, we obtain a simple, flexible, economical, and pharmaceutically "friendly" method, which allows the adaptation of the drug to different groups of patients and the obtaining of special geometries.

MESO-PP, as well as FDM, are eligible for OSDFs in pharmacies and patient care points. In the not too distant future, the use of 3DP will enable a new business model oriented towards personalization, where certain OSDFs will be produced directly at the patient's point of care. This model would incorporate the coparticipation of the pharmaceutical industry as a producer of the inks and filaments that would later be transformed into a personalized drug. This complementary production chain would be one of the most viable alternatives to creating OSDFs individualized to the needs of each patient. For this to be achievable, it is necessary to have simple, robust methods that operate with portable and inexpensive equipment and whose inks can be processed and stored in advance as "intermediate products". It will also be a requirement that the method does not require error-inducing post-processing steps (e.g., drying, polishing, removal of remaining powder). Only two techniques currently meet these requirements: FDM and MESO-PP.

Regarding the incorporation of nanocrystals in OSDFs using the MESO-PP methodology, the process will depend largely on the correct choice of the type of polymer or lipid used as ink. An optimal process would require that the NCs do not solubilize in the molten ink, but rather that they remain suspended, avoiding modifications in their morphology or crystalline characteristics. The stability and

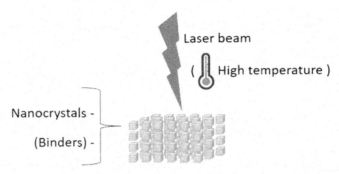

FIGURE 5.10 Printing of nanocrystals using selective laser sintering.

nonagglomeration of nanocrystals in the intermediate product and during the printing process is a determining factor to be taken into account when selecting excipients.

5.6 ELECTROMAGNETIC RADIATION-BASED PRINTING SYSTEMS

5.6.1 Selective Laser Sintering

This technique uses, similar to DSD, a bed of powdered materials consisting of a mixture of drug and excipients. To build the 3D object a thermal laser draws the specific pattern by melting the powder which then solidifies (this is called sintering) before starting the next layer (Figure 5.10). During the process, the unsintered powder acts as a support and once the printing is finished it is removed and can be reused. The objects produced with selective laser sintering (SLS) have porous structures so that the drug carrier systems obtained with this methodology have diffusion of the drug through the matrix as their main release mechanism. The degree of porosity of the matrix depends on – and can be adjusted by – the size of the powder particles and the laser energy density.

The sintering of nanoparticles with other heat sources, such as convection ovens, has been studied for a long time (Hong, 2018). In conventional sintering methods, nanoparticles are deposited on a substrate by wet processing and, then, heating of the bulk substrate to an elevated temperature occurs. The thermal sintering process usually involves high temperatures (> 200 °C) for a prolonged exposure period (> 30 min), which is limiting for many heat-sensitive drugs or polymeric materials (Awad et al., 2020; Vail et al., 1996). The main difference of selective laser sintering is that the heat source is replaced by a laser. When a laser beam is focused on the designated spot, the optical energy is directly converted into heat through a photothermal reaction, allowing to manipulate the local temperature with high controllability and selectivity. On the other hand, laser processing has certain advantages in terms of reproducibility compared to other processing techniques. The laser is an essentially massless tool, which is not subject to wear and is sterile, thus avoiding any contamination of the material being processed and increasing the reproducibility of the proposed process.

Nanoparticles have interesting properties that are advantageous for the sintering process. The depression of the melting temperature due to particle size reduction is one of the key properties in sintering, as it greatly reduces the temperature required in the process. In this sense, the thermal properties of the targeted nanoparticle are a crucial element that must be meticulously studied. On the other hand, the optical properties of the nanoparticles are also critical for laser sintering. The photothermal heating characteristics are directly related to the optical absorption of the particles. Nanoscale particles, in particular noble metal nanoparticles, exhibit unique tunable optical properties due to the surface plasmon resonance effect. This characteristic has already been extensively studied for numerous applications (Guerrero et al., 2014; Reza et al., 2012; Vio et al., 2017). If the morphology and size of the nanoparticles, as well as the strong absorption in the Rayleigh scattering regime, are tuned, it is possible to achieve local and efficient energy deposition by using a laser.

Initially, metallic nanoparticles, particularly noble metals, were the main target materials for laser sintering due to their superior chemical stability and low electrical resistance. Silver and gold nanoparticles are the most studied materials for laser sintering as microscale conductors (Yamaguchi et al., 2016; Zacharatos et al.).

This technology has the advantages of being a solvent-free process and offers faster production compared to DSD. However, the materials commonly used are powders (polystyrene, ceramic materials, glass, nylon, and metallic materials) that require high temperatures and high energy lasers to achieve sintering. This represents a drawback in the pharmaceutical field since it limits its use to active ingredients that are not degraded during the process. Also, this method has the disadvantage of requiring post-printing processing to finish the structures and remove the remaining powder. The fundamental mechanism of this method is based on the fusion of the material to be printed, so the crystalline characteristics and morphology of the nanocrystals would be modified, without being able to be redispersed as such. So far, no work has been reported using this technology for the formulation of the solid dosage form with nanocrystals. However, the melting characteristics of the nanocrystals could be used to print amorphous solids using this strategy.

5.6.2 Stereolithography

Developed in 1986, it was the first technique to be used to obtain solid dosage forms (Jain et al., 2018; Wang et al., 2016). In this technique, a UV laser is used to solidify a photopolymerizable polymer solution. The interaction of photoinitiator molecules with ultraviolet light results in the release of free radical molecules that initiate polymerization (Figure 5.11). This technique has the advantage of producing high-resolution objects at room temperature. At the moment, only a few materials can be used with this technique, as is limited to photosensitive polymers only. Due to its biocompatibility and efficient photopolymerization process, the most commonly used polymers are polyethylene glycol diacrylate (PEGDA) and gelatin methacrylate (GELMA) (Zhang et al., 2020). In contrast, many widely used pharmaceutical excipients cannot be printed or must be combined with such polymers, or

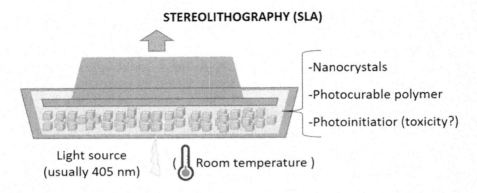

FIGURE 5.11 Obtaining a solid dosage form with nanocrystals using stereolithography.

chemically modified with photosensitive groups. For example, Shen et al. were able to develop a method for printing chitosan by digital light processing (DLP) (Shen et al., 2020). For this, they synthesized a photocurable chitosan derivative by using methacrylic anhydride.

On the other hand, to accelerate the photopolymerization process it is necessary to use photoinitiators. Among the most commonly used are diphenyl (2,4,6-trimethylbenzoyl) phosphine oxide (Irgacure TPO), 2-Hydroxy-4′-(2-hydroxyethoxy)-2-methylpropiophenone (Irgacure 2959) y Lithium phenyl-2,4,6-trimethylbenzoylphosphinate (LAP) (Bagheri and Jin, 2019). However, the safety of these compounds is unclear and some are considered toxic or unsafe (Carve and Wlodkowic, 2018). The carcinogenic risk of the light-curing material restricts its implementation in the short term and requires washing processes to ensure complete removal of the light-curing film and further research into the safety of the use of these compounds.

At the moment, no work using this technology for nanocrystal drug delivery has been reported. However, the use of the SLA 3DP approach to fabricate PEGDA composites reinforced with CNCs has been reported (Li et al., 2019). To improve the compatibility of CNCs with the PEGDA matrix, 1,3-diglycerolate diacrylate (DiGlyDA) was incorporated, which has a similar chemical structure but also has hydroxyl groups. CNCs can be used as reinforcing fillers during composite processing to improve the mechanical properties of polymers. However, CNCs dispersibility and compatibility with the polymer matrix are of vital importance, as interfacial compatibility has a significant effect on the properties of 3D-printed components. Ultimately, mechanically enhanced and structurally complex 3D-printed, CNC, composite structures were successfully printed using SLA 3DP. The addition of CNCs to the modified polymer matrix did not interfere with the 3DP process. In addition, SLA parameters, such as the thickness of the curing layer, can also be used to tune the mechanical properties and swelling behavior of the 3D-printed CNC composites. The feasibility of using CNC and SLA 3DP to fabricate composites with tunable and structurally customizable properties was confirmed. This work demonstrates that it would be possible to incorporate other types of NCs using stereolithography printing for biomedical and many other potential applications.

Although it is a fast technique that does not use temperature and allows working with a solution, some aspects still need to be improved for stereolithography to become a real alternative. The development of nontoxic photoinitiators or the synthesis of photocurable polymers that are considered GRAS are the challenges to be overcome for the technique to be used in pharmaceutical products.

5.7 CONCLUSIONS AND FUTURE PROSPECTS

Nanotechnology has revolutionized pharmaceutical sciences. Improvements in biopharmaceutical properties achieved by decreasing particle size have been widely explored in the past decades. Among the different types of nanosystems, nanocrystals are shown to be a promising alternative for the transport of APIs due to their high drug loading capacity and the scalability of their production processes. However, few developments have reached the market due to, in part, the difficulties in incorporating these systems into OSDFs. Recently, 3DP has emerged as a flexible alternative to meet this objective. This chapter described the different 3DP methods applied in pharmacy and analyzed their potential for the formulation of nanocrystals. As it could be seen, this wide range of technologies has advantages over conventional formulation methods. However, there are several barriers to overcome for 3DP methods to become an industrial alternative for nanosystems. The development of nontoxic stereolithography printing processes, the synthesis of biocompatible thermoplastic polymers that melt at low temperature, obtaining inks or polymers for 3DP that allow a correct redispersion of nanosystems, among others, are challenges to be solved if 3DP is to become the alternative of choice for the incorporation of nanosystems in OSDFs.

REFERENCES

Agrawal, A., Dudhedia, M., Deng, W., Shepard, K., Zhong L., Povilaitis, E., and Zimny, E. *AAPS PharmSciTech* 17 (2016): 214–232. https://doi.org/10.1208/s12249-015-0472-0

Amidon, G.L., Lennernäs, H., Shah, V.P., and Crison, J.R. *Pharm. Res. An Off. J. Am. Assoc. Pharm. Sci.* 12 (1995): 413–420. https://doi.org/10.1023/a:1016212804288

Awad, A., Fina, F., Goyanes, A., Gaisford, S., and Basit, A.W. *Int. J. Pharm.* 586 (2020): 119594. https://doi.org/10.1016/j.ijpharm.2020.119594

Bagheri, A., and Jin, J. *ACS Appl. Polym. Mater.* 1 (2019): 593–611. https://doi.org/10.1021/acsapm.8b00165

Baumgartner, R., Eitzlmayr, A., Matsko, N., Tetyczka, C., Khinast, J., and Roblegg, E. *Int. J. Pharm.* 477 (2014): 1–11. https://doi.org/10.1016/j.ijpharm.2014.10.008

Beck, R.C.R., Chaves, P.S., Goyanes, A., Vukosavljevic, B., Buanz, A., Windbergs, M., Basit, A.W., and Gaisford S. *Int. J. Pharm.* 528 (2017): 268–279. https://doi.org/10.1016/j.ijpharm.2017.05.074

Carve, M., and Wlodkowic, D. *Micromachines* 9 (2018): 91. https://doi.org/10.3390/mi9020091

Cunha-Filho, M., Araújo, M.R., Gelfuso, G.M., and Gratieri, T. *Ther. Deliv.* 8 (2017): 957–966. https://doi.org/10.4155/tde-2017-0067

Derby, B. *Annu. Rev. Mater. Res.* 40 (2010): 395–414. https://doi.org/10.1146/annurev-matsci-070909-104502

Eslami, F., and Elliott, J.A.W. *J. Phys. Chem. B* 118 (2014): 14675–14686. https://doi.org/10.1021/jp5063786

Flores-Ramos, M., Ibarra-Velarde, F., Jung-Cook, H., Hernández-Campos, A., Vera-Montenegro, Y., and Castillo, R. *Bioorg. Med. Chem. Lett.* 27 (2017): 616–619. https://doi.org/10.1016/j.bmcl.2016.12.004

Guerrero, A.R., Hassan, N., Escobar, C.A., Albericio, F., Kogan, M.J., and Araya, E. *Nanomedicine* 9 (2014): 2023–2039. https://doi.org/10.2217/nnm.14.126

Hong, Sukjoon (December 20th 2017). Selective Laser Sintering of Nanoparticles, Sintering of Functional Materials IgorShishkovsky, IntechOpen DOI: 10.5772/intechopen.68872. https://www.intechopen.com/chapters/55759

Hong, S. in *Sintering of Functional Materials*. InTech, 2018. https://doi.org/10.5772/65530

Jain, A., Bansal, K.K., Tiwari, A., Rosling, A., and Rosenholm, J.M. *Curr. Pharm. Des.* 24 (2018): 4979–4990. https://doi.org/10.2174/1381612825666181226160040

Junyaprasert, V.B., and Morakul, B. "*Asian J. Pharm. Sci.* 10 (2015): 13–23. https://doi.org/10.1016/j.ajps.2014.08.005

Kaur, G., Grewal, J., Jyoti, K., Jain, U.K., Chandra, R., and Madan, J. *Drug Targeting and Stimuli Sensitive Drug Delivery Systems*. Elsevier, 2018, pp. 567–626. https://doi.org/10.1016/B978-0-12-813689-8.00015-X

Keck, C.M., and Müller, R.H. *Eur. J. Pharm. Biopharm.* 62 (2006): 3–16. https://doi.org/10.1016/j.ejpb.2005.05.009

Khaled, S.A., Burley, J.C., Alexander, M.R., Yang, J., and Roberts, C.J. *Int. J. Pharm.* 494 (2015): 643–650. https://doi.org/10.1016/j.ijpharm.2015.07.067

Konta, A.A., García-Piña, M., and Serrano, D.R. *Bioeng. (Basel, Switzerland)*4, no. 4(2017). https://doi.org/10.3390/bioengineering4040079.

Lai, S.K., Wang, Y.Y., Wirtz, D., and Hanes, J. *Adv. Drug Deliv. Rev.* 61 (2009): 86–100. https://doi.org/10.1016/j.addr.2008.09.012

Li, V.C.F., Kuang, X., Mulyadi, A., Hamel, C.M., Deng, Y., and Qi, H.J. *Cellulose* 26 (2019): 3973–3985.

Marianne, A. in *Pharmaceutics. The design and manufacture of medicine*, eds. Aulton, M. and Taylor, K., London: Churchill-Livington, 2007.

Mauludin, R., Müller, R.H., and Keck, C.M. *Int. J. Pharm.* 370 (2009): 202–209. https://doi.org/10.1016/j.ijpharm.2008.11.029

Möschwitzer, J.P., and Müller, R.H. *Drug Dev. Ind. Pharm.* 39 (2013): 762–769. https://doi.org/10.3109/03639045.2012.702347

Müller, R.H., Gohla, S., and Keck, C.M. *Eur. J. Pharm. Biopharm.* 78 (2011): 1–9. https://doi.org/10.1016/j.ejpb.2011.01.007

Müller, R.H., and Junghanns, J. *Int. J. Nanomedicine* 3 (2008): 295. https://doi.org/10.2147/ijn.s595

Noyes, A.A., and Whitney, W.R. *J. Am. Chem. Soc.* 19 (1897): 930–934. https://doi.org/10.1021/ja02086a003

Paredes, A.J., Camacho, N.M., Schofs, L., Dib, A., del, M., Zarazaga, P., Literio, N., Allemandi, D.A., Sánchez Bruni, S., Lanusse, C., and Palma, S.D. *Int. J. Pharm.* https://doi.org/10.1016/j.ijpharm.2020.119501.

Paredes, A.J., McKenna, P.E., Ramöller, I.K., Naser, Y.A., Volpe-Zanutto, F., Li, M., Abbate, M.T.A., Zhao, L., Zhang, C., Abu-Ershaid, J.M., Dai, X., and Donnelly, R.F. *Adv. Funct. Mater.* 31 (2021). https://doi.org/10.1002/adfm.202005792

Patil H., Kulkarni V., Majumdar S., and Repka M.A. *Int. J. Pharm.* 471 (2014): 153–156. https://doi.org/10.1016/j.ijpharm.2014.05.024

Prasad, L.K., and Smyth, H. *Drug Dev. Ind. Pharm.* 42 (2016): 1019–1031. https://doi.org/10.3109/03639045.2015.1120743

Real, D.A., Hoffmann, S., Leonardi, D., Salomon, C., and Goycoolea, F.M. *PLoS One* 13 (2018): 115–135. https://doi.org/10.1371/journal.pone.0207625

Real, D.A., Hoffmann, S., Leonardi, D., Goycoolea, F.M., and Salomon, C.J. *J. Mol. Liq.* (2020): 114743. https://doi.org/10.1016/j.molliq.2020.114743

Real, D.A., Leonardi, D., Williams, R.O., Repka, M.A., and Salomon, C.J. *AAPS PharmSciTech* 19 (2018): 2311–2321. https://doi.org/10.1208/s12249-018-1057-5

Real, D.A., Orzan, L., Leonardi, D., and Salomon, C. *AAPS PharmSciTech.*

Real, J.P., Barberis, M.E., Camacho, N.M., Sánchez Bruni, S., and Palma, S.D. . *Int. J. Pharm.* 587 (2020): 119653. https://doi.org/10.1016/j.ijpharm.2020.119653

Repka, M.A., Majumdar, S., Battu, S.K., Srirangam, R., and Upadhye, S.B. *Expert Opin. Drug Deliv.* 5 (2008): 1357–1376. https://doi.org/10.1517/17425240802583421

Reza, A., Noor, A.S.M., and Maarof, M. *Plasmonics - Principles and Applications.* InTech, 2012.

Santos Souza, H.F., Real, D., Leonardi, D., Silber, A.M., and Salomon, C.J. *Trop. Med. Int. Heal.* 22 (2017): 1514–1522. https://doi.org/10.1111/tmi.12980

Savjani, K.T., Gajjar, A.K., and Savjani, J.K. *ISRN Pharm.* 2012 (2012): 1–10. https://doi.org/10.5402/2012/195727

Shariatnia S., Veldanda A., Obeidat S., Jarrahbashi D., and Asadi A. *Compos. Part B Eng.* 177 (2019): 107291. https://doi.org/10.1016/j.compositesb.2019.107291

Shen, Y., Tang, H., Huang, X., Hang, R., Zhang, X., Wang, Y., and Yao, X. *Carbohydr. Polym.* 235 (2020): 115970. https://doi.org/10.1016/j.carbpol.2020.115970

Tan, E.H., Parmentier, J., Low, A., and Möschwitzer, J.P. *Int. J. Pharm.* 532 (2017): 131–138. https://doi.org/10.1016/j.ijpharm.2017.08.107

Vail, N.K., Balasubramanian, B., Barlow, J.W., and Marcus, H.L. *Rapid Prototyp. J.* 2 (1996): 24–40. https://doi.org/10.1108/13552549610129764

Van Eerdenbrugh, B., Van den Mooter, G., and Augustijns, P. *Int. J. Pharm.* 364 (2008): 64–75.

Vinaud, M.C., Real, D., Fraga, C.M., Lima, N.F., De Souza Lino Junior, R., Leonardi, D., and Salomon, C.J. *Ther. Deliv.* (2020). https://doi.org/10.4155/tde-2020-0017

Vio, V., Marchant, M.J., Araya, E., and Kogan, M.K. *Curr. Pharm. Des.* 23 (2017): 1916–1926. https://doi.org/10.2174/1381612823666170105152948

Wang, J., Goyanes, A., Gaisford, S., and Basit, A.W. *Int. J. Pharm.* 503 (2016): 207–212 https://doi.org/10.1016/j.ijpharm.2016.03.016

Yamaguchi, M., Miyagi, N., Mita, M., Yamasaki, K., and Maekawa, K. *Trans. Japan Inst. Electron. Packag.* 9 (2016): E16-008-1–E16-008-8. https://doi.org/10.5104/jiepeng.9.E16-008-1

Ye, X., Patil, H., Feng, X., Tiwari, R.V., Lu, J., Gryczke, A., Kolter, K., Langley, N., Majumdar, S., Neupane, D., Mishra, S.R., and Repka, M.A. *AAPS PharmSciTech* 17 (2016): 78–88. https://doi.org/10.1208/s12249-015-0389-7

Zacharatos, F., Theodorakos, I., Karvounis, P., Tuohy, S., Braz, N., Melamed, S., Kabla, A., de la Vega, F., Andritsos, K., Hatziapostolou, A., Karnakis, D., and Zergioti, I. *Materials (Basel).* https://doi.org/10.3390/ma11112142

Zhang, J., Hu, Q., Wang, S., Tao, J., and Gou, M. *Int. J. Bioprinting* 6 (2020): 12–27. https://doi.org/10.18063/ijb.v6i1.242

6 Functional Nanomaterials

Imran Saleem[1], Yousef Rasmi[2,3], Leyla Fath-Bayati[4,5], and Zohreh Arabpour[5,6]

[1]School of Pharmacy and Biomolecular Sciences, Liverpool John Moores University
[2]Department of Biochemistry, Faculty of Medicine, Urmia University of Medical Sciences
[3]Cellular and Molecular Research Center, Urmia University of Medical Sciences
[4]Department of Tissue Engineering and Regenerative Medicine, School of Medicine, Qom University of Medical Sciences
[5]Department of Tissue Engineering and Applied Cell Sciences, School of Advanced Technologies in Medicine, Tehran University of Medical Sciences
[6]Iranian Tissue Bank and Research Center, Gen, Cell and Tissue Research Institute, Tehran University of Medical Sciences

CONTENTS

6.1 Introduction 136
6.2 Inorganic-Based Nanomaterials 137
    6.2.1 Metal and Metal Oxide-Based Nanomaterials 137
        6.2.1.1 Nanogold 139
        6.2.1.2 Nanosilver 140
        6.2.1.3 Nanoplatinum 140
        6.2.1.4 Magnetic Nanoparticles 141
    6.2.2 Semiconductor Quantum Dots (QDs) 141
6.3 Carbon-Based Nanomaterials 142
    6.3.1 Carbon Nanotubes (CNTs) 143
    6.3.2 Fullerene 144
    6.3.3 Graphene 145
    6.3.4 Graphene Oxide 146
    6.3.5 Graphene Quantum Dots (GQDs) 146
    6.3.6 Carbon Nanodiamonds (CNDs) 147
    6.3.7 Carbon Onions 147
    6.3.8 Nanoporous-Activated Carbon 148

DOI: 10.1201/9781003168584-6

6.4 Conclusion 148
References 148

6.1 INTRODUCTION

The field of nanotechnology deals with the science and engineering of designing and fabricating new constructs by controlling their size at nanoscale (Kumar 2013, Nunes et al. 2019, Silva 2004). Nanomaterials encompass a wide spectrum of materials that can be produced naturally, chemically, mechanically, or physically (De and Madhuri 2020). The shape, dimension, and synthesis method of nanomaterials have a significant impact on their properties, and functional nanomaterials (FNs) have characteristics superior to those of conventional bulk materials (Hong et al. 2019, Xu et al. 2016). Therefore, FNs are increasingly gaining popularity due to their various technological applications in different fields, including energy storage, biomedicine, bioimaging, sensors, electrochromic devices, and light-emitting diodes. Generally, FNs possess outstanding biological, optical, catalytic, electrical, mechanical, and magnetic properties originating from their nanometer size scale (Fath-Bayati et al. 2019, Fath-Bayati 2020, Hong et al. 2019, Xu et al. 2016). Nanomaterials can be organized into four groups according to their shape and dimension, including nanomaterials that have all external dimensions at nanometer size (1–100 nm) and their length equals the width categorized as zero-dimension (0D) nanomaterials, such as atomic clusters, quantum dots, and nanoparticles (Saleh 2020). Second category of nanomaterials that have two external dimensions at the nanoscale and one dimension at microscale are classified as one-dimensional (1D) nanomaterials, including nanorods, nanowires, nanotubes, and nanofibers (Sudha et al. 2018). Third category of nanomaterials is two-dimensional nanomaterials (2D) that have one external dimension at nanoscale while the other two dimensions are not in nanoscale. Examples of 2D nanomaterials include nanosheets, nanofilms, nanoplates, nanocoating, and nanoribbons. Fourth category of nanomaterials is three-dimensional (3D) nanomaterials that have external dimensions beyond 100 nm while they exhibit internal characteristics at nanoscale and include nanoporous structures such as aerogels (Dolez 2015) (Figure 6.1).

Furthermore, nanomaterials are classified into four main groups based on their composition, including: (1) inorganic-based nanomaterials (2), organic-based nanomaterials (3), carbon-based nanomaterials (4), and composite-based nanomaterials (Jeevanandam et al. 2018) (Figure 6.2). Examples of inorganic-based nanomaterials include metals (such as platinum, iron, gold, copper, and silver), metal oxides (such as titanium dioxide, zinc oxide, copper oxide, cerium oxide, and iron oxide), and semiconductors (such as silicon, nanosilicon, gallium nitride, gallium arsenide) (Hong et al. 2019, Jeevanandam et al. 2018, Sudha et al. 2018). A second chemical division of nanomaterials is composed of organic-based nanomaterials that consist of nanohydroxyapatite, lipids, nanopolysaccharides (cellulose nanomaterials, chitin nanomaterials, starch nanomaterials), protein-based bioinspired nanomaterials, and polymeric substances such as dendrimers, micelles, and liposomes. A third chemical category of nanomaterials includes carbon-based nanomaterials, such as fullerenes (C60), carbon nanotubes (CNTs), nanodiamonds, graphene (Gr), graphene oxide (GO), graphene quantum dots (GQDs), carbon nanofibers, carbon onions, and nanoporous activated carbon. A fourth

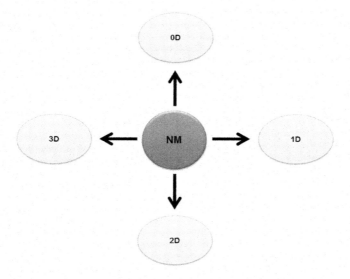

FIGURE 6.1 Classification of nanomaterials according to their dimension. Zero-dimension (0D), one-dimensional (1D), two-dimensional nanomaterials (2D), three-dimensional (3D).

chemical division of nanomaterials is composed of composite-based nanomaterials, including polymer-based nanocomposites, nonpolymer-based nanocomposites, and so on (Dolez 2015, Jeevanandam et al. 2018, Sudha et al. 2018). In the following sections, we focus on the inorganic-based nanomaterial and carbon-based nanomaterials, which are applied in various biomedical, biological, pharmaceutical, and clinical fields.

6.2 INORGANIC-BASED NANOMATERIALS

6.2.1 METAL AND METAL OXIDE-BASED NANOMATERIALS

During the past decade, metal based-nanomaterials consisting of platinum (Pt), gold (Au), silver (Ag), and copper (Cu) have been intensely assessed and results of various studies revealed that these nanoscale materials perform distinctly from their bulk counterparts (Poinern 2014), Table 6.1.

Metal oxide nanomaterials are another important class of inorganic-based nanomaterials that have excellent physical, chemical, and mechanical properties and include cerium oxide, zinc oxide, titanium dioxide, copper oxide, and iron oxides (Poinern 2014), Table 6.1. Metal and metal oxide-based nanomaterials are solid-state materials and because their application is dependent on their morphology (Asghari et al. 2016), these nanomaterials have been designed with various shapes and sizes, such as nanoparticles, quantum dots, and nanowires (Poinern 2014, Saleh 2020, Sudha et al. 2018). These nanomaterials have outstanding characteristics, such as enhanced catalytic activity and lower melting temperatures, compared to their bulk counterparts. These novel properties of nanomaterials have been

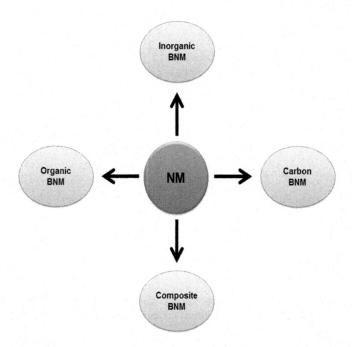

FIGURE 6.2 Classification of nanomaterials according to their composition. Inorganic-based nanomaterials (inorganic BNM), organic-based nanomaterials (organic BNM), carbon-based nanomaterials (carbon BNM), composite-based nanomaterials (composite BNM).

TABLE 6.1
Classification of Inorganic-Based Nanomaterials Based on Their Composition

Inorganic-Based Nanomaterials	Type	Reference
Metal-based Nanomaterials	Copper (Cu), platinum (Pt), gold (Au), silver (Ag),	(Poinern 2014)
Metal oxide-based Nanomaterials	Cerium oxide, titanium dioxide, zinc oxide, iron oxides, copper oxide	(Asghari et al. 2016, Poinern 2014)
Semiconductor Nanomaterials	CdSe, CdS, and HgCdTe, SiC, ZnS, ZnSe Si, Si-Ge, GaAs, InP, InGaAs, GaN, AlGaN	(Maxwell et al. 2020, Pawar et al. 2018)

attributed to their high surface-to-volume ratio, quantum-size effect, and exceptional surface chemistry (Chen et al. 2018, Poinern 2014).

In fact, as the particle size of the material decreases to the nanoscale range, the continuous band of electronic structure shifts to a separate electronic level, causing large numbers of atoms to emerge on the surface of the nanomaterial (Asghari et al. 2016).

Metal and metal oxide nanoscale materials play a critical role in various fields, such as electronics, optics, bioimaging, biosensors, chemical sensors, solar cells, and so on. The most commonly used metal or metal oxide-based FNs are in the form of nanoparticles. Therefore, in this section we introduce the metal and metal oxide-based FNs in the form of nanoparticles, followed by describing the potential applications of these nanomaterials in various fields.

6.2.1.1 Nanogold

Several gold-based nanomaterials, such as gold (Au) nanocages, gold nanoparticles, and gold nanoshells, have been used for various applications. Gold is a biocompatible metal and its properties in the nanosize range, including the melting point, considerably differ from bulk gold materials. Melting point of bulk gold materials is 1064 °C, while for gold-based nanoparticles it is approximately 600 °C. Therefore, large difference between the melting point of gold nanoparticles and bulk gold materials leads to the design and construction of gold nanostructures coated with certain molecules for adhering to tumor cells (Poinern 2014). After cell adhesion to gold nanostructures, low-power laser light was implemented to target gold-loaded tumor cells, which consequently lead to destroying of the tumor cells. In another study that takes advantages of the low melting temperature of gold nanostructures, an anticancer drug-loaded polymeric capsule coated with gold nanoparticles was irradiated with low power near-infrared laser. After irradiating and melting of gold nanoparticles, capsules ruptured and led to release of anticancer drug (Poinern 2014). In addition, gold with nanoregime helps provide a catalytic activity in chemical reactions with excellent electron transport capability (Haruta et al. 1989).

Significant work has been performed on the fabrication of gold-based construct for various applications including gold nanoparticle-based electrochemical sensor and biosensor (Maduraiveeran and Jin 2020). This was due to excellent characteristics, such as potential for surface functionalization, optical, physical, mechanical, and high surface-to-volume ratio properties (Abdel-Karim et al. 2020). The gold-based nanoparticles are regarded as durable and potential electrocatalyst in several electrochemical reactions because they have full recovery and high stability potential in chemical redox reactions (Abdel-Karim et al. 2020). Benefits of gold nanoparticle-based electrodes implementation are optimum catalytic activity, high selectivity, higher signal-to-noise ratio, and good diffusion of electroactive species (Parrilla et al. 2017). Therefore, gold nanoparticles are implemented in chemical and biological sensors owing to their excellent catalytic actvity, complete recovery in biochemical redox reactions, and chemical and optical characteristics (Maduraiveeran et al. 2018).

Electron transfer between the biomolecules and transducer is promoted due to incorporation of gold nanoparticles in modified electrodes and leading to suitable bioanalytical performance (Chen and Ostrom 2015). In addition, gold-based nanoparticle and other gold-based nanocomposites have been known as appropriate constructs for various applications in the field of biomedicine due to several properties, including simple preparation process, facile production method, high chemical durability, high biocompatibility, extensive electrochemical potential range, and favorable catalytic activity, which offer new opportunities for improvement of sensitive sensor platforms for biological, environmental, and medical applications (Chang et al. 2017, Gupta et al. 2016, Maduraiveeran and Jin 2020, Masitas et al. 2016).

6.2.1.2 Nanosilver

Silver nanomaterials have been used as antimicrobial agents to avoid bacterial or fungal contaminations in various products, including wound dressings, bandages, and clothing components (Poinern 2014). The US Food and Drug Administration (FDA) has approved nanosilver as an antimicrobial agent and consequently has been implemented in different medical devices including catheters for avoidance of infection (Poinern 2014). Previous reports revealed that hydrocarbon-coated silver nanoparticles in rat alveolar macrophages enhanced the release of cytokines in a size-independent mechanism (Carlson et al. 2008, Liu et al. 2014). In addition, other reports have revealed that silver nanoparticles could induce size-dependent toxicological and inflammatory effects on macrophages (Martínez-Gutierrez et al. 2012, Park et al. 2011, Stender et al. 2013). Another important application of silver nanoparticles is the application as an alternative strategy for antibiotics due to the development of microbial resistance overtime (Poinern 2014).

Electrochemical sensors and biosensors based on silver nanoparticles owing to their enhanced electrochemical signal, high biocompatibility, and good conductivity have made a profound impact in biomedical applications (Maduraiveeran and Jin, 2020). In the past few decades, there has been a great advance in the design and development of new platforms based on silver nanoparticles and their nanocomposites for detection of different analytes, including biological and infectious agents, disease markers, and other physiological reagents (Fekry 2017, Godfrey et al. 2017, Kumar-Krishnan et al. 2016, Maduraiveeran and Jin 2020). The designing and construction of silver nanoparticle-based nanocomposite makes use of various material matrices, including polymers, fibers, dendrimers, silicate networks, and metal oxides, which resulted in increased performance of biosensors (Liu et al. 2017, Martín-Yerga et al. 2016, Sheng et al. 2017, Yusoff et al. 2017). However, the efficiency and performance of the various biosensors rely on the distribution and avoidance of the aggregation of silver nanoparticles in the nanostructure composites. The development of an electrochemical immunosensing platform based on silver nanoparticles has exceptional properties, including simple and facile construction methods, high detection sensitivity, specificity, and rapid response. Recently, an electrochemical method was designed for tracing and detecting influenza viruses labeled with silver nanostructure (Abdel-Karim et al. 2020, Yusoff et al. 2017). The magnitude and frequency of the current, enhanced linearly either with increasing the surface coverage of the nanoparticles or boosting the concentration of influenza virus (Abdel-Karim et al. 2020).

6.2.1.3 Nanoplatinum

Platinum-based nanostructures have gained significant attention in the field of electrochemical biosensing due to their uncommon electrocatalytic and electronic properties (Chen and Chatterjee 2013, Dang et al. 2015, Deak et al. 2010, Liu et al. 2016). Material compositions, crystalline orientation, and plane surface reactive environments could affect the electron-transfer mechanism of the platinum-based nanoparticles. A study by Abellán-Llobregat, A. et al. revealed the development of a versatile electrochemical biosensor for glucose detection in humans, based on glucose oxidase (GOx) and platinum-doped graphite (Abellán-Llobregat et al. 2017). This biosensor was developed for the measurement of blood glucose concentration according to a linear scale of 0.00.9 mM and a low detection limit (6.6 mM). Furthermore, hydrogen gas sensors have been developed using Nafion electrolyte coated with Pt-C electrodes for hydrogen gas detection with suitable sensitivity, long-term durability, good performance, and wide linear range (Abdel-Karim et al. 2020, Jung et al. 2018).

6.2.1.4 Magnetic Nanoparticles

Magnetic nanoparticles are an important subset of FNs that include common magnetic elements, cobalt, nickel, manganese, iron, gadolinium, and chromium (Xue et al. 2011). Research interest towards magnetic structures continue to grow due to their significant therapeutic potential using dual-mode modification that can be controlled either by surface ligand engineering or using magnetic field. Magnetic nanomaterials are characterized using toxicity, biocompatibility, and magnetic performance (Fang and Zhang 2009, McCarthy and Weissleder 2008), because some magnetic nanomaterials such as chromium and cobalt are toxic to living body (Xue et al. 2011).

Clinical studies determined that manganese and iron are nontoxic elements, while chromium and cobalt are significantly harmful to living biological organs. Therefore, magnetic nanomaterials fabricated from nontoxic elements, such as iron and its oxidized compounds, have been largely studied for clinical applications (Jain et al. 2005, Julián-López et al. 2007, Kawai et al. 2008, Lee et al. 2007). Iron (Fe) is one of the important engineering materials in nano-state that performs distinctively from its bulk counterparts. Currently, there is great interest towards application of zero-valence iron (Fe0) or nanoiron (nano Fe) to eliminate water contaminants and impurities, and to increase the quality of water for various applications, such as drinking and agriculture. Nanoparticles such as nano Fe created noteworthy research interest due to their potential application as labeling agent in medical and in vivo molecular imaging for detection of cancers (Poinern 2014) or transplanted therapeutic cells (Fath-Bayati et al. 2019) using magnetic resonance imaging (MRI) technique. In addition, other magnetic nanostructures including gadolinium-doped nanoparticles have been implemented in clinical applications such as thermal therapy for treatment of malignant tumor owing to their unique power adsorption level (Drake et al. 2007).

6.2.2 SEMICONDUCTOR QUANTUM DOTS (QDs)

Electrical conductivity of semiconductor materials lies between a conductor and an insulator. In semiconductor materials, the highest occupied energy band, that is known as the valence band, is thoroughly occupied with electrons and the unoccupied level is the conduction band. In these materials band gap energy is less than that of insulators but more than that of conductors (Kumar 2013). When the semiconductor materials are exposed with light photons, electron that lies in the valence band is excited and jumps into the conduction band. After a short period of time, excited electrons come back to the valence band and emit fluorescence (Pawar et al. 2018).

Excitation Bohr radius is the gap between electron and hole in valence band. Decrease in size of semiconductor materials to nanoscale range results in increased surface-to-volume ratio and quantum size effect. If the dimension of the semiconductor materials is decreased to nanometer scales less than Bohr radius (a few nanometers), these nanoscale size materials are termed semiconductor quantum dots (QDs) (Pawar et al. 2018). Semiconductor materials encompass chemical elements near the commonly named "metalloid staircase" on the periodic table of elements gallium arsenide, silicon, and germanium (Saleh 2020). Generally, these semiconductor nanomaterials consist of various components from periodic table groups, including II–VI, III–V, and IV (GaAs) (Saleh 2020).

The semiconductor nanomaterials can be categorized into two types, including intrinsic semiconductors composed of pure components without doping or impurity due to presence of other metal materials. The unique properties of these materials are negative temperature coefficients of resistance that result in the enhancement of conductivity and reduction of materials resistivity by increasing temperature. Second category of semiconductor materials is named extrinsic semiconductors, in which other compounds are added to its structure for enhancement of conductivity (Saleh 2020). Distinct types of semiconductor nanoscale materials, such as CdSe, CdS, and HgCdTe, SiC, ZnS, ZnSe Si, Si-Ge, GaAs, InP, InGaAs, GaN, AlGaN, have been implemented in industrial applications due to the outstanding properties of these nanoscale materials (Maxwell et al. 2020, Pawar et al. 2018), Table 6.1.

Semiconductor nanomaterials have outstanding optical, chemical, and physical properties compared to bulk materials (Suresh 2013). Therefore, various types of semiconductor constructs exist such as solar cells, diodes, transistors, and digital and analog integrated circuits. During the past decade, there has been great interest in the designing and fabrication of nanoscale semiconductor materials that shed light on various technologies (Suresh 2013). Semiconductor quantum dots open new ways for photodynamic cancer therapy through promotion of singlet oxygen yield and suppressing of tumor cells (Kumar 2013). Also, quantum dots owing to their advanced optical and physical qualities and possibility of their surface functionalization using various biomolecules are promising candidates for biosensing (Imani et al. 2016).

6.3 CARBON-BASED NANOMATERIALS

Carbon is one of the periodic table elements that has the ability to polymerize at the atomic level and therefore can make very long chains. Carbon exists in various molecular forms and is composed of the same type of atoms but due to different structures show various features. The term "allotrope" is used for these forms, and graphite and diamond were the only known allotropes. However, at present, new types of allotropes have been described, including CBNs (Zaytseva and Neumann 2016). CBN's popularity is increasing due to their wide applications in biosensing, drug delivery, bioimaging, and tissue engineering. These interesting materials have a high surface area, a variety of surface chemistries, and unparalleled optical activity. Therefore, the new functionalized

Functional Nanomaterials

CBNs have efficient capacity of drug loading, biocompatibility, and lack of immunogenicity. However, much progress has been made in CBN's functionalization to address concerns about their health effects and improve biosafety. Recent evidence suggests that CBNs can be functionalized in conjunction with nucleic acids, active peptides, proteins, and drugs to obtain high pharmacological and low-toxic composites (Mohajeri et al. 2019). A variety of allotropes of carbon, from known allotropic phases such as amorphous carbon, graphite, and diamond to the newly discovered auspicious CNTs, fullerene, graphene, graphene quantum dots (GQDs), GO, CBNs have recently been developed (Mostofizadeh et al. 2011). Here we review the features and applications of popular functionalized CBNs.

6.3.1 CARBON NANOTUBES (CNTs)

Carbon nanotubes (CNTs) are one of the most widely used and extensively studied carbon allotropes (Iijima and Ichihashi 1993, Saito). They can be classified according to length, diameter, number of layers, and symmetry of rolled graphite sheet (chirality) (Zaytseva and Neumann 2016).

Depending on the number of walls, single-walled CNTs (SWCNT) and multi-walled CNTs (MWCNT) are the two basic forms of CNTs (Figure 6.3). Some researchers classify double-walled CNTs (DWCNTs) as a third-form of CNTs (Maiti et al. 2019, Zhang et al. 2013). SWCNTs are rolled forms of a hollow cylindrical carbon nanotube structure (single-graphene sheets) with a stable arrangement of carbon atoms linked by sp^2 bond hybridization and produced via rolling up a single graphite sheet with a high aspect ratio (length-to-diameter ratio), but MWCNTs possess several graphite layers maintained by van der Waals forces at an interlayer spacing of 3.4 A (Eatemadi et al. 2014, Hwang et al. 2020, Odom et al. 1998). In general, SWCNTs are about a few μm in length and about 1–3 nm in diameter. However, the synthesis of CNTs as long as 550 mm has recently been reported (Zhang et al. 2013). SWCNTs can be categorized into three subgroups, including (i) armchair (electrical conductivity > copper), (ii) zigzag, and (iii) chiral (Figure 6.4). While highly resistant MWCNTs consist of several layers of carbon,

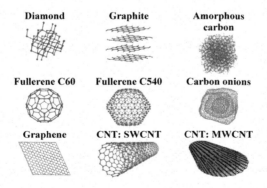

FIGURE 6.3 Schematic illustration of main carbon-based nanomaterials (Săndulescu, Tertis, et al. 2015).

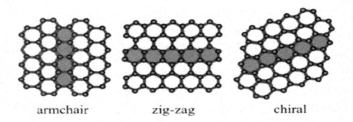

FIGURE 6.4 Different type of single-walled nanotubes: armchair, zig-zag, and chiral (Zaytseva and Neumann, 2016).

and often with variable chirality, they can reveal amazing mechanical characteristics instead of outstanding electrical features (Zaytseva and Neumann 2016).

CNTs are actively used in high-performance electron emission devices, such as electron microscopes, flat plates, and gas discharge tubes. Further, CNTs show intense luminescence from the field emission, which can be applied to lighting elements (Cha et al. 2013). Carbon nanotubes are considered to be the new generation nanoprobes (Tîlmaciu and Morris 2015). Strong conductivity, high chemical sensitivity and stability, large aspect ratio (Zhao et al. 2002), and rate of electron transfer (Lin et al. 2004) provide CNTs suitable for use in biosensors (Maiti et al. 2019). Based on their detection and transmission mechanisms, these types of biosensors are classified mainly in electronic and electrochemical CNT-based biosensors and optical biosensors (Hou et al. 2016, Jacobs et al. 2010, Kumar et al. 2015, Yang et al. 2015). CNTs have been used for composite materials such as plastic and metal alloys as reinforcing elements (Yamamoto et al.). Also, CNTs are widely applied as a weight loss enhancer and filler (Varshney 2014). In addition, intelligent functionalized CNTs show decreased or no immunogenicity per se, significant biocompatibility, and have professional photoluminescence activity, all of which amplify its increasing usage in tissue engineering, drug delivery, and imaging (Lacerda et al. 2012, Mehra et al. 2015, Mehra and Jain, 2016).

To connect different biomolecules, the nanocarrier surfaces undergo functionalization. It should be noted that the functionalization nature and kind of crystal lattices of these substances can mostly affect the entry pathway and final destruction route or excretion from cells and these factors must be considered (Elgrabli et al. 2008, Marangon et al. 2012, Raffa et al. 2010) to reduce adverse effects on tolerable biological conditions (Lacerda et al. 2008).

6.3.2 Fullerene

Fullerenes are one of the allotropes of carbon. Due to its spherical structure, the fullerenes are also called buckyballs (Notarianni et al. 2016). The family of fullerene consists of a number of atomic Cn clusters (n > 20) consisting of a spherical surface. In the structure of fullerenes, carbon atoms are usually at the vertices of pentagons and hexagons on the sphere surface (Zaytseva and Neumann, 2016). Buckminsterflern, known as formula C60, has 60 carbon atoms (each with 3 bonds) arranged in a cage-like structure that looks like a soccer ball. This type of fullerene is the most important commonly used and best researched between fullerene family (Coro et al. 2016). The spherical molecule of fullerene with 60 carbon atoms is highly symmetric, with the carbon atoms at the vertices of 12 pentagons and 12 hexagons (Figure 6.3), which makes a diameter of 0.7 nm for fullerene C60 (Kroto et al. 1985).

The C60-based nanostructures such as nanorods, CNTs, and nanoplate have found wide use in nanotechnology and nanoscience. The C60 can have various kinds and also can facilitate reactions of various substances (Coro et al. 2016). The special properties of fullerene 60 has resulted in blends with constituents to elevate their characteristic behavior. For use in the environment, covalent, endodral, and supermolecular developments permit C60 to undergo molecular manipulation and evolution of polymeric materials (Rao et al. 2020). The fullerene C60 was the first 0D allotrope of carbon to be known, but it is not the only discovered case. Indeed, large scales of C70, C76, C78, C84, and even C240 and C330 as larger clusters, have been fabricated and evaluated (Dresselhaus et al. 2011).

After diamond and graphite, fullerenes are the third carbon allotropes in the nanoscale range with strange photoelectrochemical characteristics. Fullerenes, due to their hydrophobic properties are completely insoluble in water, and appear to limit their biological applications (Durdagi et al. 2008, Horie et al. 2009, Sánchez et al. 2009). Thus, functionalization by hydroxyl, amino, or carboxyl groups has been revealed to increase the water-soluble characteristics of fullerene derivatives, thus leading to high compatibility (Afreen et al. 2015, Montellano et al. 2011). In electrochemical biosensors, fullerenes as electron mediators between the detection site and the electrode significantly increase the electron transfer rate that was caused by the biochemical or biocatalytic analytes reaction in contact with bioelements at detection sites (Hwang et al. 2020).

6.3.3 GRAPHENE

Graphene is a single layer of graphite and a 2D allotropic carbon form (Figures 6.3 (Hwang et al. 2020, Zaytseva and Neumann, 2016). In graphene structure, carbon atoms present in a 2D hexagonal crystal honeycomb network, with carbon–carbon distance of 1.42 Å and a distance between layers of 3.4 Å (Erickson et al. 2010), in which plates of graphene are held together in place via weak van der Waals forces (Hwang et al. 2020, Zaytseva and Neumann, 2016).

Graphene also represents a number of other structural elements of carbon allotropes, such as CNTs, graphite, and fullerenes. The electrical characteristics of these carbon allotropes are basically distinct from the 3D materials' characteristics. Graphene has a number of good physical characteristics, such as high mechanical strength and very high thermal stability (Zaytseva and Neumann, 2016). Graphene and CNTs have same optical, thermal, and electrical characteristics, but the 2D graphene structure allows for more diverse electronic properties (Geim and Novoselov, 2007). In addition, graphene is structurally very flexible and strong, making it attractive for the engineering of thin and flexible materials (Eda et al. 2008, Kim et al. 2009). Graphene shows a number of unique properties associated with its potential attractiveness aiding biological applications. Graphene's easy-to-functionalization perspective enriches the surface with functional groups, which actually facilitates the selective and specific identification of several biological components (Pattnaik et al. 2016, Yang et al. 2013, Zhang et al. 2013). Graphene's excellent electronic, optical, and magnetic properties make it a fast and inexpensive biosensing platform. The electron motion of the graphene layer is 100 times greater than that of silicon (Bolotin et al. 2008). In addition, chemical purity, very large surface area, and free π electron make it a good option for drug delivery (Pattnaik et al. 2016, Yang et al. 2013, Zhang et al. 2013). It is also mostly used in in vivo bioimaging, cancer diagnosis, and treatment with the aid of practical actions on a variety of fluorescent dyes, therapeutics, and other biological substances (Maiti et al. 2019).

6.3.4 Graphene Oxide

Graphene has been shown to be an fruitful additive for the production of polymer-based nanocomposites (Smith et al. 2019). However, virgin graphene faces a number of challenges, such as complex bottom-up synthesis, aggregation problems, and low solubility. Thus, compounds such as GO and GQDs are produced from carbon sources using a simple top-down method. Since graphene derivatives are easily synthesized, they are a good alternative to graphene (Yoo et al. 2014). GO is involved in the production of stimulus-responsive materials due to its unique hydrophilic, electrical, and thermal characteristics (Smith et al. 2019). The presence of functionalized oxygen groups in the GO structure enhances solubility and improves efficient surface area. In addition, GO is an important filler in polymer nanocomposites. This is due to its remarkable characteristics and high dispensability in the polymer matrix (Yoo et al. 2014), and it has been shown to be highly used as a contrast agent for MRI (Tao et al. 2012).

The presence of large π-mixed structures attached to the GO surface supplies an ultra-capacity for surface change, therefore loading a variety of aromatic drugs (Liu et al. 2010, Sun et al. 2008, Tian et al. 2011). Drug molecules can be loaded onto graphene-based nanomaterials via noncovalent interaction (Liu et al. 2008). The GO could be a roseate alternative options to gene delivery systems, because beside to evidence of GO's potential to deliver small drug molecules, the decorations with cationic polymers have highlighted its application in several other biomacromolecules (Chua et al. 2011, Feng et al. 2013, Tian et al. 2011).

6.3.5 Graphene Quantum Dots (GQDs)

Graphene quantum dots (GQD_S), which are revealed as a 0D graphene plate with lateral dimensions of less than 100 nm in one to several layers, are attractive and innovative biomaterials from the carbon family (Song L). Nontoxicity, good solubility, stable photoluminescence, and high surface grafting quality are advantages of GQD_S, which is why these materials are said to be promising candidates to replace inorganic QDs (Robinson et al. 2010). Due to quantum confinement, when converting 2D graphene sheets to GQDs, they produce excellent photoluminescence (Maiti et al. 2019, Wang et al. 2016).

Surprisingly, in comparison to other semiconductor QDs, the GQDs show superior biocompatibility and resistance to photo bleaching. In addition, the GQDs

Functional Nanomaterials 147

have prominent graphene properties, resembling a large surface area and π electrons (Chen et al. 2017, Kumawat et al. 2017, Maiti et al. 2019, Zheng et al. 2015). Since there is limited access to GQDs, applications related to them are still under development, so synthesizing GQDs from coal looks promising as it allows the production of high-quality materials on a larger scale. This will ensure the availability of high-quality GQDs in greater quantities to the scientific community, contributing to in-depth studies of the outstanding properties and accelerate the development of new applications (Robinson et al. 2010).

6.3.6 CARBON NANODIAMONDS (CNDs)

CNDs are carbon structures in nanoscale (5–100 nm) wrought of sp^3-hybridized form of carbon atoms. The exceptional electronic and optical features of NDs make them suitable for apply in specific biosensing devices (Hwang et al. 2020). Due to NDs' low toxicity, biocompatibility, refractive index, thermal conductivity, and chemical inertness of their diamond core, NDs are unique compared to other CBNs. Carbon nanodiamonds consist of a very diverse surface consisting functional groups, such as carbonyl, carboxyl, hydroxyl, and ether groups (Mochalin et al. 2012, Whitlow et al. 2017). Based on the size, NDs are classified into nanocrystalline particles (10–100 nm), ultrananocrystalline NDs (2–10 nm), and diamonds (1–2 nm).

Carbon nanodiamonds, as a delivery consignment of target molecules, can work as a substrate with a number of physical features, such as chemical functional groups and charged surface faces made by noncovalent or covalent methods on their surfaces. The noncovalent variations that use an easy and simple method are widely applied for the ND surfaces functionalization. In contrast, the covalent methods can produce stable ND surfaces, but it requires a complicated process (Hwang et al. 2020), Figure 6.3.

6.3.7 CARBON ONIONS

Carbon onions are occasionally called onion-like carbon (OLC) or carbon nano-onions (CNOs), and are composed of spherically packed carbon shells and owe their name to the onion-like concentric layer structure (Figure 6.3). Carbon onions include a variety of concentric shells, from nested fullerenes to small (<100 nm) polyhedral nanostructures (McDonough and Gogotsi 2013). Carbon onions are one of the least researched CBNs but are receiving a lot of attention due to their energy storage devices. Due to the small diameter (<10 nm), exceptional 0D structure, relatively easy dispersion and high electrical conductivity, compared to 1D nanotubes and 2D graphene, carbon onions are often applied as conductive additives in electrochemical energy storage devices, including battery and supercapacitor electrodes, in which it is necessary to increase the performance of the device (McDonough and Gogotsi 2013). However, carbon onions are more than just an additive. Devices containing carbon onion electrodes revealed a capacitance four times higher than electrolytic capacitors, which has a remarkable performance in the field of super capacitors (Pech et al. 2010). Carbon onions are also coated with redox active molecules or functionalized by electrochemical functional groups to improve overall electrochemical performance (Anjos et al. 2013). In addition, carbon onions, with ionic liquid electrolytes discovered, extend the operating voltage range and make it possible to use it in low temperature regimes (Lin et al. 2011). Because of each of these performance and qualities ranges, the carbon onions can expand into widely other applications, such as textile supercapacitors, antennas, biomedical devices, sensors, and more.

6.3.8 Nanoporous-Activated Carbon

Charcoal, or activated carbon, is a type of fine, small-pore carbon that is produced to create a large surface area for adsorption or chemical reactions. Thus, activated carbon can be applied as an adsorbent to remove contaminants in water purification processes when heavy metals, gases, and dyes are separated from water. However, the removal efficiency is not very high. Thus, the activated carbon is converted into functionalized nanoporous activated carbon to achieve the desired efficiency (Mestre and Carvalho 2017), e.g., as supercapacitor (Zhao et al. 2010).

The nanoporous structure is chiefly composed of carbon. Oxygen and hydrogen are also present in relatively small parts. Depending on the precursor, the manufacturing and processing methods after synthesis, sulfur, phosphorus, nitrogen, and minerals may also be participating. Of these, oxygen groups make the most observations on nanoporous surfaces.

6.4 CONCLUSION

This chapter introduced FNs and their significant developments and major challenges. FNs are becoming increasingly popular due to their widespread technological applications in various fields. Generally, FNs possess outstanding biological, optical, catalytic, electrical, mechanical, and magnetic properties originating from their nanometer size scale. However, to optimize the use of functional nanomaterials for various applications, a more complete understanding of their chemical and physical characteristics and functionalization methods is needed.

REFERENCES

Abdel-Karim, R., Y. Reda, and A. Abdel-Fattah. "*Nanostructured Materials-Based Nanosensors.*" *Journal of The Electrochemical Society* 167, no. 3(2020): 037554.

Abellán-Llobregat, A., I. Jeerapan, A. Bandodkar, et al. "*A stretchable and screen-printed electrochemical sensor for glucose determination in human perspiration.*" *Biosensors and Bioelectronics* 91 (2017): 885–891.

Afreen, S., K. Muthoosamy, S. Manickam, and U. Hashim. "*Functionalized fullerene (C60) as a potential nanomediator in the fabrication of highly sensitive biosensors.*" *Biosensors and Bioelectronics* 63 (2015): 354–364.

Anjos, D.M., J.K. McDonough, E. Perre, et al. "*Pseudocapacitance and performance stability of quinone-coated carbon onions.*" *Nano Energy* 2, no. 5(2013): 702–712.

Asghari, F., Z. Jahanshiri, M. Imani, M. Shams-Ghahfarokhi, and M. Razzaghi-Abyaneh. "*Antifungal nanomaterials: synthesis, properties, and applications*", in *Nanobiomaterials in Antimicrobial Therapy*. Elsevier, 2016: 343–383.

Bolotin, K.I., K.J. Sikes, Z. Jiang, et al. *"Ultrahigh electron mobility in suspended graphene."* Solid State Communications 146, no. 9(2008): 351–355.
Carlson, C., S.M. Hussain, A.M. Schrand, et al. *"Unique cellular interaction of silver nanoparticles: size-dependent generation of reactive oxygen species."* The Journal of Physical Chemistry B 112, no. 43(2008): 13608–13619.
Cha, C., S.R. Shin, N. Annabi, M.R. Dokmeci, and A. Khademhosseini. *"Carbon-based nanomaterials: Multifunctional materials for biomedical engineering."* ACS Nano 7, no. 4(2013): 2891–2897.
Chang, Z., Y. Zhou, L. Hao, et al. *"Simultaneous determination of dopamine and ascorbic acid using β-cyclodextrin/Au nanoparticles/graphene-modified electrodes."* Analytical Methods 9, no. 4(2017): 664–671.
Chen, A., and S. Chatterjee. *"Nanomaterials based electrochemical sensors for biomedical applications."* Chemical Society Reviews 42, no. 12(2013): 5425–5438.
Chen, A., and C. Ostrom. *"Palladium-based nanomaterials: Synthesis and electrochemical applications."* Chemical Reviews 115, no. 21(2015): 11999–12044.
Chen, F., W. Gao, X. Qiu, et al. *"Graphene quantum dots in biomedical applications: Recent advances and future challenges."* Frontiers in Laboratory Medicine 1, no. 4(2017): 192–199.
Chen, Y., Z. Fan, Z. Zhang, et al. *"Two-dimensional metal nanomaterials: Synthesis, properties, and applications."* Chemical Reviews 118, no. 13(2018): 6409–6455.
Chua, J.Y., A.V. Pendharkar, N. Wang, et al. *"Intra-arterial injection of neural stem cells using a microneedle technique does not cause microembolic strokes."* Journal of Cerebral Blood Flow & Metabolism 31, no. 5(2011): 1263–1271.
Coro, J., M. Suárez, L.S.R. Silva, K.I.B. Eguiluz, and G.R. Salazar-Banda. *"Fullerene applications in fuel cells: A review."* International Journal of Hydrogen Energy 41, no. 40(2016): 17944–17959.
Dang, X., H. Hu, S. Wang, and S. Hu. *"Nanomaterials-based electrochemical sensors for nitric oxide."* Microchimica Acta 182, no. 3-4(2015): 455–467.
De, S., and R. Madhuri. *"Functionalized nanomaterials for electronics and electrical and energy industries"*, in Handbook of Functionalized Nanomaterials for Industrial Applications. Elsevier, 2020: 269–296.
Deak, E., B. Rüster, L. Keller, et al. *"Suspension medium influences interaction of mesenchymal stromal cells with endothelium and pulmonary toxicity after transplantation in mice."* Cytotherapy 12, no. 2(2010): 260–264.
Dolez, P.I. *"Nanomaterials definitions, classifications, and applications"*, in Nanoengineering. Elsevier, 2015: 3–40.
Drake, P., H.-J. Cho, P.-S. Shih, et al. *"Gd-doped iron-oxide nanoparticles for tumour therapy via magnetic field hyperthermia."* Journal of Materials Chemistry 17, no. 46(2007): 4914–4918.
Dresselhaus, M.S., G. Dresselhaus, and P.C. Eklund. *"Fullerenes."* Journal of Materials Research 8, no. 8(2011): 2054–2097.
Durdagi, S., T. Mavromoustakos, N. Chronakis, and M.G. Papadopoulos. *"Computational design of novel fullerene analogues as potential HIV-1 PR inhibitors: Analysis of the binding interactions between fullerene inhibitors and HIV-1 PR residues using 3D QSAR, molecular docking and molecular dynamics simulations."* Bioorganic and Medicinal Chemistry 16, no. 23(2008): 9957–9974.
Eatemadi, A., H. Daraee, H. Karimkhanloo, et al. *"Carbon nanotubes: Properties, synthesis, purification, and medical applications."* Nanoscale Research Letters 9, no. 1(2014): 393.
Eda, G., G. Fanchini, and M. Chhowalla. "Large-area ultrathin films of reduced graphene oxide as a transparent and flexible electronic material." Nature Nanotechnology 3, no. 5(2008): 270–274.

Elgrabli, D., M. Floriani, S. Abella-Gallart, et al. *"Biodistribution and clearance of instilled carbon nanotubes in rat lung."* Particle and Fibre Toxicology 5, no. 1(2008): 20.

Erickson, K., R. Erni, Z. Lee, et al. *"Determination of the local chemical structure of graphene oxide and reduced graphene oxide."* Advanced Materials 22, no. 40(2010): 4467–4472.

Fang, C., and M. Zhang. *"Multifunctional magnetic nanoparticles for medical imaging applications."* Journal of Materials Chemistry 19, no. 35(2009): 6258–6266.

Fath-Bayati, L., J. Ai, S.-J. Mowla, S. Ebrahimi-Barough, and E. Motevaseli. *"Tracking of intraperitoneally and direct intrahepatic administered mesenchymal stem cells expressing miR-146a-5p in mice hepatic tissue."* Health Biotechnology and Biopharma 4, no. 3(2020): 56–76.

Fath-Bayati, L., M. Vasei, and E. Sharif-Paghaleh. *"Optical fluorescence imaging with shortwave infrared light emitter nanomaterials for in vivo cell tracking in regenerative medicine."* Journal of Cellular and Molecular Medicine 23, no. 12(2019): 7905–7918.

Fekry, A.M. *"A new simple electrochemical moxifloxacin hydrochloride sensor built on carbon paste modified with silver nanoparticles."* Biosensors and Bioelectronics 87, (2017): 1065–1070.

Feng, L., X. Yang, X. Shi, et al. *"Polyethylene glycol and polyethylenimine dual-functionalized nano-graphene oxide for photothermally enhanced gene delivery."* Small 9, no. 11(2013): 1989–1997.

Geim, A.K., and K.S. Novoselov. *"The rise of graphene."* Nature Materials 6, no. 3(2007): 183–191.

Godfrey, I.J., A.J. Dent, I.P. Parkin, S. Maenosono, and G. Sankar. *"Structure of gold–silver nanoparticles."* The Journal of Physical Chemistry C 121, no. 3(2017): 1957–1963.

Gupta, A., D.F. Moyano, A. Parnsubsakul, et al. *"Ultrastable and biofunctionalizable gold nanoparticles."* ACS Applied Materials & Interfaces 8, no. 22(2016): 14096–14101.

Haruta, M., N. Yamada, T. Kobayashi, and S. Iijima. *"Gold catalysts prepared by coprecipitation for low-temperature oxidation of hydrogen and of carbon monoxide."* Journal of Catalysis 115, no. 2(1989): 301–309.

Hong, L., S.-H. Luo, C.-H. Yu, et al. *"Functional nanomaterials and their potential applications in antibacterial therapy."* Pharmaceutical Nanotechnology 7, no. 2(2019): 129–146.

Horie M.F.A., Saito Y., Yoshida Y., Sato H., Ohi H. et al. "Antioxidant action of sugar-pendant C60 fullerenes." *Bioorganic and Medicinal Chemistry Letters* 19, no. 20(2009 Oct 15): 5902–5904. 10.1016/j.bmcl.2009.08.067.

Hou, G., L. Zhang, V. Ng, Z. Wu, and M. Schulz. *"Review of recent advances in carbon nanotube biosensors based on field-effect transistors."* Nano LIFE 06, no. 03n04(2016): 1642006.

Hwang, H.S., J.W. Jeong, Y.A. Kim, and M. Chang. *"Carbon nanomaterials as versatile platforms for biosensing applications."* Micromachines 11, no. 9(2020): 814.

Iijima, S., and T. Ichihashi. *"Single-shell carbon nanotubes of 1-nm diameter."* Nature 363, (1993): 603.

Imani, S., A.J. Bandodkar, A.V. Mohan, et al. *"A wearable chemical–electrophysiological hybrid biosensing system for real-time health and fitness monitoring."* Nature Communications 7, no. 1(2016): 1–7.

Jacobs, C.B., M.J. Peairs, and B.J. Venton. *"Review: Carbon nanotube based electrochemical sensors for biomolecules."* Analytica Chimica Acta 662, no. 2(2010): 105–127.

Jain, T.K., M.A. Morales, S.K. Sahoo, D.L. Leslie-Pelecky, and V. Labhasetwar. *"Iron oxide nanoparticles for sustained delivery of anticancer agents."* Molecular Pharmaceutics 2, no. 3(2005): 194–205.

Jeevanandam, J., A. Barhoum, Y.S. Chan, A. Dufresne, and M.K. Danquah. *"Review on nanoparticles and nanostructured materials: history, sources, toxicity and regulations."* Beilstein Journal of Nanotechnology 9, no. 1(2018): 1050–1074.

Julián-López, B., C. Boissière, C. Chanéac, et al. "*Mesoporous maghemite–organosilica microspheres: a promising route towards multifunctional platforms for smart diagnosis and therapy.*" Journal of Materials Chemistry 17, no. 16(2007): 1563–1569.
Jung, S.-W., E.K. Lee, and S.-Y. Lee. "*Communication—Concentration-cell-type nafion-based potentiometric hydrogen sensors.*" ECS Journal of Solid State Science and Technology 7, no. 12(2018): Q239.
Kawai, N., M. Futakuchi, T. Yoshida, et al. "*Effect of heat therapy using magnetic nanoparticles conjugated with cationic liposomes on prostate tumor in bone.*" The Prostate 68, no. 7(2008): 784–792.
Kim, K.S., Y. Zhao, H. Jang, et al. "*Large-scale pattern growth of graphene films for stretchable transparent electrodes.*" Nature 457, no. 7230(2009): 706–710.
Kroto, H.W., J.R. Heath, S.C. O'Brien, R.F. Curl, and R.E. Smalley. "*C60: Buckminsterfullerene.*" Nature 318, no. 6042(1985): 162–163.
Kumar-Krishnan, S., A. Hernandez-Rangel, U. Pal, et al. "*Surface functionalized halloysite nanotubes decorated with silver nanoparticles for enzyme immobilization and biosensing.*" Journal of Materials Chemistry B 4, no. 15(2016): 2553–2560.
Kumar, A. "*Functional nanomaterials: From basic science to emerging applications.*" Trans Tech Publ (2013): 1–19.
Kumar, S., W. Ahlawat, R. Kumar, and N. Dilbaghi. "*Graphene, carbon nanotubes, zinc oxide and gold as elite nanomaterials for fabrication of biosensors for healthcare.*" Biosensors and Bioelectronics 70, (2015): 498–503.
Kumawat, M.K., M. Thakur, R.B. Gurung, and R. Srivastava. "*Graphene quantum dots for cell proliferation, nucleus imaging, and photoluminescent sensing applications.*" Scientific Reports 7, no. 1(2017): 15858.
Lacerda, L., H. Ali-Boucetta, M.A. Herrero, et al. "*Tissue histology and physiology following intravenous administration of different types of functionalized multiwalled carbon nanotubes.*" Nanomedicine (Lond) 3, no. 2(2008): 149–161.
Lacerda, L., J. Russier, G. Pastorin, et al. "*Translocation mechanisms of chemically functionalised carbon nanotubes across plasma membranes.*" Biomaterials 33, no. 11(2012): 3334–3343.
Lee, H., M.K. Yu, S. Park, et al. "*Thermally cross-linked superparamagnetic iron oxide nanoparticles: synthesis and application as a dual imaging probe for cancer in vivo.*" Journal of the American Chemical Society 129, no. 42(2007): 12739–12745.
Lin, R., P.-L. Taberna, S. Fantini, et al. "*Capacitive energy storage from −50 to 100 °c using an ionic liquid electrolyte.*" The Journal of Physical Chemistry Letters 2, no. 19(2011): 2396–2401.
Lin, Y., F. Lu, Y. Tu, and Z. Ren. "*Glucose biosensors based on carbon nanotube nanoelectrode ensembles.*" Nano Letters 4, no. 2(2004): 191–195.
Liu, Y., X. Liu, Y. Liu, et al. "*Construction of a highly sensitive non-enzymatic sensor for superoxide anion radical detection from living cells.*" Biosensors and Bioelectronics 90, (2017): 39–45.
Liu, Y., Y. Xu, Y. Tian, et al. "*Functional nanomaterials can optimize the efficacy of vaccines.*" Small 10, no. 22(2014): 4505–4520.
Liu, Y., D. Yu, C. Zeng, Z. Miao, and L. Dai. "*Biocompatible graphene oxide-based glucose biosensors.*" Langmuir 26, no. 9(2010): 6158–6160.
Liu, Z., H. Forsyth, N. Khaper, and A. Chen. "*Sensitive electrochemical detection of nitric oxide based on AuPt and reduced graphene oxide nanocomposites.*" Analyst 141, no. 13(2016): 4074–4083.
Liu, Z., J.T. Robinson, X. Sun, and H. Dai. "*PEGylated nanographene oxide for delivery of water-insoluble cancer drugs.*" Journal of the American Chemical Society 130, no. 33(2008): 10876–10877.

Maduraiveeran, G., and W. Jin. *"Functional nanomaterial-derived electrochemical sensor and biosensor platforms for biomedical applications"*, in *Handbook of Nanomaterials in Analytical Chemistry.* Elsevier, 2020: 297–327.

Maduraiveeran, G., M. Kundu, and M. Sasidharan. *"Electrochemical detection of hydrogen peroxide based on silver nanoparticles via amplified electron transfer process."* *Journal of Materials Science* 53, no. 11(2018): 8328–8338.

Maiti, D., X. Tong, X. Mou, and K. Yang. *"Carbon-based nanomaterials for biomedical applications: A recent study."* *Frontiers in Pharmacology* 9, no. 1401(2019): 1401.

Marangon, I., N. Boggetto, C. Ménard-Moyon, et al. *"Intercellular carbon nanotube translocation assessed by flow cytometry imaging."* *Nano Letters* 12, no. 9(2012): 4830–4837.

Martín-Yerga, D., E.C. Rama, and A. Costa-García. *"Electrochemical study and applications of selective electrodeposition of silver on quantum dots."* *Analytical Chemistry* 88, no. 7(2016): 3739–3746.

Martínez-Gutierrez, F., E.P. Thi, J.M. Silverman, et al. *"Antibacterial activity, inflammatory response, coagulation and cytotoxicity effects of silver nanoparticles."* *Nanomedicine: Nanotechnology, Biology and Medicine* 8, no. 3(2012): 328–336.

Masitas, R.A., S.L. Allen, and F.P. Zamborini. *"Size-dependent electrophoretic deposition of catalytic gold nanoparticles."* *Journal of the American Chemical Society* 138, no. 47(2016): 15295–15298.

Maxwell, T., M.G.N. Campos, S. Smith, et al. Quantum Dots, in *Nanoparticles for Biomedical Applications.* Elsevier, 2020: 243–265.

McCarthy, J.R., and R. Weissleder. *"Multifunctional magnetic nanoparticles for targeted imaging and therapy."* *Advanced Drug Delivery Reviews* 60, no. 11(2008): 1241–1251.

McDonough, J.K., and Y. Gogotsi. *"Carbon onions: Synthesis and electrochemical applications."* *Interface Magazine* 22, no. 3(2013): 61–66.

Mehra, N.K., K. Jain, and N.K. Jain. *"Design of multifunctional nanocarriers for delivery of anti-cancer therapy."* *Current Pharmaceutical Design* 21, no. 42(2015): 6157–6164.

Mehra, N.K., and N.K. Jain. *"Multifunctional hybrid-carbon nanotubes: New horizon in drug delivery and targeting."* *Journal of Drug Targeting* 24, no. 4(2016): 294–308.

Mestre, A., and A. Carvalho. "Nanoporous carbon synthesis: An old story with exciting new chapters. name." *Materials Science* (2017). https://doi.org/10.5772/intechopen.72476.

Mochalin, V.N., O. Shenderova, D. Ho, and Y. Gogotsi. *"The properties and applications of nanodiamonds."* *Nature Nanotechnology* 7, no. 1(2012): 11–23.

Mohajeri, M., B. Behnam, and A. Sahebkar. *"Biomedical applications of carbon nanomaterials: Drug and gene delivery potentials."* *Journal of Cellular Physiology* 234, no. 1(2019): 298–319.

Montellano, A., T. Da Ros, A. Bianco, and M. Prato. *"Fullerene C60 as a multifunctional system for drug and gene delivery."* *Nanoscale* 3 (2011): 4035–4041.

Mostofizadeh, A., Y. Li, B. Song, and Y. Huang. *"Synthesis, properties, and applications of low-dimensional carbon-related nanomaterials."* *Journal of Nanomaterials* 2011 (2011): 685081.

Notarianni, M., J. Liu, K. Vernon, and N. Motta. *"Synthesis and applications of carbon nanomaterials for energy generation and storage."* *Beilstein Journal of Nanotechnology* 7, (2016): 149–196.

Nunes, D., A. Pimentel, A. Gonçalves, et al. *"Metal oxide nanostructures for sensor applications."* *Semiconductor Science and Technology* 34, no. 4(2019): 043001.

Odom, T.W., J.-L. Huang, P. Kim, and C.M. Lieber. *"Atomic structure and electronic properties of single-walled carbon nanotubes."* *Nature* 391, no. 6662(1998): 62–64.

Park, J., D.-H. Lim, H.-J. Lim, et al. *"Size dependent macrophage responses and toxicological effects of Ag nanoparticles."* *Chemical Communications* 47, no. 15(2011): 4382–4384.

Parrilla, M., R. Cánovas, and F.J. Andrade. *"Based enzymatic electrode with enhanced potentiometric response for monitoring glucose in biological fluids."* Biosensors and Bioelectronics 90, (2017): 110–116.

Pattnaik, S., K. Swain, and Z. Lin. *"Graphene and graphene-based nanocomposites: biomedical applications and biosafety."* Journal of Materials Chemistry B 4, no. 48(2016): 7813–7831.

Pawar, R.S., P.G. Upadhaya, and V.B. Patravale. *"Quantum dots: Novel realm in biomedical and pharmaceutical industry"*, in Handbook of Nanomaterials for Industrial Applications. Elsevier, 2018: 621–637.

Pech, D., M. Brunet, H. Durou, et al. *"Ultrahigh-power micrometre-sized supercapacitors based on onion-like carbon."* Nature Nanotechnology 5, no. 9(2010): 651–654.

Poinern, G.E.J. A Laboratory Course in Nanoscience and Nanotechnology. CRC Press, 2014.

Raffa, V., G. Ciofani, O. Vittorio, C. Riggio, and A. Cuschieri. *"Physicochemical properties affecting cellular uptake of carbon nanotubes."* Nanomedicine (Lond) 5, no. 1(2010): 89–97.

Rao, N., R. Singh, and L. Bashambu. *"Carbon-based nanomaterials: Synthesis and prospective applications."* Materials Today: Proceedings 44 (2020): 608–614.

Robinson, J.T., K. Welsher, S.M. Tabakman, et al. "High performance in vivo near-IR (> 1 μm) imaging and photothermal cancer therapy with carbon nanotubes." *Nano Research* 3, no. 11(2010): 779–793.

Saito, R.A.D., Dresselhaus, G., and Dresselhaus, M.S. Physical Properties of Carbon Nanotubes. Imperial College Press.

Saleh, T.A. *"Nanomaterials: Classification, properties, and environmental toxicities."* Environmental Technology & Innovation 20 (2020): 101067. https://doi.org/10.1016/j.eti.2020.101067.

Sánchez, L., R. Otero, J.M. Gallego, R. Miranda, and N. Martín. *"Ordering fullerenes at the nanometer scale on solid surfaces."* Chemical Reviews 109, no. 5(2009): 2081–2091.

Săndulescu, R., Tertis, M., Cristea, C., and E. Bodoki *"New materials for the construction of electrochemical biosensors."* Micro and Nanoscale Applications (2015): 1–36.

Sheng, Q., Y. Shen, J. Zhang, and J. Zheng. *"Ni doped Ag@ C core–shell nanomaterials and their application in electrochemical H_2O_2 sensing."* Analytical Methods 9, no. 1(2017): 163–169.

Silva, G.A. *"Introduction to nanotechnology and its applications to medicine."* Surgical Neurology 61, no. 3(2004): 216–220.

Smith, A.T., A.M. LaChance, S. Zeng, B. Liu, and L. Sun. *"Synthesis, properties, and applications of graphene oxide/reduced graphene oxide and their nanocomposites."* Nano Materials Science 1, no. 1(2019): 31–47.

Song L.S.J., Lu J., Lu C. "Structure observation of graphene quantum dots by single-layered formation in layered confinement space." *Chemical Science* 6, (2015): 4846–4850.

Stender, A.S., K. Marchuk, C. Liu, et al. *"Single cell optical imaging and spectroscopy."* Chemical Reviews 113, no. 4(2013): 2469–2527.

Sudha, P.N., K. Sangeetha, K. Vijayalakshmi, and A. Barhoum. *"Nanomaterials history, classification, unique properties, production and market"*, in Emerging Applications of Nanoparticles and Architecture Nanostructures. Elsevier, 2018: 341–384.

Sun, X., Z. Liu, K. Welsher, et al. *"Nano-graphene oxide for cellular imaging and drug delivery."* Nano Research 1, no. 3(2008): 203–212.

Suresh, S. *"Semiconductor nanomaterials, methods and applications: A review."* Nanoscience and Nanotechnology 3, no. 3(2013): 62–74.

Tao, H., K. Yang, Z. Ma, et al. *"In vivo NIR fluorescence imaging, biodistribution, and toxicology of photoluminescent carbon dots produced from carbon nanotubes and graphite."* Small 8, no. 2(2012): 281–290.

Tian, B., C. Wang, S. Zhang, L. Feng, and Z. Liu. "*Photothermally enhanced photodynamic therapy delivered by nano-graphene oxide.*" ACS Nano 5, no. 9(2011): 7000–7009.

Tîlmaciu, C.-M., and M.C. Morris. "*Carbon nanotube biosensors.*" Frontiers in Chemistry 3, no. 59(2015). https://doi.org/10.3389/fchem.2015.00059.

Varshney, K. "*Carbon nanotubes: A review on synthesis, properties and applications.*" International Journal of Engineering Research and General Science 2, no. 4(2014): 660–677.

Wang, J., S. Cao, Y. Ding, et al. "*Theoretical investigations of optical origins of fluorescent graphene quantum dots.*" Scientific Reports 6, no. 1(2016): 24850.

Whitlow, J., S. Pacelli, and A. Paul. "*Multifunctional nanodiamonds in regenerative medicine: Recent advances and future directions.*" Journal of Controlled Release 261 (2017): 62–86.

Xu, Z., Y. Liu, F. Ren, F. Yang, and D. Ma. "*Development of functional nanostructures and their applications in catalysis and solar cells.*" Coordination Chemistry Reviews 320 (2016): 153–180.

Xue, X., F. Wang, and X. Liu. "*Emerging functional nanomaterials for therapeutics.*" Journal of Materials Chemistry 21, no. 35(2011): 13107–13127.

Yamamoto, T.W.K., Hernandez, E.R. "Mechanical properties, thermal stability and heat, a.d. transport in carbon nanotubes." In: Jorio, G., Dresselhaus, M.S., editors. *Carbon, S. Nanotubes: Advanced Topics in the Synthesis, Properties and Applications.* Springer and B, 2008.

Yang, K., L. Feng, X. Shi, and Z. Liu. "*Nano-graphene in biomedicine: Theranostic applications.*" Chemical Society Reviews 42, no. 2(2013): 530–547.

Yang, N., X. Chen, T. Ren, P. Zhang, and D. Yang. "*Carbon nanotube based biosensors.*" Sensors and Actuators B: Chemical 207, (2015): 690–715.

Yoo, B.M., H.J. Shin, H.W. Yoon, and H.B. Park. "*Graphene and graphene oxide and their uses in barrier polymers.*" Journal of Applied Polymer Science 131, no. 1(2014). https://doi.org/10.1002/app.39628.

Yusoff, N., P. Rameshkumar, M.S. Mehmood, et al. "*Ternary nanohybrid of reduced graphene oxide-nafion@ silver nanoparticles for boosting the sensor performance in nonenzymatic amperometric detection of hydrogen peroxide.*" Biosensors and Bioelectronics 87, (2017): 1020–1028.

Zaytseva, O., and G. Neumann. "*Carbon nanomaterials: production, impact on plant development, agricultural and environmental applications.*" Chemical and Biological Technologies in Agriculture 3, no. 1(2016): 17.

Zhang, H., G. Grüner, and Y. Zhao. "*Recent advancements of graphene in biomedicine.*" Journal of Materials Chemistry B 1, no. 20(2013): 2542–2567.

Zhang, R., Y. Zhang, Q. Zhang, et al. "*Growth of half-meter long carbon nanotubes based on Schulz-Flory distribution.*" ACS Nano 7, no. 7(2013): 6156–6161.

Zhao, L., L.-Z. Fan, M.-Q. Zhou, et al. "*Nitrogen-containing hydrothermal carbons with superior performance in supercapacitors.*" Advanced Materials 22, no. 45(2010): 5202–5206.

Zhao, Q., Z. Gan, and Q. Zhuang. "*Electrochemical sensors based on carbon nanotubes.*" Electroanalysis 14, no. 23(2002): 1609–1613.

Zheng, X.T., A. Ananthanarayanan, K.Q. Luo, and P. Chen. "*glowing graphene quantum dots and carbon dots: Properties, syntheses, and biological applications.*" Small 11, no. 14(2015): 1620–1636.

7 Trigger-Sensitive Nanoparticle for Drug Delivery

Hadiqa Nazish Raja, Basalat Imran, and Fakhar Ud Din
Quaid-i-Azam University

CONTENTS

- 7.1 Introduction 156
  - 7.1.1 Modes of Targeted Drug Delivery by Nanoparticles 156
    - 7.1.1.1 Passive Targeting 156
    - 7.1.1.2 Active Targeting 157
    - 7.1.1.3 Trigger-Sensitive Targeting 158
- 7.2 Trigger-Sensitive Nanoparticles for Drug Delivery 158
- 7.3 Modes of Drug Release from Trigger-Sensitive Nanoparticles 159
- 7.4 Types of Triggering Stimuli 159
- 7.5 Endogenous Trigger-Sensitive Nanoparticles 161
  - 7.5.1 pH-Sensitive Nanoparticles 161
    - 7.5.1.1 Mechanism of Drug Release from the pH-Sensitive DDS 162
  - 7.5.2 Redox-Sensitive Nanoparticles 164
  - 7.5.3 Enzyme-Sensitive Nanoparticles 164
  - 7.5.4 Ionic Microenvironment-Sensitive Nanoparticles 165
- 7.6 Exogenous Trigger-Sensitive Nanoparticles 166
  - 7.6.1 Photosensitive Nanoparticles 166
  - 7.6.2 Temperature-Sensitive Nanoparticles 167
  - 7.6.3 Magnetic Field Nanoparticles 167
  - 7.6.4 Electric Field-Sensitive Nanoparticles 169
  - 7.6.5 Ultrasound-Sensitive Nanoparticles 170
- 7.7 Dual/Multi-Responsive DDS 170
- 7.8 Applications of Trigger-Sensitive Nanoparticulate Drug Delivery System 171
- 7.9 Limitations of Trigger-Sensitive Nanoparticulate Drug Delivery System 171
- 7.10 Conclusion 173
- References 173

7.1 INTRODUCTION

Drug delivery is meant for the formulation and development of a technique, carrier system, or a technology with an intention of delivery of the drug in the body to achieve its therapeutic and pharmacological response in the body (Gundloori et al., 2019; Mishra et al., 2019; Nayak et al., 2018) Drug delivery system is carrier system or formulation containing the drug to execute the drug delivery inside the body. An ideal drug delivery system meant for the effective drug delivery needs to be designed in such a way that it delivers the drug to the specific desired pathological site in a controlled manner over an enhanced time period. The conventional drug delivery systems pose a lot of problems (Nayak et al., 2018). The uncontrolled nonspecific drug release and biodistribution associated with the conventional drug delivery system may lead to the reduced efficacy, repeated drug intake, patient incompliance, and enhanced toxicity at the sites other than the pathological site of action (Bhagwat and Vaidhya, 2013). The concept of "magic bullet" was first introduced by Paul Ehrlich about 100 years ago. The "magic bullet" conception involved an idea to target particular microbes without targeting the healthy parts (Stirland et al., 2013). The researchers had been trying to develop an ideal drug delivery system since very long ago to overcome the limitations of the conventional drug delivery system and to develop such magic bullets (Stayton et al., 2013). The concept of magic bullets laid the foundation of the targeted delivery. Abundant of drug delivery systems carrying the therapeutic substances in the body are incapable of differentiating the pathological cells from the healthy cells, thus such system exerts the deleterious effects on the healthy cells as well. The need to overcome drawbacks of such drug delivery systems led the foundation of the targeted drug delivery approach (Singh et al., 2019). Targeted drug delivery is meant for the target-specific delivery of the therapeutic moiety at the pathological site to enhance the efficacy and reduce the toxic effects of the therapeutic moiety (Wang et al., 2015). By targeting the specific diseased parts of the body, the drug is ensured to be released in those pathological parts to exert the therapeutic response without harming the nonpathological parts (Dahiya et al., 2018). The site-specific and controlled release of the drug at the target site enhances efficacy of low dose that also increases the patient compliance. Nanotechnology has brought revolution in the field of targeted drug delivery. The targeted drug delivery by utilizing nanocarriers has made it possible to allow targeting not only at tissue and organ level, but also at the cellular, molecular, and organelle level (Garcia et al., 2019).

7.1.1 MODES OF TARGETED DRUG DELIVERY BY NANOPARTICLES

The nanoparticulate drug delivery system allows the site specific release by different targeting mechanism as shown in Figure 7.1, but among all of them the trigger-sensitive targeting is the smartest approach towards targeted drug delivery (Li et al., 2018; Wang et al., 2015).

7.1.1.1 Passive Targeting

The passive targeting occurs due to the EPR effect (enhanced permeability and retention effect) in which the nanocarriers ensure the deposition of the drug in

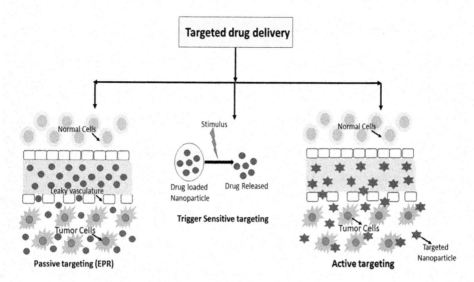

FIGURE 7.1 Modes of targeted drug delivery by nanoparticles.

diseased tissues having leaky vasculature (Maeda, 2015; Shi et al., 2017). Passive targeting is a size-dependent phenomenon. The EPR effect is phenomenon of the extravasation of the particles from blood vessels at the tumor sites due enhanced permeability and porosity of the blood vessels, through which nanoparticles of size between 30 and 200 nm can easily permeate leading to their enhanced retention and accumulation at the tumor site (Albanese et al., 2012; Longo et al., 2015; Matsumura and Maeda, 1986). The extravasated nanoparticles may be detected as foreign particles and get opsonized by the reticuloendothelial system. The stealth liposomes have been produced to overcome this issue by coating nanoparticles by PEG (polyethylene glycol) that ensure escape of nanoparticles from being recognized as foreign particles (Pillai, 2014). The nanoparticles having passive mode of targeting are being utilized for the clinical applications; but it has been claimed that such passively targeted nanoparticles can deliver only 0.7% of the administered dose to target the tumor sites (Berndt et al., 2006). The substantial variability of the EPR effect in humans leads to limited passive targeting and enhances the interest researchers to develop nanoparticles meant for active targeting (Kobayashi et al., 2014).

7.1.1.2 Active Targeting

Active targeting involves the immobilization of the targeting ligands on the surface of nanocarriers that interact with targets on the targeted cells or receptors. This targeted binding leads to site-specific drug release and pharmacological effect (Verhoef and Anchordoquy, 2013; Yang et al., 2015). But the active targeting may also involve the nonspecific off-target side effects as the receptors targeted by ligands may be present in healthy cells along with the diseased cells (Dvir et al., 2010; Shuhendler et al., 2012). In an effort to optimize selectivity for targeted delivery, trigger-sensitive nanoparticles as smart drug delivery system have been developed.

7.1.1.3 Trigger-Sensitive Targeting

Trigger-sensitive drug release mechanism is the most efficient selective mechanism involved in targeted drug delivery (Arrue and Ratjen, 2017; Wang et al., 2016). Various endogenous and exogenous triggers can be utilized for targeted drug delivery. The trigger-sensitive nanoparticles are capable to allow the drug release triggered by a trigger at targeted sites. A smart drug delivery system is defined as the delivery system that inhibits the drug release before reaching the target sites and particularly releases the drug at a controlled rate at the targeted sites (Liu et al., 2016). The trigger-sensitive nanoparticles form the basis of such smart drug delivery system capabilities to respond the particular triggers to allow the drug release at the particular sites at a controlled rate without causing harmful effects on the healthy tissues (Lopes et al., 2018). The trigger-sensitive nanoparticles have attained particular attention as magic bullets in recent years that have unique characteristic to turn on or turn off the therapeutic response of the drug dependent on the existence of the particular trigger to which the smart nanoparticle is sensitive (Banerjee, 2011). This chapter mainly focuses on these trigger-responsive, smart nanoparticles based on various stimuli acting as triggers, their release mechanisms, and the limitations associated with them.

7.2 TRIGGER-SENSITIVE NANOPARTICLES FOR DRUG DELIVERY

Several novel techniques and the smart technologies have been employed to overcome the limitations associated with the conventional drug delivery systems in order to have controlled drug release and to make the targeted drug delivery approach possible (Lopes et al., 2013; Tiwari et al., 2017). The targeted drug delivery is the main requirement of a smart and an ideal drug delivery system to optimize the therapeutic efficacy and to reduce the associated side effects on the healthy tissues (Ding et al., 2016; Kojima, 2010). Among the several such smart approaches invented for the sake of targeted drug delivery, trigger-responsive nanoparticulate drug delivery system is the smartest technology to ensure the targeted delivery and minimizing the adverse effects (Cheng et al., 2014; Kurisawa and Yui, 1998). The trigger-sensitive nanoparticles are tuned in a way that they become responsive to a particular stimulus or multiple stimuli at the same time and the release of drug loaded as cargo in nanoparticulate carrier is dependent on presence or absence of stimuli (Bawa et al., 2018; Puoci et al., 2008). The trigger-sensitiveness is mainly controlled by the tuning and the composition of the nanoparticles to make them responsive to particular triggers (Cheng et al., 2014; Bawa et al., 2009; Bawa 2018). The smartly designed trigger-sensitive nanoparticles enhance the drug efficacy by controlling the drug release, individual pathological variations, and target-specific drug delivery and can be engineered to variate the drug release and the cellular intake of the drug (Baeza et al., 2015; Castillo et al., 2019; Yang et al., 2016).

7.3 MODES OF DRUG RELEASE FROM TRIGGER-SENSITIVE NANOPARTICLES

The nanoparticles are broadly classified as organic, inorganic and hybrid nanoparticles. The organic nanoparticles are mainly composed of organic materials that have high biocompatibility and are biodegradable. The examples of organic nanoparticles mainly include liposomes, polymersomes, nano-emulsions, synthetic organic systems, organo-gels, glutamic acid derivatives, dendrimers, and micelles (Huo et al., 2016). The drug release is controlled by the slow destruction of the organic nanocarrier. Due to decreased physiological stability of organic nanoparticles, the triggered drug release occurs by the breakdown of the organic carrier in the presence of trigger. This triggered release mechanism involves the nanostructure destruction induced by particular trigger (Ma et al., 2014). Owing to the high biocompatibility and biodegradability, there are few limitations associated with the organic nanoparticles that mainly include: reduced drug loading capability, decreased chemical and thermal stability, enhanced chances to be taken up by immune system (Huo et al., 2016).

The inorganic nanoparticle is composed of the inorganic materials. Such particles are nonbiodegradable. The examples of inorganic nanoparticles include mesoporous silica NPs, iron sulfate NPs, gold NPs, and carbonaceous NPs. The drug release from such nanocarriers occurs by diffusion (Huo et al., 2016). The trigger-responsive drug release from such nanoparticle occurs in a specific way. As mesoporous silica is highly porous, having several holes to load drug and are capped or sealed by nanovalves on the entrance or surface (Bawa et al., 2018; Zheng et al., 2018). The trigger stimulates the nanovalve opening that allows the control drug release while keeping the inorganic nanocarrier intact. This triggered release mechanism involves opening of the nanovalve for controllable drug releasing (Cao et al., 2019; Luo et al., 2011). Inorganic nanoparticles have several advantages like having enhanced stability and being multifunctional but issues related to their biocompatibility, biodegradability, and biosafety are problematic (Huo et al., 2016). The hybrid nanocarriers have both organic and inorganic components in order to utilize the benefits and overcome the limitations of both of them (Crayton and Tsourkas, 2011; Montes and Maleki, 2020). The compounds are linked by linker bonds and those intermediate linking bonds break down in these nanoparticles being triggered by a stimulus, thus releasing the drug at the site under the influence of the trigger (Chen et al., 2014; Liu et al., 2018; Wu et al., 2018). The different mechanisms involved in the drug release from the trigger-sensitive nanoparticulate drug delivery system are shown in Figure 7.2.

7.4 TYPES OF TRIGGERING STIMULI

The trigger-sensitive nanoparticulate drug delivery system releases the drug in response to the presence of a particular trigger or multiple triggers to which the nanoparticulate system is sensitive. The particular stimulus to which the sensitive nanoparticulate carrier responds is called triggers (Chen et al., 2016; Cho et al., 2015). These triggers are classified as endogenous and exogenous triggers as shown in Figure 7.3.

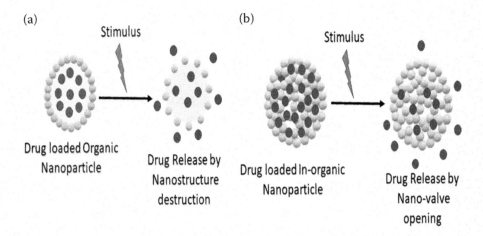

FIGURE 7.2 Modes of drug release from trigger-sensitive nanoparticles: A. Drug release by nanostructure destruction in organic nanoparticles. B. Drug release by nano-valve opening and diffusion in inorganic nanoparticles. Both A and B mechanisms are found in hybrid nanoparticles.

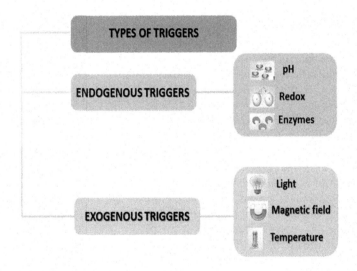

FIGURE 7.3 Types of triggering stimuli.

The internal physiological or pathological stimuli that are the characteristic feature of the pathological parts marking them changed from the normal parts are named as the endogenous triggers (Chen et al., 2016). These triggers mainly include the change in pH, redox reactions, ionic microenvironment, enzyme concentration, etc. Such stimuli efficiently trigger the drug release from the trigger-sensitive nanoparticle at particular pathological sites and protecting the healthy sites from the deleterious effects of the drug (Mura et al., 2013). The stimuli applied outside the body externally meant for the drug release inside the body are called exogenous triggers (Bhatnagar and Venuganti, 2015; Mura et al., 2013). The exogenous triggers may include changes in temperature, externally applied magnetic or electric field, ultrasound, or light (Kong et al., 2013).

The other classification of such triggers includes physical, chemical, and biological triggers (Raza, Rasheed, et al., 2019). The physical triggers include electric or magnetic field, temperature, ultrasound, and osmotic pressure, etc. The examples of chemical triggers include glucose level, redox reaction, and change in pH. The biological triggers may include the enzymes and endogenous receptors (Lopes et al., 2013).

7.5 ENDOGENOUS TRIGGER-SENSITIVE NANOPARTICLES

The endogenous triggering stimuli of biological and chemical origin that are characteristic features of the pathological sites can be utilized to trigger the on-demand drug release from the functionalized nanocarriers to attain the enhanced therapeutic effect at the pathological sites (Hatakeyama, 2017). For this purpose the nanoparticles acting as drug carrier must be smart enough to produce the response to such endogenous stimuli that will activate the nanoparticles to trigger the drug release at the particular site. Such smart nanoparticles may be sensitive to the ionic microenvironment within the tissues, acidic environment at the tumor sites, overproduced glutathione, pH at the specific sites, over expansion of particular enzymes, increased level of hydrogen peroxide or reactive oxygen species, etc., that will trigger the drug release (Raza, Rasheed, et al., 2019). The endogenous environment of pathological tissues is quite varying from that of normal tissues, which provides effective basis of the associated triggers for on-demand drug release (Huo et al., 2016). The nanoparticulate system sensitive to endogenous triggers that can be utilized for targeted drug delivery mainly include

- pH-sensitive nanoparticles
- Redox-sensitive nanoparticles
- Enzyme-ensitive nanoparticles
- Ionic microenvironment-sensitive nanoparticles

7.5.1 pH-Sensitive Nanoparticles

Among several endogenous stimuli, the variation in pH is notably employed for the smart trigger-sensitive drug delivery system because of its intrinsic variation based on the anatomical, physiological, and pathological basis (Gupta et al., 2002; Sonawane et al., 2017). The pH-sensitive drug delivery systems composed of smart polyanionic or polycationic polymers are capable of sensing the minute changes in pH within the body. The existing pH variation in the body, ranging from highly acidic or basic pH to the normal physiological pH at organ, tissue, and even at cellular and subcellular level, provides the smart basis of the triggering intrinsic stimulus to ensure the targeted drug delivery (Sonawane et al., 2017). Thus the anatomical and pathological variation in pH can be utilized to formulate the pH-controlled drug delivery system that enables to release the drug at the particular site (Schmaljohann, 2006). The tumor tissues having pH 6.5–7.2 and inflamed tissues having pH <7.4 can be differentiated because of their pH variation from normal physiological pH that is 7.4 (Schmaljohann, 2006). The particular segment within GIT can be targeted because of existence of natural pH variations in different segments of GIT like stomach has pH 1.5–3.5, small intestine pH is 5.5–6.8, and large intestine has pH of 6.4–7. Cellular targeting is also even possible as the pH of organelles like endosomes having pH 5.5–6 and lysosomes having pH 4.5–5 is also variant from pH of cytosol that is 7.4 (Schmaljohann, 2006). Thus the cellular and even subcellular targeting can be achieved by utilizing the pH-sensitive drug delivery system. Such smart systems have wide applications in cancer therapy, gene delivery, biomedical purposes, tumor targeting, and bioimaging, etc. (Porta-i-Batalla et al., 2016; Xifré-Pérez et al., 2016).

7.5.1.1 Mechanism of Drug Release from the pH-Sensitive DDS

The drug is released at the controlled rate at particular sites based on pH variation by two main mechanisms (Ratemi, 2018):

1. **Protonation or Deprotonation**

The pH-sensitive smart polymers used in the formation of the pH-sensitive nanoparticulate system are mainly polyelectrolytes either polyacidic polymers or polybasic polymers (Huh et al., 2012). The basic mechanism of protonation and deprotonation of the polyelectrolyte polymers is mainly based on their particular functional groups (York et al., 2008). Polyacidic or anionic polymers have acidic functional groups like carboxyl group that undergo protonation by accepting proton in an acidic medium, so they remain unionized and unswollen in an acidic pH (Ratemi, 2018). Such polymers on reaching the basic intestinal pH, undergo protonation by losing proton and get ionized and charged in the basic medium. Like charges repel each other leading to the swelling of the polymer that allows the drug release at basic pH (Ratemi, 2018; Yoshida et al., 2013). The mechanism of drug release from the polyacidic nanoparticulate DDS is summarized in Figure 7.4.

Polybasic or cationic polymers having basic cationic functional groups remain ionized in an acidic environment (Ratemi, 2018). Like charges repel each other, so the polymer undergoes swelling in acidic pH leading to the drug release at particular pH (Yoshida et al., 2013). While in basic pH, these polymers undergo deprotonation and get unionized at the basic pH (Schmaljohann, 2006). The mechanism of drug release from the polyacidic nanoparticulate DDS is summarized in Figure 7.5.

2. **Cleavage of Acid-Labile Bonds**

pH-responsive drug delivery systems may respond to the particular pH on the basis of cleavable linkages in polymers or drug linked with the polymer through such linkages (Javadzadeh and Hamedeyazdan, 2012). Most acid-labile linkages like oxime, acetal, hydrazone, imine, orthoester, ketal, and the cis-aconityl amide bonds undergo cleavage at low pH and lead to the drug release as shown in Figure 7.6.

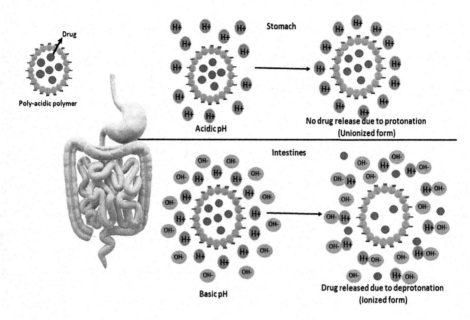

FIGURE 7.4 Mechanism of drug release from pH-sensitive nanpoparticulate DDS composed of polyacidic polymer.

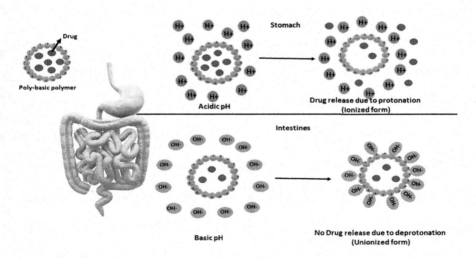

FIGURE 7.5 Mechanism of drug release from pH-sensitive nanpoparticulate DDS composed of polybasic polymer.

These linkages may be present in the polymer itself or the drug can be linked with the polymer through such linkages to ensure the targeted drug delivery at particular pH like acidic pH at the tumor sites (Javadzadeh and Hamedeyazdan, 2012; Ratemi 2018).

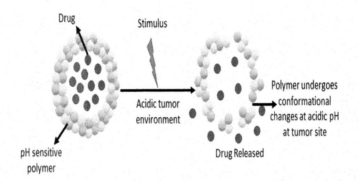

FIGURE 7.6 Cleavage of acid-labile linkages.

The only limitation associated with the pH-sensitive nanoparticulate drug delivery system is because of nonbiodegradability of the pH-sensitive polymers. Most of the pH-sensitive polymers are not biodegradable because the body lacks the enzymes to degrade most of these polymers used in such smart drug delivery system (Balamuralidhara et al., 2011).

7.5.2 REDOX-SENSITIVE NANOPARTICLES

Redox-responsive nanoparticulate DDS has high stability during circulation and only responds to the reducing microenvironment of the targeted tumor sites (Mintzer and Simanek, 2009). The reducing microenvironment at the tumor sites is mainly due to reduced or oxidized states of glutathione and NADP. The concentration of glutathione (GSH) is two times higher at the tumor site as compared to the normal tissues that generate the reducing environment at the tumor site (Schafer and Buettner, 2001). The redox-responsive nanoparticulate system stays stable and intact within the normal cells but undergoes redox reaction due to the reduced microenvironment at the tumor site, thus releasing the drug only at the tumor site. This type of system not only enhances the accumulation of the drug at the target site but also reduces the toxicity and enhances the efficacy of the drug (Fleige et al., 2012).

7.5.3 ENZYME-SENSITIVE NANOPARTICLES

Biological enzymes being natural catalysts (like lipases, proteases, glycosidases, and phospholipases, etc.) are particularly involved in almost all the metabolic and biological events occurring in the body. The enzymes are specific in action, which means they act on the specific substrate that undergoes catalysis (Rasheed et al., 2018). The unique characteristics of enzymes like biorecognition, specificity, catalytic efficacy, process efficiency, selectivity, and enzyme-related regulation and dysregulation in the extracellular and intracellular environment are highly advantageous to produce the enzyme-sensitive nanoparticulate drug delivery system that allows the enzyme-triggered drug release at a particular site. It is a revolutionary basis of biomedicine (Liu et al., 2016; Raza, Rasheed, et al., 2019). The enzyme-responsive nanoparticulate system is a smart drug delivery system that undergoes the variation in its physicochemical properties upon the catalytic action of the enzyme (Hu et al., 2013; Raza, Rasheed, et al., 2019). Among several enzymes, proteases, trypsin, and oxidoreductase-based enzyme-responsive drug delivery systems are of prime importance. Proteases are overexpressed in cancer and inflammation, thus can be utilized to target these diseases. Oxidoreductases can also be targeted due to their involvement in cancer and Alzheimer's disease (Kundu and Surh, 2010). Trypsin is the most important enzyme that regulates the pancreatic secretion that is itself involved in the secretion of several other enzymes (Basel et al., 2011). The enzyme-related disturbance in the diseased cells allows the basis of the smart drug delivery system to identify, regulate, and monitor the different pathological conditions (Raza, Rasheed, et al., 2019).

The mechanism of drug release from enzyme-responsive drug delivery system involves the enzyme-triggered catalytic degradation of the polymeric carrier in which drug has been loaded. The polymeric nanocarrier carrying the drug stays intact in OFF state when there is no catalytic action of enzyme. The catalytic action of the enzyme on the polymeric nanocarrier switches on the ON state leading to the drug release as shown in Figure 7.7 (Radhakrishnan et al., 2014; Raza, Rasheed, et al., 2019).

A major challenge in enzyme-responsive DDSs is to precisely control the initial response time of the systems (Liu et al., 2016).

7.5.4 Ionic Microenvironment-Sensitive Nanoparticles

The natural anatomic and pathological ionic microenvironment variation in the body is an interesting stimulus to formulate the trigger-sensitive drug delivery system. This system is just like pH-sensitive drug delivery system (Zhao and Moore, 2001). This system is formed by the incorporation of acidic or basic functional groups to the polymeric backbone to allow the on-demand targeted drug release. The incorporation of such functional groups actually effect the degree of ionization of polymer with respect to the surrounding microenvironment, thus regulating the release of drug loaded (Karewicz et al., 2010). The degree of

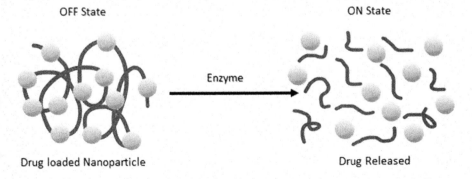

FIGURE 7.7 Mechanism of drug release from enzyme-sensitive nanoparticles.

ionization of such smart polymers varies in different environment, thus the drug is only released from the ionized form of polymer based on the pH of the surrounding medium of the polymeric carrier carrying the drug (Zhang et al., 2005). The microenvironment of tumor cells is different from the normal physiological microenvironment. The tumor site is more acidic, so such variations are useful and innovative to design the smart microenvironment-sensitive nanoparticulate system (Furyk et al., 2006).

The drug release mechanism is based on the degree of ionization of the polymeric nanocarrier based on the pH of the surrounding microenvironment as shown in Figures 7.4 and 7.5 (Schmaljohann, 2006). The incorporation of negatively charged acidic functional group like carboxyl group causes an increased repulsion in surrounding negatively charged basic microenvironment. Thus, such polymers undergo swelling at high pH leading to drug release (Ratemi, 2018). The incorporation of positively charged basic functional groups like amine causes an increased repulsion in the surrounding positively charged microenvironment. Such polymers undergo swelling at low pH leading to drug release (Ratemi, 2018; Yoshida et al., 2013). All such polymers are utilized as functionalized smart polymeric nanocarriers carrying drug that release the drug dependent on the surrounding microenvironment (Zhang et al., 2005).

7.6 EXOGENOUS TRIGGER-SENSITIVE NANOPARTICLES

The exogenous triggering stimuli of external origin can be utilized to trigger the drug release from the functionalized nanocarriers to attain the enhanced therapeutic effect at the pathological sites (Hatakeyama, 2017). For this purpose the nanoparticles acting as drug carrier must be smart enough to produce the response to such external stimuli that will activate the nanoparticles to trigger the drug release at the particular site. Such smart nanoparticles may be sensitive to the external triggers like magnetic field, temperature, radiations, pressure, ultrasound that will trigger the drug release (Raza, Rasheed, et al., 2019). The extrinsic exogenous triggers can be utilized in executing a more controllable drug release by artificially controlling the tunable parameters of the extrinsic triggers like magnetic field, light, ultrasound, temperature, electric field, etc. (Huo et al., 2016). The nanoparticulate system sensitive to exogenous triggers that can be utilized for targeted drug delivery mainly include:

- Photosensitive nanoparticles
- Temperature-sensitive nanoparticles
- Magnetic field-sensitive nanoparticles
- Electrical field-sensitive nanoparticles
- Ultrasound-sensitive nanoparticles

7.6.1 Photosensitive Nanoparticles

Light-sensitive nanoparticles are of prime importance in the pharmaceutical field for sake of targeted drug delivery (Lino and Ferreira, 2018). The targeted drug release can be achieved by utilizing the wide range of wavelengths of light ranging from ultraviolet to near infrared (NIR) (Brown et al., 2009; Chen et al., 2016; Hossion et al., 2013). The light of ultraviolet and visible region have less penetration to be utilized for in vivo drug delivery purposes but NIR is the best exogenous light source for the targeted drug delivery because of the associated high penetration power and high safety profile (Xiang et al., 2018). There are mainly three mechanisms of drug release from the photosensitive nanoparticulate system that include photothermal effect, two photon activation, and upconverting nanoparticle as shown in Figure 7.8. Drug is released by photothermal effect when the photothermal agent in the composition of nanoparticulate carrier converts light to heat and the heat produced induces the drug release from the thermoresponsive material in the carrier system (Xiang et al., 2018). Photon activation phenomenon and upconverting nanoparticles are two basic mechanisms to overcome the issue of using UV light as an exogenous stimulus. Two photon activation involves the redox reaction (gain and loss of electrons) stimulated by two photons of NIR light (that is equivalent to 1 UV photon) leading to the targeted drug release at the particular site (Guardado-Alvarez et al., 2014; Yang et al., 2016). Upconverting nanoparticulate system is highly specialized nanoparticulate system that converts NIR beam to high energy beam of ultraviolet light that causes the conformational changes in the UV-sensitive material of the nanocarrier to trigger the drug release (Gwon et al., 2018).

7.6.2 TEMPERATURE-SENSITIVE NANOPARTICLES

Temperature is the most widely used stimulus for the targeted drug delivery. Temperature may act as endogenous or exogenous trigger to stimulate the drug release (Gu et al., 2018; Sánchez-Moreno et al., 2018). As the temperature of the tumor cells is higher (40–42°C) as compared to the normal physiological temperature (37 °C), it acts as an endogenous trigger for the drug release at the tumor site (Liu et al., 2017; Raza, Hayat, et al., 2019). Temperature may also be utilized as an external trigger to induce drug release. The thermosensitive polymers are used in the formulation of this smart nanoparticulate system. There are basic two mechanisms of drug release from the thermosensitive nanoparticulate DDS under influence of temperature. One mechanism involves the burst release from the thermosensitive carrier at an elevated temperature (Khoee and Karimi, 2018). Other mechanism involves the utilization of external stimulus like light, electric field, etc. that raises the temperature at the particular site leading to the burst drug release at that site (Yang et al., 2018).

7.6.3 MAGNETIC FIELD NANOPARTICLES

The major disadvantage of the conventional drug delivery systems is the generalized systemic effect that exhibits the undesired side effects at sites other than the pathological sites (Pankhurst et al., 2009). The chemotherapeutic agents have more side effects than the therapeutic advantages, due to which scientists are struggling to overcome these issues. The objective of targeted drug delivery can be achieved by utilizing the external triggers, such as magnetic field (Pankhurst et al., 2003). The

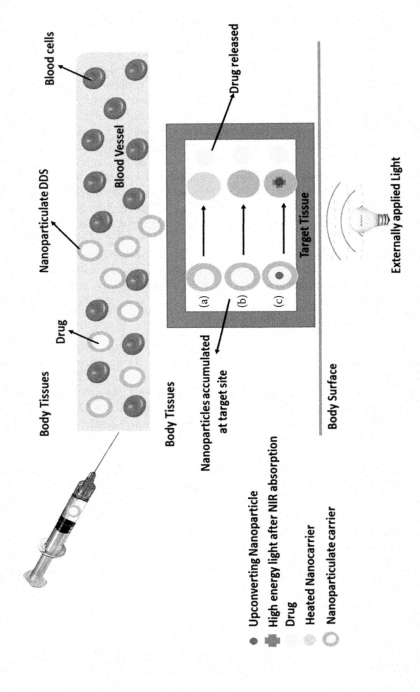

FIGURE 7.8 Mechanism of drug release from photosensitive nanoparticles, A. Photothermal effect. B. Two photon activation phenomenon. C. Upconverting nanoparticles.

Trigger-Sensitive Nanoparticle

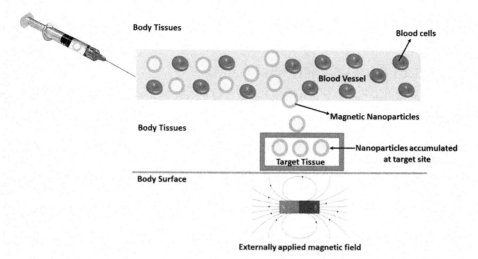

FIGURE 7.9 Mechanism of drug release from magnetic field-sensitive nanoparticles.

utilization of the metal nanoparticles as drug carriers is surely a smart approach to deliver the drug at the targeted site by applying a magnetic field externally. The drug-loaded metallic nanoparticle is first administered into the body that gets accumulated at the target site by applying an external magnetic field as shown in Figure 7.9 (Pankhurst et al., 2016). Once the metallic nanoparticles get accumulated at the tumor site, the drug gets released through enzymatic activity or due to physiological changes like pH, temperature, etc. at the pathological site (Wu et al., 2010). In this way the drug gets concentrated at the pathological site to ensure the targeted pharmacological effect. Moreover, the magnetic field can be used for diagnostic purposes to diagnose different diseases (Frimpong and Hilt, 2010).

7.6.4 ELECTRIC FIELD-SENSITIVE NANOPARTICLES

The conductive nanoparticulate carriers are responsive to the externally applied electric field. Only weak electric field can be utilized as an external trigger to stimulate the on-demand, targeted drug release (Samanta et al., 2016). The weak electric field is applied at the target site. The heat produced, redox reaction stimulated, and nanodestruction induced by electric field are mainly involved in drug release from electroresponsive nanoparticulate carrier (Ge et al., 2012; Jeon et al., 2011; Wang et al., 2016). After the administration of the electroresponsive drug delivery system, the heat, cleavage, redox reaction, or any other physiological changes induced by the externally applied electric field induce the drug release at a particular site (Puiggalí-Jou et al., 2018). The release can be stopped by removing the external electric field, thus providing control over the drug release in better way (Petcharoen and Sirivat, 2016).

The only limitation associated with such system is that there are several parameters like voltage, temperature, pH, time, current, etc. are to be controlled and optimized while utilizing electric field as a trigger for an efficient targeted drug delivery (Samanta et al., 2016).

7.6.5 ULTRASOUND-SENSITIVE NANOPARTICLES

Ultrasound waves being safer, noninvasive, and having better control over the tissue penetration are of great interest as an exogenous trigger for targeted drug delivery (Paris et al., 2015). The mechanical, thermal, and radiation properties of the ultrasound waves trigger the effective drug release from the nanoparticulate drug carrier at the target site (Luo et al., 2017). Different nanoparticles carrier can be utilized that are responsive to the thermal, mechanical, or radiation properties of the waves (Aryal et al., 2010). The nanoparticulate carrier of the thermoresponsive material may undergo cleavage by the heat produced by the ultrasound waves (Paris et al., 2018). The vibrations produced by ultrasound waves may also cause the conformational changes in the nanocarrier leading to the drug release at the particular site as shown in Figure 7.10 (Xin et al., 2017).

7.7 DUAL/MULTI-RESPONSIVE DDS

With advancements in nanotechnology, nowadays, dual and multi-responsive DDS are preferred over utilizing a single stimulus-responsive DDS for the targeted drug delivery (An et al., 2016). Multi stimuli used may be endogenous, exogenous, or both (Zhang et al., 2018). The combination of multiple triggers like pH, temperature, light, magnetic field, electric field, redox potential, etc. in the composition of the multi-responsive nanoparticulate DDS will be effective enough to ensure and regulate the targeted drug release at the specific sites (Li et al., 2020; Luo et al., 2019; Wen et al., 2018; Xin et al., 2017). Such multi-responsive DDS act as a smart carrier system to

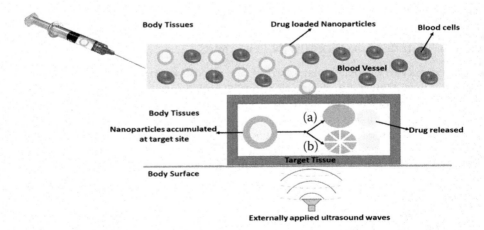

FIGURE 7.10 Mechanism of drug release from ultrasound-sensitive nanoparticles, A. Thermal response of ultrasound waves. B. Nanodestruction by vibrations induced by ultrasound waves.

release drug at the target site under the influence of multiple stimuli, thus enhancing the drug release at a site when triggered by multiple stimuli at the same time (Chen et al., 2017; Guragain et al., 2015; Lei et al., 2020). The examples of different stimuli-sensitive nanoparticulate systems have been tabulated in Table 7.1.

7.8 APPLICATIONS OF TRIGGER-SENSITIVE NANOPARTICULATE DRUG DELIVERY SYSTEM

The trigger-sensitive nanoparticles can serve the diverse applications in prevention, treatment, mitigation, and diagnosis of diseases or the pathological condition. The chemotherapy given in cancer has more adverse effects as compared to the beneficial aspects. The nanoparticles serving the targeted delivery can be optimized to prevent the side effects at the normal sites other than the targeted pathological sites. The nanoparticles can not only target at tissue and organ level, but the site-specific targeting can be achieved at cellular and subcellular level through their functionalization. In case of brain diseases, the blood brain barrier does not allow the drug to reach site of action. But nanoparticles can cross the blood brain barrier to reach the target site (De Jong and Borm, 2008). The stimuli-sensitive nanoparticles are of prime importance in analytical and diagnostic fields. Biosensors involve the utilization of the biorecognition component along with transducer that converts the biological event into the measurable electrical signal (Avella-Oliver et al., 2016). Thermosensitive nanoparticles responsive to thermal changes are used in the thermosensors. Trigger-sensitive nanoparticles are also utilized for noninvasive imaging purposes. Such smart nanoparticles may also serve as platform for theragnostic purposes in which the nanosystem serves the purpose of both, diagnosis and therapy (Herynek et al., 2016; Karimi et al., 2016; Shen et al., 2017). The photosensitive nanoparticles have certain applications in photodynamic therapy (Juarranz et al., 2008). The variation of pH is found at organ and tissue level, even at cellular and subcellular level. This variation is also found at the pathological tumor sites, which favors the targeted delivery of pH-sensitive nanoparticles (Liu et al., 2014). Several nanoparticulate drug delivery systems have been developed by the researchers, but still efforts are being made to get their approval for being used clinically. However, some nanomedicines have been approved by FDA for their clinical use (Kawasaki and Player, 2005; Pogodin et al., 2012).

7.9 LIMITATIONS OF TRIGGER-SENSITIVE NANOPARTICULATE DRUG DELIVERY SYSTEM

Trigger-sensitive nanoparticles are highly advantageous in several biomedical applications due to many beneficial aspects of surface modification, biocompatibility, enhanced drug loading, biodegradability, and targeted drug release properties in response to particular endogenous or exogenous stimulus. Despite of several advantages, there are few limitations associated with them (Pham et al., 2020). The major limitation of nanoparticles is poor target site specificity due to which nanoparticles poorly distinguish the target sites from the nontarget sites. It not only leads to the side effects at the nontarget sites but also reduces the drug availability at

TABLE 7.1
Recently Designed Stimuli-Sensitive Nanoparticles

Stimulus	Delivery System	Composition	Drug Loaded	Mechanism	Application	References
Endogenous Stimuli						
pH/ Ionic microenvironment	Micelles	Hyperbranched double hydrophilic block copolymer of poly (ethylene oxide)-hyperbranched polyglycerol	Doxorubicin	pH dependent drug release	Treatment of cancer	(Lee et al., 2012)
Enzymes	Nanocapsules	Biopolymers	Doxorubicin	Enzyme-sensitive drug release	Treatment of cervical cancer	(Radhakrishnan et al., 2014)
Exogenous Stimuli						
Temperature	Nanogel	Chitosan grafted PNIPAM based. nanogel assembly	Curcumin		Targeted drug delivery	(Luckanagul et al., 2018)
Electric field	Nanocomposite film	Polypyrrole/graphene oxide nanocomposite film	Dexamethasone	Electrochemical reduction	On-demand drug delivery without the passive release of the drug.	(Weaver et al., 2014)
Magnetic field	Nanoparticles	Manganese ferrite (MnFe2O4) nanoparticles functionalized with mono/multilayers of chitosan and alginate sodium	Curcumin	Magnetic hyperthermia	Targeted drug delivery against the tumor, Imaging.	(Jardim et al., 2018)
Ultrasound	Nanoparticle aggregates (NPA)	Drug-loaded PLGA nanoparticles were. transformed into nanoparticles aggregates	Doxorubicin	Ultrasonic vibrations stimulated. NPA dissociation promoting. enhanced tumor uptake	Targeted drug delivery to the disease site	(Papa et al., 2017)
Light (NIR)	Nanoparticles	Hollow mesoporous Prussian blue nanoparticles filled with phase change material (1-tetradecanol) loaded with two drugs.	Doxorubicin and Camptothecin.	Melting of 1-tetradecanol resulted in the escape of 1-TD and drugs from the carrier.	Tumor treatment through synergistic photo-thermal and chemotherapy.	(Chen et al., 2017)

the target sites. The need is to modify the surfaces of nanoparticles that they surely deliver the drug at the target sites (Figueroa et al., 2019). Other limitations associated are toxicity and biocompatibility issues. The need is to overcome the cytotoxicity by utilizing the biocompatible and biodegradable materials in synthesis of the nanoparticles (Vakili-Ghartavol et al., 2020). The shape and size of nanoparticles also determine their toxicity. A study revealed that the spherical nanoparticles of size lying between 1 μm and 300 μm exhibit the least toxicity (Woźniak et al., 2017). The functional groups may also determine the cytotoxic effects of the nanoparticles (Aisida et al., 2019). The biodegradable nature of nanoparticles can surely reduce the toxic effects of nanoparticles due to biocompatibility issue (Parmar et al., 2018; Tapeinos et al., 2018). Another limitation associated with trigger-sensitive nanoparticles is nonspecific identification of the linkages and the functional groups that will allow release of the drug at the particular site under the influence of the particular stimulus (Blum et al., 2015). Researchers are still making efforts to overcome these issues to utilize the nanoparticles clinically. Once these limitations are addressed, nanoparticulate systems will bring clinical revolution.

7.10 CONCLUSION

Nanotechnology has a wide range of applications and with recent advancements nanoparticles are very smartly being used as biomedicines. Nanoparticles are of prime importance in pharmaceutical and biomedical applications due to associated potential benefits like enhanced drug loading, targeted drug delivery, biodegradability, and control over drug release accomplished by responding to various endogenous and exogenous stimuli. All these characteristics are promising to enhance the efficacy of the drug delivery systems that will help in disease prevention and control effectively. However, owing to so many benefits, there are also few limitations associated with them, such as toxicity, size and shape control, biocompatibility with the biological membranes, which are required to be addressed in order for them to be used clinically. Several efforts are still being made to overcome these issues to make their clinical use possible and to take advantages of these smart carriers clinically.

ABBREVIATIONS

UV   Ultraviolet
DDS   Drug delivery system
NIR   Near infrared
EPR   Enhanced permeability and retention
PEG   Polyethylene glycol
GIT   Gastrointestinal tract

REFERENCES

Aisida, S.O., et al. "Incubation period induced biogenic synthesis of PEG enhanced Moringa oleifera silver nanocapsules and its antibacterial activity." *Journal of Polymer Research* 26, no. 9(2019): 1–11.

Albanese, A., P.S. Tang, and W.C.W. Chan. "The effect of nanoparticle size, shape, and surface chemistry on biological systems." *Annual Review of Biomedical Engineering* 14 (2012): 1–16.

An, X., et al. "Rational design of multi-stimuli-responsive nanoparticles for precise cancer therapy." *ACS Nano* 10, no. 6(2016): 5947–5958.

Arrue, L., and L.J.C. Ratjen. "Internal targeting and external control: Phototriggered targeting in nanomedicine."*ChemMedChem* 12, no. 23(2017): 1908–1916.

Aryal, S., C.-M.J. Hu, and L. Zhang "Polymer – cisplatin conjugate nanoparticles for acid-responsive drug delivery." *ACS Nano* 4, no. 1(2010): 251–258.

Avella-Oliver, M., et al. "Towards photochromic and thermochromic biosensing." *TrAC Trends in Analytical Chemistry* 79 (2016): 37–45.

Baeza, A., M. Colilla, and M. Vallet-Regí. "Advances in mesoporous silica nanoparticles for targeted stimuli-responsive drug delivery." *Expert Opinion on Drug Delivery* 12, no. 2(2015): 319–337.

Balamuralidhara, V., et al. "pH sensitive drug delivery systems: a review." *American Journal of Drug Discovery and Development* 1, no. 1(2011): 25.

Banerjee, R.J.N. "Trigger-responsive nanoparticles: Control switches for cancer therapy." *Nanomedicine* 6, no. 10(2011): 1657–1660.

Basel, M.T., et al. "Protease-sensitive, polymer-caged liposomes: a method for making highly targeted liposomes using triggered release." *ACS Nano* 5, no. 3(2011): 2162–2175.

Bawa, P., et al. "Stimuli-responsive polymers and their applications in drug delivery." *Biomedical Materials* 4, no. 2(2009): 022001.

Bawa, K.K., et al. "PLA-based triblock copolymer micelles exhibiting dual acidic ph/reduction responses at dual core and core/corona interface locations." *Macromolecular Rapid Communications* 39, no. 24(2018): 1800477.

Berndt, I., et al. "Mechanics versus thermodynamics: Swelling in multiple-temperature-sensitive core–shell Microgels." *Angewandte Chemie International Edition* 45, no. 7(2006): 1081–1085.

Bhagwat, R., and I. Vaidhya. "Novel drug delivery systems: An overview." *International Journal of Pharmaceutical Sciences and Research* 4, no. 3(2013): 970.

Bhatnagar, S., and V.V. Venuganti. "Cancer targeting: Responsive polymers for stimuli-sensitive drug delivery." *Journal of Nanoscience and Nanotechnology* 15, no. 3(2015): 1925–1945.

Blum, A.P., et al. "Stimuli-responsive nanomaterials for biomedical applications." *Journal of the American Chemical Society* 137, no. 6(2015): 2140–2154.

Brown, A.A., O. Azzaroni, and W.T. Huck. "Photoresponsive polymer brushes for hydrophilic patterning." *Langmuir* 25, no. 3(2009): 1744–1749.

Cao, Z., et al. "pH-and enzyme-triggered drug release as an important process in the design of anti-tumor drug delivery systems." *Biomedicine and Pharmacotherapy* 118 (2019): 109340.

Castillo, R.R., et al. "Advances in mesoporous silica nanoparticles for targeted stimuli-responsive drug delivery: An update." *Expert Opinion on Drug Delivery* 16, no. 4(2019): 415–439.

Chen, Y., et al. "Hollow mesoporous organosilica nanoparticles: A generic intelligent framework-hybridization approach for biomedicine." *Journal of the American Chemical Society* 136, no. 46(2014): 16326–16334.

Chen, S., et al. "Photo, pH and redox multi-responsive nanogels for drug delivery and fluorescence cell imaging." *Polymer Chemistry* 8, no. 39(2017): 6150–6157.

Chen, H., et al. "Multifunctional phase-change hollow mesoporous Prussian blue nanoparticles as a NIR light responsive drug co-delivery system to overcome cancer therapeutic resistance." *Journal of Materials Chemistry B* 5, no. 34(2017): 7051–7058.

Chen, H., D. Liu, and Z.J.C.L. Guo. "Endogenous stimuli-responsive nanocarriers for drug delivery." *Chemistry Letters* 45, no. 3(2016): 242–249.
Cheng, W., et al. "Stimuli-responsive polymers for anti-cancer drug delivery." *Materials Science and Engineering* 45 (2014): 600–608.
Cho, H.J., et al. "Engineered photo-responsive materials for near-infrared-triggered drug delivery." *Journal of Industrial and Engineering Chemistry* 31 (2015): 15–25.
Crayton, S.H., and A. Tsourkas. "pH-titratable superparamagnetic iron oxide for improved nanoparticle accumulation in acidic tumor microenvironments." *ACS Nano* 5, no. 12(2011): 9592–9601.
Dahiya, M.S., V.K. Tomer, and S. Duhan. "Metal–ferrite nanocomposites for targeted drug delivery." in *Applications of Nanocomposite Materials in Drug Delivery*. Elsevier (2018), 737–760.
De Jong, W.H., and P.J. Borm "Drug delivery and nanoparticles: applications and hazards." *International Journal of Nanomedicine* 3, no. 2(2008): 133.
Ding, C., et al. "Recent advances in stimuli-responsive release function drug delivery systems for tumor treatment." *Molecules* 21, no. 12(2016): 1715.
Dvir, T., et al. "Photo-targeted nanoparticles." *Nano Letters*. 10, no. 1(2010): 250–254.
Figueroa, S.M., et al. "Influenza A virus mimetic nanoparticles trigger selective cell uptake." *Proceedings of the National Academy of Sciences* 116, no. 20(2019): 9831–9836.
Fleige, E., M.A. Quadir, and R. Haag. "Stimuli-responsive polymeric nanocarriers for the controlled transport of active compounds: concepts and applications." *Advanced Drug Delivery Reviews*. 64, no. 9(2012): 866–884.
Frimpong, R.A., and J.Z. Hilt. "Magnetic nanoparticles in biomedicine: synthesis, functionalization and applications." *Nanomedicine* 5, no. 9(2010): 1401–1414.
Furyk, S., et al. "Effects of end group polarity and molecular weight on the lower critical solution temperature of poly (N-isopropylacrylamide)." *Journal of Polymer Science Part A: Polymer Chemistry* 44, no. 4(2006): 1492–1501.
Garcia, E., et al. "Cell-line-based studies of nanotechnology drug-delivery systems: A brief review." in *Nanocarriers for Drug Delivery*. Elsevier (2019), 375–393.
Ge, J., et al. "Drug release from electric-field-responsive nanoparticles." *ACS Nano* 6, no. 1(2012): 227–233.
Gu, M., et al. "Applications of stimuli-responsive nanoscale drug delivery systems in translational research." *Drug Discovery Today* 23, no. 5(2018): 1043–1052.
Guardado-Alvarez, T.M., et al. "Photo-redox activated drug delivery systems operating under two photon excitation in the near-IR." *Nanoscale* 6, no. 9(2014): 4652–4658.
Gundloori, R.V., A. Singam, and N. Killi. "Nanobased intravenous and transdermal drug delivery systems." in *Applications of Targeted Nano Drugs and Delivery Systems*. Elsevier (2019), 551–594.
Gupta, P., K. Vermani, and S. Garg. "Hydrogels: from controlled release to pH-responsive drug delivery." *Drug Discovery Today* 7, no. 10(2002): 569–579.
Guragain, S., et al. "Multi-stimuli-responsive polymeric materials." *Chemistry–A European Journal* 21, no. 38(2015): 13164–13174.
Gwon, K., et al. "Improved near infrared-mediated hydrogel formation using diacrylated Pluronic F127-coated upconversion nanoparticles." *Materials Science and Engineering: C* 90 (2018): 77–84.
Hatakeyama, H. Recent advances in endogenous and exogenous stimuli-responsive nanocarriers for drug delivery and therapeutics." *Chemical and Pharmaceutical Bulletin* 65, no. 7(2017): 612–617.
Herynek, V., et al. "Using ferromagnetic nanoparticles with low Curie temperature for magnetic resonance imaging-guided thermoablation." *International Journal of Nanomedicine* 11 (2016): 3801.

Hossion, A.M., et al. "Visible light controlled release of anticancer drug through double activation of prodrug." *ACS Medicinal Chemistry Letters* 4, no. 1(2013): 124–127.

Hu, X., et al. "Photo-triggered release of caged camptothecin prodrugs from dually responsive shell cross-linked micelles." *Macromolecules* 46, no. 15(2013): 6243–6256.

Huh, K.M., et al. "pH-sensitive polymers for drug delivery." *Macromolecular Research* 20, no. 3(2012): 224–233.

Huo, M., Y. Chen, and J. Shi. "Triggered-release drug delivery nanosystems for cancer therapy by intravenous injection: where are we now?" *Expert Opinion on Drug Delivery* 13, no. 9(2016): 1195–1198.

Jardim, K.V., et al. "Novel magneto-responsive nanoplatforms based on MnFe2O4 nanoparticles layer-by-layer functionalized with chitosan and sodium alginate for magnetic controlled release of curcumin." *Materials Science and Engineering: C* 92 (2018): 184–195.

Javadzadeh, Y., and S. Hamedeyazdan. "Novel drug delivery systems for modulation of gastrointestinal transit time." in *Recent Advances in Novel Drug Carrier Systems*. IntechOpen, (2012).

Jeon, G., et al. "Electrically actuatable smart nanoporous membrane for pulsatile drug release." *Nano Letters* 11, no. 3(2011): 1284–1288.

Juarranz Á. , et al. "Photodynamic therapy of cancer. Basic principles and applications." *Clinical and Translational Oncology* 10, no. 3(2008): 148–154.

Karewicz, A., et al. ""Smart" alginate–hydroxypropylcellulose microbeads for controlled release of heparin." *International Journal of Pharmaceutics* 385, no. 1-2(2010): 163–169.

Karimi, M., et al. "Temperature-responsive smart nanocarriers for delivery of therapeutic agents: applications and recent advances." *ACS Applied Materials & Interfaces* 8, no. 33(2016): 21107–21133.

Kawasaki, E.S., and A. Player. "Nanotechnology, nanomedicine, and the development of new, effective therapies for cancer." *Nanomedicine: Nanotechnology, Biology and Medicine* 1, no. 2(2005): 101–109.

Khoee, S., and M.R. Karimi "Dual-drug loaded Janus graphene oxide-based thermoresponsive nanoparticles for targeted therapy." *Polymer* 142 (2018): 80–98.

Kobayashi, H., R. Watanabe, and P.L. Choyke. "Improving conventional enhanced permeability and retention (EPR) effects; what is the appropriate target?" *Theranostics* 4, no. 1(2014): 81.

Kojima, C. "Design of stimuli-responsive dendrimers." *Expert Opinion on Drug Delivery* 7, no. 3(2010): 307–319.

Kong, S.D., et al. "Magnetic field activated lipid–polymer hybrid nanoparticles for stimuli-responsive drug release." *Acta Biomaterialia* 9, no. 3(2013): 5447–5452.

Kundu, J.K., and Y.-J. Surh "Nrf2-Keap1 signaling as a potential target for chemoprevention of inflammation-associated carcinogenesis." *Pharmaceutical Research* 27, no. 6(2010): 999–1013.

Kurisawa, M., and N. Yui. "Dual-stimuli-responsive drug release from interpenetrating polymer network-structured hydrogels of gelatin and dextran." *Journal of Controlled Release* 54, no. 2(1998): 191–200.

Lee, S., et al. "Hyperbranched double hydrophilic block copolymer micelles of poly (ethylene oxide) and polyglycerol for pH-responsive drug delivery." *Biomacromolecules* 13, no. 4(2012): 1190–1196.

Lei, B., et al. "Double security drug delivery system DDS constructed by multi-responsive (pH/redox/US) microgel." *Colloids and Surfaces B: Biointerfaces* (2020): 111022.

Li, J., et al. "Near-infrared light and magnetic field dual-responsive porous silicon-based nanocarriers to overcome multidrug resistance in breast cancer cells with enhanced efficiency." *Journal of Materials Chemistry B* 8, no. 3(2020): 546–557.

Li, Y., Y. Zhang, and W.J.N.R. Wang. "Phototriggered Targeting of Nanocarriers for Drug Delivery." *Nano Research* 11, no. 10(2018): 5424–5438.
Lino, M.M., and L. Ferreira. "Light-triggerable formulations for the intracellular controlled release of biomolecules." *Drug Discovery Today* 23, no. 5(2018): 1062–1070.
Liu, J., et al. "pH-sensitive nano-systems for drug delivery in cancer therapy." *Biotechnology Advances* 32, no. 4(2014): 693–710.
Liu, D., et al. "The smart drug delivery system and its clinical potential." *Theranostics* 6, no. 9(2016): 1306.
Liu, M., et al. "Internal stimuli-responsive nanocarriers for drug delivery: Design strategies and applications." *Materials Science and Engineering: C* 71 (2017): 1267–1280.
Liu, W., et al. "Hypoxia-activated anticancer prodrug for bioimaging, tracking drug release, and anticancer application." *Bioconjugate Chemistry* 29, no. 10(2018): 3332–3343.
Longo, J., et al. "Preventing metastasis by targeting lymphatic vessels with photodynamic therapy based on nanostructured photosensitizer." *Journal of Nanomedicine & Nanotechnology* 6, no. 5(2015): 1.
Lopes, J.R., et al. "Physical and chemical stimuli-responsive drug delivery systems: targeted delivery and main routes of administration." *Current Pharmaceutical Design* 19, no. 41(2013): 7169–7184.
Lopes, C.M., P. Barata, and R. Oliveira. "Stimuli-responsive nanosystems for drug-targeted delivery." in *Drug Targeting and Stimuli Sensitive Drug Delivery Systems*. Elsevier (2018), 155–209.
Luckanagul, J.A., et al. "Chitosan-based polymer hybrids for thermo-responsive nanogel delivery of curcumin." *Carbohydrate Polymers* 181 (2018): 1119–1127.
Luo, Z., et al. "Mesoporous silica nanoparticles end-capped with collagen: redox-responsive nanoreservoirs for targeted drug delivery." *Angewandte Chemie International* 50, no. 3(2011): 640–643.
Luo, Z., et al. "On-demand drug release from dual-targeting small nanoparticles triggered by high-intensity focused ultrasound enhanced glioblastoma-targeting therapy." *ACS Applied Materials & Interfaces* 9, no. 37(2017): 31612–31625.
Luo, Y., et al. "Dual pH/redox-responsive mixed polymeric micelles for anticancer drug delivery and controlled release." *Pharmaceutics* 11, no. 4(2019): 176.
Ma, X., et al. "Ultra-pH-sensitive nanoprobe library with broad pH tunability and fluorescence emissions." *Journal of the American Chemical Society* 136, no. 31(2014): 11085–11092.
Maeda, H. "Toward a full understanding of the EPR effect in primary and metastatic tumors as well as issues related to its heterogeneity." *ScienceDirect* 91 (2015): 3–6.
Matsumura, Y., and H. Maeda. "A new concept for macromolecular therapeutics in cancer chemotherapy: Mechanism of tumoritropic accumulation of proteins and the antitumor agent smancs." *Cancer Research* 46, 12 Part 1(1986): 6387–6392.
Mintzer, M.A., and E.E. Simanek. "Nonviral vectors for gene delivery." *Chemical Reviews* 109, no. 2(2009): 259–302.
Mishra, R.K., et al. "Efficient nanocarriers for drug-delivery systems: Types and fabrication." in *Nanocarriers for Drug Delivery*. Elsevier (2019), 1–41.
Montes, S., and H. Maleki. "Aerogels and their applications." in *Colloidal Metal Oxide Nanoparticles*. Elsevier (2020), 337–399.
Mura, S., J. Nicolas, and P.J.N.m. Couvreur. "Stimuli-responsive nanocarriers for drug delivery." *Nature Materials* 12, no. 11(2013): 991–1003.
Nayak, A.K., et al. "Drug delivery: Present, past, and future of medicine." in *Applications of Nanocomposite Materials in Drug Delivery*. Elsevier (2018), 255–282.
Pankhurst, Q.A., et al. "Applications of magnetic nanoparticles in biomedicine." *Journal Of Physics D: Applied Physics* 36, no. 13(2003): R167.

Pankhurst, Q., et al. "Progress in applications of magnetic nanoparticles in biomedicine." *Journal of Physics D: Applied Physics* 42, no. 22(2009): 224001.

Pankhurst, Q., S. Jones, and J. Dobson. "Applications of magnetic nanoparticles in biomedicine: The story so far." *Journal of Physics D: Applied Physics* 49, no. 50(2016): 501002.

Papa, A.-L., et al. "Ultrasound-sensitive nanoparticle aggregates for targeted drug delivery." *Biomaterials* 139 (2017): 187–194.

Paris, J.L., et al. "Polymer-grafted mesoporous silica nanoparticles as ultrasound-responsive drug carriers." *ACS Nano* 9, no. 11(2015): 11023–11033.

Paris, J.L., et al. "Mesoporous silica nanoparticles engineered for ultrasound-induced uptake by cancer cells." *Nanoscale* 10, no. 14(2018): 6402–6408.

Parmar, A., et al. "Anti-proliferate and apoptosis triggering potential of methotrexate-transferrin conjugate encapsulated PLGA nanoparticles with enhanced cellular uptake by high-affinity folate receptors." *Artificial Cells, Nanomedicine, and Biotechnology* 46, sup2(2018): 704–719.

Petcharoen, K., and A. Sirivat. "Magneto-electro-responsive material based on magnetite nanoparticles/polyurethane composites." *Materials Science and Engineering: C* 61 (2016): 312–323.

Pham, S.H., Y. Choi, and J. Choi. "Stimuli-Responsive Nanomaterials for Application in Antitumor Therapy and Drug Delivery." *Pharmaceutics* 12, no. 7(2020): 630.

Pillai, G. "Nanomedicines for cancer therapy: An update of FDA approved and those under various stages of development." *SOJ Pharmacy and Pharmaceutical Sciences* 1, no. 2(2014): 13.

Pogodin, S., et al. "Nanoparticle-induced permeability of lipid membranes." *ACS Nano* 6, no. 12(2012): 10555–10561.

Porta-i-Batalla, M., et al. "Sustained, controlled and stimuli-responsive drug release systems based on nanoporous anodic alumina with layer-by-layer polyelectrolyte." *Nanoscale research letters* 11, no. 1(2016): 1–9.

Puiggalí-Jou, A., et al. "Smart Drug Delivery from Electrospun Fibers through electro-responsive polymeric nanoparticles." *ACS Applied Bio Materials* 1, no. 5(2018): 1594–1605.

Puoci, F., F. Iemma, and N. Picci. "Stimuli-responsive molecularly imprinted polymers for drug delivery: a review." *Current Drug Delivery* 5, no. 2(2008): 85–96.

Radhakrishnan, K., et al. "Dual enzyme responsive and targeted nanocapsules for intracellular delivery of anticancer agents." *RSC Advances* 4, no. 86(2014): 45961–45968.

Rasheed, T., et al. "The smart chemistry of stimuli-responsive polymeric carriers for target drug delivery applications." in *Stimuli Responsive Polymeric Nanocarriers for Drug Delivery Applications*, 1. Elsevier (2018), 61–99.

Ratemi, E. *Stimuli Responsive Polymeric Nanocarriers for Drug Delivery Applications 1, Types and Triggers*. Woodhead Publishing (2018), 121.

Raza, A., Rasheed, T., et al. "Endogenous and exogenous stimuli-responsive drug delivery systems for programmed site-specific release." *Molecules* 24, no. 6(2019): 1117.

Raza, A., Hayat, U., et al. "'Smart' materials-based near-infrared light-responsive drug delivery systems for cancer treatment: a review." *Journal of Materials Research and Technology* 8, no. 1(2019): 1497–1509.

Samanta, D., N. Hosseini-Nassab, and R.N. Zare. "Electroresponsive nanoparticles for drug delivery on demand." *Nanoscale* 8, no. 17(2016): 9310–9317.

Sánchez-Moreno, P., et al. "Thermo-sensitive nanomaterials: recent advance in synthesis and biomedical applications." *Nanomaterials* 8, no. 11(2018): 935.

Schafer, F.Q., and G.R. Buettner. "Redox environment of the cell as viewed through the redox state of the glutathione disulfide/glutathione couple." *Free Radical Biology and Medicine*. 30, no. 11(2001): 1191–1212.

Schmaljohann, D. "Thermo-and pH-responsive polymers in drug delivery." *Advanced Drug Delivery Reviews* 58, no. 15(2006): 1655–1670.

Shen, S., et al. "Near-infrared light-responsive nanoparticles with thermosensitive yolk-shell structure for multimodal imaging and chemo-photothermal therapy of tumor." *Nanomedicine: Nanotechnology, Biology and Medicine* 13, no. 5(2017): 1607–1616.

Shi, J., et al. "Cancer nanomedicine: Progress, challenges and opportunities." *Nature Reviews Cancer* 17, no. 1(2017): 20.

Shuhendler, A.J., et al. "A novel solid lipid nanoparticle formulation for active targeting to tumor $\alpha v \beta 3$ integrin receptors reveals cyclic RGD as a double-edged sword." *Advanced Healthcare Materials* 1, no. 5(2012): 600–608.

Singh, A.K., et al. "Engineering nanomaterials for smart drug release: Recent advances and challenges." in *Applications of Targeted Nano Drugs and Delivery Systems*. Elsevier (2019), 411–449.

Sonawane, S.J., R.S. Kalhapure, and T. Govender. "Hydrazone linkages in pH responsive drug delivery systems." *European Journal of Pharmaceutical Sciences* 99 (2017): 45–65.

Stayton, P.S., B. Ghosn, and J.T. Wilson. "Targeting." in *Biomaterials Science*. Elsevier (2013), 1028–1036.

Stirland, D., et al. "Targeted drug delivery for cancer therapy." in *Biomaterials for Cancer Therapeutics*. Elsevier (2013), 31–56.

Tapeinos, C., et al. "Functionalised collagen spheres reduce H2O2 mediated apoptosis by scavenging overexpressed ROS." *Nanomedicine: Nanotechnology, Biology and Medicine* 14, no. 7(2018): 2397–2405.

Tiwari, A., et al. "Curcumin encapsulated zeolitic imidazolate frameworks as stimuli responsive drug delivery system and their interaction with biomimetic environment." *Scientific Reports* 7, no. 1(2017): 1–12.

Vakili-Ghartavol, R., et al. "Toxicity assessment of superparamagnetic iron oxide nanoparticles in different tissues." *Artificial Cells, Nanomedicine, and Biotechnology* 48, no. 1(2020): 443–451.

Verhoef, J.J.F., and T.J.A. Anchordoquy. "Questioning the use of PEGylation for drug delivery." *Drug Delivery and Translational Research* 3, no. 6(2013): 499–503.

Wang, C.E., et al. "Polymer nanostructures synthesized by controlled living polymerization for tumor-targeted drug delivery." *Science Direct* 219 (2015): 345–354.

Wang, S., P. Huang, and X.J.A.M. Chen. "Hierarchical targeting strategy for enhanced tumor tissue accumulation/retention and cellular internalization." *Advanced Materials* 28, no. 34(2016): 7340–7364.

Weaver, C.L., et al. "Electrically controlled drug delivery from graphene oxide nanocomposite films." *ACS Nano* 8, no. 2(2014): 1834–1843.

Wen, K., et al. "Near-infrared/pH dual-sensitive Nanocarriers for enhanced intracellular delivery of doxorubicin." *ACS Biomaterials Science & Engineering* 4, no. 12(2018): 4244–4254.

Woźniak, A., et al. "Size and shape-dependent cytotoxicity profile of gold nanoparticles for biomedical applications." *Journal of Materials Science: Materials in Medicine* 28, no. 6(2017): 92.

Wu, J., et al. "Chemodrug-gated biodegradable hollow mesoporous organosilica nanotheranostics for multimodal imaging-guided low-temperature photothermal therapy/chemotherapy of cancer." *ACS Applied Materials & Interfaces* 10, no. 49(2018): 42115–42126.

Wu, A., P. Ou, and L. Zeng. "Biomedical applications of magnetic nanoparticles." *Nano* 5, no. 05(2010): 245–270.

Xiang, J., et al. "Near-infrared light-triggered drug release from UV-responsive diblock copolymer-coated upconversion nanoparticles with high monodispersity." *Journal of Materials Chemistry B* 6, no. 21(2018): 3531–3540.

Xifré-Pérez, E., et al. "Sustained, controlled and stimuli-responsive drug release systems based on nanoporous anodic alumina with layer-by-layer polyelectrolyte." *Nanoscale Research Letters* (2016).

Xin, Y., et al. "PLGA nanoparticles introduction into mitoxantrone-loaded ultrasound-responsive liposomes: in vitro and in vivo investigations." *International Journal of Pharmaceutics* 528, no. 1-2(2017): 47–54.

Yang, G., et al. "Near-infrared-light responsive nanoscale drug delivery systems for cancer treatment." *Coordination Chemistry Reviews* 320 (2016): 100–117.

Yang, J., et al. "NIR-controlled morphology transformation and pulsatile drug delivery based on multifunctional phototheranostic nanoparticles for photoacoustic imaging-guided photothermal-chemotherapy." *Biomaterials* 176 (2018): 1–12.

Yang, Q., and S.K. Lai. "Anti-PEG immunity: Emergence, characteristics, and unaddressed questions." *Wiley Interdisciplinary Reviews* 7, no. 5(2015): 655–677.

Yang, K., L. Feng, and Z. Liu. "Stimuli responsive drug delivery systems based on nano-graphene for cancer therapy." *Advanced Drug Delivery Reviews* 105 (2016): 228–241.

York, A.W., S.E. Kirkland, and C.L. McCormick. "Advances in the synthesis of amphiphilic block copolymers via RAFT polymerization: stimuli-responsive drug and gene delivery." *Advanced Drug Delivery Reviews* 60, no. 9(2008): 1018–1036.

Yoshida, T., et al. "pH-and ion-sensitive polymers for drug delivery." *Expert Opinion On Drug Delivery* 10, no. 11(2013): 1497–1513.

Zhang, R., et al. "A novel pH-and ionic-strength-sensitive carboxy methyl dextran hydrogel." *Biomaterials* 26, no. 22(2005): 4677–4683.

Zhang, L., et al. "Dual pH/reduction-responsive hybrid polymeric micelles for targeted chemo-photothermal combination therapy." *Acta Biomaterialia* 75 (2018): 371–385.

Zhao, B., and J.S. Moore. "Fast pH-and ionic strength-responsive hydrogels in microchannels." *Langmuir* 17, no. 16(2001): 4758–4763.

Zheng, L., et al. "Fabrication of acidic pH-cleavable polymer for anticancer drug delivery using a dual functional monomer." *Biomacromolecules* 19, no. 9(2018): 3874–3882.

8 Metal Organic Frameworks for Drug Delivery

Saima Zulfiqar[1], Shahzad Sharif[2], and Muhammad Yar[3]
[1]Government College University Lahore (GCUL) and COMSATS University Islamabad, Lahore Campus
[2]Government College University Lahore (GCUL)
[3]COMSATS University Islamabad, Lahore Campus

CONTENTS

8.1 Introduction	182
8.2 Synthesis of MOFs and MOF Nanoparticles (nMOFs)	183
8.3 Characteristics of Metal Organic Frameworks	183
8.3.1 Pore Size of MOFs	184
8.3.2 Particle Size	184
8.3.3 Rationality and Biodegradability of MOFs	184
8.3.4 Toxicological Compatibility of MOFs	185
8.4 Functionalization of MOFs	185
8.4.1 Surface Modification	185
8.4.2 Covalent Modification	185
8.4.3 Coordination Modulation	186
8.4.4 Noncovalent Modification	186
8.4.5 External Modification	187
8.5 Drug Delivery through Stimuli Based MOFs	187
8.5.1 Single-Stimulus-Based MOFs for Drug Delivery	187
8.5.1.1 pH-Controlled MOFs	187
8.5.1.2 Magnetic MOFs	188
8.5.1.4 Ion-Receptive MOFs	189
8.5.1.5 Temperature-Sensitive MOFs	189
8.5.2 Multiple Stimuli-Responsive MOFs	191
8.6 Multifunctional MOFs	194
8.6.1 Polymer Mantle on MOFs	194
8.6.2 Magnetic Core-Shell MOFs	196
8.6.3 Core-Shell MOF nanocomposites	197
8.6.4 Flexible MOF as Theranostic Nanoplatform	197

8.7 Drug Delivery to Organs 198
8.8 Challenges and Future Perspectives 200
8.9 Conclusions 201
References 202

8.1 INTRODUCTION

At the present time, a large number of active ingredients (AIs) are being used as therapeutics, but these have disadvantages, such as instability, rapid biodegradation, low specificity, toxicities, and evident after effects. Keeping in view all these constraints, it is necessary to institute novel ways for delivery of drugs to the specific site, reducing adverse effects, their sustained drug release in time and at a regulated rate, and enhancing the therapeutic efficiency (Kaelin 2005, Keith et al. 2005, Vickers 2017). Simultaneously, diagnosis, therapy, and therapeutic response are controlled by using one assimilated system consisting of therapy and diagnosis (Ryu et al. 2012). As a result, drug-delivery systems (DDS) have been evolved. These avoid the use of conventional medicine and therapies.

The optimized concentration, increased therapeutic effects, and low after effects of drugs help in targeting drug to the tissues selectively, called as targeting drug delivery system (TDDS) (Gao et al. 2014). TDDS is being used for treatment of cancer. For this, liposomes, microspheres, nanoparticles, microemulsion, albumin, lipoproteins, emulsion, and polymer conjugates are being used as carriers for drug. The composition of nanocarrier materials helps in evolution of TDDS (Baeza et al. 2015, Gul et al. 2018). As there are many new polymers and materials being prepared rapidly, drug nanocarrier materials have got much responsiveness. Metal organic frameworks (MOFs), also known as porous coordination polymers (PCPs) are porous organic-inorganic hybrid extended networks, elicited from inorganic elements (cations, clusters, chains, etc.), and organic poly-dentate ligands (Della Rocca et al. 2011).

MOFs are preferred over traditional porous material because of having distinctive characteristics: several categories (Farrusseng et al. 2009) manifold functions, substantial pores and specific surface area, negligible density of crystal, reproducible pore dimensions, good bionic catalytic properties, and biocompatibility (Lee et al. 2009). These are mostly used in separations, gas storage, nonlinear optics, catalysis, sensing, and light-harvesting as DDS (Zhu and Ren 2014) and as contrast agents (Lin et al. 2006).

Especially, as drug delivery system, MOFs are used due to having large volumes and pore surfaces corresponding high drug sorption ability, functionalization of the surface through unique host-guest interactions enabling either optimized drug release, adsorption/desorption of drug or surface modification to enhance their biodistribution and in vivo stability, stable to perform their functions, and, then, being discharged from the body, averting endogenous accumulation and related toxicity.

Due to loading of large dosage of drug, the scientists have emphasized on their use for delivery of anticancer drugs. In this chapter, we discuss the recent advances on how MOFs help in achieving targeted drug delivery systems.

8.2 SYNTHESIS OF MOFS AND MOF NANOPARTICLES (NMOFS)

When a therapeutic molecule is injected intravenously, it must have particle size smaller than 200 nm in aqueous medium, so that it moves through capillaries without agglomeration (Horcajada et al. 2012). Particle size is the controlling factor in drug delivery. Nanoparticles of MOFs have gained attraction of scientists due to their innate characteristics as reproducible shapes and sizes, adaptive surface-volume ratios, and easy external surface reorganization (modifications) (Doane and Burda 2012). It is important to synthesize stable and reproducible MOFs. In the following, some synthetic approaches are enlisted as reverse microemulsion, microwave irradiation, solvothermal method, and ultrasonic synthesis.

In the most common solvo-thermal method, numerous parameters (concentration, pressure, reaction time, stoichiometry, temperature, pH) are taken into account. It is quite easy to adjust these parameters in order to either adjust reaction kinetics or mitigate the nucleation process; as a result MOFs of different sizes are obtained (Hermes et al. 2007, Cravillon et al. 2009, Horcajada et al. 2009, Chalati et al. 2011). The MOF size can be controlled by adding some blocking agent, e.g., pyridine that can tune particle size of indium BDC from 4 μm to 900 nm (Cho et al. 2008).

Reverse microemulsions help in adjustment of growth kinetics as well as nucleation of MOFs. The particle size of gadolinium BDC nanorods synthesized from microemulsion of [NMeH$_3$]$_2$[BDC] and GdCl$_3$ in cetyltrimethyl ammonium bromide (CTAB)/water/1-hexanol/isooctane system for 2 hours was governed by changing the molar ratio of water and surfactant microemulsion. At low molar ratio of water and surfactant microemulsion, size of the particle was 100–125 nm, while at high molar ratio, 1–2 μm (Rieter et al. 2006).

Ultrasonic method is low cost, environment friendly and gives high yield. Qiu et al. synthesized different sized (50–100 nm) microporous MOF (Zn$_3$(BTC)$_2$.12H$_2$O) under different reaction environments. It was observed that reaction for 30 minutes produced crystals of 100–200 nm diameter, while up to 90 minutes, the diameter increased to 700–900 nm. Consequently, this method could be used preferred to change the structural properties of MOFs (size and shape) under different reaction conditions (Qiu et al. 2008)

Synthesis of MOFs through microwave radiation has many benefits, such as phase selectivity, narrow particle size distributions, and fast crystallization along with facile morphology control. In it, nucleation as well as growth of crystal is controlled by temperature and time. For example, MIL-101(Cd) nMOF has particle size 40–90 nm at 210 °C between 1 and 40 minutes. It is proved that microwave method is the best among all (Jhung et al. 2007).

8.3 CHARACTERISTICS OF METAL ORGANIC FRAMEWORKS

MOFs are suitable exporters of drug due to unique characteristic, such as a large pore diameter, large specific surface area, nontoxicity, and good biocompatibility to human body and easy metabolism. The properties, such as pore size, particle size, stability of MOFs, can be adjusted rationally to enhance the drug encapsulation, control rate of drug liberation, and carry drug to a specific point. However, some basic characteristics of MOFs are being discussed in the following sections.

8.3.1 PORE SIZE OF MOFs

Porosity of MOFs is modified to increase their capacity of drug encapsulation. In recent years, researchers have prepared the porous MOFs that can be utilized for the enhanced loading capacity as well as multi-functional targeted delivery of drugs. As Gao et al. prepared ZIF-8 that had 51% drug encapsulation capacity after modification (Gao et al. 2016). Furthermore, pore size can be tuned by using a suitable organic ligand. Anyways, there are few studies regarding the influence of pore sizes of MOFs on loading of drug.

8.3.2 PARTICLE SIZE

Particle size of drug-loaded MOFs helps in not only transferring of drug effectively but also at the specific site. The drug delivery system having size 100 nm targets the cancerous tissues comparatively easily. The particle size of 100 nm helps in achieving multifunctional targeting. As a result, size distribution of MOFs could be controlled. For instance, Duan et al. synthesized zeolitic imidazolate-based AZIF-8 and investigated effect of its particle diameter on tumor treatment (Duan et al. 2018). They controlled the particle size of AZIF-8 using nontoxic poly-allylamine hydrochloride (PAH) that also altered nucleation rate of AZIF-8. Larger particle size of AZIF-8 was obtained by using more PAH. In vitro as well as in vivo studies proved that AZIF-8 of 60 nm had remarkable biocompatibility against tumors. Gao investigated active targeting, light sensitivity, magnetic sensitization, and load chemotherapeutic drugs along with factors affecting Fe-MIL-53-NH_2 simultaneously (Gao et al. 2017). It was observed that large particle size was achieved when low concentration of reactants was used. Moreover, Gao et al. examined the effect of benzoic acid on pore size of UIO-66-NH_2 [4]. In this case, small particle size was obtained by using low concentration of benzoic acid. However, it could be concluded that particle size of MOFs can be obtained depending upon requirement through concentration of reactants. Currently, MOFs as nanocarriers of drugs having different particle size are being used.

8.3.3 RATIONALITY AND BIODEGRADABILITY OF MOFs

Rationality of MOFs is the principle condition for their use in drug distribution in body. These loaded with drug must be intact prior to reach the target site to make sure their effectiveness and safety in vitro and release drug at suitable position. Rachel prepared Zn-MTX (methotrexate) and Zr-MTX(methotrexate) nanoscale coordination polymeric materials at elevated temperature through surfactant-assisted and the microwave heating methods. Both were unstable due to polymerization of particles in water to disrupt the liposomal surface. Thereafter, they synthesized Gd-MTX(methotrexate) nanoscale coordination polymeric material via microwave heating method followed by wrapping in phospholipid bilayer. This drug loaded system was stable. Hence, in order to develop drug-loading system, the effect of every component on the stability must be taken into account. On the other hand, in medical applications, matrix material recommended for is required to be chemically instable and its byproducts should directly approach the body's metabolic system to avoid endogenous accumulation. This property helps in release of drug effectively: even though, it is affected by hydrophobicity/hydrophilicity property, pore size and host-guest interactions (Gao, Cui et al. 2018).

8.3.4 Toxicological Compatibility of MOFs

For drug delivery system, MOFs need toxicologically compatibility owing to nontoxic building blocks. Suitable MOFs can be prepared by blending different precursors, as metal nodes and organic linkers. Among metal nodes, iron being a component of hemoglobin is the most appropriate: its concentration in blood plasma is almost 22 mM. Ti, Mn, Zr, Zn, Fe, Ca, Cu, and Mg are the most suitable metals (Horcajada et al. 2012).

The organic linkers used for the synthesis are prepared from natural compounds. The most commonly used linkers for drug delivery are imidazolates, polycarboxylates, amines, pyridyl, etc. Terephthalic acid, trimesic acid, 2,6 napthalenedicarboxylic acid, 1-methylimidazole, 2-methylimidazole, isonicotinic acid, and 5-aminoisophthalic acid have toxicity statistics (LD_{50}) of 5 g/kg, 8.4 g/kg, 5 g/kg, 1.13 g/kg, 1.4 g/kg, 5 g/kg, and 1.6 g/kg, respectively, that confirmed their use in various bioapplications (Horcajada et al. 2010). It was deduced that this system could be used for drug delivery system.

Another suitable preference to use endogenous organic spacers as linkers is that they have essential components of body configuration. These could be reutilized in lowering the destructive effects significantly. Till now, many such MOFs are integrated, such as amino acid-based and nucleo-based MOFs (Mizutani et al. 1998; An et al. 2009).

8.4 FUNCTIONALIZATION OF MOFS

8.4.1 Surface Modification

A large number of nano-based emulsions, lipid/polymer-based fibrils based on lipid, as well as polymer and metal particles (Liong et al. 2008, O'Hanlon et al. 2012, Zhu et al. 2012, Leung et al. 2013) are being used in biomedical field. To deliver bioactive molecules at the specific site with least side effects, surfaces of MOFs and nMOFs are modified to use them for biomedical applications. The surfaces of MOFs are functionalized to control stability and/or the targeting abilities (Simon-Yarza et al. 2017). The different methods for surface modifications are: (i) covalent modification, (ii) coordination modulation, (iii) noncovalent modification, and (iv) external surface modulations. Here below, these methods have been discussed in detail with their advantages and limitations.

8.4.2 Covalent Modification

Covalent modification takes place after synthesis of MOFs and through interaction of drugs and biomolecules with organic ligands and frameworks in the MOFs via covalent interactions. At first, this was investigated by Kiang and his coworkers in 1999 (Kiang et al. 1999). Afterwards, Canivet et al. conjugated peptides with free amino groups on the organic precursor of MOF ((In) MIL-68-NH$_2$), while having stable framework during modification (Canivet et al. 2011). Till now, this method is considered favorable for biomacromolecules (proteins and nucleic acid). Morris et al. conjugated framework of MOF with DNA by click chemistry (Morris et al. 2014). Kahn et al. proclaimed a unique DNA-MOF system based on stimuli via covalent inter reaction between EDC/NHS amide and amine-incorporated cytosine-rich DNA and carboxylic acid-functionalized MOFs (Morris et al. 2014). Moreover, linkers of MOFs are conjugated with nucleophiles of proteins (amines, hydroxyls, and thiols) through nucleophilic substitution. Lin et al. used nucleophilic substitution to functionalized carboxylate groups in MOFs by its interaction with amine groups in trypsin (Liu et al. 2013). Jung et al. conjugated green fluorescent protein (EGFP) and MOFs by coupling with either DCC or EDC (Jung et al. 2011). However, electrophiles present in MOFs may induce covalent conjugation of proteins.

8.4.3 COORDINATION MODULATION

For biomedical applications, MOFs are modified during synthesis or post synthesis, based upon coordination modulation through a modulator like carboxylate-containing molecule. Abánades Lázaro et al. modified UiO-66 during its synthesis by using biomolecules biotin and folic acid as chemical modulators (Abánades Lázaro, Haddad et al. 2018). The same procedure was used by Dong et al. to tailor surface of TCCP@UiO-66 with the porphyrin; the resultant product had excellent porosity (Kan et al. 2018). The association of nucleophilic functional groups (thiols, phosphates, imidazoles, and carboxylates) to metal cluster nodes on MOFs' surface can be performed post-synthetically (Zhu, Gu et al. 2014, Park et al. 2016). This phenomenon was exchange of surface ligand. For example, Wang et al. incorporated carboxyl-functionalized diiodo-substituted BODIPY on to the surface of UiO-66 after its synthesis (Wang, Wang et al. 2016). Lin et al. molded the UiO-68-NH$_2$ through bonding of phosphate groups of small interfering RNA (siRNA) to the surface Zr^{4+} of this MOF at its surface (He et al. 2014).

8.4.4 NONCOVALENT MODIFICATION

MOFs can be modified by adsorbing the biomacromolecule on to their surface through supramolecular interactions. Such an example is radiolabeled model of UiO-66 reported by Hong et al. (Chen et al. 2017). Pyrene-derived PEG was used to functionalize the UiO-66 via π-π stacking interactions as well as reaction of cysteine residues of nucleolin-peptide with maleimide residue of pyrene-derived PEG chains. Protein-coated MOF system has been developed to control changes in glucose levels (Ma et al. 2013). Biomacromolecules as methylene green (MG) and glucose dehydrogenase (GDH) were imbibed on to ZIFs to functionalize the surface of the bulky molecules (MOF).

8.4.5 EXTERNAL MODIFICATION

To modify the external surface of MOF nanoparticles, silica coating is used to bind free drugs or biomolecules to the exterior of MOF nanoparticles. Yang et al. and Taylor et al. used silica to insert silyl-cyclic oligopeptide c(RGDfK) on to Fc-Gd and MIL-101 MOF nanoparticles to target tumor respectively (Ma et al. 2013). In the same manner, Zhang et al. attached SGDEVDK oligopeptide on the exterior of nMOFs *through* coating of silica (Zhang et al. 2015). Huang et al. changed the surface of Fe(bbi) MOFs through coupling of folic acid and silica dusting (Gao et al. 2013). This method has limitation for biomedical applications due to toxicity of silica. It is required to substitute the characteristics of MOFs and MOF nanoparticles during or after their synthesis for biological applications.

8.5 DRUG DELIVERY THROUGH STIMULI BASED MOFS

New substances have gained great attention particularly in the area of bioapplications (Lu et al. 2016). Along with this, stimuli-responsive MOFs are among the considerable agents for controlled drug discharge. Basically, stimuli-responsive MOFs may be separated in two classes: multiple stimuli and single stimuli-responsive MOFs respectively. Here, we investigated miscellaneous stimuli-responsive MOFs involving single and multi-responsive MOFs which are capable of achieving controlled discharge of encapsulated drugs by effect of different stimuli like pH, magnetic field, ions, temperature, etc.

8.5.1 SINGLE-STIMULUS-BASED MOFs FOR DRUG DELIVERY

8.5.1.1 pH-Controlled MOFs

Among all the permeable MOFs activated by outer stimuli, pH-controlled MOFs have proved the most vast study, particularly for cancer treatment, due to the acidic tumor surrounding (Bai et al. 2016). As yet, various studies ensure pH-stimulated characteristics of MOFs for therapy for malignant tumor and drug discharge and. For instance, Hunag and group synthesized a Fe(bbi) MOF derived from 1,1'-'-(1,4-butanediyl) bis (imidazole) as well as iron ions through precipitation method (Gao et al. 2013). In situ coating of DOX was done by merely augmenting DOX into metal solution (Figure 8.1a) and, as estimated, the encapsulated DOX can be discharged on decreasing the pH (Figure 8.1b). Hence, MOF surface was covered with silica to avoid quick degradation of the polymeric material to govern efficient DOX discharge. More significantly, folic acid was joined to the freshly prepared nanocomposites for targeted drug discharge. Resultantly, the MOFs based drug deliverer exhibited pH-controlled gradual discharge behavior and better anticancer activity.

Recently, Qian and team explained an interesting positively charged MOF ZJU-101 for the delivery of sodium salt of diclofenac, a negatively charged dose (Yang et al. 2016). The positively charged host substance was made up of Zr and 2,2'-bipyridine-5,5'-dicarboxylate (BPYDC) having large upholding capability of 0.546g g^{-1} of sodium salt of diclofenac. This salt was discharged rapidly in affected tissues than in healthy tissues due to exchange of ions at low pH that reduce the

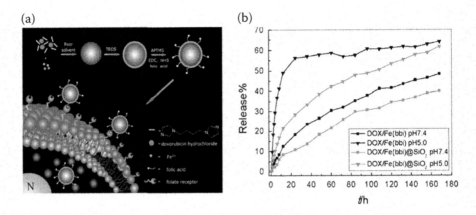

FIGURE 8.1 (a) Diagrammatic working of DOX/Fe(bbi)@SiO2–FA. (b) In vitro release of DOX•HCl from DOX/Fe(bbi) and DOX/Fe(bbi)@SiO2 on pH change. Based on (Gao, Zheng et al. 2013).

Coulombic interactions amid negatively charged drug and positively charged ZJU-101. Hence, pH-responsive diclofenac sodium@ZJU-101 was confirmed an active anticancer drug carrier.

More interestingly, the team proposed PCN-221 MOFs as an oral drug deliverer (Lin, Hu et al. 2016). PCN-221 was synthesized by combining 5,10,15,20-tetrakis (4-carboxyphenyl) porphyrin (TCPP) as well as $ZrCl_4$ followed by entrapping of an anticancer drug methotrexate (MTX) via saturating the MOF in MTX. As a result, the MTX-PCN-221 having 0.40g g^{-1} holding capability was soaked in PBS solution at physiological temperature, pH 7.4 and 2.0. Test data showed that approximately 40% of MTX was discharged at pH 2.0 (stomach pH) following 3 days. On the other hand, 100% of MTX was discharged at pH 7.4 (intestinal pH). Hence, it was proved that pH responsive PCN-221 could be used as an oral drug deliverer.

8.5.1.2 Magnetic MOFs

Magnetic response-based platforms offer a variety of options in drug delivery owing to their promising applications in separation of molecules magnetically, MRI, magnetic directing, as well as magnetic hyperthermia. Among all these properties, magnetically controlled drug delivery has been found to be a unique platform to develop drug-enclosed probes in cancer to enhance treatment efficiency; subsequently they were introduced by Watson and team (Freeman et al. 1960). Candidate nanocarriers derived from MOFs for such purpose are ordinarily core-shell NPs, for instance Fe_3O_4 was sometimes utilized as a magnetic center with a MOF shell. Zhang and group explained a targeted drug carrier, magnetic MOF, $Fe_3O_4/Cu_3(BTC)_2$ (HKUST-1, BTC=benzene-1,3,5-tricarboxylate), and nanocomposites (Ke et al. 2011). Nimesulide (NIM), an NSAID drug, was encapsulated on to magnetic nanocarriers by immersing $Fe_3O_4/Cu_3(BTC)_2$ in the NIM trichloromethane solution. Such assessments indicated that the compound had appreciable magnetic property for targeted drug release as well as separation.

Guan and team put forward a single step in situ pyrolysis technique to form γ-Fe_2O_3@MOFs. In this study, γ-Fe_2O_3@MIL-53(Al) showed its ability for drug discharge (Wu et al. 2014). As supposed, the magnetic nanocomposites demonstrated regulated discharge in saline environment at 37 °C, i.e., entrapped IBU was entirely discharged after 7 days in three phases: first, around one-third of the drug was discharged quickly in the first 3 hours; then half drug was discharged in the time span of 2 days; and last, the remaining drug was discharged in next 5 days. This data approved that magnetic γ-Fe_2O_3@MIL-53(Al) was easy for delivery of different drugs.

8.5.1.4 Ion-Receptive MOFs

Ion-based MOFs carry drugs through strong association of drugs with frameworks, controlling their dissemination and delivery. When ionic drugs interact electrostatically with ionic frameworks, ionic drug is released via ion exchange (a chemical stimuli-responsive process). For example, Rosi et al. synthesized an anionic $Zn_8(ad)_4(BPDC)_6O\cdot 2Me_2NH_2\cdot 8DMF\cdot 11H_2O$ (bioMOF-1) by adding biphenyl dicarboxylate to the solution of zinc acetate dehydrate and adenine (An et al. 2009). It was employed to for discharge of procainamide HCl, a cationic antiarrythmic drug. This has a drug loading capacity of up to 0.22 g/g and it was observed that drug procainamide HCl was released due to presence of the cations in the biological fluids. Controlled experiment in PBS buffer (pH = 7.4) and nanopure water confirmed release of procainamide HCl stimulated by buffer cations.

Yang et al. formulated a drug carrier from neutral MOF-74-Fe(II) via oxidation (Hu et. 2013). It showed 15.9 wt% encapsulation of ibuprofen-negative ions (Figure 8.2a). During synthesis of this MOF, hydroxide might be produced due to two particular Ibu-anions entrapped within MOF channels and representing two separate drug release manners. When drug-encapsulated MOF was immersed in PBS solution, coordinated free anions and sodium ibuprofen were discharged due to either ion exchange or diffusion method. The release of dose was activated by phosphate anions via competitive adsorption (Figure 8.2b).

8.5.1.5 Temperature-Sensitive MOFs

These respond to physiological temperature (37 °C). Poly(N-isopropyl acrylamide) (PNIPAM) is mostly promising for development of thermosensitive drug nanocarriers even at lower critical solution temperature (Gebeyehu et al. 2018). At low temperature or below its cloud point (Tc, at 32 °C), it gets dissolved in water due to hydrophilic nature. Thus, it also develops aggregates (Figure 8.3a).

Therefore, Sada et al. investigated loading capacity of drugs, caffeine, procainamide, and resorufin of a switchable UiO-66- PNIPAM nanocarrier through soaking in guest solution (Nagata et al. 2015). The release of these drugs was assessed at 25 °C or 40 °C (Figure 8.3b). Maximum release was observed at standard temperature and blocked at 40 °C; it exhibited controlled drug release via change in temperature. Qian et al. loaded anticancer drug MTX on to zinc-based MOFs: $Zn_{16}(ad)_8(TP)_8O_2(H_2O)_8\cdot 4HTP\cdot 36DMF\cdot 16H_2O$ (ad=adenine, H_2TP= [1,1′:4′,1″-terphenyl]-4,4″- dicarboxylic acid for ZJU-64, and 2′,5′-dimethyl-[1,1′:4′,1″-terphenyl]-4,4″-dicarboxylic acid for ZJU-64-CH_3) (Nagata et al. 2015)

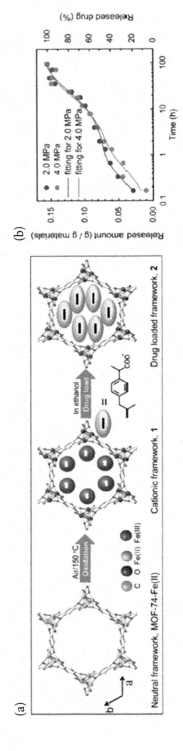

FIGURE 8.2 (a) Encapsulation of ibuprofen within MOF-74-Fe(II) via oxidation. (b) Ibuprofen release profiles in phosphate saline buffer at 2.0 and 4.0 MPa, respectively. Based on (Hu et. 2013).

Metal Organic Frameworks for Drug Delivery

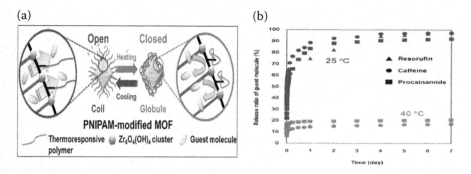

FIGURE 8.3 (a) Schematic loading of PNIPAM on to UiO-66-PNIPAM. (b) Release profile drug from drug-encapsulated UiO-66-PNIPAM at 25 °C and 40 °C in aqueous environment for a week. Based on (Nagata et al. 2015).

via impregnation method (Lin, Hu et al. 2016). Loading capacities of ZJU-64 and ZJU-64-CH$_3$ were evaluated up to 13.45 and 10.63 wt%, respectively, while same amount was released at physiological temperature of 37 °C for 72 hours. Thus, it was deduced that these could be used as temperature-responsive drug carriers.

8.5.2 Multiple Stimuli-Responsive MOFs

MOFs as drug carriers could be used in both ways – MOF response towards particular stimulus for specific drug delivery in complex human body and multiple stimuli-responsive MOFs for an extended period of time in specified conditions for achieving efficiency in chemotherapy (Lin, Hu et al. 2016). First discovered supramolecular host, pillararenes, was comprehensively studied for so many years due to its simple and advantageous functioning (Strutt et al. 2014). Supramolecular chemistry and material sciences developed so many pillararenes-based host-guest systems, such as nanovalves, which recently are being used for drug delivery (Li et al. 2013). Different research groups have been working to use MOF for bioapplications by designing nanocarriers (Li et al. 2013). Guest molecules are stored by host molecules as they also participating as dual stimuli-responsive gatekeepers after modification such as UMCM-1-NH-Py and CP5. A human body with acidic tumor cells, when treated with CP5, which act as neutral, results in enervation of noncovalent binding interaction of CP5 and stalk. Temperature also plays major role in weakening the noncovalent interaction of CP5 with stalk. This may disengage the MOF that leads to release of cargo. Methyl viologen salts added as combative binding agent as it is were found to be compatible with CP5 as compared to Py. A nanocarrier is linked and coordinated with pH as well as competitive binding agent within one drug that cause minor toxicity with biocompatibility (Figure 8.4).

UiO-66-NH$_2$ 5-Fu is another example of CP5-capped nanocarrier that is further reconstructed with quaternary ammonium salt which is attached with CP5 ring to form pseudorotaxanes. A combative binding agent needed to activate the discharge of 5-Fu, Zn^{2+}, was found to be compatible to release 5-Fu and CP5; thus, showing

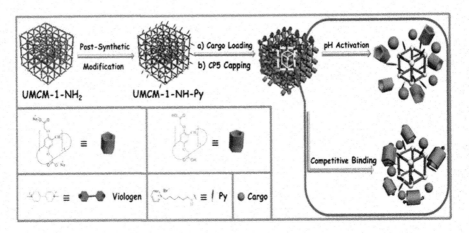

FIGURE 8.4 Symbolic representation of binate-stimuli receptive DDS due to UMCM-1-NH$_2$ NMOF locked via pillararenes. Based on (Tan et al. 2015).

natural linking for Zn2+. A therapy for central nervous system disease can be carried out by dual stimuli-responsive nanocarrier CP5-based MOF. CP5-gated Zr-MOFs lately developed with triply responsive diagnostic radioactive drug (Tan et al. 2016). Quaternary ammonium salt secured with UiO-66-NH$_2$ after synthesis and negatively charged CP5 was enclosed on the stalk which is present on the surface of MOF that made a complicated host-guest combination (Tan et al. 2015). A supramolecular interaction between MOF and nanovalves of CP5 reduced by pH, increased temperature, and binding agents that manage selective release of drug (Figure 8.5). This is how a triple modal relating to the healing of disease was headed under one system.

Tumor cells for bone cancer therapy treatment and bone regeneration in lower pH and high Ca^{2+} concentration in hard tissues create a response for the liberation of therapeutics. High temperature encourages the release of therapeutic agents to destroy cancer cells by weakening the host-guest interaction. Nanocarrier with triply responsive drugs acquires biocompatibility, negligible toxicity and advantageous properties. Supramolecular macrocycles when tied with MOF showed successive results for specified regulation of drug, such as β-cyclodextrin (Wang et al. 2015, Gao, Wang et al. 2018).

Multifunctional MOF working as a drug carrier was modified by a research group into MIL-101 post synthesis. Host-guest interaction tied with nanocarrier to confine the movement, encapsulated with β-CD derivative and peptide polymer on the surface of MIL-101-N$_3$, DOX, and linking of β-CD on the MOF possessed redox responsiveness in the existence of benzoic imine bond in disulfide bond and K (ad) RGDS-PEG1900 (Gao, Wang et al. 2018). DDS has shown efficient results in increased absorption of tumor cell followed by release of drug in cancer therapy. Cell toxicity of DOX reduced by surface modification as observed in vitro experiments and DDS found highly effective in vivo experiments. By supramolecular

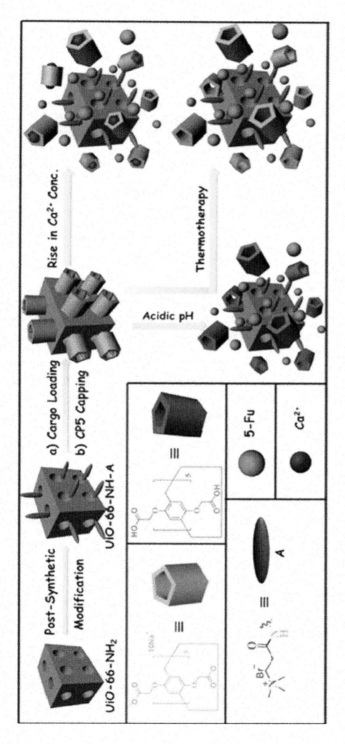

FIGURE 8.5 Schematic illustration of DDS due to trio stimuli developed on CP5-capped UiO-66. Based on (Tan et al. 2016).

complexation of UiO-68-azobenzene encapsulated with β-CD released as a result of light effect and binding agent.

Photo-controllable isomerization of azobenzene enhanced un-tying of the gating β-CD to synchronize the cargo release. Amantadine, a combative binding agent, shows more affinity with β-CD and activates the detachment of the gatekeeper from stalk.

Timely release of encapsulated β-CD–capped UiO-68-azo in dual stimulus response came from nanocarrier with rhodamine B. One pot enclosed DOX with biodegradable MOFs, Fe-BTC, and Zn-BTC, analyzed at room temperature. MOFs can be interfered to activate the release of drug by biostable liposomes at low pH as the coordination between metal center and carboxylic acid group by protonation in acidic medium get collapsed. Moreover, release of drug is also encouraged by electrostatic interaction of liposomes with DOX. At start MOFs nanocarriers were made to release drug in low acidic conditions but it provides remarkable possibility in stimuli responsive drugs.

8.6 MULTIFUNCTIONAL MOFS

Despite the rapid and wide development of various MOF-based treatment facilities, many challenges are still unaddressed due to the limited nature (e.g., traditional chemotherapy) and resisted therapeutic efficacy (multidrug resistance) of current cancer treatments. Much effort is devoted in the development of multi-tasking theranostic nanomedical platforms, e.g., versatile diagnostic or treatment systems based on MOFs to rule out the already discussed issues in gaining superior antitumor efficiency. Among these platforms, core-shell MOFs nanocomposites have been found to demonstrate excellent potential as nanomedical systems. This chapter discusses core-shell structures of MOFs nanocarriers, involving magnetic core-shell MOFs, polymer-coated MOFs, as well as nanocomposites based on other core-shell MOFs Moreover, some multitasking therapeutic nanosystems fabricated from MOFs will be explained as well.

8.6.1 POLYMER MANTLE ON MOFS

As Lin's research group revealed tge first ever demonstration of surface variation of a MOF structure having a silica shell to control the discharge of metallic ions and photoluminescence sensing of dipicolinic acid (Rieter et al. 2007). Various multi-tasking core-shell MOF nanocomposites are constructed for uses in biomedical sciences in order to provide peculiar advantages constituting increased stability, biocompatibility, tunable functionality, as well as water dispersibility. For example, Lin's research group developed Pt-constituting nanocoordination polymers (NCPs) fabricated with Tb^{3+} cations and c,c,t-(diamminedichlorodisuccinatio), and covered their surface with a silica to boost their biocompatibility as well as functionality (Figure 8.6) (Rieter et al. 2008). This controlled release from these NCPs proposed that these nanocomposites can effectively govern the discharge of Pt species by changing density of silica shell. Silyl-derived c(RGDfK) had been pasted on to exterior of already formed nanocomposites to increase intake of cells. In vitro

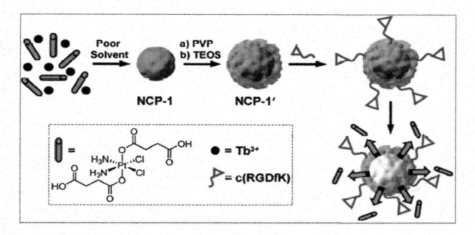

FIGURE 8.6 Illustrative assembly of silica-coated MOF to deliver drug into cancerous cells. Based on (Rieter et al. 2008).

synthesis demonstrated favorable anticancer efficiencies as well as better biocompatibility of Pt (IV) NCPs.

Although all the silica-coated MOFs showed good drug-loading ability, such nanoplatforms were restricted by a rapid drug discharge. Hence, Lin et al. constructed a novel cisplatin faced MOF synthesized from disuccinatocisplatin (DSCP) as well as $ZrCl_4$ more sustained by lipid (Huxford-Phillips et al. 2013). An anisamide (AA) which showed attraction for σ-receptor cells was activated on nanocomposites surface to increase cellular uptake. It substantiated better anticancer efficiency as well as allowed the development of lipid-covered NCPs for nanomedical applications.

Horcajada et al. modified nanoscale MIL-100(Fe) using heparin to provide MOF having considerable biological characteristics through a one-pot methodology (Giménez-Marqués et al. 2016). In comparison with unaltered MIL-100(Fe), heparin covered nanocomposites demonstrated their colloidal stability as well as saved their efficiency to encapsulate increased caffeine amount involving a better discharge. Moreover, the heparin-covered MOF exhibited better stealth characteristics for the cell uptake and toxicity, functionalization of the alternate approach, macrophage uptake, cytokines production, and cytokine profile along with reactive oxygen species (ROS) generation. Obviously, the surface alteration process made a new insight into the construction of theranostics MOFs.

Moreover, Tendeloo et al. constructed a poly(ϵ-caprolactone) layered zirconium MOF for delivery of anticancer drug (Filippousi et al. 2016). Here, poly(ϵ-caprolactone)-tocopheryl poly(ethylene glycol)-succinate (PCL-TPGS) copolymer was constructed on to the surface of UiO-67 and UiO-66 MOFs because of its nontoxic nature and biocompatibility. Further, two standard drugs, hydrophobic taxol and hydrophilic cisplatin, were utilized to determine the drug-encapsulation capability and drug-discharge characteristic of the drug-containing nanocarriers. In vitro drug discharge determinations showed a gradual delivery of drug and diminished burst effects. Furthermore, the fabricated polymer-layered MOFs exhibited decreased toxicity as a consequence of MTT analysis utilizing cell lines (HSC-3 and U-87 MG). All the data proposed that such polymer-coated MOFs fabrications were considerable for more biomedical applications.

8.6.2 Magnetic Core-Shell MOFs

Practice of iron compounds as a backbone for construction of MOF-based theranostic devices to diagnose as well as treat cancer have gained appreciable considerations due to their high magnetic properties for instance, magnetic hyperthermia, magnetic separation and MRI. For example, Fu et al. investigated a theranostic effect of Fe_3O_4@UiO-66 core-shell nanocomposite to be utilized in vitro and in vivo MRI and distribution of drug simultaneously (Filippousi et al. 2016). The external surface of UiO-66 was utilized to entrap DOX and powerful superparamagnetic core Fe_3O_4 can be rapidly amassed in an outer magnetic field as well as ascertained its capability for MRI.

Sahu et al. also constructed multitasking magnetic core-shell MOFs as theranostic-based nanoplatforms. It was isoreticular MOF (IRMOF-3) covered Fe_3O_4 NPs for drug (paclitaxel) discharge, continued by alteration with folic acid and rhodamine B isothiocyanate (Chowdhuri, Bhattacharya et al. 2016). This work showed that targeted drug discharge, cell imaging, and chemotherapy along with MRI were considered via multitasking porous nanoplatform. Consequently, an analogous IRMOF-3 covered Fe_3O_4 nanocomposite was constructed as a targeted anticancer nanoplatform (Chowdhuri, Singh et al. 2016). Particularly, a biodegradable O-carboxymethyl chitosan (OCMC) was attached to Fe_3O_4 core, rather than oxide itself, to enhance loading ability and pH sensitivity. Significantly, carbon dots (CDs) had been incorporated in magnetic nanocomposites to understand photosensitive imaging beforehand. The mixed drug delivered showed high dose of 1.63 g DOX g^{-1} magnetic MOF having pH-controlled release of drug, cancer cell targeting, MRI, magnetic-guided drug delivery, and light imaging. Efficient theranostic efficacy, facile functionality, as well as biocompatibility render the magnetic MOF fabrication a considerable candidate in biomedicine.

Wang et al. synthesized multitasking Au nanoclusters/Fe_3O_4@poly(acrylic acid)/zeolitic imidazolate framework-8 (ZIF-8) NPs (Fe_3O_4@PAA/AuNCs/ZIF-8 NPs) for diagnosis of cancer along with visualization therapy (Fig. 11A) (Filippousi et al. 2016). The resulted nanomedical platform exhibited many healing characteristics like high DOX loading ability, dual pH responsive drug discharge, facile magnetic separation, tri-model imaging, as well as high ability of in vivo inhibition of tumor (Fig. 9B). Tri-model imaging of magnetic resonance, optical imaging, and X-ray CT scan were used to join the properties and reduce the limitations of various imaging techniques. Hence, these multitasking core-shell nanocarriers exhibited high potential for effective theranostic applications.

A different thought-provoking research on multitasking magnetic core-shell MOF was performed by Chen et al. (Wang, Zhou et al. 2016). Multitasking iron nanocomposites [Fe_3O_4@C@MIL-100(Fe) (FCM)] had been developed for dihydroartemisinin (DHA) delivery and Fe (III) as cancer theranostic system at the same time. Fe (III) in acidic cancerous tissues induced production of ROS that trigger death of tumor cells, through reduction. Moreover, magnetic backbone of Fe_3O_4@C, having characteristics of MRI entrapped CDs for fluorescence-based photosensitive imaging (FOI). In vivo and in vitro trials showed double-modal image-guided, pH-controlled chemotherapy to be better in diagnostics as well as efficient treatment of cancer.

8.6.3 Core-Shell MOF Nanocomposites

A great deal of struggle has been put in designing multitasking core-shell MOFs as effective theranostic nanosystems. In addition to the already discussed core-shell MOFs, theranostic platforms and innumerable unique core-shell MOF nanocomposites exhibited appreciable prospective for nanoscale uses (Chowdhuri, Laha et al. 2016). For instance, Wang et al. reported nanocube $Mn_3[Co(CN)_6]_2$@SiO_2@Ag for cancer therapy (Wang et al. 2015). Internal core of the Mn_3[Co$(CN)_6]_2$ MOF structure worked as a T_1-T_2 double modal MRI imaging tool and coupled photon fluorescence imaging tool; external shell of Ag NPs empower the platform with better TPF imaging and photothermal ability simultaneously. DOX had been loaded in the multitasking nanoplatform with an increased loading capability of 600 mg g^{-1}, and drug discharge could be initiated by light because of surface plasmon resonance (SPR) of unreactive Ag NPs. Conditioned analysis showed that the cooperative therapeutic effects of photothermal therapy and chemotherapy were higher than that of chemotherapy or photothermal therapy. All this data proposed $Mn_3[Co(CN)_6]_2$@SiO_2@Ag core-shell nanocomposites as considerable agents for an imaging-directed cancer therapeutic system. Significantly, this multitasking MOF based nanoplatform exhibited no exact adverse effects to healthy organs and could be removed from body via excretory system of modal rat.

Moreover, advances in core-shell MOFs nanocomposites in lieu of easy, precise and effective identification have also gained enhanced consideration. For example, Yang et al. designed a multitasking MOF for double-modal imaging of cancer cells in vivo as well as in vitro. (Yang et al. 2014) Internal core of the MOF was synthesized from 1,1'-dicarboxyl ferrocene (Fc) and magnetic Gd (III) ions, and external shell of silica remained for auxiliary conjugation of fluorescent dye RBITC as well as targeting ligand RGD to give their powerful luminescence and targeting abilities correspondingly. The already prepared multitasking Fc-Gd@SiO_2(RBITC)-RGD nanocomposites having features like better water dispersion, excellent stability, reduced cytotoxicity, and enabled double-modal T_1- and T_2-weighted magnetic resonance imaging in vivo, influencing a unique system for imaging and targeting of detection schemes.

8.6.4 Flexible MOF as Theranostic Nanoplatform

Latest advances in MOF synthesis methodologies have made the approach towards multi-targeted nanomedical applications in diagnosis as well as treatment because of the surface integration of multitasking constituents through inclusion and after synthesis modifications of MOFs combined with precise synthesis of MOFs.

Moreover, drug used as building units of MOFs are a direct method to initiate the progress of imminent theranostics or multimodal NPs. For instance, Dastidar et al. synthesized different Mn^{II} MOFs by combining nonsteroidal anti-inflammatory drugs (NSAIDs) as well as ligand 1,2-bis(4-pyridyl)ethylene (L) as scaffold elements for drug delivery as well as cell imaging (Bera et al. 2020). Both these MOFs exhibited ability for drug delivery and cell imaging; drug moieties being constituent of the assembly. In addition, better biodegradability as well as biocompatibility and anti-inflammatory responses of MOFs render these constructions in parallel with great possibilities for bioapplications.

8.7 DRUG DELIVERY TO ORGANS

Successful incorporation of cells as well as release of intracellular drug is key factors for effective delivery of drugs (Canton and Battaglia 2012). Nanoparticles are either assimilated via active transport system like endocytosis, or passive transport when there is insufficient size (Treuel et al. 2013), allowing release of drug in cytosol directly. MOFs and nano-MOFs (nMOFs) (Furukawa et al. 2013), having large pore size due to multidentate organic linkers, provide opportunity of loading large amount of drugs along with adjustable structural characteristics (He et al. 2014, Zhu, Gu et al. 2014). These have potential to carry drug to the specific organ or site (Chen et al. 2017).

Lazaro et al. reported Zr-MOFs as drug carriers to cancerous breast cells (Lázaro et al. 2018). These DCA-loaded terephthalate Zr-MOFs were modified through adsorption of drugs that could be used against cancer, such as 5-fluorouracil (5-FU) on to their surface. These had loading capacity of 5-fluorouracil (5-FU) up to 4.3% w/w. Graph between cell viability and MOF concentration (mg/mL) clearly explored the high level of cytotoxicity of 5- FU@DCA@MOFs against MCF-7 cells, indicating effective intracellular delivery of 5-FU.

Slight increase in toxicity of 5-FU@DCA@Zr-L2 was observed, while other MOFs had significant cell viability up to 33±8 on incubation with 0.5 mgmL^{-1} of MOFs. On the other hand, higher concentration of DCA-loaded terephthalate Zr-MOFs exhibited prominent cytotoxicity, killing all cells at all concentrations, putting forward the clue that it is phenomena of surface chemistry affecting uptake of cells as well as significant cytotoxic behavior. From this experiment, it has been deduced that post-synthetical adsorption of DCA at defects sites has control over porosity up to 20 nm nanoparticles of hierarchical MOFs, and loading of anticancer drug as DCA up to 15–25% w/w. MCF-7 cancerous breast cells were killed by small size of 20 nm DCA-loaded particles. Moreover, it was observed that DCA@Zr-L5 and DCA@Zr-L6 released DCA through passive transport system effectively into cytosol, even though these had larger size than others (Orellana-Tavra, Haddad et al. 2017). Simultaneous delivery of 5-FU from 5-FU@DCA@MOFs augments more cytotoxicity than DCA@MOFs and the free drugs.

Wang et al. explored the in vitro cytotoxic effect of small-sized (121±17 nm) MIL- 88B-on-UIO-66 for MCF-7 breast cancer cells via the MTT assay (Wang et al. 2020). Drugs, 5-fluorouracil (5-FU), and Alendronate were loaded on to this MOF and as a result, five different drug delivery systems (16.6%5-Fu@MOF,

Metal Organic Frameworks for Drug Delivery

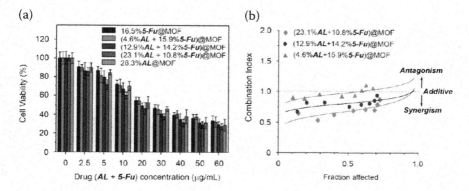

FIGURE 8.7 (a) MCF-7 cells viability towards different doses of MIL-88B-on-UIO-66 (b) Effect of combination index-fraction. Based on (Wang et al. 2020).

4.6%AL+15.9%5-Fu@MOF, 12.9%AL+14.2%5-Fu@MOF, 23.2%AL+10.8%5-Fu@MOF, and 28.3%Al@MOF) were obtained. At low concentration, the hybrid drug delivery systems exhibited cooperative effect, while at high concentration, the dual-drug delivery systems had antagonistic effect and single drug delivery system represented cooperative effect. Overall, delivery systems based on hybrid drug have excellent synergistic effect irrespective of dose (Figure 8.7).

However, these nanocarriers and the drugs have negligible cytotoxic behavior even at high dose against MCF-7 cells. It is proved that loading different concentrations of these dual-drug delivery systems have different kinetics of these drug releases in hybrid-phase MOF.

To investigate the cytotoxic behavior against hepatic and breast cancerous cells, El-Bindary et al. (2020) used ZIF-8 nanoparticles through crystal violet dye binding assay. Cell sustainability of MCF-7 and HepG-2 cell lines was reduced up to 47 and 53.3% by using free DOX, while reduced up to 13.5% and 16.8%, respectively by enhancing the dose up to 100 μgmL of DOX in DOX@ZIF-8 NPs. When DOX@ZIF-8 (2.9–62.5 μg/mL) was incubated for 24 hours against MCF-7 and HepG-2 cell lines, these cancerous cells were remained only 23.9 and 30.2%, respectively. On using 500 μg/ml dose of DOX, cell sustainability against of MCF-7 and HepG-2 cell lines was reduced to 3.5% and 5.3%, respectively. At acidic pH of endosomes and lysosomes, DOX@ZIF-8 nanoparticles could disseminate DOX rapidly within the cells (Wang et al. 2019).

It could be proposed that hepatic and breast cancer cells could internalize large number of nanoparticles and result in enhanced effectiveness of anticancer drug DOX.

Abanades et al. investigated the cytotoxic effect of, Zr-fum-FA (PS), Zr-fum-L1-PEG2000 (s), and Zr-fum (s) against HeLa cervical cancer cells through MTS assay [12]. More than 100% proliferation was observed after 72 hours incubation at 1 mgmL^{-1} of the MOFs, due to cellular metabolism of the autogenous fumarate linker. It, with DCA@Zr-fum-FA (CM) and DCA@Zr-fum (s) after 72 hours, showed no effect on proliferation of cells, while DCA@Zr-fum-L1-PEG2000 (s) had slight effect on

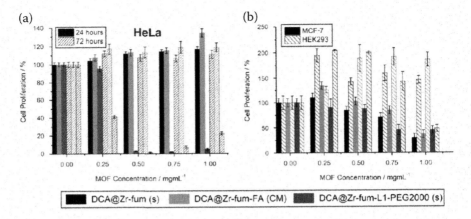

FIGURE 8.8 Cytotoxicity measurement of DCA-loaded Zr-fum composite via MTS assay against (a) HeLa cells incubation for 24 and 72 hours; (b) MCF-7 and HEK293 cells to 72 hours. Reproduced with permission (Abanades, Haddad et al. 2018).

incubation with 0.5 mgmL^{-1} of MOF; perhaps, it is due to partial internalization by cervical cell lines via caveolae-mediated endocytosis [13]. On the other hand, HEK293 human embryonic kidney cell lines and MCF-7 human breast carcinoma were utilized to investigate the cytotoxic behavior of DCA@Zr-fum (s), DCA@Zr-fum- FA (CM) as well as DCA@Zr-fum-L1-PEG2000 (s). At 1 mgmL^{-1} dosage, these have low cytotoxicity to MCF-7 breast cancer cells, whereas negligible cytotoxic effect to HEK293 kidney cells, as depicted in Figure 8.8.

Hence, MOFs as well as MOF nanoparticles are the potential aspirants that can deliver the drugs to their specific site addressing the issue that nanomedicine is presently encountering.

8.8 CHALLENGES AND FUTURE PERSPECTIVES

Even though use of MOFs is being appreciated in drug delivery and biomedicines, there are many complications unresolved.

Keeping in view the need of MOFs for nanomedicines development, efforts are recommended to address, how to effectively load drugs and then achieve desired drugs release, and also degradation of DDSs is required to further explored systematically. Although few articles have been reported regarding mechanisms for drug release and payloads (He et al. 2019, Li et al. 2019), as well as perversion of MOF (Guo et al. 2017), but detailed investigations are needed. It is necessary to understand the fundamental processes that elucidate in proper way (Guo, Su et al. 2017, Wang et al. 2018, Zhang, Chen et al. 2018). Through reviewing the positions of molecules and atoms, new and more effective MOF-based DDSs could be developed. Precise and controlled drug release via MOF at lower temperature than physiological temperatures (37 °C) can be achieved by using MOF constituted of lanthanide cations and the luminescent ligands (Silva et al. 2019). Here, the release of drug is much more problematical. To overcome this issue, novel DDSs (thermally triggered decarboxylation mechanism) could be developed. Besides, deficiencies of different mechanisms should be controlled and addressed.

Furthermore, it is still problem to enhance the efficiency of cancer drugs and reduce their after effects. When MOF-based drug is administered in acidic conditions, ligand gets protonated resulting in weak bond between metal node as well as ligand and release of drug not occurred at its specific site (premature drug release). The profile of drug release from MOFs differs according to different degree of protonation of different MOFs. It is difficult to control this problematic issue due to their diverse in vivo properties (physical and chemical) in distinct organs. To overcome the drawbacks like, short flow, untimely discharge and low biocompatibility, protective shells, such as glucose (Zhang, Shang et al. 2018), PEG (Gupta et al. 2019), chitosan (Karakeçili et al. 2019), pH gelatin polymer (Javanbakht et al. 2019), erythrocyte membrane (Zhang, Wang et al. 2018), phospholipid bilayer (PLB) (Sheno et al. 2019), polymer brush (Sheno et al. 2019), GSH polymer (Liu et al. 2019), hydrogel, pectin, and chitosan and sodium alginate (Vahed et al. 2019), should be used to cover the surface of the carriers. By using these, metabolic process and early drug release of nMOFs in blood could be receded as well as high effect of drug could be obtained. Anyhow, some novel approaches, as "click to release", in situ fabrication of effective drugs from chemically inactive pro-drugs in subcellular organelles (Wang 2019), introducing 'sacrificial' interface (McHugh et al. 2018), supra molecular nanovalves (Wu et al. 2018) should be analyzed. Till now, in vivo long-term toxicity of none of the MOFs has been probed (Chen et al. 2019); which will enable them to use in drug delivery and biomedical applications. To use them in translation into the practical clinical application, it is compulsory to assess chronic and acute toxicities as well as fates of MOFs in living bodies. As unique characteristics of biomaterials play an important role in drug release when and where needed, new functional materials should be proposed for construction and modification of MOFs.

8.9 CONCLUSIONS

In conclusion, since past few decades, various applications of MOFs have been accounted mainly for their high stability and surface area, well-defined structure, toxicological compatibility, tunable pore size, high porosity, biodegradability, and facile functionalization. During these recurring years, MOFs have been extensively explored in biomedical applications and drug delivery. The use of MOFs has been very well investigated for the delivery of a variety of therapeutic agents including proteins, nucleic acid, and anticancer drugs.

In this chapter, we have summed up how MOFs are functionalized through pore encapsulation, surface adsorption, functional molecules as the building block, and covalent binding. In functionalization approaches, hydrogen bonding, π–π interaction and van der Waals interactions play an important role. Here, we have discussed how structure, properties, and principal release mechanisms play an important role in the evolution of MOF-based drug delivery systems as well as contribution of MOFs in drug delivery towards specific organs for treatment and diagnosis of critical diseases, such as cancer and other diseases. Despite this, at present, research products are not accessible for clinical use due to lack of valid knowledge about mechanism and detailed toxicological investigations. On the whole, uses of MOFs in drug delivery systems have promising future.

REFERENCES

Abanades, I., S. Haddad, J. Rodrigo-Munoz, R. Marshall, B. Sastre, V. del Pozo, D. Fairen-Jimenez, and R. Forgan. "Surface-functionalisation of Zr-Fumarate MOF for selective cytotoxicity and immune system compatibility in nanoscale drug delivery." (2018). https://doi.org/10.5525/gla.researchdata.656.

Abánades Lázaro, I., S. Haddad, J.M. Rodrigo-Muñoz, C. Orellana-Tavra, V. Del Pozo, D. Fairen-Jimenez, and R.S. Forgan. "Mechanistic investigation into the selective anticancer cytotoxicity and immune system response of surface-functionalized, dichloroacetate-loaded, UiO-66 nanoparticles." *ACS Applied Materials & Interfaces* 10, no. 6(2018): 5255–5268.

An, J., S.J. Geib, and N.L. Rosi. "Cation-triggered drug release from a porous zinc– adeninate metal– organic framework." *Journal of the American Chemical Society* 131, no. 24(2009): 8376–8377.

Baeza, A., M. Colilla, and M. Vallet-Regí. "Advances in mesoporous silica nanoparticles for targeted stimuli-responsive drug delivery." *Expert Opinion on Drug Delivery* 12, no. 2(2015): 319–337.

Bai, Y., Y. Dou, L.-H. Xie, W. Rutledge, J.-R. Li, and H.-C. Zhou. "Zr-based metal–organic frameworks: design, synthesis, structure, and applications." *Chemical Society Reviews* 45, no. 8(2016): 2327–2367.

Bera, S., A. Chowdhury, K. Sarkar, and P. Dastidar. "Design and synthesis of znii-coordination polymers anchored with NSAIDs: Metallovesicle formation and multi-drug delivery." *Chemistry–An Asian Journal* 15, no. 4(2020): 503–510.

Canivet, J., S. Aguado, G. Bergeret, and D. Farrusseng. "Amino acid functionalized metal–organic frameworks by a soft coupling–deprotection sequence." *Chemical Communications* 47, no. 42(2011): 11650–11652.

Canton, I., and G. Battaglia. "Endocytosis at the nanoscale." *Chemical Society Reviews* 41, no. 7(2012): 2718–2739.

Chalati, T., P. Horcajada, R. Gref, P. Couvreur, and C. Serre. "Optimisation of the synthesis of MOF nanoparticles made of flexible porous iron fumarate MIL-88A." *Journal of Materials Chemistry* 21, no. 7(2011): 2220–2227.

Chen, D., D. Yang, C.A. Dougherty, W. Lu, H. Wu, X. He, T. Cai, M.E. Van Dort, B.D. Ross, and H. Hong. "In vivo targeting and positron emission tomography imaging of tumor with intrinsically radioactive metal–organic frameworks nanomaterials." *ACS Nano* 11, no. 4(2017): 4315–4327.

Chen, G., X. Leng, J. Luo, L. You, C. Qu, X. Dong, H. Huang, X. Yin, and J. Ni. "In Vitro Toxicity Study of a Porous Iron (III) Metal–Organic Framework." *Molecules* 24, no. 7(2019): 1211.

Cho, W., H.J. Lee, and M. Oh. "Growth-controlled formation of porous coordination polymer particles." *Journal of the American Chemical Society* 130, no. 50(2008): 16943–16946.

Chowdhuri, A.R., D. Bhattacharya, and S.K. Sahu. "Magnetic nanoscale metal organic frameworks for potential targeted anticancer drug delivery, imaging and as an MRI contrast agent." *Dalton Transactions* 45, no. 7(2016): 2963–2973.

Chowdhuri, A.R., D. Laha, S. Pal, P. Karmakar, and S.K. Sahu. "One-pot synthesis of folic acid encapsulated upconversion nanoscale metal organic frameworks for targeting, imaging and pH responsive drug release." *Dalton Transactions* 45, no. 45(2016): 18120–18132.

Chowdhuri, A.R., T. Singh, S.K. Ghosh, and S.K. Sahu. "Carbon dots embedded magnetic nanoparticles@ chitosan@ metal organic framework as a nanoprobe for pH sensitive targeted anticancer drug delivery." *ACS Applied Materials & Interfaces* 8, no. 26(2016): 16573–16583.

Cravillon, J., S. Münzer, S.-J. Lohmeier, A. Feldhoff, K. Huber, and M. Wiebcke. "Rapid room-temperature synthesis and characterization of nanocrystals of a prototypical zeolitic imidazolate framework." *Chemistry of Materials* 21, no. 8(2009): 1410–1412.

Della Rocca, J., D. Liu, and W. Lin. "Nanoscale metal–organic frameworks for biomedical imaging and drug delivery." *Accounts of Chemical Research* 44, no. 10(2011): 957–968.

Doane, T.L., and C. Burda. "The unique role of nanoparticles in nanomedicine: imaging, drug delivery and therapy." *Chemical Society Reviews* 41, no. 7(2012): 2885–2911.

Duan, D., H. Liu, M. Xu, M. Chen, Y. Han, Y. Shi, and Z. Liu. "Size-controlled synthesis of drug-loaded zeolitic imidazolate framework in aqueous solution and size effect on their cancer theranostics in vivo." *ACS Applied Materials & Interfaces* 10, no. 49(2018): 42165–42174.

El-Bindary, A.A., E.A. Toson, K.R. Shoueir, H.A. Aljohani, and M.M. Abo-Ser. "Metal–organic frameworks as efficient materials for drug delivery: Synthesis, characterization, antioxidant, anticancer, antibacterial and molecular docking investigation." *Applied Organometallic Chemistry* 34, no. 11(2020): e5905.

Farrusseng, D., S. Aguado, and C. Pinel. "Metal–organic frameworks: opportunities for catalysis." *Angewandte Chemie International Edition* 48, no. 41(2009): 7502–7513.

Filippousi, M., S. Turner, K. Leus, P.I. Siafaka, E.D. Tseligka, M. Vandichel, S.G. Nanaki, I.S. Vizirianakis, D.N. Bikiaris, and P. Van Der Voort. "Biocompatible Zr-based nanoscale MOFs coated with modified poly (ε-caprolactone) as anticancer drug carriers." *International Journal of Pharmaceutics* 509, no. 1-2(2016): 208–218.

Freeman, M., A. Arrott, and J. Watson. "Magnetism in medicine." *Journal of Applied Physics* 31, no. 5(1960): S404–S405.

Furukawa, H., K.E. Cordova, M. O'Keeffe, and O.M. Yaghi. "The chemistry and applications of metal-organic frameworks." *Science* 341, no. 6149(2013). http://doi.org/10.1126/science.1230444.

Gao, P.F., L.L. Zheng, L.J. Liang, X.X. Yang, Y.F. Li, and C.Z. Huang. "A new type of pH-responsive coordination polymer sphere as a vehicle for targeted anticancer drug delivery and sustained release." *Journal of Materials Chemistry B* 1, no. 25(2013): 3202–3208.

Gao, X., R. Cui, G. Ji, and Z. Liu. "Size and surface controllable metal–organic frameworks (MOFs) for fluorescence imaging and cancer therapy." *Nanoscale* 10, no. 13(2018): 6205–6211.

Gao, X., X. Hai, H. Baigude, W. Guan, and Z. Liu. "Fabrication of functional hollow microspheres constructed from MOF shells: Promising drug delivery systems with high loading capacity and targeted transport." *Scientific Reports* 6, (2016): 37705.

Gao, X., Y. Wang, G. Ji, R. Cui, and Z. Liu. "One-pot synthesis of hierarchical-pore metal–organic frameworks for drug delivery and fluorescent imaging." *CrystEngComm* 20, no. 8(2018): 1087–1093.

Gao, X., M. Zhai, W. Guan, J. Liu, Z. Liu, and A. Damirin. "Controllable synthesis of a smart multifunctional nanoscale metal–organic framework for magnetic resonance/optical imaging and targeted drug delivery." *ACS Applied Materials & Interfaces* 9, no. 4(2017): 3455–3462.

Gao, Y., J. Xie, H. Chen, S. Gu, R. Zhao, J. Shao, and L. Jia. "Nanotechnology-based intelligent drug design for cancer metastasis treatment." *Biotechnology Advances* 32, no. 4(2014): 761–777.

Gebeyehu, B.T., S.-Y. Huang, A.-W. Lee, J.-K. Chen, J.-Y. Lai, D.-J. Lee, and C.-C. Cheng. "Dual stimuli-responsive nucleobase-functionalized polymeric systems as efficient tools for manipulating micellar self-assembly behavior." *Macromolecules* 51, no. 3(2018): 1189–1197.

Giménez-Marqués, M., T. Hidalgo, C. Serre, and P. Horcajada. "Nanostructured metal–organic frameworks and their bio-related applications." *Coordination Chemistry Reviews* 307, (2016): 342–360.

Gul, R., N. Ahmed, K.U. Shah, G.M. Khan, and A.U. Rehman. "Functionalised nanostructures for transdermal delivery of drug cargos." *Journal of Drug Targeting* 26, no. 2(2018): 110–122.

Guo, C., F. Su, Y. Song, B. Hu, M. Wang, L. He, D. Peng, and Z. Zhang. "Aptamer-templated silver nanoclusters embedded in zirconium metal–organic framework for bifunctional electrochemical and SPR aptasensors toward carcinoembryonic antigen." *ACS Applied Materials & Interfaces* 9, no. 47(2017): 41188–41199.

Gupta, V., S. Mohiyuddin, A. Sachdev, P. Soni, P. Gopinath, and S. Tyagi. "PEG functionalized zirconium dicarboxylate MOFs for docetaxel drug delivery in vitro." *Journal of Drug Delivery Science and Technology* 52, (2019): 846–855.

He, C., K. Lu, D. Liu, and W. Lin. "Nanoscale metal–organic frameworks for the co-delivery of cisplatin and pooled siRNAs to enhance therapeutic efficacy in drug-resistant ovarian cancer cells." *Journal of the American Chemical Society* 136, no. 14(2014): 5181–5184.

He, Y., W. Zhang, T. Guo, G. Zhang, W. Qin, L. Zhang, C. Wang, W. Zhu, M. Yang, and X. Hu. "Drug nanoclusters formed in confined nano-cages of CD-MOF: dramatic enhancement of solubility and bioavailability of azilsartan." *ACTA Pharmaceutica Sinica B* 9, no. 1(2019): 97–106.

Hermes, S., T. Witte, T. Hikov, D. Zacher, S. Bahnmüller, G. Langstein, K. Huber, and R.A. Fischer. "Trapping metal-organic framework nanocrystals: an in-situ time-resolved light scattering study on the crystal growth of MOF-5 in solution." *Journal of the American Chemical Society* 129, no. 17(2007): 5324–5325.

Horcajada, P., T. Chalati, C. Serre, B. Gillet, C. Sebrie, T. Baati, J.F. Eubank, D. Heurtaux, P. Clayette, and C. Kreuz. "Porous metal–organic-framework nanoscale carriers as a potential platform for drug delivery and imaging." *Nature Materials* 9, no. 2(2010): 172–178.

Horcajada, P., R. Gref, T. Baati, P.K. Allan, G. Maurin, P. Couvreur, G. Ferey, R.E. Morris, and C. Serre. "Metal–organic frameworks in biomedicine." *Chemical Reviews* 112, no. 2(2012): 1232–1268.

Horcajada, P., C. Serre, D. Grosso, C. Boissiere, S. Perruchas, C. Sanchez, and G. Férey. "Colloidal route for preparing optical thin films of nanoporous metal–organic frameworks." *Advanced Materials* 21, no. 19(2009): 1931–1935.

Huxford-Phillips, R.C., S.R. Russell, D. Liu, and W. Lin. "Lipid-coated nanoscale coordination polymers for targeted cisplatin delivery." *RSC Advances* 3, no. 34(2013): 14438–14443.

Javanbakht, S., P. Nezhad-Mokhtari, A. Shaabani, N. Arsalani, and M. Ghorbani. "Incorporating Cu-based metal-organic framework/drug nanohybrids into gelatin microsphere for ibuprofen oral delivery." *Materials Science and Engineering: C* 96, (2019): 302–309.

Jhung, S.H., J.H. Lee, J.W. Yoon, C. Serre, G. Férey, and J.S. Chang. "Microwave synthesis of chromium terephthalate MIL-101 and its benzene sorption ability." *Advanced Materials* 19, no. 1(2007): 121–124.

Jung, S., Y. Kim, S.-J. Kim, T.-H. Kwon, S. Huh, and S. Park. "Bio-functionalization of metal–organic frameworks by covalent protein conjugation." *Chemical Communications* 47, no. 10(2011): 2904–2906.

Kaelin, W.G. "The concept of synthetic lethality in the context of anticancer therapy." *Nature Reviews Cancer* 5, no. 9(2005): 689–698.

Kan, J.-L., Y. Jiang, A. Xue, Y.-H. Yu, Q. Wang, Y. Zhou, and Y.-B. Dong. "Surface decorated porphyrinic nanoscale metal–organic framework for photodynamic therapy." *Inorganic Chemistry* 57, no. 9(2018): 5420–5428.

Karakeçili, A., B. Topuz, S. Korpayev, and M. Erdek. "Metal-organic frameworks for on-demand pH controlled delivery of vancomycin from chitosan scaffolds." *Materials Science and Engineering: C* 105, (2019): 110098.

Ke, F., Y.-P. Yuan, L.-G. Qiu, Y.-H. Shen, A.-J. Xie, J.-F. Zhu, X.-Y. Tian, and L.-D. Zhang. "Facile fabrication of magnetic metal–organic framework nanocomposites for potential targeted drug delivery." *Journal of Materials Chemistry* 21, no. 11(2011): 3843–3848.

Keith, C.T., A.A. Borisy, and B.R. Stockwell. "Multicomponent therapeutics for networked systems." *Nature Reviews Drug Discovery* 4, no. 1(2005): 71–78.

Kiang, Y.-H., G.B. Gardner, S. Lee, Z. Xu, and E.B. Lobkovsky. "Variable pore size, variable chemical functionality, and an example of reactivity within porous phenylacetylene silver salts." *Journal of the American Chemical Society* 121, no. 36(1999): 8204–8215.

Lázaro, I.A., S.A. Lázaro, and R.S. Forgan. "Enhancing anticancer cytotoxicity through bimodal drug delivery from ultrasmall Zr MOF nanoparticles." *Chemical Communications* 54, no. 22(2018): 2792–2795.

Lee, J., O.K. Farha, J. Roberts, K.A. Scheidt, S.T. Nguyen, and J.T. Hupp. "Metal–organic framework materials as catalysts." *Chemical Society Reviews* 38, no. 5(2009): 1450–1459.

Leung, K.C.-F., C.-H. Wong, X.-M. Zhu, S.-F. Lee, K.W. Sham, J.M. Lai, C.-P. Chak, Y.-X.J. Wang, and C.H. Cheng. "Ternary hybrid nanocomposites for gene delivery and magnetic resonance imaging of hepatocellular carcinoma cells." *Quantitative Imaging in Medicine and Surgery* 3, no. 6(2013): 302.

Li, H., D.-X. Chen, Y.-L. Sun, Y.B. Zheng, L.-L. Tan, P.S. Weiss, and Y.-W. Yang. "Viologen-mediated assembly of and sensing with carboxylatopillar [5] arene-modified gold nanoparticles." *Journal of the American Chemical Society* 135, no. 4(2013): 1570–1576.

Li, Z., S. Zhao, H. Wang, Y. Peng, Z. Tan, and B. Tang. "Functional groups influence and mechanism research of UiO-66-type metal-organic frameworks for ketoprofen delivery." *Colloids and Surfaces B: Biointerfaces* 178, (2019): 1–7.

Lin, W., Q. Hu, K. Jiang, Y. Yang, Y. Yang, Y. Cui, and G. Qian. "A porphyrin-based metal–organic framework as a pH-responsive drug carrier."*Journal of Solid State Chemistry* 237, (2016): 307–312.

Lin, W., Q. Hu, J. Yu, K. Jiang, Y. Yang, S. Xiang, Y. Cui, Y. Yang, Z. Wang, and G. Qian. "Low cytotoxic metal-organic frameworks as temperature-responsive drug carriers." *ChemPlusChem* 81, no. 8(2016): 804.

Lin, W., W. Lin, H. An, K. Taylor, and W. Rieter. "Nanoscale metal organic frameworks as potential multimodal contrast agents." *Journal of the American Chemical Society* 128, (2006): 9024–9025.

Liong, M., J. Lu, M. Kovochich, T. Xia, S.G. Ruehm, A.E. Nel, F. Tamanoi, and J.I. Zink. "Multifunctional inorganic nanoparticles for imaging, targeting, and drug delivery." *ACS Nano* 2, no. 5(2008): 889–896.

Liu, W.-L., S.-H. Lo, B. Singco, C.-C. Yang, H.-Y. Huang, and C.-H. Lin. "Novel trypsin–FITC@ MOF bioreactor efficiently catalyzes protein digestion." *Journal of Materials Chemistry B* 1, no. 7(2013): 928–932.

Liu, Y., C.S. Gong, Y. Dai, Z. Yang, G. Yu, Y. Liu, M. Zhang, L. Lin, W. Tang, and Z. Zhou. "In situ polymerization on nanoscale metal-organic frameworks for enhanced physiological stability and stimulus-responsive intracellular drug delivery." *Biomaterials* 218, (2019): 119365.

Lu, Y., A.A. Aimetti, R. Langer, and Z. Gu. "Bioresponsive materials." *Nature Reviews Materials* 2, no. 1(2016): 1–17.

Ma, W., Q. Jiang, P. Yu, L. Yang, and L. Mao. "Zeolitic imidazolate framework-based electrochemical biosensor for in vivo electrochemical measurements." *Analytical Chemistry* 85, no. 15(2013): 7550–7557.

McHugh, L.N., M.J. McPherson, L.J. McCormick, S.A. Morris, P.S. Wheatley, S.J. Teat, D. McKay, D.M. Dawson, C.E. Sansome, and S.E. Ashbrook. "Hydrolytic stability in hemilabile metal–organic frameworks." *Nature Chemistry* 10, no. 11(2018): 1096–1102.

Mizutani, M., N. Maejima, K. Jitsukawa, H. Masuda, and H. Einaga. "An infinite chiral single-helical structure formed in Cu (II)-L-/D-glutamic acid system." *Inorganica Chimica Acta* 283, no. 1(1998): 105–110.

Morris, W., W.E. Briley, E. Auyeung, M.D. Cabezas, and C.A. Mirkin. "Nucleic acid–metal organic framework (MOF) nanoparticle conjugates." *Journal of the American Chemical Society* 136, no. 20(2014): 7261–7264.

Nagata, S., K. Kokado, and K. Sada. "Metal–organic framework tethering PNIPAM for ON–OFF controlled release in solution." *Chemical Communications* 51, no. 41(2015): 8614–8617.

O'Hanlon, C.E., K.G. Amede, R. Meredith, and J.M. Janjic. "NIR-labeled perfluoropolyether nanoemulsions for drug delivery and imaging." *Journal of Fluorine Chemistry* 137, (2012): 27–33.

Orellana-Tavra, C., S. Haddad, R.J. Marshall, I. Abánades Lázaro, G. Boix, I. Imaz, D. Maspoch, R.S. Forgan, and D. Fairen-Jimenez. "Tuning the endocytosis mechanism of Zr-based metal–organic frameworks through linker functionalization." *ACS Applied Materials & Interfaces* 9, no. 41(2017): 35516–35525.

Park, J., Q. Jiang, D. Feng, L. Mao, and H.-C. Zhou. "Size-controlled synthesis of porphyrinic metal–organic framework and functionalization for targeted photodynamic therapy." *Journal of the American Chemical Society* 138, no. 10(2016): 3518–3525.

Qiu, L.-G., Z.-Q. Li, Y. Wu, W. Wang, T. Xu, and X. Jiang. "Facile synthesis of nanocrystals of a microporous metal–organic framework by an ultrasonic method and selective sensing of organoamines." *Chemical Communications* 31, (2008): 3642–3644.

Rieter, W.J., K.M. Pott, K.M. Taylor, and W. Lin. "Nanoscale coordination polymers for platinum-based anticancer drug delivery." *Journal of the American Chemical Society* 130, no. 35(2008): 11584–11585.

Rieter, W.J., K.M. Taylor, H. An, W. Lin, and W. Lin. "Nanoscale metal– organic frameworks as potential multimodal contrast enhancing agents." *Journal of the American Chemical Society* 128, no. 28(2006): 9024–9025.

Rieter, W.J., K.M. Taylor, and W. Lin. "Surface modification and functionalization of nanoscale metal-organic frameworks for controlled release and luminescence sensing." *Journal of the American Chemical Society* 129, no. 32(2007): 9852–9853.

Ryu, J.H., H. Koo, I.-C. Sun, S.H. Yuk, K. Choi, K. Kim, and I.C. Kwon. "Tumor-targeting multi-functional nanoparticles for theragnosis: new paradigm for cancer therapy." *Advanced Drug Delivery Reviews* 64, no. 13(2012): 1447–1458.

Sheno, N.N., S. Farhadi, A. Maleki, and M. Hamidi. "A novel approach for the synthesis of phospholipid bilayer-coated zeolitic imidazolate frameworks: preparation and characterization as a pH-responsive drug delivery system." *New Journal of Chemistry* 43, no. 4(2019): 1956–1963.

Silva, J.Y.R., Y.G. Proenza, L.L. da Luz, S. de Sousa Araújo, M.A. Gomes Filho, S.A. Junior, T.A. Soares, and R.L. Longo. "A thermo-responsive adsorbent-heater-thermometer nanomaterial for controlled drug release:(ZIF-8, EuxTby)@ AuNP core-shell." *Materials Science and Engineering: C* 102, (2019): 578–588.

Simon-Yarza, T., S. Rojas, P. Horcajada, and C. Serre. "4.38 The Situation of Metal-Organic Frameworks in Biomedicine." *Comprehensive Biomaterials II*; Elsevier 4, (2017): 719–749.

Strutt, N.L., H. Zhang, and J.F. Stoddart. "Enantiopure pillar [5] arene active domains within a homochiral metal–organic framework." *Chemical Communications* 50, no. 56(2014): 7455–7458.

Tan, L.-L., H. Li, Y.-C. Qiu, D.-X. Chen, X. Wang, R.-Y. Pan, Y. Wang, S.X.-A. Zhang, B. Wang, and Y.-W. Yang. "Stimuli-responsive metal–organic frameworks gated by pillar [5] arene supramolecular switches." *Chemical Science* 6, no. 3(2015): 1640–1644.

Tan, L.-L., N. Song, S.X.-A. Zhang, H. Li, B. Wang, and Y.-W. Yang. "Ca 2+, pH and thermo triple-responsive mechanized Zr-based MOFs for on-command drug release in bone diseases." *Journal of Materials Chemistry B* 4, no. 1(2016): 135–140.

Tan, L.L., H. Li, Y. Zhou, Y. Zhang, X. Feng, B. Wang, and Y.W. Yang. "Zn2+-triggered drug release from biocompatible zirconium MOFs equipped with supramolecular gates." *Small* 11, no. 31(2015): 3807–3813.

Treuel, L., X. Jiang, and G.U. Nienhaus. "New views on cellular uptake and trafficking of manufactured nanoparticles." *Journal of the Royal Society Interface* 10, no. 82(2013): 20120939.

Vahed, T.A., M.R. Naimi-Jamal, and L. Panahi. "Alginate-coated ZIF-8 metal-organic framework as a green and bioactive platform for controlled drug release." *Journal of Drug Delivery Science and Technology* 49, (2019): 570–576.

Vickers, N.J. "Animal communication: when i'm calling you, will you answer too?" *Current Biology* 27, no. 14(2017): R713–R715.

Wang, D., Z. Guo, J. Zhou, J. Chen, G. Zhao, R. Chen, M. He, Z. Liu, H. Wang, and Q. Chen. "Novel Mn3 [Co (CN) 6] 2@ SiO2@ Ag Core–Shell Nanocube: Enhanced Two-Photon Fluorescence and Magnetic Resonance Dual-Modal Imaging-Guided Photothermal and Chemo-therapy." *Small* 11, no. 44(2015): 5956–5967.

Wang, D., J. Zhou, R. Chen, R. Shi, G. Xia, S. Zhou, Z. Liu, N. Zhang, H. Wang, and Z. Guo. "Magnetically guided delivery of DHA and Fe ions for enhanced cancer therapy based on pH-responsive degradation of DHA-loaded Fe3O4@ C@ MIL-100 (Fe) nanoparticles." *Biomaterials* 107, (2016): 88–101.

Wang, F., Y. Zhang, Z. Liu, Z. Du, L. Zhang, J. Ren, and X. Qu. "A Biocompatible Heterogeneous MOF–Cu Catalyst for In Vivo Drug Synthesis in Targeted Subcellular Organelles." *Angewandte Chemie* 131, no. 21(2019): 7061–7066.

Wang, H., T. Li, J. Li, W. Tong, and C. Gao. "One-pot synthesis of poly (ethylene glycol) modified zeolitic imidazolate framework-8 nanoparticles: Size control, surface modification and drug encapsulation." *Colloids and Surfaces A: Physicochemical and Engineering Aspects* 568, (2019): 224–230.

Wang, W., L. Wang, Z. Li, and Z. Xie. "BODIPY-containing nanoscale metal–organic frameworks for photodynamic therapy." *Chemical Communications* 52, no. 31(2016): 5402–5405.

Wang, X.-G., Z.-Y. Dong, H. Cheng, S.-S. Wan, W.-H. Chen, M.-Z. Zou, J.-W. Huo, H.-X. Deng, and X.-Z. Zhang. "A multifunctional metal–organic framework based tumor targeting drug delivery system for cancer therapy." *Nanoscale* 7, no. 38(2015): 16061–16070.

Wang, X.G., L. Xu, M.J. Li, and X.Z. Zhang. "Construction of Flexible-on-Rigid Hybrid-Phase Metal–Organic Frameworks for Controllable Multi-Drug Delivery." *Angewandte Chemie International Edition* 59, no. 41(2020): 18078–18086.

Wang, Z.-C., Y. Zhang, and Z.-Y. Li. "A low cytotoxic metal–organic framework carrier: pH-responsive 5-fluorouracil delivery and anti-cervical cancer activity evaluation." *Journal of Cluster Science* 29, no. 6(2018): 1285–1290.

Wu, M.-X., H.-J. Yan, J. Gao, Y. Cheng, J. Yang, J.-R. Wu, B.-J. Gong, H.-Y. Zhang, and Y.-W. Yang. "Multifunctional supramolecular materials constructed from polypyrrole@ UiO-66 nanohybrids and pillararene nanovalves for targeted chemophotothermal therapy." *Acs Applied Materials & Interfaces* 10, no. 40(2018): 34655–34663.

Wu, Y. n., M. Zhou, S. Li, Z. Li, J. Li, B. Wu, G. Li, F. Li, and X. Guan. "Magnetic metal–organic frameworks: γ-Fe2O3@ MOFs via confined in situ pyrolysis method for drug delivery." *Small* 10, no. 14(2014): 2927–2936.

Yang, H., C. Qin, C. Yu, Y. Lu, H. Zhang, F. Xue, D. Wu, Z. Zhou, and S. Yang. "RGD-Conjugated Nanoscale Coordination Polymers for Targeted T1-and T2-weighted Magnetic Resonance Imaging of Tumors in Vivo." *Advanced functional materials* 24, no. 12(2014): 1738–1747.

Yang, Y., Q. Hu, Q. Zhang, K. Jiang, W. Lin, Y. Yang, Y. Cui, and G. Qian. "A large capacity cationic metal–organic framework nanocarrier for physiological pH responsive drug delivery." *Molecular Pharmaceutics* 13, no. 8(2016): 2782–2786.

Zhang, H., Y. Shang, Y.-H. Li, S.-K. Sun, and X.-B. Yin. "Smart metal–organic framework-based nanoplatforms for imaging-guided precise chemotherapy." *ACS Applied Materials & Interfaces* 11, no. 2(2018): 1886–1895.

Zhang, L., Y. Chen, R. Shi, T. Kang, G. Pang, B. Wang, Y. Zhao, X. Zeng, C. Zou, and P. Wu. "Synthesis of hollow nanocages MOF-5 as drug delivery vehicle to solve the load-bearing problem of insoluble antitumor drug oleanolic acid (OA)." *Inorganic Chemistry Communications* 96, (2018): 20–23.

Zhang, L., J. Lei, F. Ma, P. Ling, J. Liu, and H. Ju. "A porphyrin photosensitized metal–organic framework for cancer cell apoptosis and caspase responsive theranostics." *Chemical Communications* 51, no. 54(2015): 10831–10834.

Zhang, L., Z. Wang, Y. Zhang, F. Cao, K. Dong, J. Ren, and X. Qu. "Erythrocyte membrane cloaked metal–organic framework nanoparticle as biomimetic nanoreactor for starvation-activated colon cancer therapy." *ACS Nano* 12, no. 10(2018): 10201–10211.

Zhu, G., and H. Ren. *Porous Organic Frameworks: Design, Synthesis and Their Advanced Applications*. Springer, 2014.

Zhu, X.-M., J. Yuan, K.C.-F. Leung, S.-F. Lee, K.W. Sham, C.H. Cheng, D.W. Au, G.-J. Teng, A.T. Ahuja, and Y.-X.J. Wang. "Hollow superparamagnetic iron oxide nanoshells as a hydrophobic anticancer drug carrier: intracellular pH-dependent drug release and enλhanced cytotoxicity." *Nanoscale* 4, no. 18(2012): 5744–5754.

Zhu, X., J. Gu, Y. Wang, B. Li, Y. Li, W. Zhao, and J. Shi. "Inherent anchorages in UiO-66 nanoparticles for efficient capture of alendronate and its mediated release." *Chemical Communications* 50, no. 63(2014): 8779–8782.

9 Graphene and Graphene-Based Nanomaterials
Current Applications and Future Perspectives

Abid Hussain, Yuhua Weng, and Yuanyu Huang
Beijing Institute of Technology

CONTENTS

9.1 Introduction and History of Graphene	209
9.2 Structure and Composition of Graphene Oxide	210
9.3 Preparation of Graphene Oxide (GO)	211
9.4 Functionalization of Graphene Oxide	214
9.5 Graphene Oxide in Nanomedicines	216
9.6 Graphene-Based Materials in Cancer Therapy	218
9.7 Graphene-Based Materials in Cancer Diagnosis	221
9.8 Conclusion	222
9.9 Future Perspective	222
Acknowledgments	223
References	223

9.1 INTRODUCTION AND HISTORY OF GRAPHENE

Graphene is a two-dimensional (2D) carbon sheet having sp2 hybridization arranged in a honeycomb lattice with a molecular weight that exceeds 106–107 g/mol. Technically, graphene is nonmetal but due to its properties being like that of semiconducting metal, graphene is often termed as quasi-metal (Yang et al. 2018; Zhang et al. 2020).

Back in 2004, Andre Geim and Kostya Novoselov at the University of Manchester isolated graphene for the first time. They separated graphite into distinct carbon layers using adhesive tape, which led them to win the Nobel Prize in physics in 2010. A decade later from the discovery, graphene picked up headlines due to the applications in cryogenic pressure vessels, aerospace composite tooling, and electric vehicle batteries, such as aramid nanofiber reinforced supercapacitors. The importance of graphene is obvious from the fact that more than 2,300 graphene-related patents have been approved in a single calendar year, which shows the significance of industrialization and commercialization of graphene (Alvial-Palavicino and Konrad 2019).

In this chapter, we explain the evolving technologies of graphene in various fields, such as industrial, medical, biomedical, nanomaterials, cancer diagnosis, and treatment. A similar structure to graphene was first discovered in 1859 by a chemist named Benjamin C. Brodie and a further developing method was introduced in 1957 by Hummers and Offeman. The developed method is known as Hammer's method (Brisebois and Siaj 2020, Tiwari et al. 2018).

9.2 STRUCTURE AND COMPOSITION OF GRAPHENE OXIDE

Various spectroscopic as well as microscopic techniques have been developed to investigate the in-depth structural features of graphene oxide. For example, thickness of a single layer of graphene oxide is often investigated by atomic force microscopy. This technique can also evaluate the number of layers, while the electrical defects can be also investigated by conductive atomic force microscopy. Similarly, various other scientific techniques have been used to characterize graphene and graphene based-materials, such as high-resolution transmission electron microscope, scanning tunneling microscope, and nano-contact atomic force microscopy. Regarding different forms and structure elucidation of graphene, many analytical tools, such as Fourier transform infrared spectroscopy (FTIR), X-ray photoelectron spectroscopy (XPS), X-ray absorption near-edge spectroscopy, solid-state nuclear magnetic resonance (NMR), and Raman spectroscopy are generally used.

Taking advantage of the abovementioned scientific techniques, Lerf-Klinovski's model has explained the best description of graphene oxide, which was further elaborated by Gao et al. (2009). According to the above model, the hydroxyl and epoxy functional groups are attached to the basal plane of each graphene oxide particle. These functional groups at the basal planes of sp3 hybridized carbon atoms are nearby and there is no direct correlation between the size of graphene oxide and the content of each functional group (Lerf et al. 1998). Notwithstanding, the structural motifs of graphene oxide are still in investigational stages, thus the precise atomic structure of graphene oxide remains unclear.

The structure of graphene oxide is largely derived from graphite oxide. The graphite oxide was first prepared in 1859 (Brodie 1859) and its structure and composition are still debatable due to nonstoichiometric composition as well as a high level of hygroscopicity. There are some important but still conflicting models regarding graphitic oxide that are proposed, such as Hofmann model in 1936 stated that the graphitic oxide consist of modified planar carbon layers of epoxy with C_2O molecular formula (Hofmann and Holst 1939). Further suggested by Ruess in 1946 that the carbon layers are wrinkled or puckered, while oxygen bridges consisting of hydroxyl and ether are present at carbon atom 1 and 3 (Ruess 1947). Furthermore, the acidic property of graphite oxide was evaluated by Hofmann using enol and keto-type structure integration into their model and found ether bridges at 1 and 3 carbon atoms. Later on, the ether and epoxide groups were replaced by Scholz and Boehm to hydroxyl and carbonyl groups and proposed a novel structure with corrugated carbon (Scholz and Boehm 1969). A very different model was proposed by Nakajima et al., which stated that the two carbon atoms are sp3 hybridized and present perpendicularly to the layers. The amount of hydroxyl and carbonyl groups are in relative amounts proportional to the level of hydration (Nakajima and Matsuo 1994). Moreover, Lerf and his team proposed a model based on NMR studies. Lerf suggested that the flat aromatic regions are randomly distributed with un-oxidized benzene rings and puckered regions of alicyclic six-membered rings containing ether, C=C, and C-OH (Lerf et al. 1998).

All of the above studies gave us an illustration of the fundamental structure and different compositional features of graphite oxide. However, it is also obvious that a refined structure of graphene oxide is imperative, while the above discussion gave some idea about the structure of graphene oxide, which is further explained below.

Each sheet of graphene oxide consists of a network of hexagonal carbon rings consisting of mainly sp2 hybridized carbon atom and sp3 hybridized carbon having oxygen groups. The sp3 hybridized oxygen-bearing groups, such as hydroxyl, epoxide, and carboxyl, shall be considered as oxidized regions and they generally disrupt the honeycomb lattice structured sheet. The honeycomb lattice sheet is an extended sp2 hybridized and shall be considered as un-oxidized regions. The oxygen-bearing groups such as hydroxyl, epoxide, and carboxyl are uniformly but randomly displaced on the graphene plane either slightly below or above (Figure 9.1).

9.3 PREPARATION OF GRAPHENE OXIDE (GO)

The structure evaluation suggested that graphene oxide is a nonstoichiometric chemical structure comprising carbon, oxygen, as well as hydrogen. The three functional groups are present in different ratios, and the ratios are considered to mainly dependent on the processing procedures. The oxygen functional groups are introduced abundantly during chemical exfoliation to the flat carbon grid. Furthermore, the oxygen functional groups can be introduced as bridging oxygen groups, such as epoxide groups, hydroxyl, and carbonyl as well as phenol. These functional groups bring unique properties to graphene oxide and thus enhance its acceptability in various fields. These unique properties made graphene oxide accessible and scalable for industrial use, such as graphene-derived materials, polymer composites, graphene-based sensors, photovoltaics, membranes, and purification materials.

As from the literature and studies available to date, the comprehensive structure of graphene oxide is debatable and ambiguous, still, vast efforts are made for the synthesis of graphene oxide in the past decade. The reason behind the efforts regarding the synthesis is the environmental acceptability and compassion and, of course, the effectiveness of graphene oxide in various applied industries. In this section, we try to critically discuss the synthesis of graphene oxide with respect to its structure.

The Brodie method can be considered as the earliest and breakthrough in the importance of graphene oxide. Although he termed the product as graphic acid, which was later known to be graphite oxide and graphite oxide is the precursor of graphene oxide. Brodie reported the first change of graphite while blending with strong oxidants. Due to scientific advancements, the chemistry of graphite oxide was studied

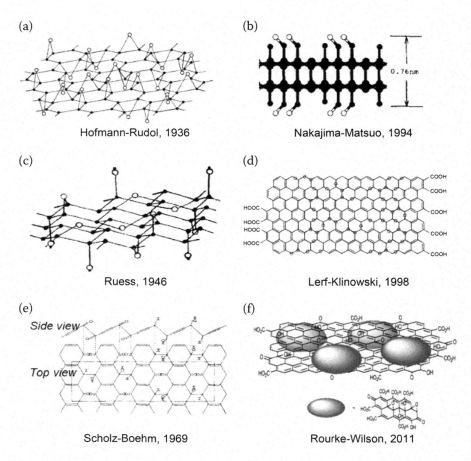

FIGURE 9.1 The figure elaborates different structures of graphite or graphene oxide. Reprinted from (Sun 2019) with permission from Elsevier.

extensively that it is still believed that graphite oxide is capable to transform into graphene oxide and graphite. Mostly, graphite is the key starting material which is extensively used in experimental research among scientist for industrial production of graphene which leads to various applications (Brodie 1859).

Moreover, the pioneering method termed as Hummer's method is still acceptable and widely used to oxidize graphite. The Hummers method developed in 1958 has many advantages and presenting enhanced reproducibility as compared to Brodie and earlier scientists. The hummer's method mainly depends on a blend of an acid such as sulfuric acid and potassium permanganate as shown in Figure 9.2. Sulfuric acid and potassium permanganate are used in the first two steps, while the whole procedure comprises three main steps. The first couple of steps involve a period for the intercalation of graphite, which is further followed by immediate or subsequent oxidization of the obtained graphite compounds. The final step is to obtain the

Graphene and Graphene-Based Nanomaterials

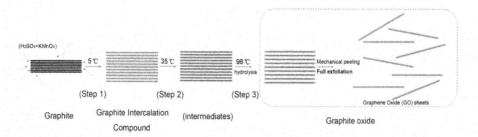

FIGURE 9.2 A general schematic presentation of graphene oxide preparation using Hummers–Offeman method. Reprinted with permission from (Hummers Jr and Offeman 1958) Copyright (1958) American Chemical Society.

homogeneous graphite oxide solution. The above obtained graphite oxide is then hydrolyzed, which is then exfoliated forming single sheet using sonication, mechanical peeling, or by some other shearing stress. To obtain graphene oxide sheets, various researchers completely break the stacked structure of graphite oxide using desired strength ultrasonication (Beckett and Croft 1952).

According to Brodie's method in 1859, the oxidation is brought via fuming nitric acid and potassium chlorate. This method was believed to be very tiresome and nonsignificant because oxidation usually takes a long time. The method was modified by Staudenmaier by the addition of sulfuric acid. This adjustment of the reaction components brought fast completion in a single vessel leading to enhanced processing outcomes and yield (Staudenmaier 1898). Later on, the scientific community focused on cutting the processing time as well as limiting the exchanges of toxic gases. Finally, a time-saving and nonhazardous method was presented by Hammers and Offeman by changing the oxidant from nitric acid, potassium chlorate, and potassium permanganate to a water-free mixture. This effective, time-consuming and hazardous method takes not more than 2 hours for completion at a low temperature.

Recently, modern and advanced ways were established to oxidize graphite by replacing chlorate and permanganate-based oxidizers with potassium chromate or in combination with nitric acid. Furthermore, potassium ferrate in sulfuric acid is also used to prepare graphene oxide at room temperature. The potassium ferrate is considered less toxic as compared to chlorate but the suitability of ferrate ions is still debatable because ferrate ions show low stability in acidic solution while less oxidative in a neutral and basic environment. The oxidation of graphite was also brought using benzoyl peroxide (BPO) (Shen et al. 2009). It usually acts as a main oxidizing agent to oxidize graphene oxide in a metal-free environment.

Recently, to overcome or replace the hazardous and toxic method, efforts are made in the development of graphene oxide. The old and traditional methods include chlorate and permanganate oxidation that are more toxic. The new methods mostly involved novel oxidants, which are considered more scalable but still limited due to the cooling and handling of large volume of hot solutions during the manufacturing process.

9.4 FUNCTIONALIZATION OF GRAPHENE OXIDE

The physicochemical characteristics imparted under its 2D structure, graphene oxide becomes highly selective and very sensitive when it comes to functionalization with additional chemical groups and/or linker molecules. This functionalization makes the graphene oxide a very versatile and promising candidate for ample biomedical and bio-sensing applications. As discussed above, the Hummer's method is an easy and cost-effective method and frequently used for the preparation of graphene oxide. Due to its high conductivity after reduction, selective targeting after proper functionalization, graphene oxide gained high levels of attention in the scientific community related to nanomaterials.

At the basal planes, the graphene oxide bears epoxy and hydroxyl groups. While the edges of the sheets are bind with other functional groups, such as lactone, carboxyl, phenol, carbonyl, and quinone groups via covalent bond. This bonding creates oxidized regions of sp3 hybridized carbon atoms, which, in turn, disrupt nonoxidized regions of the original sp2 honeycomb linkage. Generally, the water solubility of graphene sheets is limited, which depends on strong $\pi-\pi$ bonds between layers. It is the functional group that give the polarity and make graphene oxide hydrophilic and water-soluble. This water solubility in turn makes the graphene oxide important for chemical derivatization and further processing (Compton and Nguyen 2010, Du 2008, Loh et al. 2010).

Regardless of the sensitivity and careful considerations, graphene oxide can be easily modified via chemical functionalization. The said modification has aided in the application of graphene oxide in biomedical sensors, diagnostic and therapeutical applications, and electrochemistry. Numerous methods have already been developed to maximize the advantages and quality while minimizing the disadvantages of graphene oxide. Chemical functionalization of graphene oxide maximizes the stability in water, selectivity for sensing application in electrochemistry, and increase in the surface area can lead to enhanced sensitivity of graphene oxide. Some of the useful functionalization methods are discussed further.

The hydroxyl groups of the GO are mostly functionalized using silanization and esterification. At the basal planes of the GO sheets, the hydroxyl groups are the ideal and more effective sites for chemical modification. This kind of modification is termed as silanization, where organofunctional alkoxysilane molecules are coated on various surfaces, especially carbon nanomaterial surfaces to obtain desired physical and chemical properties (Xu et al. 2008). Carpio and coworkers reported enhanced multifunctionality of GO when functionalized with N-(trimethoxysilylpropyl) ethylenediamine tri-acetic acid (EDTA-silane). This aids in the metal adsorption and antimicrobial characteristics of GO (Reference). In comparison, GO functionalized with EDTA-silane showed an improvement in the antimicrobial property of GO. Moreover, GO functionalized with EDTA-silane showed reduced human cell toxicity making its way for future clinical applications as a promising carbon nanomaterial (Bekyarova et al. 2009).

To enhance the water solubility, Hu et al. combined silanization with a reduction to functionalize graphene oxide. In this study, 3-methacryloxypropyltrimethoxysilane (MPS) was used as a silanized agent on the hydroxyl groups of GO and reduced-GO to get MPS-GO and MPS-rGO as shown in Figure 9.3. They noticed an increase in

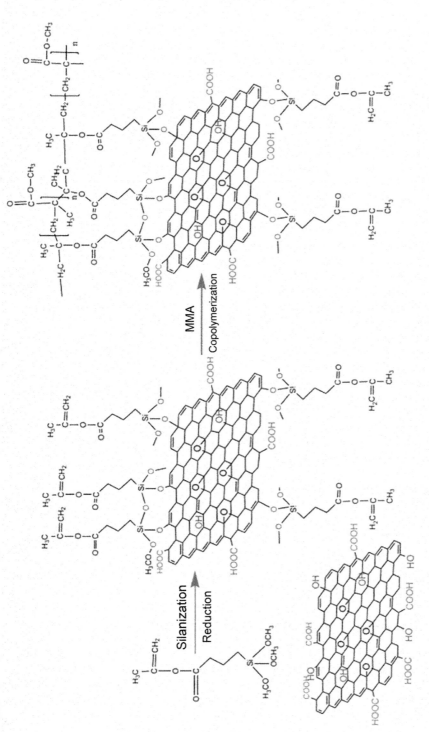

FIGURE 9.3 3-methacryloxypropyltrimethoxysilane used for silanization of graphene oxide, followed by copolymerization of graphene oxide functionalization. Reproduced with permission from reference (Hu et al. 2014) copyright 2014 Elsevier.

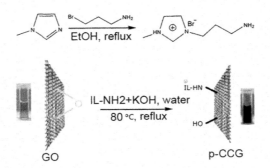

FIGURE 9.4 Functionalization of GO to prepare p-CCG. Reproduced with permission from (Yang, Shan, et al. 2009) copyright 2009 Royal Society of Chemistry.

the water solubility of MPS-GO and MPS-rGO as compared to nonfunctionalized GO and reduced-GO. Furthermore, they also used poly(methyl methacrylate) to functionalize GO and reduced-GO derivatives via in situ copolymerization to increase the tensile strength of the polymer (Hu et al. 2014). The nanosheets displayed stable dispersion in the polymer matrix leading to maximum mechanical strength demonstrating an importance of GO in industrial applications.

Usually, the α-carbon of the epoxide is attacked to functionalize GO. An ionic liquid, 1-(3-aminopropyl)-3-methylimidazolium bromide was used at the same ring-opening mechanism to functionalize GO by Yang et al. as shown in Figure 9.4 (Yang, Li, et al. 2009). The chemically converted graphene oxide sheets showed better stability in polar solvents as well as excellent potential for electrochemical sensor applications. Similarly, GO-platelets were functionalized by Yang et al. using 3-aminopropylt-riethoxy-silane (APTS) via the nucleophilic reaction between the amino groups of APTS and epoxide. The resulted functionalized GO displayed increased mechanical strength, better poly-dispersity and solubility in most polar solvents, which can lead to enhanced encapsulation of therapeutics and biosensor applications.

Mostly, polymer chemical linkers and small molecules are functionalized on GO via esterification or amidation and activation. Bovine serum albumin (BSA) was functionalized on GO by Shen et al. in 2010 using a popular method termed diimide-activated amidation (Shen et al. 2010). This method is often used to link proteins and materials together. The GO-conjugated BSA survived the protein from denaturation as well exhibited excellent solubility in water.

9.5 GRAPHENE OXIDE IN NANOMEDICINES

Graphene, a single-layered carbon atom arranged in a two-dimensional (2D) honeycomb lattice, belongs to a basic building block for other graphitic materials, such as carbon nanotubes and graphite (Lee et al. 2008; Meyer et al. 2007). The extraordinary physicochemical characteristics of graphene, which include excellent electrical conduction and electron transport abilities, high surface area of about 2630 m^2/g for single-layer graphene, exceptional thermal and electrical conductivities, durable mechanical strength, unprecedent flexibility and impermeability, and ease of functionalization have led the foundation for its importance in the field of catalysis, sensors, biotechnology, and energy technology (Chen, Feng, and Li 2012; Shao et al. 2010, Xu et al. 2011). Mostly, biomolecules are adsorbing through π–π stacking and electrostatic interaction, thus enhancing the application of graphene in constructing biosensors and drug loading. The inherent chemical and physical characteristics of graphene, such as bulky sp2 hybridized carbon area as well as an ultra-high surface area made graphene-constructed nanomaterials encouraging carriers for effective as well as efficient gene and drug delivery, especially after versatile surface modification (Pattnaik, Swain, and Lin 2016, Tonelli et al. 2015).

Graphene, bearing a short history from 2004 till now, with graphene oxide (GO) gaining the prime importance among its various derivatives (Katsnelson 2007). Like other allotropes of carbon, pristine graphene, the nonmodified graphene is hydrophobic and therefore demeanor dispersibility problems in physiological media (Konkena and Vasudevan 2012, Sasidharan et al. 2011). That is the reason that graphene is modified to less toxic and more dispersible oxidized forms, such as GO (Prato, Kostarelos, and Bianco 2008). The edges of GO can be functionalized with hydroxyl, carboxyl, and carbonyl groups, while the basal plane can be decorated with hydroxyl and epoxy groups which are hydrophobic and can adsorb drugs and other biomolecules due to π–π interactions (Gao 2015). The sheet edges and hydrophilic basal of GO gives a structure that stabilizes hydrophobic molecules by presenting its hydrophilic sheet, which act as a surfactant (Lotya et al. 2010).

GO, the oxidized derivative of graphite having numerous properties such as amphiphilicity, easy surface functionalization, fluorescence quenching ability (Hong et al. 2012; Wu et al. 2017), and SERS property (Jiang et al. 2017), is gaining much more importance as a promising material for the biomedical applications. Importantly, graphene oxide can be easily and effectively functionalized with targeting ligands to facilitate targeted drug and gene delivery as well as targeted imaging (Biju 2014, Goenka, Sant, and Sant 2014, Liu, Cui, and Losic 2013, Loh et al. 2010, Ma et al. 2012, Sun 2008, Wu et al. 2014, Yan et al. 2013; Yang, Zhang, et al. 2009; Zhang 2010).

However, considering the extended structure of graphene, an observable concern is its toxicity after administration (Bhushan et al. 2014). The surface functionalization of graphene oxide plays a critical role in better biocompatibility. Mostly, surface functionalization decreases the strong hydrophobicity nature of graphene and hence reduces the interaction of graphene and related materials, such as GO with cells and tissues. This reduced interaction further leads to a reduction in singlet oxygen generation that can facilitate apoptosis via the caspase-3 activation pathway (Gurunathan and Kim 2016; Zhang and Gurunathan 2016).

Mostly, Hummer's method is used to oxidize the graphite (Guerrero-Contreras and Caballero-Briones 2015, Paulchamy, Arthi, and Lignesh 2015, Xiong et al. 2013, Dimiev and Tour 2014). Briefly, the graphite is treated with potassium permanganate and sulfuric acid resulting in graphitic salt, followed by exfoliation in solvents with sonication. The graphitic salt serves as a precursor for GO (Posudievsky et al. 2012). Recently, Chun-Hu Chen reported rapid and high quantity oxidation of graphene by modifying the Hummer's method, includes "PAOM" (preformed acidic oxidizing medium) as a modified version. Before the addition of graphitic flakes, KMnO4, and concentrated H_2SO_4 were mixed to get oxidized graphene oxide in a high yield (Chen et al. 2017). This type of high oxidization of pristine graphene presents reactive sites for further derivatization.

Furthermore, the degree of oxidization also affects the physicochemical properties of GO, which further impact the modification of GO for its application (Drewniak et al. 2015). Gongkai Wang and his group reported a chemical reaction for the oxidation of graphene to form high-quality GO resulting in a less exothermic reaction as compared to the Hummers method, which emits toxic gases. The degree of oxidation is well-monitored and controlled with strong oxidizing agents and the chemical reaction presented was an inspiration from unzipping carbon nanotubes (Wang et al. 2011).

As intense research has been done on GO, the actual understanding of the mechanism of formation is been lacking. The concept of the formation of C=O covalent bonds with the introduction of oxygen atoms into graphene lattice is a theoretical report. However, Ayrat M. Dimiev suggested a three-step procedure for the formation of GO, to better understand the mechanism. Starting from graphitic intercalation compound (GIC) to be the first step, while GIC converts to oxidized pristine GO in the second step involving diffusion of the oxidizing agent into preoccupied graphitic galleries and become an ordered form of graphite. The last step involves the hydrolysis of pristine GO and finally the GO is formed upon exposure to water (Dimiev and Tour 2014).

Mostly, two factors greatly affect the surface chemistry of GO, such as size and surface charge of the sheet. By adjusting the pH, and varying oxygen-containing groups, the surface properties of GO can be greatly tuned. The change in pH causing variability in the dispersion stability of GO, and this change in pH can be simply achieved with NaOH and HCl titration (Abbasia et al. 2017).

9.6 GRAPHENE-BASED MATERIALS IN CANCER THERAPY

As we know, the obvious issue of graphene-based nanomaterials is to increase their biocompatibility and their behavioral control in biological systems. Addressing this, surface functionalization is an attractive strategy (Urbas et al. 2014). As discussed in the above sections of this chapter, GO holds chemically reactive-oxygen functional groups, such as hydroxyl and carboxylic acid, and utilizing these reactive groups covalently conjugated to hydrophilic polymers, such as PEG and DEX, and react with a small molecule, which leads to improved biocompatibility as well as a reduction in their nonspecific binding to cells, biological molecules, and tissues (Alibolandi 2017, Hassanpour, Ghorbanpour, and Tehrani 2016, Layek and Nandi 2013, Wang, Colombi Ciacchi, and Wei 2017). Besides the mostly used covalent chemical reactions, graphene can also be modified noncovalently to different polymers and/or biomolecules using electrostatic binding, hydrophobic interactions, or π–π stacking for enhanced stability especially aqueous medium (Georgakilas et al. 2016, Xie et al. 2016).

Surface area (2630 m^2/g) and sp2 hybridized carbon area of GO has always been advantageous to high drug/gene loading to achieve drug/gene targeting in high amount to the site of action (Yang, Asiri, et al. 2013). For example, Dai et al. discovered the first nano-sized graphene oxide (NGO) as a novel and efficient nano-delivery system for water-insoluble aromatic drugs into cells for the treatment of cancer (Liu et al. 2008). They loaded SN38, a water-insoluble anticancer drug on to modified NGO-PEG by simple noncovalent adsorption via π–π stacking, the resulted nGO–PEG–SN38 displayed 1000-folds high cell toxicity in HCT-116 cells as compared to its counterpart CPT-11.

The same research group further explored targeting graphene oxide to a specific type of cells by loading doxorubicin to as-synthesized NGO-PEG conjugated with CD20+ antibody called Rituxan for specifically killing of B-cell lymphoma [40].

Similarly, folic acid (FA)-modified graphene oxide has been vastly used as an effective nanocarrier for targeted drug delivery. For example, Zhang et al. synthesized sulfonic acid-modified graphene oxide conjugated FA to target cells overexpressed folate receptor (FR). Moreover, to achieve a synergistic cytotoxic effect, the FA-conjugated GO delivery system was further loaded with DOX and CPT. The two drugs loaded FA conjugated GO show specific targeting and increase cell toxicity compared to the single leaded drug (Zhang et al. 2010).

As far as an environmental and external stimulus is concerned, several researchers also studied GO-based smart drug delivery systems, which can behave according to the external stimuli, such as temperature, pH, and light, etc. Lately, Pan and coworkers designed a thermoresponsive drug delivery system such as graphene sheet grafted with poly(N-isopropyl acrylamide) and loaded with CPT. The resultant nanocomplex demonstrated higher in vitro cytotoxicity compared to CPT dissolved in DMSO (Pan et al. 2011). Besides, to enhance the anticancer effect, GO-based nanocomposites have also been designed by Yang et al. They synthesized $GO-Fe_3O_4$ nanoparticle hybrid, a dual-targeting drug delivery cargo comprising magnetic and biomolecule. The resultant dual targeting hybrid nanoparticle exhibited specific targeting in vitro in SK3 human breast cancer cells (Pan et al. 2011).

Graphene oxide can also be used as gene carriers, such as functionalized GO modified by polyethyleneimine (PEI) and chitosan (CS) has been demonstrated to be effective in small interfering RNA (siRNA) as well as plasmid DNA delivery. For example, Feng et al. studied gene delivery by modifying graphene oxide with polyethyleneimine (having a molecular weight of 1.2 and 10 kDa) (Shi et al. 2014). GO modified with PEI nanoparticles (GO-PEI) were also loaded subsequently with plasmid DNA to target EGFP expression in HeLa cells. As a result, the GO–PEI having a molecular weight of -10k showed excellent cytotoxicity compared to GO-PEI having a molecular weight of 10k. Although both GO-PEI (having a molecular weight of 1.2 and 10 kDa) demonstrated similar EGFP transfection efficiency. These results proposed that the PEI-modified GO is a promising and effective candidate for targeted gene delivery. Similarly, another research group recently prepared GO-sheets functionalized with chitosan (CS) loaded with plasmid DNA (pDNA). Using a certain nitrogen/phosphate ratio, the pDNA was successfully and efficiently delivered to HeLa cells. Furthermore, the surface of GO was conjugated PEI via an amide bond by Zhang and coworkers (Zhang, Chen, et al. 2011). The resulting GO-PEI as shown below was successfully loaded with DOX and siRNA. The results revealed that the nanocomplex showed maximum cytotoxicity and effective knockdown of Bcl-2 protein expression due to a strong synergistic effect.

As photothermal therapy (PTT) generates heat by taking advantage of an optical-absorbing agent under light irradiation. The generated heat results in cytotoxicity via localized high temperature. GO has been extensively used as a PTT agent in the form of PEG-GO, which exhibits high near-infrared absorbance, thereby showing promise in PTT treatment of cancer. For example, Yang and his research team designed and prepared GO-PEG. They studied in vivo tumor uptake and PTT of GO-PEG in xenograft tumor mouse models. Taking advantage of enhanced permeability and retention effect (EPR), the resultant nanocarrier demonstrates excellent tumor uptake of GO-PEG due to efficient tumor passive targeting of GO (Yang et al. 2010). Furthermore, taking advantage of the strong absorbance of graphene oxide in the near-infrared region, excellent tumor destruction was also achieved at low-power near-infrared (NIR) laser irradiation. Similarly, Zhang et al. achieved dual chemo-photothermal therapy by preparing NGO-PEG-DOX. Due to the excellent NIR absorbance of PEGylated graphene oxide, chemotherapy combined with photothermal therapy was simultaneously carried out. The combined chemo-photothermal treatment resulted in a much higher therapeutic efficacy compared to either chemotherapy or photothermal therapy alone in terms of in vivo tumor treatment in a mouse model (Zhang, Guo, et al. 2011) (Figure 9.5).

Photodynamic therapy (PDT), depending on the generation of reactive oxygen species is a novel and effective method to treat cancer. The reactive oxygen species are generally produced as a result of near-infrared irradiation of photosensitizer (PS) molecules, which leads to the destruction of cancer cells. According to a study,

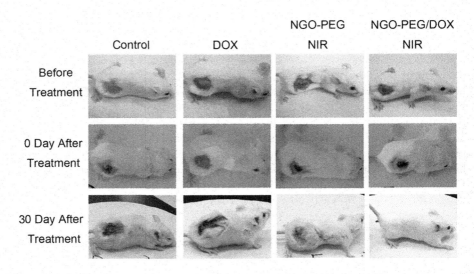

FIGURE 9.5 Representative photos of tumors on mice after treatments indicated. The laser irradiated tumor on NGO-PEG-DOX injected mouse was completely destructed. Reprinted from (Zhang, Guo, et al. 2011) with permission from Elsevier.

Chlorin e6, a strong photosensitizer loaded on to graphene oxide to achieve photodynamic therapy. Huang et al. conjugated GO with folic acid and loaded it with Chlorin e6 for targeted delivery to cancer cells. The resultant nanocomplex using a 633 nm He–Ne laser irradiation effectively destroyed cancer cells photodynamically (Yang, Feng, et al. 2013).

9.7 GRAPHENE-BASED MATERIALS IN CANCER DIAGNOSIS

Over the past decade, graphene-based nanomaterials have been extensively studied with vast molecular imaging techniques. These include photo-acoustic, magnetic resonance imaging (MRI), and optical and positron emission tomography [PET] imaging.

According to a research study, the graphene oxide and PEGylated graphene oxide both exhibited photoluminescence in the visible and infrared range. Welsher et al. [31] covalently conjugated GO–PEG to both Rituxan and anti-CD20 antibody specific labeling of B-cell specific. Due to the intrinsic photoluminescence of GO, they observed strong fluorescent signals from the B cells (Yang, Feng, et al. 2013).

Although the intrinsic fluorescence of graphene oxide has been extensively studied and found to be useful in biomedical imaging, the detectible fluorescence of graphene oxide is affected by the auto-fluorescence interference of living tissues, which has partial use in biomedical imaging particularly in vivo animal imaging. Therefore, external fluorescent dyes have been utilized to label graphene oxide for imaging.

However, the fluorescence signal by GO can be quenched if fluorophores are directly labeled (Yang, Feng, et al. 2013). To avoid quenching of GO fluorescence, PEG is often used as a linker between fluorescein and GO for intracellular optical imaging. The PEGylated GO-conjugated fluorescein demonstrates excellent fluorescence signal, better biocompatibility, as well as pH tunable fluorescence in vitro. This shows that the direct interactions between fluorescein and GO can be reduced via a PEG linker. In the past several years, most of the research community label PEGylated GO utilizing external fluorescent dyes for in vitro and in vivo imaging. Furthermore, PEGlyated GO was conjugated with Cy7, and NIR dye by Yang and his team, and evaluate fluorescence imaging in mouse tumor models. The Cy7 was conjugated at the terminal site of PEG and the resultant GO-PEG-Cy7 was administered to tumor-bearing mic. A strong fluorescent signal was detected over a long period, suggesting efficient passive targeting by GO modified with PEG and conjugated with MIR dye (Chang et al. 2011). Similarly, another group [34] labeled the same GO modified with PEG with 125I. The labeling was achieved by anchoring iodine atoms on the defects and edges of GO. The resultant material showed that radiolabeling is an exceptionally delicate method and could precisely track the labeled molecule in vivo in a quantitative model. Recently, the same group (Yuan et al. 2012) invented GO–PEG labeled with 64Cu for in vivo positron emission tomography (PET) imaging, and recognized active tumor targeting by GO–PEG conjugated with TRC105, to target vasculatures in 4T1 tumors. This is the first ever in vivo demonstration of graphene nanomaterials.

9.8 CONCLUSION

Briefly, GO sheets are carbon-based materials with important functionalities, such as epoxy, carboxyl, hydroxyl groups, and further addition of other active groups at the basal planes and edges. As discussed, several functionalization methods and procedures have been developed to maximize the mechanical, biomedical, imaging, chemical, and therapeutic properties of graphene and graphene-based materials. In regards to the functionalization of GO using salinization of graphene oxide, hydroxyl groups, and reduced GO with MPS and other molecules are widely used. The above methods can improve tensile strength as well as increase solubility. Similarly, many methods, such as functionalization of graphene oxide with epoxy groups, which involve attacking the α-carbon of the epoxide are also gaining interest. The obtained graphene nanosheets from such functionalization presents maximum dispersibility especially in polar solvent as well as the nanosheets also has the capability of high gas barrier characteristics. Moreover, activation coupled esterification and amidation are commonly used methods to functionalize the carboxylic groups of GO. These two methods bring solubilization of GO and help the easy functionalization of graphene oxide with different biomolecules. The capability to improve the properties of graphene oxide demonstrates the diverse range of potential applications, particularly in the biomedicine field, hydrophobic drug delivery, medical diagnostics, and precise DNA-sequence detection. Regarding noncovalent functionalization, the electrical properties of graphene oxide can be improved by π–π stacking and, as well as reduction in internal resistance, refining the electron transfer, maximizing the specific capacitance. Consequently, the development of noncovalent functionalization of graphene oxide is obviously an efficient and tunable method for electrochemical sensors assembly as well as precise detection of biomolecules. In summary, modified graphene oxide nanosheets are currently used extensively in device development, which are used in different fields, such as advanced materials industries and biomedical and diagnostics industries around the globe. However, the complete prospective of graphene and graphene-based materials is still in the development.

9.9 FUTURE PERSPECTIVE

From stagnant to vibrant model, with theoretical assumption to experimental findings, the time dependence has been considered for inherent property exploration of GO. Interesting methods and characterization techniques are on the way from research towards industrial development. Consequently, comprehensive properties and characteristics of the graphene oxide shall likely become entirely unraveled and widely acceptable, and the resulting characteristics are advantageous and leading to the practical applications of graphene and graphene-based materials.

Scientific research regarding the synthesis of graphene and graphene-based materials is now at such a plateau, where the Hummer's method involving the use of concentrated sulfuric acid, is still considered the most acceptable approach. However, effective innovations are much needed, which requires more contributions from the scientists around the globe with intelligence.

Moreover, graphene and graphene oxide materials have been broadly applied in various biomedical and nanotechnology field. Currently, a majority of studies are focused on toxicities in vital organs of the body, such as liver and lungs. However, it is imperative to explore the toxicity of graphene-based materials in brain or neurological aspect in the future. Generally, graphene-based materials have shown toxicities in many biological applications; however, a mechanistic and in-depth investigation is needed in future.

Finally, much more focus should be given to the uniform size and properties to further enrich the application of graphene oxide. It is impractical for downstream industries to embrace graphene and graphene-based materials from upstream entities having unstable presentations.

ACKNOWLEDGMENTS

This work was supported by the Natural Science Foundation of Guangdong Province (2019A1515010776), the Hu-Xiang Young Talent Program (2018RS3094), the Hunan Provincial Natural Science Foundation of China (2018JJ1019), the National Natural Science Foundation of China (31871003, 32001008), the Beijing Nova Program from Beijing Municipal Science & Technology Commission (Z201100006820005), the Beijing-Tianjin-Hebei Basic Research Cooperation Project (19JCZDJC64100), the National Key R&D Program of China (2021YFE0106900), and the Young Elite Scientist Sponsorship Program of Beijing Association for Science and Technology (2020–2022). We thank Biological & Medical Engineering Core Facilities (Beijing Institute of Technology) for providing advanced equipment.The authors declare no conflict of interest.

REFERENCES

Abbasia, F., J. Karimi-Sabetb, C. Ghotbia, Z. Abbasia, S.A. Mousavia, and N. Aminia. "The effect of pH on lateral size and surface chemistry of graphene oxide." *Scientia Iranica* 24, no. 6 (2017): 3554–3559.

Alibolandi, Mona, Marzieh Mohammadi, Seyed Mohammad Taghdisi, Mohammad Ramezani, and Khalil Abnous. "Fabrication of aptamer decorated dextran coated nano-graphene oxide for targeted drug delivery." *Carbohydrate Polymers* 155 (2017): 218–229.

Alvial-Palavicino, Carla, and Kornelia Konrad. "The rise of graphene expectations: Anticipatory practices in emergent nanotechnologies." *Futures* 109, (2019): 192–202.

Beckett, R.J., and R.C. Croft. "The structure of graphite oxide." *The Journal of Physical Chemistry* 56, no. 8(1952): 929–935.

Bekyarova, Elena, Mikhail E. Itkis, Palanisamy Ramesh, Claire Berger, Michael Sprinkle, Walt A de Heer, and Robert C Haddon. "Chemical modification of epitaxial graphene: Spontaneous grafting of aryl groups." *Journal of the American Chemical Society* 131, no. 4(2009): 1336–1337.

Bhushan, Bharat, Dan Luo, Scott R Schricker, Wolfgang Sigmund, and Stefan Zauscher. *Handbook of Nanomaterials Properties*. Springer Science & Business Media, 2014.

Biju, Vasudevanpillai. "Chemical modifications and bioconjugate reactions of nanomaterials for sensing, imaging, drug delivery and therapy." *Chemical Society Reviews* 43, no. 3(2014): 744–764.

Brisebois, P.P., and Mohamed Siaj. "Harvesting graphene oxide–years 1859 to 2019: a review of its structure, synthesis, properties and exfoliation." *Journal of Materials Chemistry C* 8, no. 5(2020): 1517–1547.

Brodie, Benjamin Collins. XIII. "On the atomic weight of graphite." *Philosophical transactions of the Royal Society of London*, no. 149(1859): 249–259.

Chang, Yanli, Sheng-Tao Yang, Jia-Hui Liu, Erya Dong, Yanwen Wang, Aoneng Cao, Yuanfang Liu, and Haifang Wang. "In vitro toxicity evaluation of graphene oxide on A549 cells." *Toxicology Letters* 200, no. 3(2011): 201–210.

Chen, Chun-Hu, Shin Hu, Jyun-Fu Shih, Chang-Ying Yang, Yun-Wen Luo, Ren-Huai Jhang, Chao-Ming Chiang, and Yung-Jr Hung. "Effective synthesis of highly oxidized graphene oxide that enables wafer-scale nanopatterning: preformed acidic oxidizing medium approach." *Scientific Reports* 7, no. 1(2017): 3908.

Chen, Da, Hongbin Feng, and Jinghong Li. "Graphene oxide: Preparation, functionalization, and electrochemical applications." *Chemical Reviews* 112, no. 11(2012): 6027–6053.

Compton, Owen C, and SonBinh T Nguyen. "Graphene oxide, highly reduced graphene oxide, and graphene: Versatile building blocks for carbon-based materials." *Small* 6, no. 6(2010): 711–723.

Dimiev, Ayrat M, and James M Tour. "Mechanism of graphene oxide formation." *ACS Nano* 8, no. 3(2014): 3060–3068.

Drewniak, Sabina Elżbieta, Tadeusz Piotr Pustelny, Roksana Muzyka, and Agnieszka Plis. "Studies of physicochemical properties of graphite oxide and thermally exfoliated/reduced graphene oxide." *Polish Journal of Chemical Technology* 17, no. 4(2015): 109–114.

Du, X. "X.; Skachko, I.; Barker, A.; Andrei, EY." *Nature Nanotechnology* 3 (2008): 491.

Gao, Wei. "The chemistry of graphene oxide." In *Graphene Oxide*, 61–95. Springer, 2015.

Gao, Wei, Lawrence B Alemany, Lijie Ci, and Pulickel M Ajayan. "New insights into the structure and reduction of graphite oxide." *Nature Chemistry* 1, no. 5(2009): 403–408.

Georgakilas, Vasilios, Jitendra N Tiwari, K. Christian Kemp, Jason A Perman, Athanasios B Bourlinos, Kwang S Kim, and Radek Zboril. "Noncovalent functionalization of graphene and graphene oxide for energy materials, biosensing, catalytic, and biomedical applications." *Chemical Reviews* 116, no. 9(2016): 5464–5519.

Goenka, Sumit, Vinayak Sant, and Shilpa Sant. "Graphene-based nanomaterials for drug delivery and tissue engineering." *Journal of Controlled Release* 173 (2014): 75–88.

Guerrero-Contreras, Jesus, and Felipe Caballero-Briones. "Graphene oxide powders with different oxidation degree, prepared by synthesis variations of the Hummers method." *Materials Chemistry and Physics* 153 (2015): 209–220.

Gurunathan, Sangiliyandi, and Jin-Hoi Kim. "Synthesis, toxicity, biocompatibility, and biomedical applications of graphene and graphene-related materials." *International Journal of Nanomedicine* 11 (2016): 1927.

Hassanpour, Akbar, Khatereh Ghorbanpour, and Abbas Dadkhah Tehrani. "Covalent and Non-covalent Modification of Graphene Oxide Through Polymer Grafting." In *Advanced 2D Materials*, John Wiley & Sons (2016): 287–351. http://dx.doi.org/10.1002/9781119242635.ch8

Hofmann, Ulrich, and Rudolf Holst. "Über die Säurenatur und die Methylierung von Graphitoxyd." *Berichte der deutschen chemischen Gesellschaft (A and B Series)* 72, no. 4(1939): 754–771.

Hong, Bong Jin, Zhi An, Owen C Compton, and SonBinh T Nguyen. "Tunable Biomolecular Interaction and Fluorescence Quenching Ability of Graphene Oxide: Application to "Turn-on" DNA Sensing in Biological Media." *Small* 8, no. 16(2012): 2469–2476.

Hu, Xinjun, Enqi Su, Bochao Zhu, Junji Jia, Peihong Yao, and Yongxiao Bai. "Preparation of silanized graphene/poly (methyl methacrylate) nanocomposites in situ copolymerization and its mechanical properties." *Composites Science and Technology* 97, (2014): 6–11.

Hummers Jr, William S, and Richard E Offeman. "Preparation of graphitic oxide." *Journal of the American Chemical Society* 80, no. 6(1958): 1339–1339.

Jiang, Tao, Xiaolong Wang, Shiwei Tang, Jun Zhou, Chenjie Gu, and Jing Tang. "Seed-mediated synthesis and SERS performance of graphene oxide-wrapped Ag nanomushroom." *Scientific Reports* 7, no. 1(2017): 9795.

Katsnelson, Mikhail I. "Graphene: Carbon in two dimensions." *Materials Today* 10, no. 1-2(2007): 20–27.

Konkena, Bharathi, and Sukumaran Vasudevan. "Understanding aqueous dispersibility of graphene oxide and reduced graphene oxide through p K a measurements." *The Journal of Physical Chemistry Letters* 3, no. 7(2012): 867–872.

Layek, Rama K, and Arun K Nandi. "A review on synthesis and properties of polymer functionalized graphene." *Polymer* 54, no. 19(2013): 5087–5103.

Lee, Changgu, Xiaoding Wei, Jeffrey W Kysar, and James Hone. "Measurement of the elastic properties and intrinsic strength of monolayer graphene." *Science* 321, no. 5887(2008): 385–388.

Lerf, Anton, Heyong He, Michael Forster, and Jacek Klinowski. "Structure of graphite oxide revisited." *The Journal of Physical Chemistry B* 102, no. 23(1998): 4477–4482.

Liu, Jingquan, Liang Cui, and Dusan Losic. "Graphene and graphene oxide as new nanocarriers for drug delivery applications." *Acta Biomaterialia* 9, no. 12(2013): 9243–9257.

Liu, Zhuang, Joshua T Robinson, Xiaoming Sun, and Hongjie Dai. "PEGylated nanographene oxide for delivery of water-insoluble cancer drugs." *Journal of the American Chemical Society* 130, no. 33(2008): 10876–10877.

Loh, Kian Ping, Qiaoliang Bao, Goki Eda, and Manish Chhowalla. "Graphene oxide as a chemically tunable platform for optical applications." *Nature Chemistry* 2, no. 12(2010): 1015.

Lotya, Mustafa, Paul J King, Umar Khan, Sukanta De, and Jonathan N Coleman. "High-concentration, surfactant-stabilized graphene dispersions." *ACS Nano* 4, no. 6(2010): 3155–3162.

Ma, Xinxing, Huiquan Tao, Kai Yang, Liangzhu Feng, Liang Cheng, Xiaoze Shi, Yonggang Li, Liang Guo, and Zhuang Liu. "A functionalized graphene oxide-iron oxide nanocomposite for magnetically targeted drug delivery, photothermal therapy, and magnetic resonance imaging." *Nano Research* 5, no. 3(2012): 199–212.

Meyer, Jannik C, Andre K Geim, Mikhail I Katsnelson, Konstantin S Novoselov, Tim J Booth, and Siegmar Roth. "The structure of suspended graphene sheets." *Nature* 446, no. 7131(2007): 60–63.

Nakajima, Tsuyoshi, and Yoshiaki Matsuo. "Formation process and structure of graphite oxide." *Carbon* 32, no. 3(1994): 469–475.

Pan, Yongzheng, Hongqian Bao, Nanda Gopal Sahoo, Tongfei Wu, and Lin Li. "Water-soluble poly (N-isopropylacrylamide)–graphene sheets synthesized via click chemistry for drug delivery." *Advanced Functional Materials* 21, no. 14(2011): 2754–2763.

Pattnaik, Satyanarayan, Kalpana Swain, and Zhiqun Lin. "Graphene and graphene-based nanocomposites: biomedical applications and biosafety." *Journal of Materials Chemistry B* 4, no. 48(2016): 7813–7831.

Paulchamy, B., G. Arthi, and B.D. Lignesh. "A simple approach to stepwise synthesis of graphene oxide nanomaterial." *Journal of Nanomedicine & Nanotechnology* 6, no. 1(2015): 1.

Posudievsky, Oleg Yu, Oleksandra A Khazieieva, Vyacheslav G Koshechko, and Vitaly D Pokhodenko. "Preparation of graphene oxide by solvent-free mechanochemical oxidation of graphite." *Journal of Materials Chemistry* 22, no. 25(2012): 12465–12467.

Prato, Maurizio, Kostas Kostarelos, and Alberto Bianco. "Functionalized carbon nanotubes in drug design and discovery." *Accounts of Chemical Research* 41, no. 1(2008): 60–68.

Ruess, G. "Über das graphitoxyhydroxyd (graphitoxyd)." *Monatshefte für Chemie und verwandte Teile anderer Wissenschaften* 76, no. 3-5(1947): 381–417.

Sasidharan, Abhilash, L.S. Panchakarla, Parwathy Chandran, Deepthy Menon, Shantikumar Nair, C.N.R. Rao, and Manzoor Koyakutty. "Differential nano-bio interactions and toxicity effects of pristine versus functionalized graphene." *Nanoscale* 3, no. 6(2011): 2461–2464.

Scholz, Werner, and H.P. Boehm. "Untersuchungen am graphitoxid. VI. Betrachtungen zur struktur des graphitoxids." *Zeitschrift für anorganische und allgemeine Chemie* 369, no. 3-6(1969): 327–340.

Shao, Yuyan, Jun Wang, Hong Wu, Jun Liu, Ilhan A Aksay, and Yuehe Lin. "Graphene based electrochemical sensors and biosensors: A review." *Electroanalysis: An International Journal Devoted to Fundamental and Practical Aspects of Electroanalysis* 22, no. 10(2010): 1027–1036.

Shen, Jianfeng, Yizhe Hu, Min Shi, Xin Lu, Chen Qin, Chen Li, and Mingxin Ye. "Fast and facile preparation of graphene oxide and reduced graphene oxide nanoplatelets." *Chemistry of Materials* 21, no. 15(2009): 3514–3520.

Shen, Jianfeng, Min Shi, Bo Yan, Hongwei Ma, Na Li, Yizhe Hu, and Mingxin Ye. "Covalent attaching protein to graphene oxide via diimide-activated amidation." *Colloids and Surfaces B: Biointerfaces* 81, no. 2(2010): 434–438.

Shi, Sixiang, Feng Chen, Emily B Ehlerding, and Weibo Cai. "Surface engineering of graphene-based nanomaterials for biomedical applications." *Bioconjugate Chemistry* 25, no. 9(2014): 1609–1619.

Staudenmaier, L. "Verfahren zur darstellung der graphitsäure." *Berichte der deutschen chemischen Gesellschaft* 31, no. 2(1898): 1481–1487.

Sun, Ling. "Structure and synthesis of graphene oxide." *Chinese Journal of Chemical Engineering* 27, no. 10(2019): 2251–2260.

Sun, Xiaoming, Zhuang Liu, Kevin Welsher, Joshua Tucker Robinson, Andrew Goodwin, Sasa Zaric, and Hongjie Dai. "Nano-graphene oxide for cellular imaging and drug delivery." *Nano Research* 1, no. 3(2008): 203–212.

Tiwari, Santosh K, Raghvendra Kumar Mishra, Sung Kyu Ha, and Andrzej Huczko. "Evolution of graphene oxide and graphene: From imagination to industrialization." *ChemNanoMat* 4, no. 7(2018): 598–620.

Tonelli, Fernanda MP, Vânia AM Goulart, Katia N Gomes, Marina S Ladeira, Anderson K Santos, Eudes Lorençon, Luiz O Ladeira, and Rodrigo R Resende. "Graphene-based nanomaterials: biological and medical applications and toxicity." *Nanomedicine* 10, no. 15(2015): 2423–2450.

Urbas, Karolina, Malgorzata Aleksandrzak, Magdalena Jedrzejczak, Malgorzata Jedrzejczak, Rafal Rakoczy, Xuecheng Chen, and Ewa Mijowska. "Chemical and magnetic functionalization of graphene oxide as a route to enhance its biocompatibility." *Nanoscale Research Letters* 9, no. 1(2014): 656.

Wang, Gongkai, Xiang Sun, Changsheng Liu, and Jie Lian. "Tailoring oxidation degrees of graphene oxide by simple chemical reactions." *Applied Physics Letters* 99, no. 5(2011): 053114.

Wang, Zhuqing, Lucio Colombi Ciacchi, and Gang Wei. "Recent advances in the synthesis of graphene-based nanomaterials for controlled drug delivery." *Applied Sciences* 7, no. 11(2017): 1175.

Wu, Huixia, Haili Shi, Yapei Wang, Xiaoqing Jia, Caizhi Tang, Jiamin Zhang, and Shiping Yang. "Hyaluronic acid conjugated graphene oxide for targeted drug delivery." *Carbon* 69 (2014): 379–389.

Wu, Xu, Yuqian Xing, Kevin Zeng, Kirby Huber, and Julia Xiaojun Zhao. "Study of fluorescence quenching ability of graphene oxide with a layer of rigid and tunable silica spacer." *Langmuir* 34, no. 2(2017): 603–611.

Xie, Meng, Hailin Lei, Yufeng Zhang, Yuanguo Xu, Song Shen, Yanru Ge, Huaming Li, and Jimin Xie. "Non-covalent modification of graphene oxide nanocomposites with chitosan/dextran and its application in drug delivery." *RSC Advances* 6, no. 11(2016): 9328–9337.

Xiong, Bin, Yingke Zhou, Yuanyuan Zhao, Jie Wang, Xia Chen, Ryan O'Hayre, and Zongping Shao. "The use of nitrogen-doped graphene supporting Pt nanoparticles as a catalyst for methanol electrocatalytic oxidation." *Carbon* 52, (2013): 181–192.

Xu, Bin, Shufang Yue, Zhuyin Sui, Xuetong Zhang, Shanshan Hou, Gaoping Cao, and Yusheng Yang. "What is the choice for supercapacitors: graphene or graphene oxide?" *Energy & Environmental Science* 4, no. 8(2011): 2826–2830.

Xu, Yuxi, Hua Bai, Gewu Lu, Chun Li, and Gaoquan Shi. "Flexible graphene films via the filtration of water-soluble noncovalent functionalized graphene sheets." *Journal of the American Chemical Society* 130, no. 18(2008): 5856–5857.

Yan, Liang, Ya-Nan Chang, Lina Zhao, Zhanjun Gu, Xiaoxiao Liu, Gan Tian, Liangjun Zhou, Wenlu Ren, Shan Jin, and Wenyan Yin. "The use of polyethylenimine-modified graphene oxide as a nanocarrier for transferring hydrophobic nanocrystals into water to produce water-dispersible hybrids for use in drug delivery." *Carbon* 57 (2013): 120–129.

Yang, Gao, Lihua Li, Wing Bun Lee, and Man Cheung Ng. "Structure of graphene and its disorders: a review." *Science and technology of advanced materials* 19, no. 1(2018): 613–648.

Yang, Huafeng, Fenghua Li, Changsheng Shan, Dongxue Han, Qixian Zhang, Li Niu, and Ari Ivaska. "Covalent functionalization of chemically converted graphene sheets via silane and its reinforcement." *Journal of Materials Chemistry* 19, no. 26(2009): 4632–4638.

Yang, Huafeng, Changsheng Shan, Fenghua Li, Dongxue Han, Qixian Zhang, and Li Niu. "Covalent functionalization of polydisperse chemically-converted graphene sheets with amine-terminated ionic liquid." *Chemical Communications* 45, no. 26(2009): 3880–3882.

Yang, Kai, Liangzhu Feng, Xiaoze Shi, and Zhuang Liu. "Nano-graphene in biomedicine: Theranostic applications." *Chemical Society Reviews* 42, no. 2(2013): 530–547.

Yang, Kai, Shuai Zhang, Guoxin Zhang, Xiaoming Sun, Shuit-Tong Lee, and Zhuang Liu. "Graphene in mice: ultrahigh in vivo tumor uptake and efficient photothermal therapy." *Nano Letters* 10, no. 9(2010): 3318–3323.

Yang, Xiaoying, Xiaoyan Zhang, Yanfeng Ma, Yi Huang, Yinsong Wang, and Yongsheng Chen. "Superparamagnetic graphene oxide–Fe 3 O 4 nanoparticles hybrid for controlled targeted drug carriers." *Journal of Materials Chemistry* 19, no. 18(2009): 2710–2714.

Yang, Yuqi, Abdullah Mohamed Asiri, Zhiwen Tang, Dan Du, and Yuehe Lin. "Graphene based materials for biomedical applications." *Materials Today* 16, no. 10(2013): 365–373.

Yuan, Jifeng, Hongcai Gao, Jianjun Sui, Hongwei Duan, Wei N Chen, and Chi B Ching. "Cytotoxicity evaluation of oxidized single-walled carbon nanotubes and graphene oxide on human hepatoma HepG2 cells: An iTRAQ-coupled 2D LC-MS/MS proteome analysis." *Toxicological Sciences* 126, no. 1(2012): 149–161.

Zhang, Liming, Jingguang Xia, Qinghuan Zhao, Liwei Liu, and Zhijun Zhang. "Functional graphene oxide as a nanocarrier for controlled loading and targeted delivery of mixed anticancer drugs." *Small* 6, no. 4(2010): 537–544.

Zhang, Wei, Changchun Chai, Qingyang Fan, Yanxing Song, and Yintang Yang. "PBCF-graphene: A 2D Sp2 hybridized honeycomb carbon allotrope with a direct band gap." *ChemNanoMat* 6, no. 1(2020): 139–147.

Zhang, Wen, Zhouyi Guo, Deqiu Huang, Zhiming Liu, Xi Guo, and Huiqing Zhong. "Synergistic effect of chemo-photothermal therapy using PEGylated graphene oxide." *Biomaterials* 32, no. 33(2011): 8555–8561.

Zhang, Xi-Feng, and Sangiliyandi Gurunathan. "Biofabrication of a novel biomolecule-assisted reduced graphene oxide: an excellent biocompatible nanomaterial." *International Journal of Nanomedicine* 11, (2016): 6635.

Zhang, Yi, Biao Chen, Liming Zhang, Jie Huang, Fenghua Chen, Zupei Yang, Jianlin Yao, and Zhijun Zhang. "Controlled assembly of Fe_3O_4 magnetic nanoparticles on graphene oxide." *Nanoscale* 3, no. 4(2011): 1446–1450.

10 Nanomaterials for Lungs Targeting

Keerti Mishra and Akhlesh Kumar Jain
Guru Ghasidas Central University

CONTENTS

10.1 Introduction 229
    10.1.1 Anatomy of Lungs and Drug Deposition 231
    10.1.2 Fate of Inhaled Particles 231
    10.1.3 History of Inhaled or Targeted Chemotherapies in Lungs 233
10.2 Methods of Drug Delivery in Lungs 233
    10.2.1 Advantages of Lung Targeting 234
    10.2.2 Challenges in Lung Targeting 235
10.3 Micro- and Nanocarrier-Mediated Lung Targeting 236
    10.3.1 Microspheres-Mediated Lung Targeting 236
    10.3.2 Nanoparticle-Mediated Lung Targeting 243
    10.3.3 Liposome-Mediated Lung Targeting 245
    10.3.4 Niosome-Mediated Lung Targeting 247
    10.3.5 Dendrimer-Mediated Lung Targeting 248
    10.3.6 Ligand-Mediated Lung Targeting 249
10.4 Conclusion and Future Perspective 250
References 251

10.1 INTRODUCTION

Since the past two decades, numerous approaches through nasal and repiratory tract have been revealed and exploited for the delivery of drug candidates, which have enabled the use of different drugs via the aforementioned route. Several research studies have advocated a substantial increase in the absorption as well as bioavailability of the drugs aside from the least side effects possible with inhalational therapy. Occasionally, systemic drug delivery of antimicrobials with high doses causes serious adverse/side effects, which called out for the necessity of inhalational formulations that could also provide targeted drug delivery in the airways following less drug exposure systematically (Zhou, et al. 2015). For instance, inhalation of protein therapeutics is a great illustration for demonstrating high targeted potential towards the application of protein in a particular area of the lungs (Hertel, et al. 2015, Kargozar and Mozafari 2018). Anatomical features and permeability factors play the vital role in administration of therapeutics via pulmonary or nasal route. Further, physical and chemical features of the drug, such as particle size and shape, the surface charge of drug, drug entrapment efficiency, drug loading capacity, muco-adhesivity, and biocompatibility, are all important considerations for efficient drug delivery. As conventional therapy bears a bunch of limitations, the development and use of nanotechnology is receiving more consideration in respect to the treatment of numerous critical ailments and disorders. In addition, out of innumerable approaches of nanotechnology, drug targeting through pulmonary route has attracted more attention as it targets the drug directly into lungs exclusive of local or systemic treatment (Servatan, et al. 2020, Vishali, et al. 2019). Furthermore, the pulmonary route bypasses first-pass metabolism, resulting in reduced therapeutic wastage and improved treatment for a variety of disorders (Liang, et al. 2015).

The huge surface area of the lungs, thin epithelial layer, and copious blood supply assure quick absorption of medication therapies, making them unrivaled for drug delivery in the treatment of lung disease or systemic disease. As stated above, drug delivery through the lungs chiefly avoids the first-pass metabolism, since lungs offer a lower degree of drug metabolism in comparison to liver and gastrointestinal tract. Because of ease in self-administration and noninvasiveness, drug delivery through the lungs has become more striking. Delivery of drugs through lung targeting bears both, a bunch of rewards and few shortcomings (Figure 10.1). Drug delivery through inhalation is being popular for more than five decades, but few limitations such as formulation development, dosing, and drug stability have restricted the clinical success and use of drug delivery through inhalation (Neumiller and Campbell 2010). In this chapter, drug delivery through the pulmonary route is discussed along with the various nanomaterials used for lung targeting. Drug therapies can go to a variety of places after entering the lungs, including interactions with air–liquid interfaces, diffusion pathways, transit via respiratory epithelia, and mucus clearance (Todoroff and Vanbever 2011). Thus, in order to distribute drug therapeutics successfully by the lungs with supreme advantages, it becomes essential to understand the anatomy of the lungs.

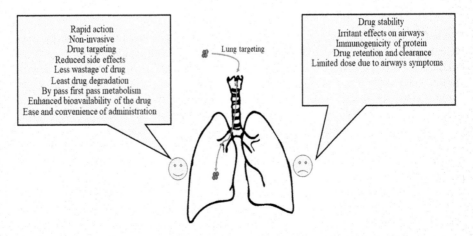

FIGURE 10.1 Depiction of advantages and disadvantages of lung targeting.

Nanomaterials for Lungs Targeting

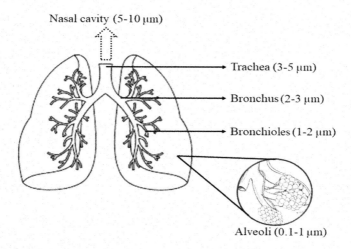

FIGURE 10.2 The size-dependent deposition of particles in the lungs.

10.1.1 ANATOMY OF LUNGS AND DRUG DEPOSITION

Lungs are extremely vascularized organs with a widely large surface area, which helps in improving the drug bioavailability up to a higher extent (Fattal, et al. 2014). Terminal bronchioles of the lungs differentiate into respiratory bronchioles and later turns up to small alveoli (Figure 10.2) (Knudsen and Ochs 2018). Alveoli, when covered with epithelial cells, are called as granular or membranous pneumocytes, the former one secretes surfactant phospholipids, while later one is involved in gaseous exchanges. The route of administration for medication delivery into the bloodstream is through the alveoli epithelium, which is accompanied by tight junctions of 0.5–0.6 nm gap (Dias, et al. 2018, Hamlington, et al. 2018). Alveolar ducts also contain macrophages engulfing inhaled particles (Wachtel 2016).

The route of administration for medication delivery into the bloodstream is through the alveoli epithelium, which is accompanied by tight junctions of 0.5–0.6 nm gap. The fate of an inhaled medicine in terms of deposition and distribution is complicated, as it is impacted by a variety of parameters such as breathing rate and lung volume in different people. Furthermore, the size of particles decides the air flow, anticipated location in the lung, distribution, and deposition of particles, which befalls through quite a few diverse mechanisms.

10.1.2 FATE OF INHALED PARTICLES

Inertial impaction, sedimentation, and diffusion are three key methods for depositing particles in the lungs (Hinds 1999). The aerodynamic diameter of the particle, d_a, is the area where the particles will deposit; it can be hypothesized by assuming a spherical particle falling under gravity; the diameter of this spherical particle with a density of 1 g/cm³ has the sedimentation rate as the particle of interest; and the aerodynamic diameter of this spherical particle with a density of 1 g/cm³ has the same sedimentation rate as the particle of interest (Hinds 1999), which is defined by the equation:

$$d_a = \sqrt{\frac{\rho}{\rho_a}} d_g$$

Where, ρ = mass density of the particle, ρ_a = unit density (1 g/cm^3), and d_g = geometric diameter.

Particles having aerodynamic size greater than 5 μm are deposited in the mouth and upper airways by inertial impaction, while particles having size between 1 and 5 μm are deposited in the interior lung tissues by inertial impaction and settling. Particles with an aerodynamic diameter less than 1 μm are driven by diffusion, with the majority remaining suspended and aroused (Heyder, et al. 1986). The alveolar area of the lungs is the best place for particles having size of 1–3 μm. Particles having size of 100 nm are deposited in the alveolar region of the lungs through diffusion, but small particles are not feasible drug delivery methods since they demand a large amount of energy to create as liquid droplets or disaggregate as a dry powder (Heyder, et al. 1986).

Drug particles having size of 1–5 μm are deposited deeply in the interiors of the lungs, while drug particles having size greater than 10 μm are often accumulated in the oropharyngeal area of the lungs, according to certain research (Chow, et al. 2007, Malcolmson and Embleton 1998). Despite this fact, a couple of studies have demonstrated the deposition of nano-size ranged particles in the lungs, suggesting higher and uniform deposition of nano-size ranged particles in lungs than particles in micron-size range (Ahmad, et al. 2009). The size of the drug particle determines the rate of clearance in the alveolar area. Even though various characteristics regarding the interactions of macrophages and drug particles are still unfamiliar, numerous researchers agreed on the succeeding statements (Carvalho, et al. 2011):

i. Particles larger than 6 μm are typically exhaled rather than phagocytosed
ii. Macrophages are capable of taking up particles with a size range of 1–5 μm
iii. Macrophages phagocytosed particles smaller than 200 nm once they pass through the cellular barrier without using any energy

These findings advocate that smaller particles when accumulate deeply in the interiors of the lungs, easily escape out from macrophage clearance. As a result, it is determined that drug particles in the nanoscale size range have superior deposition capabilities for drug delivery in the lungs' alveolar areas. The downsides associated with the formulation development of inhalable drugs and difficulties linked with drug deposition may be typically avoided through the means of nanomaterials, which will penetrate more deeply in the lungs and will reach the alveolar region more easily. Moreover, nanomaterials can avoid macrophage clearance without any such difficulty and penetrate the epithelium. Parallelly, the surface of nanomaterials may be altered effortlessly to improve the bioavailability, to enhance the penetrability in the mucus layers, or to increase target functionalities to support the delivery of the desired drug.

10.1.3 HISTORY OF INHALED OR TARGETED CHEMOTHERAPIES IN LUNGS

A large number of data has been published on the delivery aspects of chemotherapeutic agents through aerosols in preclinical studies despite the limited oncological use of inhalational agents in humans (Sharma, et al. 2001). In vitro models (Azarmi, et al. 2006, Tseng, et al. 2009) and animal models (Dames, et al. 2007, Hershey, et al. 1999, Rao, et al. 2003) were later developed, as well as phase I/II human trials, to support the feasibility of aerosol-based drug delivery for the treatment of various cancers (Lemarie, et al. 2011, Otterson, et al. 2007, Otterson, et al. 2010, Tatsumura, et al. 1993, Zarogoulidis, et al. 2013).

In 1968, 5-fluorouracil (5-FU), a chemotherapeutic drug, was tried out for inhalational therapy (Tatsumura, et al. 1983). When patients inhaled 5-FU before surgery, the concentration of 5-FU in lung tumors was higher than in other residual tissue. Tatsumura et al. found a high concentration of 5-FU in the major bronchus and nearby lymph nodes for approximately 4 hours after 5-FU was administered (Tatsumura, et al. 1993).

Furthermore, 5-FU-loaded lipid-coated nanoparticles showed increased cytotoxic capabilities and prolonged drug release (Hitzman, et al. 2006). In a phase I/II clinical trial, Otterson and colleagues evaluated and established the promising efficacy of breathed doxorubicin (Otterson, et al. 2007, Otterson, et al. 2010). In a small number of individuals, nontargeted aerosol-based medication administration has resulted in a reasonable decline in pulmonary function and dose-limiting pulmonary toxicity. Targeting inhaled chemotherapeutics into tumors while protecting healthy lung tissue becomes more difficult as a result. There hasn't been a mechanism for drug targeting in the lungs described yet. Azarmi et al. later developed doxorubicin-loaded poly (butyl cyanocrylate) nanoparticles for targeted medication administration to the lungs by inhalation. Despite passive diffusion, doxorubicin-loaded poly (butyl cyanocrylate) nanoparticles successfully displayed in vitro absorption into the H460 and A549 cell lines via endocytosis (Azarmi, et al. 2008).

10.2 METHODS OF DRUG DELIVERY IN LUNGS

Particles larger than 6 μm are not phagocytosed and are deposited in the upper respiratory tract, whereas particles between 1 and 5 μm reach the small airways and are competently taken up by macrophages, and particles less than 1 μm are usually exhaled out and are phagocytosed by macrophages. However, there is substantial evidence that nanoparticles have penetrated the alveoli and may have even entered the bloodstream.

As a result, nanoparticles are a viable choice for delivering drugs in a systematic manner (Deng, et al. 2019, Gaul, et al. 2018). Nanomaterials, which have a diameter of fever than 200 nm, can easily pass through cell membranes and other biological barriers, and are thus utilized as a vector for the delivery of various genes (Anton, et al. 2012, Puisney, et al. 2018, Xiong, et al. 2019). Some unique devices, such as pressurized metered-dose inhalers (PMDIs), nebulizers, dry powder inhalers (DPIs), or Respimat Soft Mist Inhaler (SMI), are used to deliver medications into the respiratory system (Ibrahim, et al. 2015, Islam and Richard 2019, Mozafari, et al. 2018). PMDIs have a pressurized canister in which the medicine is suspended in a propellant, similar to deodorant or perfume bottles (Bensch, et al. 2018). In order to deliver drugs properly from PMDIs into nasal cavities, patient synchronization is desired which is tough for children and elderly people. Since drugs are more efficient in their dry forms than the liquid forms, hence, dry powder inhalers enhance the stability as well as the shelf life of the drug (Duke, et al. 2019), but at the same time it is hard to deliver drugs as dry powder in patients with poor inhaling capacity or with poor lung function (Ibrahim, et al. 2015). In nebulizers, the drug is converted into aerosol ranged between 1 and 5 μm, which is later inhaled by patients. Nebulizers are often preferred over PMDIs, since in nebulizers, no synchronization is needed from patients. Another way for delivering drugs to the lungs is SMI, which combines the benefits of both PMIDs and nebulizers (Vincken, et al. 2018). Drugs are used in SMI in the form of a solution, which is commonly water or ethanol, and a metered dose is released on actuation, which is recently created into a fine mist, which may be breathed easily (Vallorz, et al. 2019).

In order to deliver drugs accurately through nanomaterials, appropriate particle size and proper inhaling devices are required as per the situation of patient and disease (Perriello and Sobieraj 2016). Figure 10.2 shows a schematic illustration of the respiratory tract as well as particle sizes that can be used to target certain areas.

10.2.1 Advantages of Lung Targeting

Lung targeting over conventional therapy offers numerous advantages including:

1. Through lung targeting, drugs are directly delivered into the desired diseased site, which requires a lower drug dose and less systemic side effects as compared to systemic treatment. In addition, lung targeting bypasses hurdles to the drug's therapeutic efficacy, such as poor gastrointestinal absorption and first-pass metabolism (Labiris and Dolovich 2003, Wolff 1998).
2. Pulmonary medication delivery is frequently noninvasive and needle-free, allowing a wide spectrum of substances to be delivered, from tiny molecules to big proteins (Labiris and Dolovich 2003, Wolff 1998).
3. Lungs have a vast surface area for absorption of about 100 m^2 and a highly permeable membrane in the alveolar region with an estimated thickness of 0.2 to 0.7 μm. (Patton and Byron 2007).
4. Bulky drug molecules with low gastrointestinal absorption can be considerably absorbed because mucociliary clearance in the lung periphery is slow, resulting in a protracted stay in the lung (Agnew, et al. 1981).
5. Drugs that are manufactured as dry powders have a number of advantages, including increased chemical and physical stability, longer shelf lives, and no need for refrigeration during storage (McBride, et al. 2013).
6. Targeted medication delivery to the lungs may result in a lower overall medication dose and fewer side effects, which are generally associated with systemic drug exposure at high concentrations. Otherwise, drug administration to the alveolar region can be targeted for systemic drug administration, in which the drug is absorbed by the thin epithelial layer of cells and then circulated throughout the body (Labiris and Dolovich 2003, Wolff 1998).

10.2.2 CHALLENGES IN LUNG TARGETING

Some of the major hurdles in the targeting of drugs therapeutics in the lungs are:

i. In order to improve the bioavailability of inhaled medications, the drug's water solubility must be adequate (Hastedt, et al. 2016). It is highlighted because the pulmonary lining fluid has a limited overall capacity of 150 mL and is circulated as a thin liquid film with a thickness of lesser than 30 m across a wide epithelium surface with an area of roughly 140–160 m^2 (Fröhlich, et al. 2016, Wauthoz and Amighi 2015). Drugs with a low potency need comparatively higher doses, i.e., more than 100 mg, because of which water solubility is essential and highlights the necessity to fabricate various meaningful approaches for the enhancement of drug water solubility (Hastedt, et al. 2016).

ii. Despite the availability of several effective drugs, the demand for desired clinical aids, and aggressive exertions is to formulate new therapies and drugs, as from numerous potent antibiotics, only a couple of antibiotics are permitted for use in inhalational therapy to treat different pulmonary infections (Velkov, et al. 2015, Wenzler, et al. 2016). Only four antibiotics have been licenced for inhalational therapy in Europe for serious lung infections due to cystic fibrosis: Aztreonam, Tobramycin, Colistin, and Levofloxacin (Elborn, et al. 2016, Velkov, et al. 2015, Wenzler, et al. 2016). Furthermore, a few medications are currently being tested in clinical trials for the treatment of lung infections (Lewis 2013). Unfortunately, the success rate of these medications is low, even after putting in a lot of effort and time in their clinical trials. In conclusion, it is still a challenge to treat pulmonary associated infections by utilizing those certain antibiotics, as pathogens develop resistant to an individual drug (Velkov, et al. 2015).

iii. Based on known clinical research, it is expected that currently approved inhalational formulations can only be used to slow the progress of the virus and reduce the destruction of airway tissues (Tolker-Nielsen 2014), not to completely eradicate the infection. For the reasons stated above, *Pseudomonas aeruginosa*, a gram-negative bacterium, develops resistance to many antibiotics quickly. Furthermore, *Pseudomonas aeruginosa* is well known for causing complex and complicated pulmonary infections due to the formation of *Pseudomonas aeruginosa* biofilms, which are multi-cellular surface-attached spatially oriented bacterial communities having bacterial cells in high metabolic external regions and low metabolic/persister central regions that are significantly responsible for the development of complex and complicated pulmonary infections (Branda, et al. 2005, Costerton, et al. 1999, Schachter 2003). Antibiotics, paradoxically, demonstrate their mechanisms of action with exceptional expertise on metabolically active bacterial cells (Lewis 2013); as a result, bacterial cells remain dormant in biofilms, where biofilm increases survival and infection recurrence. Furthermore, the extracellular matrix of biofilms, which is made up of lipids, DNA, alginate, and extracellular polymers, acts as a significant barrier to antimicrobial agent penetration (Branda, et al. 2005, Mah, et al. 2003). Strong interactions between the drug and biofilm components, e.g., have reduced the efficacy of Tobramycin (an aminoglycoside broadly reported as first-line treatment in cystic fibrosis-associated infections), resulting in slow and ineffective drug penetration into the biofilm matrix (Müller, et al. 2018, Tseng, et al. 2013). Furthermore, the low pH of the infected tissues and biofilm can protonate medications like ciprofloxacin, boosting charge interaction with alginate and lowering the concentration of free drug at the site of action (Suci, et al. 1994). As a result, antibiotic concentrations do not reach the minimum inhibitory concentration (MIC), which promotes microenvironmental pressure and biofilm development, as well as the emergence of drug-resistant bacterial subpopulations. As a result, germs that develop mucosal biofilms are more difficult to remove with traditional inhalation treatments.

iv. It is important to keep the drug concentration in the pulmonary lining fluid compartment higher than the MIC for as long as possible (Rodvold, et al. 2011, Zhou, et al. 2015), which may be difficult to achieve due to systemic absorption of drugs across the air-blood barrier or efficient pulmonary clearance mechanisms, such as mucociliary (Hardy, et al. 2013, Pérez, et al. 2014). Furthermore, continuous use of high-dose antibiotics without regulated release and selective targeting efficacy to the desired region may produce damage in healthy lung cells (d'Angelo et al. 2014, Hoffmann et al. 2002).

Researchers have proposed and pursued numerous techniques to avoid or slow the development of antibiotic resistance, thereby enhancing the patient's quality of life, after recognizing the predicted hurdles and knowledge gaps in overcoming antimicrobial resistance.

10.3 MICRO- AND NANOCARRIER-MEDIATED LUNG TARGETING

In the past few years, several approaches including nanotechnology to provide an edge to lung targeting have been discovered. Innumerable studies have indicated that numerous drugs are encapsulated in nanomaterials or micromaterials to improve their bioavailability, selectivity, shelf-life, cytotoxic potential, and to reduce the undesirable side effects or toxic effects. Some of those nanomaterials or micromaterials used to encapsulate the drug for targeting lungs are briefly discussed below.

10.3.1 MICROSPHERES-MEDIATED LUNG TARGETING

Microspheres may be defined as globular particles of size less than 200 µm usually composed of natural or synthetic polymer. Microspheres are widely preferred as the drug carrier for targeting to diverse places. Inhalation of encapsulated pharmaceuticals in microspheres in the form of aerosolized pharmaceuticals can be regarded as the most effective method of medication delivery to the systemic circulation (Ventura, et al. 2008). Tilmicosin-gelatine microspheres (TMS-GMS) were created (Table 10.1) to effectively treat pulmonary infections in animals, as frequent

TABLE 10.1
List of Various Nanomaterials Used for the Lung-Mediated Delivery; Key Characterization Aspects and Vital Outcome

Drug	Objective of the Study	Method of Preparation	Size Range/ Average Size	% Drug Loading	% Entrapment Efficiency	Conclusion	Ref.
Microspheres							
Tilmicosin	Lung-targeting and prolongation of elimination half-life	Emulsion-chemical cross-linking	5.0–5.0 μm	23.25–39.83	54.33–70.47	TMS-GMS showed preferentially lung targeting compared to other organs with increased drug half-life.	(Yang, et al. 2019)
Enoxaparin	Selective accumulation of drug in to lungs and prolonged enoxaparin release	Emulsion crosslinking method	12–15 μm	15–30	82.42–99.70	Albumin microspheres showed an efficient and prolonged antithrombotic response via passive lung targeting in rats.	(Ibrahim, et al. 2017)
Sophoridine	Efficient treatment of lung cancer	Emulsion-solvent extraction	12–24 μm	1–10	4–65	The formulation demonstrated two prolonged way effect by preferential lung accumulation and prolonged drug kinetics, indicating potential of formulated microspheres in treating lung cancer.	(Wang, et al. 2016)
Cefquinome	Lung targeting	Atomization and spray drying	7–30 μm	18.3 ± 1.3	91.6 ± 2.6	The formulation demonstrated preferential and highest distribution of drug into organ of interest for prolonged period of time.	(Qu, et al. 2017)

(Continued)

TABLE 10.1 (Continued)
List of Various Nanomaterials Used for the Lung-Mediated Delivery; Key Characterization Aspects and Vital Outcome

Drug	Objective of the Study	Method of Preparation	Size Range/ Average Size	% Drug Loading	% Entrapment Efficiency	Conclusion	Ref.
Carboplatin	Anti-tumor effect	Emulsion method	5.0–28.6 μm	23.76	-	Greater antitumor response by parenteral administration of the formulation.	(Lu, et al. 2003)
Nanoparticles							
Docetaxel	Effective treatment with fewer side effects	Modified pro-liposome	720 ± 0.23 nm	–	95.46 ± 0.29	Nanoparticle dramatically improved accumulation of docetaxel to the lungs in comparison to injection of plain drug. Further, nanoparticles decreased hazardous effects on RBC and tissue damages.	(Zhang, et al. 2017)
Liposomes							
Docetaxel	Enhanced delivery of drug to lung with fewer side effects	Thin-film hydration method	100–120 nm	–	85.1 ± 6.5-88.6 ± 7.1	Liposomes effectively enhanced targeting potential of drug to lungs tissue after CD133 conjugation.	(Ma, et al. 2018)
Baicalin	Developed liposomes of nanometre size as a site-specific carrier system	Effervescent dispersion and	131.7 ± 11.7 nm	–	81.9±1.29- 83.8±1.35	Liposome was proven as an admirable carrier for Chinese drug baicalin for lung disease.	(Wei, et al. 2017)

(Continued)

TABLE 10.1 (Continued)
List of Various Nanomaterials Used for the Lung-Mediated Delivery; Key Characterization Aspects and Vital Outcome

Drug	Objective of the Study	Method of Preparation	Size Range/ Average Size	% Drug Loading	% Entrapment Efficiency	Conclusion	Ref.
Pirfenidone	Prolongation of drug kinetics and enhanced distribution of drug in desired organ	lyophilized techniques Film hydration method	582.3 ± 21.6 nm	—	87.2	Liposomes prominently maintained the prolonged drug kinetics with improved drug accumulation into the lungs.	(Meng and Xu 2015)
Rifampicin	Fine mist inhalation of rifampicin	Lipid thin film hydration technique	78.3 ± 3.2 to 4381 ± 73 nm		Up to ≈ 42	An increased bio-adhesive ability chitosan coated liposomes and lower side effects of rifampicin against A549 epithelial cells.	(Zaru, et al. 2009)
Niosomes							
Cisplatin	To increase the potential of treatment and decrease the toxicity of carboplatin (CBP).	Hand shaking	3.72 μm	—	29.2%	Cisplatin entrapped in niosomes protected against weight loss and bone marrow toxicity along with greatly decreased secondary growth of lung tumors.	(Gude, et al. 2002)
Carboplatin	To increase the therapeutic benefits niosomal carboplatin	Lipid film hydration method	5.78 μm	—	31.983	Niosomal carboplatin has showed greatly enhanced antimetastatic effect with fewer side effects.	(Zhang and Lu 2001)

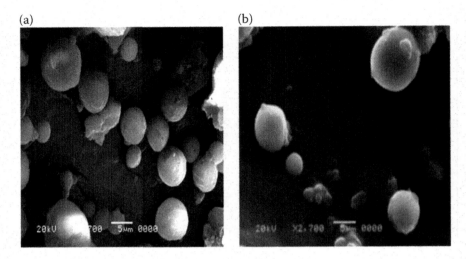

FIGURE 10.3 SEM micrographs of freeze-dried (a) plain and (b) Enox-Alb MS of formula F8. Magnification ×2700. Reproduced under Creative Commons License 4.0 from Ibrahim, S.S., Osman, R., Mortada, N.D., Geneidy, A.S., & Awad, G.A. 2017. Passive targeting and lung tolerability of enoxaparin microspheres for a sustained antithrombotic activity in rats. *Drug delivery* 24(1): 243–251.

administration of Tilmicosin was sought due to the short elimination half-life. When compared to crude TMS, TMS-GMS prolonged the in vitro kinetics of TMS. TMS-GMS was found to have an 8.48 times higher absorption of TMS in the lungs than TMS-injection, indicating that TMS-GMS is a sustained-release formulation with an extended Tilmicosin elimination half-life that could be used for lung targeting in clinical veterinary medications (Yang, et al. 2019). Similarly, Enoxaparin albumin microspheres (Enox-Alb MS) (Figure 10.3, Table 10.1) were developed for the proficient treatment of pulmonary embolism, where animal studies have advocated high lung targeting proficiency with no sign of potential tissue toxicity. Albumin microspheres also maintained the anticoagulant activity up to 38 hours, while the commercial product was able to do the same only up to 5 hours. Based on the findings, it was determined that in the future, compounded Enox-Alb MS might be used as a carrier for regulated and targeted drug delivery of Enoxaparin to the lungs for effective pulmonary embolism treatment (Ibrahim, et al. 2017).

Furthermore, poly(lactide-co-glycolide) (PLGA) microspheres loaded with sophoridine were prepared (Table 10.1) and evaluated for possible treatment of lung cancer through direct targeting. Formulated microspheres exhibited a sustained-release profile for 14 days (Figure 10.4). The concentration of medicine in the lungs was significantly higher than in the other essential organs, implying that formed microspheres may improve the medicine's therapeutic efficiency against lung cancer (Wang, et al. 2016).

Poly lactic-co-glycolic acid (PLGA) microspheres loaded with cefquinome were also manufactured and examined with a similar goal (Figure 10.5, Table 10.1). In vivo studies suggested that maximum drug release was in the lung, identifying it as

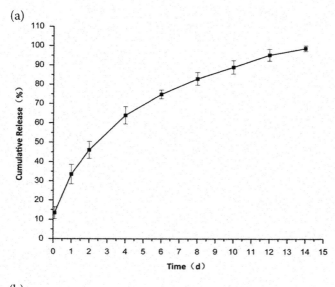
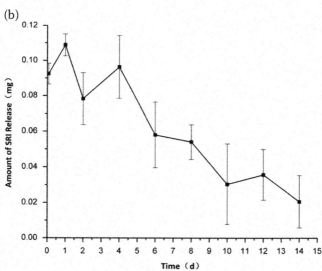

FIGURE 10.4 (a) Cumulative and (b) amount in vitro release profiles of sophoridine-loaded poly(lactide-co-glycolide) microspheres (x ± SD, n = 3). Reproduced under Creative Commons License 4.0 from Wang, W., Cai, Y., Zhang, G., Liu, Y., Sui, H., Park, K., & Wang, H. 2016. Sophoridine-loaded PLGA microspheres for lung targeting: preparation, in vitro, and in vivo evaluation. *Drug delivery* 23(9): 3674–3680.

a target tissue. Prepared microspheres revealed a sustained release profile for 36 hours; nevertheless, in vivo data argued that the greatest drug release was in the lung, designating it as a target tissue. The potential use of prepared microspheres as a drug delivery carrier with improved drug concentration at the desired targeted site

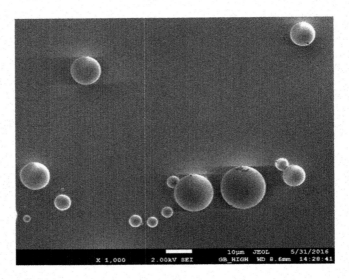

FIGURE 10.5 Scanning electron microscopy of cefquinome-loaded poly lactic-co-glycolic acid microspheres. Reproduced under Creative Commons License 4.0 from Qu, S., Zhao, L., Zhu, J., Wang, C., Dai, C., Guo, H., & Hao, Z. 2017. Preparation and testing of cefquinome-loaded poly lactic-co-glycolic acid microspheres for lung targeting. *Drug delivery* 24(1): 745–751.

and the least effect on normal or healthy tissues has been demonstrated (Qu, et al. 2017).

Acute lung injury (ALI), a dangerous ailment deprived of any such fruitful results, is mainly caused due to oxidative stress, since mitochondria-targeted antioxidant is a problematic approach, sialic acid (SA)-modified microspheres (MS) loaded with TPP-modified curcumin (Cur-TPP) were developed for accurately targeting the mitochondria and enhancing the antioxidant activity. The distribution of microspheres into the lungs of murine mice has increased, whereas SA has efficiently targeted E-selectin, which is strongly expressed on inflamed tissues, and has shown optimal lung-targeted characteristics in ALI model mice following modification with SA. When compared to the H_2O_2 group, SA/Cur-TPP/MS with greater antioxidative activity, reduced intracellular ROS formation, and enhanced mitochondrial membrane potential adds upto to a lower degree of apoptosis in the endothelial cells of the human umbilical vein. In vivo investigations have demonstrated that SA/Cur-TPP/MS effectively reduced inflammation and reduced oxidative stress, suggesting that SA/Cur-TPP/MS could be a potential drug delivery carrier for the treatment of ALI (Jin, et al. 2018). In addition, carboplatin (CPt) microspheres with a high in vivo lung-targeting effectiveness were developed (Table 10.1). It was reported that CPt microspheres had more cytotoxic effects than CPt solution (a half dose of the microspheres equivalent to the pure medication had comparable effects) (Lu, et al. 2003).

10.3.2 NANOPARTICLE-MEDIATED LUNG TARGETING

Nanoparticles (NPs), a submicron colloidal system with a wide range of biological features, can be employed as nanomedicine to treat a variety of ailments (Sahoo, et al. 2007). It has been proven that nanoparticles had improved the pharmacokinetic as well as pharmacodynamic properties of several drugs and had inhibited or decreased multidrug resistance, and other major problems associated with lung targeting (Wicki, et al. 2015). Nanoparticles are made from a variety of materials and are then employed to solubilize, entrap, or encapsulate the target medicine to increase its in vivo delivery. Nanoparticles also have unique magnetic, electrical, and photosensitive capabilities that are necessary for illness diagnosis and treatment (Gindy and Prud'homme 2009). Likewise, docetaxel-lecithoid nanoparticles (DTX-LN) were formulated (Table 10.1) to accomplish a higher delivery of docetaxel to the lungs with the least systemic side effects. During biodistribution studies, it was observed that nanoparticles had intensely enhanced the biodistribution of docetaxel in lungs compared to the DTX-INJ (Figure 10.6). In addition, DTX-LN indicated a decrease in haematotoxicity and the least tissue damage, proving themselves as an alternative carrier having improved drug delivery efficacy in lungs with the least systemic side effects (Zhang, et al. 2017, Gindy and Prud'homme 2009).

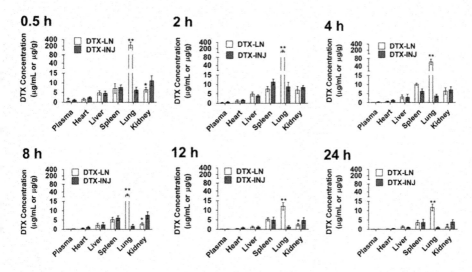

FIGURE 10.6 Biodistribution of Docetaxel-lecithoid nanoparticles and Docetaxel injection in rabbit at a dose of 2 mg/kg Docetaxel equivalence. Each value represents the mean ± S.D. (n = 3). $^*P<0.05$ and $^{**}P<0.01$ indicate a significant difference between Docetaxel-lecithoid nanoparticles and Docetaxel injection. Reproduced under Creative Commons License 4.0 from Zhang, L., Liu, Z.H., Cheng, X.G., Xia, Z., Liu, Y., & Yu, Y. 2017. Docetaxel-Loaded Lecithoid Nanoparticles with Enhanced Lung Targeting Efficiency and Reduced Systemic Toxicity: Developed by Solid Dispersion and Effervescent Techniques. *Chemical and Pharmaceutical Bulletin* 65(10): 959–966.

Radiation-induced lung toxicity is still a major issue, despite many breakthroughs in pulmonary radiotherapy. To circumvent this, noninvasive lung imaging utilizing super-paramagnetic iron oxide nanoparticles decorated with the neutral natural dietary amino acid glycine (GSPIONs) was developed. Further evaluation through pulmonary magnetic resonance imaging suggested that designed nanoparticles are excellent imaging tools as well as diagnostic tools which may be later used for precised targeted radiotherapy for lung cancer, which has significant potential to decrease the pulmonary complications of radiation (Chakraborty, et al. 2020). Curcumin micellar nanoparticles (Cur-NPs) were developed for effective pulmonary administration of the anticancer chemical curcumin in the treatment of lung cancer. Cur-NPs' size-dependent activity against malignant lung cells was proven in an in vitro cytotoxicity assay. In addition, when compared to raw curcumin, Cur-NPs were more efficient in suppressing the inflammatory marker interleukin-8 (IL8). Along with this, Cur-NPs demonstrated excellent aerosolization characteristics, suggesting that Cur-NPs when inhaled could be competent in treating lung cancer with minimal side effects (Lee, et al. 2016). The lung is an appealing target for gene therapy treatment of numerous monogenetic disorders, including cystic fibrosis. Even after numerous clinical trials, there are only a few vectors available for enhancing lung function through gene therapy, indicating the need for the creation of a formulation with improved gene transfer efficacy and the requisite physicochemical features for successful therapy. As a result, a novel cell penetrating peptide (CPP)-based nonviral vector was developed, which uses GAG-binding enhanced transduction (GET) for very successful gene transfer. GET peptides directly pair with DNA for the production of nanoparticles through electrostatic interactions (NPs). PEGylated variants of the GET peptide were created, and DNA NPs with varying densities of PEG coatings were generated for efficient in vivo administration using a novel approach. When compared to non-PEGylated complexes, PEGylated nanoparticles showed improved biodistribution, improved safety profiles, and effective gene transfer of a reporter luciferase plasmid (Figure 10.7). Furthermore, when compared to polyethyleneimine (PEI), a widely used nonviral gene carrier in preclinical settings, gene expression was significantly increased, describing an advanced approach combining novel GET peptides for enhanced transfection with a tuneable PEG coating for effective lung gene therapy (Osman, et al. 2018)

It was also shown that endothelial cells play a critical role in numerous physiological processes, making endothelium-targeting systems a promising therapy option for diseases. As a result, Fernandez *et al.* used span 80 nanoparticles of xanthan gum (XG), a natural polysaccharide, to create a negatively charged hydrophilic surface with stabilizing qualities and the ability to target endothelial cells. After that, a plasmid encoding enhanced green fluorescent protein (pEGFP) was added to the nanosystem, and the system's protection and stability were proven. The cytotoxicity and transfection capacity of nanoparticles into human umbilical vein endothelial cells were used to measure their in vivo biocompatibility. GFP expression was found in the vascular endothelium of the lung, liver, and kidney in biodistribution assays, suggesting the gene targeting ability of XG-functionalized span 80 nanoparticles to endothelial cells (Fernandez-Piñeiro et al. 2018).

Nanomaterials for Lungs Targeting 245

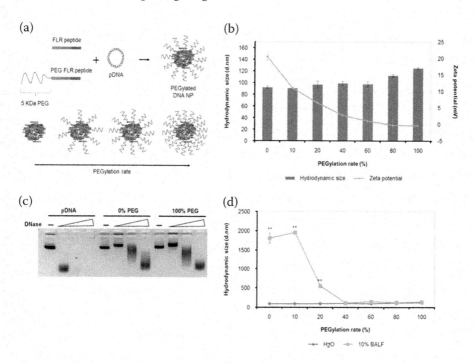

FIGURE 10.7 The effect of PEGylation on the physiochemical properties of DNA NPs. (a) Schematic of non-PEGylated and PEGylated FLR peptides blended at different molar ratios to form DNA NPs with a tuneable coating of PEG on the outer surface. (b) To confirm the formation of DNA NPs we measured the diameter and charge of complexes at 0%, 10%, 20%, 40%, 60%, 80%, and 100% PEGylation rates. The hydrodynamic size and zeta-potentials were measured in ultrapure water and 10 mM NaCl at pH 7.0, respectively. (c) We tested the ability of complexes to protect DNA from nucleases. Naked pDNA, 0% PEG DNA NPs and 100% PEG DNA NPs were incubated with DNase I for 30 minutes, digested using proteinase K, and visualized using a gel shift assay. Degradation of DNA was evidenced by a shift, smear, or loss of the DNA band. (d) The stability of PEGylated DNA complexes was investigated following 1 h incubation in ultrapure water or 10% (v/v) BALF in PBS. The hydrodynamic diameter of respective NPs was measured by DLS. Error bars indicate SD, n = 3. Two-tailed Student's t-test, *$P<.05$, **$P<.01$. Reproduced under Creative Commons License 4.0 from Osman, G., Rodriguez, J., Chan, S.Y., Chisholm, J., Duncan, G., Kim, N., Tatler, A.L., Shakesheff, K.M., Hanes, J., Suk, J.S., & Dixon, J.E. 2018. PEGylated enhanced cell penetrating peptide nanoparticles for lung gene therapy. *Journal of Control Release* 10(285): 35–45.

10.3.3 LIPOSOME-MEDIATED LUNG TARGETING

Liposomes, spherical vesicles with one or more than one lipid bilayer enclosed in an aqueous compartment, corresponding to cell membrane structure are an example of nanomedicines used in cancer treatment. Liposomes, initially fuse with the cell lipid bilayer, afterward delivers the drug payload straight to the cells (Etheridge, et al. 2013). Liposomes may be utilized as a carrier for drug therapies or genes due to their high loading capacity for numerous compounds, which allows for easy chemical modification for a wide range of applications; ability to reduce adverse effects of encapsulated drugs and improve drug stability (Hattori, et al. 2006, Kawakami, et al. 2000, Opanasopit, et al. 2002). A CD133 aptamer modified docetaxel liposomal formulation (CD133-DTX-LP) was made and studied further (Table 10.1). The presence of CD133 aptamer on the outer surface of CD133-DTX-LP has shown a slower drug release, which is justified by the presence of CD133 aptamer on the outer surface, which could have inhibited the drug release. CD133-DTX-LP has expressively inhibited the proliferation of cells in the cytotoxicity studies and enhanced the therapeutic efficiency. Appreciable tumor targeting and significant antitumor activity of CD133-DTX-LP was confirmed with minimum systemic side effects (Ma, et al. 2018). Baicalin-loaded nanoliposomes (Table 10.1) demonstrated a sustained release profile of the drug for 24 hours and stability for at least 1 year. Further, biodistribution studies advocated a higher concentration of Baicalin from nanoliposomes than from Baicalin solution into the lungs (Figure 10.8). Afterward, no injury was reported to the lungs by Baicalin-loaded nanoliposomes. Prominently, improved antitumor activity in the nude mice having orthotopic lung cancer was exhibited, suggesting that Baicalin-loaded nanoliposomes could express themselves as outstanding drug delivery nanocarrier for the treatment of lung cancer with an exceptionally high lung-targeting potency and therapeutic efficacy (Wei, et al. 2017).

Likewise, with the same purpose, pirfenidone (PFD)-loaded liposomes were formulated (Table 10.1), which exhibited distinctive sustained and prolonged

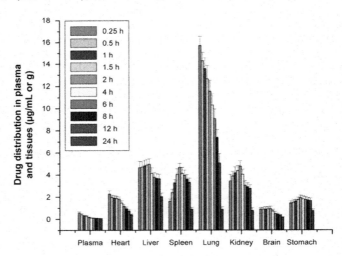

FIGURE 10.8 Distribution of baicalin liposomes in various rabbit tissues after intravenous administration ($\bar{x} \pm SD$, n = 5; plasma: μg/mL, tissues: μg/g). Reproduced under Creative Commons License 4.0 from Wei, Y., Liang, J., Zheng, X., Pi, C., Liu, H., Yang, H., Zou, Y., Ye, Y., & Zhao, L. 2016. Lung-targeting drug delivery system of baicalin-loaded nanoliposomes: development, biodistribution in rabbits, and pharmacodynamics in nude mice bearing orthotopic human lung cancer. *International Journal of Nanomedicine* 12: 251–261.

release of PFD. Drug from liposome was highly distributed in the lungs because liposomes were substantially entrapped in the dense physical vascular network of the lungs. Also, from the histopathological results, it was confirmed that the prepared formulation could significantly treat pathological injury of the lung tissue. From the study, it was concluded that PFD-loaded liposomes can prominently maintain the sustained release of the drug, with increased lungs targeting potential. (Meng and Xu 2015). For the drug delivery to the lungs through nebulization, Rifampicin (RIF)-based chitosan (CHT)-coated liposomes were developed and evaluated (Table 10.1). Developed liposomes had advocated increased mucoadhesive properties and negligible toxicity towards A549 epithelial cells. Because CHT-coated liposomes have a strong mucoadhesive property, they can be employed to transfer medications to the lungs for the treatment of a variety of lung and systemic disorders (Zaru, et al. 2009). The clinical efficacy and safety of a liposomal formulation of cisplatin (Lipoplatin) and gemcitabine were compared to a combination of cisplatin and gemcitabine. The liposomal formulation was found to be more effective in the treatment of advanced non-small cell lung cancer, with no associated life-threatening side effects or death (Mylonakis, et al. 2010).

10.3.4 NIOSOME-MEDIATED LUNG TARGETING

Niosomes are nonionic surfactant-based spherical vesicles in the microscopic size range that are commonly utilized as medication carriers. Niosomes are vesicles that can be unilamellar or multilamellar. These are often made up of nonionic surfactant and cholesterol, and they can successfully entrap hydrophilic pharmaceuticals in the aqueous compartment and lipophilic medications in the vesicular membrane, which is made up of cholesterol or lipid components. Due to the presence of nonionic surfactants, better tumor targeting to the liver and brain can be achieved. Moreover, niosomes are more stable than liposomes. Further, it is also reported that cytotoxic activities of cytotoxic drugs in tumor-bearing mice have been improved and drug toxicities have been decreased when entrapped in the niosomes (Gude, et al. 2002). The effect of cisplatin niosomes and theophylline on activated macrophages was studied using a murine B16F10 melanoma model (Table 10.1). Cisplatin niosomes showed notable antimetastatic action on subsequent lung tumors and reduced toxicity. Furthermore, theophylline, either alone or in conjunction with cisplatin or activated macrophages, was found to have no antimetastatic impact (Gude, et al. 2002). For the improvement in the treatment efficacy along with the minimized side effects of the drug carboplatin, niosomes of the same (CBS-NS) were prepared (Table 10.1). In vitro release studies were illustrated by a bi-exponential equation, and decent targeting efficiency of niosomes was observed. A significant increase in antitumor effect of CBS-NS was observed when injected on mice having lung carcinoma (Zhang and Lu 2001). Rifampicin-loaded niosomes with Span-85 and cholesterol demonstrated up to 65 % of the drug localization into the lungs. Further, such nanocarriers could help in overcoming drug resistance as is often noted once the drug has reached its subtherapeutic level in the body (Jain and Vyas 1995).

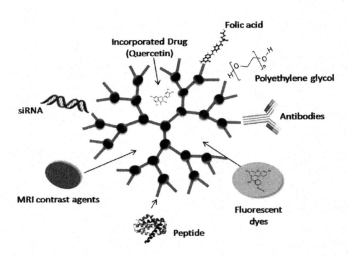

FIGURE 10.9 Dendrimers as multifunctional nanoplatforms. Reproduced under Creative Commons License 4.0 from Madaan, K., Kumar, S., Poonia, N., Lather, V., & Pandita, D. 2014. Dendrimers in drug delivery and targeting: Drug-dendrimer interactions and toxicity issues. *Journal of Pharmacy & Bioallied Sciences* 6(3): 139–150.

10.3.5 DENDRIMER-MEDIATED LUNG TARGETING

Dendrimer, a spherical highly branched nanocarrier, has shown promising in vitro and in vivo findings (Wolinsky and Grinstaff 2008), and as a multifunctional nanoplatforms (Figure 10.9), dendrimer has been used to treat a variety of cancers, including lung cancer (Khandare, et al. 2006, Ly, et al. 2013). Dendrimers are highly branching, monodispersed, flexible nanoscale structures with a changeable functional group that facilitates multivalent ligand conjugation (Chittasupho, et al. 2017). Dendrimers consist of several layers of dendrons or concentric branched units emerging from a central initiator core, a number of layers of dendrons or concentric branched units present in dendrimer defines generation (G) of dendrimer (Scott, et al. 2005). Dendrimers are less likely to be identified and taken up by the reticuloendothelial system due to their lower size range, which typically ranges from 1 to 100 nm.

The growth-inhibitory activity of doxorubicin-PAMAM dendrimer attached to liposomes was high in NCI-H460 and DMS114 lung cell lines, but it was reduced when the dendrimer was complexed to egg yolk phosphatidylcholine: stearyl amine (EPC:SA) liposomes. However, the addition of hexadecyl-phosphocholine (HePC) in the complex helped to increase the cytotoxic potential (Papagiannaros, et al. 2005). When fluorescence microscopy and flow cytometry were used to assess the impact of functional groups attached to the surface of dendrimers in A549 lung epithelial cells, it was discovered that dendrimers with anionic groups are primarily taken up by caveolae-mediated endocytosis, whereas dendrimers with cationic or neutral groups are taken up by non-clathrin, non-caveolae-mediated endocytosis (Perumal, et al. 2008).

Synthetic lung surfactants, exosurf, and tyloxapol were utilized to see how they affected dendrimer-mediated transfection in eukaryotic cells. The results imply that Exosurf boosted the rate of dendrimer luciferase plasmid transfection in human T-cell leukemia, but tyloxapol, a nonionic surfactant, improved dendrimer-mediated gene transfer due to enhanced porosity in the cell membrane and DNA uptake (Kukowska-Latallo, et al. 1999). Lactoferrin receptors (LfRs) were found in abundance on bronchial epithelial (BEAS-2B) cells but not on alveolar epithelial (A549) cells, indicating lactoferrin's targeting selectivity as a targeting ligand for bronchial epithelial cells. Lf-conjugated polyethylenimine branching polymer has recently been employed to transport genes to the lungs (Elfinger, et al. 2007). Rifampicin (RIF) was successfully loaded in the mannosylated dendritic architecture for the discriminating distribution of RIF in alveolar macrophages. At pH 7.4, drug release, hemolytic toxicity, and cytotoxicity were significantly reduced, but drug release and alveolar macrophage uptake were significantly boosted at pH 5.0. In addition, the device was biocompatible and has demonstrated RIF delivery to particular sites (Kumar, et al. 2006). Methylprednisolone-polyamidoamine dendrimer (PAMAM G4-OH) complex was prepared for the effective treatment of asthma, which showed a significant increase in dendrimer residence times when compared to the free drug. Furthermore, after 24 hours of intranasal delivery, more than 35% of the total dendrimer was recovered from the lung (Kannan, et al. 2005).

10.3.6 Ligand-Mediated Lung Targeting

Great strategies have been discovered in recent years associated with the targeting of lungs through systemic administration, which has demonstrated excellent results in order to treat several lung related diseases including asthma, cystic fibrosis, lung cancer or pulmonary infection. Out of those discovered strategies, nanocarriers have indicated the most appreciable results, however, complex structure and low stability are still a concern (Vega-Villa, et al. 2008). To drive nanocarriers to tumors, various targeting ligands have been used (Balyasnikova, et al. 2002). Lung targeting strategies based on the macromolecular carriers like peptides or antibody drug conjugated systems have limited use because of the insufficient potency and immunogenicity (Teicher and Chari 2011). As a result, using a prodrug method, tiny ligands with well-defined structures, minimal immunogenicity, and high targeting efficiency could readily overcome the above-mentioned problem. Asthma is one of the most serious inflammatory lung illnesses, and it is difficult to treat well since existing treatments include frequent corticosteroid inhalation, which can cause systemic side effects or ineffective treatment due to medication tissue specificity and poor compliance. A phenolic propanediamine derivative, N, N, N'-trimethyl-N'-(2-hydroxyl-3-methyl-5-^{123}Iiodobenzyl)-1,3-propanediamine (HIPD), has been extensively utilized as an imaging agent for the diagnosis of localized pulmonary diseases. A new analogue N, N, N'-trimethyl-N'-(4-hydroxyl-benzyl)-1,3-propanediamine (TPD), based on HIPD was discovered to be used as a lung targeting ligand. TPD was later covalently conjugated with Rhein, an anti-inflammatory chemical used in asthma treatment. During the research, it was shown that when TPD-Rhein was precisely disseminated to rats' lungs, cellular absorption efficiency was greatly improved, resulting in a 13-fold rise in C_{max} and a 103-fold rise in AUC_{0-t} for lung compared to Rhein alone. In the lungs of asthmatic rats, serum histamine, serum IL-5, and bronchoalveolar lavage fluid IL-5 levels all decreased significantly, demonstrating that the TPD-Rhein combination is an effective and safe therapeutic option for asthma (Li, et al. 2018). The targeting and imaging efficiency of ^{99m}Tc-(tricine)-HYNIC-(Ser)$_3$-J18 for non-small cell lung cancer (NSCLC) was studied in A-549 xenografted nude mice, where (Ser)$_3$-J18 peptide was conjugated with HYNIC and labelled with ^{99m}Tc using tricine. ^{99m}Tc-(tricine)-HYNIC-(Ser)$_3$-J18 had high labelling efficiency at room temperature and has demonstrated considerable stability in saline and human plasma. The radiolabelled peptide had specifically targeted NSCLC tumor and had indicated a high target uptake with tolerable low circumstantial activity for tumor imaging in mice. ^{99m}Tc-(tricine)-HYNIC-(Ser)$_3$-J18 advocated itself as a better radiolabelled peptide for NSCLC targeting and imaging when compared with previously reported ^{99m}Tc-(EDDA/tricine)-HYNIC-(Ser)$_3$-J18 radiolabelled peptide (Shaghaghi, et al. 2018).

10.4 CONCLUSION AND FUTURE PERSPECTIVE

Formulation scientist has been playing a vital role on drugs to explore various materials alone or in combination to be used as carriers for medicine, and to additional modifications to provide superior characteristics, which could target the specific organ of interest that could lead to therapeutic benefits. Nanocarrier-based carriers have the limitless ability of site-specific applications, hence constantly being tested for a role in lung-specific targeting. The number of questions were raised around the topic of nanocarrier-mediated lung targeting. Varied solutions in effort to discuss all the concern aspects, indicating substitutes and designing appropriate systems to meet the necessities. Delivery of therapeutics to lungs seems to be a lucrative choice because of the huge surface area associated with the alveolar region; which offers wonderful prospects to increase the efficacy of treatment not only systemically but also locally. However, the destiny of the medication in lung tissue is always a serious worry. For the treatment of asthma and other pulmonary disorders, such as tuberculosis, chronic obstructive pulmonary disease, and lung cancer, local application to the lungs is ideal. The use of nano- and microcarriers, such as microspheres, nanoparticles, liposomes, niosomes, and dendrimers for local and site-specific drug delivery is discussed in this chapter. Further, carriers could provide controlled drug delivery at a predetermined rate for the prolonged period of time at a minimum therapeutic dose with greater patient compliance that ultimately decreases the potential toxicities. However, there is still a necessity to recognize cell-specific receptors, which might smooth up the targeting to deep interior tissues of the lungs. The chapter also deals with other relevant topics such as selection of materials for selective targeting, role of size of in targeting, challenges and limitations of the drug carriers, and selective targeting of nanocarriers to the lungs and their constructive growth in terms of drug kinetics. Nevertheless, restricting the phagocytic uptake or suppressing the macrophages uptake by designing suitable particle size of nanoparticles made of appropriate polymer in order to achieve interior lung targeting. Pulmonary targeting of nanomaterials has to still overcome numerous limitations, such as stability of formulation or poor delivery efficiency.

REFERENCES

Agnew, J., Pavia, D., and Clarke, S. "Airways Penetration of Inhaled Radioaerosol: An Index to Small Airways Function?" *Eur J Respir Dis* 62, no. 4(1981): 239–255.

Ahmad, F.J., Mittal, G., Jain, G.K., et al. "Nano-Salbutamol Dry Powder Inhalation: A New Approach for Treating Broncho-Constrictive Conditions." *Eur J Pharm Biopharm* 71, no. 2(2009): 282–291.

Anton, N., Jakhmola, A., and Vandamme, T.F. "Trojan Microparticles for Drug Delivery." *Pharmaceutics* 4, no. 1(2012): 1–25.

Azarmi, S., Löbenberg, R., Roa, W.H., Tai, S., and Finlay, W. "Formulation and in Vivo Evaluation of Effervescent Inhalable Carrier Particles for Pulmonary Delivery of Nanoparticles." *Drug Dev Ind Pharm* 34, no. 9(2008): 943–947.

Azarmi, S., Tao, X., Chen, H., et al. "Formulation and Cytotoxicity of Doxorubicin Nanoparticles Carried by Dry Powder Aerosol Particles." *Int J Pharm* 319, no. 1-2(2006): 155–161.

Balyasnikova, I., Yeomans, D., McDonald, T., and Danilov, S. "Antibody-Mediated Lung Endothelium Targeting: In Vivo Model on Primates." *Gene Ther* 9, no. 4(2002): 282–290.

Bensch, G., Parker, E.D., Ariely, R., Stoyanov, S., and Ramakrishnan, K. "Real-World Study Characterizing Patients Prior to Receiving Albuterol Multidose Dry Powder Inhaler or Short-Acting B2-Agonist Via Pressurized Metered-Dose Inhalers for Asthma and Copd in the United States." *J Allergy Clin Immunol* 141, no. 2(2018): AB208.

Branda, S.S., Vik, S., Friedman, L., and Kolter, R. "Biofilms: The Matrix Revisited." *Trends Microbiol* 13, no. 1(2005): 20–26.

Carvalho, T.C., Peters, J.I., and Williams III, R.O. "Influence of Particle Size on Regional Lung Deposition–What Evidence Is There?" *Int J Pharm* 406, no. 1-2(2011): 1–10.

Chakraborty, A., Royce, S.G., Selomulya, C., and Plebanski, M. "A Novel Approach for Non-Invasive Lung Imaging and Targeting Lung Immune Cells." *Int J Mol Sci* 21, no. 5(2020): 1613.

Chittasupho, C., Anuchapreeda, S., and Sarisuta, N. "Cxcr4 Targeted Dendrimer for Anti-Cancer Drug Delivery and Breast Cancer Cell Migration Inhibition." *Eur J Pharm Biopharm* 119, (2017): 310–321.

Chow, A.H., Tong, H.H., Chattopadhyay, P., and Shekunov, B.Y. "Particle Engineering for Pulmonary Drug Delivery." *Pharm Res* 24, no. 3(2007): 411–437.

Costerton, J.W., Stewart, P.S., and Greenberg, E.P. "Bacterial Biofilms: A Common Cause of Persistent Infections." *Science* 284, no. 5418(1999): 1318–1322.

d'Angelo, I., Conte, C., La Rotonda, M.I., et al. "Improving the Efficacy of Inhaled Drugs in Cystic Fibrosis: Challenges and Emerging Drug Delivery Strategies." *Adv Drug Deliv Rev* 75, (2014): 92–111.

Dames, P., Gleich, B., Flemmer, A., et al. "Targeted Delivery of Magnetic Aerosol Droplets to the Lung." *Nat Nanotechnol* 2, no. 8(2007): 495.

Deng, Q., Ou, C., Shen, Y.-M., et al. "Health Effects of Physical Activity as Predicted by Particle Deposition in the Human Respiratory Tract." *Sci Total Environ* 657, (2019): 819–826.

Dias, S.A., Planus, E., Angely, C., et al. "Perfluorocarbon Induces Alveolar Epithelial Cell Response through Structural and Mechanical Remodeling." *Biomech Model Mechanobiol* 17, no. 4(2018): 961–973.

Duke, D.J., Scott, H.N., Kusangaya, A.J., et al. "Drug Distribution Transients in Solution and Suspension-Based Pressurised Metered Dose Inhaler Sprays." *Int J Pharm* 566, (2019): 463–475.

Elborn, J.S., Vataire, A.-L., Fukushima, A., et al. "Comparison of Inhaled Antibiotics for the Treatment of Chronic Pseudomonas Aeruginosa Lung Infection in Patients with Cystic Fibrosis: Systematic Literature Review and Network Meta-Analysis." *Clin Ther* 38, no. 10(2016): 2204–2226.

Elfinger, M., Maucksch, C., and Rudolph, C. "Characterization of Lactoferrin as a Targeting Ligand for Nonviral Gene Delivery to Airway Epithelial Cells." *Biomaterials* 28, no. 23(2007): 3448–3455.

Etheridge, M.L., Campbell, S.A., Erdman, A.G., et al. "The Big Picture on Nanomedicine: The State of Investigational and Approved Nanomedicine Products." *Nanomedicine* 9, no. 1(2013): 1–14.

Fattal, E., Grabowski, N., Mura, S., et al. "Lung Toxicity of Biodegradable Nanoparticles." *J Biomed Nanotechnol* 10, no. 10(2014): 2852–2864.

Fernandez-Piñeiro, I., Alvarez-Trabado, J., Márquez, J., Badiola, I., and Sanchez, A. "Xanthan Gum-Functionalised Span Nanoparticles for Gene Targeting to Endothelial Cells." *Colloids Surf, B* 170 (2018): 411–420.

Fröhlich, E., Mercuri, A., Wu, S., and Salar-Behzadi, S. "Measurements of Deposition, Lung Surface Area and Lung Fluid for Simulation of Inhaled Compounds." *Front Pharmacol* 7 (2016): 181.

Gaul, R., Ramsey, J.M., Heise, A., Cryan, S.-A., and Greene, C.M. "Nanotechnology Approaches to Pulmonary Drug Delivery: Targeted Delivery of Small Molecule and Gene-Based Therapeutics to the Lung." In *Design of Nanostructures for Versatile Therapeutic Applications*. 221–253: Elsevier, 2018.

Gindy, M.E., and Prud'homme, R.K. "Multifunctional Nanoparticles for Imaging, Delivery and Targeting in Cancer Therapy." *Expert Opin Drug Delivery* 6, no. 8(2009): 865–878.

Gude, R., Jadhav, M., Rao, S., and Jagtap, A. "Effects of Niosomal Cisplatin and Combination of the Same with Theophylline and with Activated Macrophages in Murine B16f10 Melanoma Model." *Cancer Biothe Radiopharm* 17, no. 2(2002): 183–192.

Hamlington, K.L., Smith, B.J., Dunn, C.M., et al. "Linking Lung Function to Structural Damage of Alveolar Epithelium in Ventilator-Induced Lung Injury." *Respir Physiol Neurobiol* 255, (2018): 22–29.

Hardy, C.L., LeMasurier, J.S., Mohamud, R., et al. "Differential Uptake of Nanoparticles and Microparticles by Pulmonary Apc Subsets Induces Discrete Immunological Imprints." *J Immunol* 191, no. 10(2013): 5278–5290.

Hastedt, J.E., Bäckman, P., Clark, A.R., et al. "Scope and Relevance of a Pulmonary Biopharmaceutical Classification System Aaps/Fda/Usp Workshop March 16-17th, 2015 in Baltimore, Md." Springer (2016).

Hattori, Y., Kawakami, S., Nakamura, K., Yamashita, F., and Hashida, M. "Efficient Gene Transfer into Macrophages and Dendritic Cells by in Vivo Gene Delivery with Mannosylated Lipoplex Via the Intraperitoneal Route." *J Pharmacol Exp Ther* 318, no. 2(2006): 828–834.

Hershey, A.E., Kurzman, I.D., Forrest, L.J., et al. "Inhalation Chemotherapy for Macroscopic Primary or Metastatic Lung Tumors: Proof of Principle Using Dogs with Spontaneously Occurring Tumors as a Model." *Clin Cancer Res* 5, no. 9(1999): 2653–2659.

Hertel, S.P., Winter, G., and Friess, W. "Protein Stability in Pulmonary Drug Delivery Via Nebulization." *Adv Drug Deliv Rev* 93, (2015): 79–94.

Heyder, J., Gebhart, J., Rudolf, G., Schiller, C.F., and Stahlhofen, W. "Deposition of Particles in the Human Respiratory Tract in the Size Range 0.005–15 Mm." *J Aerosol Sci* 17, no. 5(1986): 811–825.

Hinds, W.C. *Aerosol Technology: Properties, Behavior, and Measurement of Airborne Particles*. John Wiley & Sons, 1999.

Hitzman, C.J., Elmquist, W.F., Wattenberg, L.W., and Wiedmann, T.S. "Development of a Respirable, Sustained Release Microcarrier for 5-Fluorouracil I: In Vitro Assessment of Liposomes, Microspheres, and Lipid Coated Nanoparticles." *J Pharm Sci* 95, no. 5(2006): 1114–1126.

Hoffmann, I.M., Rubin, B.K., Iskandar, S.S., et al. "Acute Renal Failure in Cystic Fibrosis: Association with Inhaled Tobramycin Therapy." *Pediatr Pulmonol* 34, no. 5(2002): 375–377.

Ibrahim, M., Verma, R., and Garcia-Contreras, L. "Inhalation Drug Delivery Devices: Technology Update." *Medical Devices* 8, (2015): 131.

Ibrahim, S.S., Osman, R., Mortada, N.D., Geneidy, A.-S., and Awad, G.A. "Passive Targeting and Lung Tolerability of Enoxaparin Microspheres for a Sustained Antithrombotic Activity in Rats." *Drug Delivery* 24, no. 1(2017): 243–251.

Islam, N., and Richard, D. "Inhaled Micro/Nanoparticulate Anticancer Drug Formulations: An Emerging Targeted Drug Delivery Strategy for Lung Cancers." *Curr Cancer Drug Targets* 19, no. 3(2019): 162–178.

Jain, C., and Vyas, S. "Preparation and Characterization of Niosomes Containing Rifampicin for Lung Targeting." *J Microencapsulation* 12, no. 4(1995): 401–407.

Jin, F., Liu, D., Yu, H., et al. "Sialic Acid-Functionalized Peg–PLGA Microspheres Loading Mitochondrial-Targeting-Modified Curcumin for Acute Lung Injury Therapy." *Mol Pharmaceutics* 16, no. 1(2018): 71–85.

Kannan, R.M., Inagopalla, R., Kannan, S., et al. "Dendrimer-Based Nanodevices for Asthma Drug Delivery: Synthesis, in-Vitro and in Vivo Studies." Paper presented at the 05AIChE: 2005 AIChE Annual Meeting and Fall Showcase (2005).

Kargozar, S., and Mozafari, M. "Nanotechnology and Nanomedicine: Start Small, Think Big." *Mater Today: Proc* 5, no. 7(2018): 15492–15500.

Kawakami, S., Wong, J., Sato, A., et al. "Biodistribution Characteristics of Mannosylated, Fucosylated, and Galactosylated Liposomes in Mice." *Biochim Biophys Acta Gen Subj* 1524, no. 2-3(2000): 258–265.

Khandare, J.J., Jayant, S., Singh, A., et al. "Dendrimer Versus Linear Conjugate: Influence of Polymeric Architecture on the Delivery and Anticancer Effect of Paclitaxel." *Bioconjugate Chem* 17, no. 6(2006): 1464–1472.

Knudsen, L., and Ochs, M. "The Micromechanics of Lung Alveoli: Structure and Function of Surfactant and Tissue Components." *Histochem Cell Biol* 150, no. 6(2018): 661–676.

Kukowska-Latallo, J.F., Chen, C., Eichman, J., Bielinska, A.U., and Baker Jr, J.R. "Enhancement of Dendrimer-Mediated Transfection Using Synthetic Lung Surfactant Exosurf Neonatal in Vitro." *Biochem Biophys Res Commun* 264, no. 1(1999): 253–261.

Kumar, P.V., Asthana, A., Dutta, T., and Jain, N.K. "Intracellular Macrophage Uptake of Rifampicin Loaded Mannosylated Dendrimers." *J Drug Targeting* 14, no. 8(2006): 546–556.

Labiris, N.R., and Dolovich, M.B. "Pulmonary Drug Delivery. Part I: Physiological Factors Affecting Therapeutic Effectiveness of Aerosolized Medications." *Br J Clin Pharmacol* 56, no. 6(2003): 588–599.

Lee, W.-H., Loo, C.-Y., Ong, H.-X., et al. "Synthesis and Characterization of Inhalable Flavonoid Nanoparticle for Lung Cancer Cell Targeting." *J Biomed Nanotechnol* 12, no. 2(2016): 371–386.

Lemarie, E., Vecellio, L., Hureaux, J., et al. "Aerosolized Gemcitabine in Patients with Carcinoma of the Lung: Feasibility and Safety Study." *J Aerosol Med Pulm Drug Delivery* 24, no. 6(2011): 261–270.

Lewis, K. "Platforms for Antibiotic Discovery." *Nat Rev Drug Discovery* 12, no. 5(2013): 371–387.

Li, J., Yang, Y., Wan, D., Peng, Y., and Zhang, J. "A Novel Phenolic Propanediamine Moiety-Based Lung-Targeting Therapy for Asthma." *Drug Delivery* 25, no. 1(2018): 1117–1126.

Liang, Z., Ni, R., Zhou, J., and Mao, S. "Recent Advances in Controlled Pulmonary Drug Delivery." *Drug Discovery Today* 20, no. 3(2015): 380–389.

Lu, B., Zhang, J., and Yang, H. "Lung-Targeting Microspheres of Carboplatin." *Int J Pharm* 265, no. 1-2(2003): 1–11.

Ly, T.U., Tran, N.Q., Hoang, T.K.D., et al. "Pegylated Dendrimer and Its Effect in Fluorouracil Loading and Release for Enhancing Antitumor Activity." *J Biomed Nanotechnol* 9, no. 2(2013): 213–220.

Ma, J., Zhuang, H., Zhuang, Z., et al. "Development of Docetaxel Liposome Surface Modified with Cd133 Aptamers for Lung Cancer Targeting." *Artif Cells, Nanomed, Biotechnol* 46, no. 8(2018): 1864–1871.

Mah, T.-F., Pitts, B., Pellock, B., et al. "A Genetic Basis for *Pseudomonas Aeruginosa* Biofilm Antibiotic Resistance." *Nature* 426, no. 6964(2003): 306–310.

Malcolmson, R.J., and Embleton, J.K. "Dry Powder Formulations for Pulmonary Delivery." *Pharm Sci Technol Today* 1, no. 9(1998): 394–398.

McBride, A.A., Price, D.N., Lamoureux, L.R., et al. "Preparation and Characterization of Novel Magnetic Nano-in-Microparticles for Site-Specific Pulmonary Drug Delivery." *Mol Pharmaceutics* 10, no. 10(2013): 3574–3581.

Meng, H., and Xu, Y. "Pirfenidone-Loaded Liposomes for Lung Targeting: Preparation and in Vitro/in Vivo Evaluation." *Drug Des, Dev Ther* 9, (2015): 3369.

Mozafari, M., Rajadas, J., and Kaplan, D. *Nanoengineered Biomaterials for Regenerative Medicine.* Elsevier, 2018.

Müller, L., Murgia, X., Siebenbürger, L., et al. "Human Airway Mucus Alters Susceptibility of Pseudomonas Aeruginosa Biofilms to Tobramycin, but Not Colistin." *J Antimicrob Chemother* 73, no. 10(2018): 2762–2769.

Mylonakis, N., Athanasiou, A., Ziras, N., et al. "Phase Ii Study of Liposomal Cisplatin (Lipoplatin™) Plus Gemcitabine Versus Cisplatin Plus Gemcitabine as First Line Treatment in Inoperable (Stage Iiib/Iv) Non-Small Cell Lung Cancer." *Lung Cancer* 68, no. 2(2010): 240–247.

Neumiller, J.J., and Campbell, R.K. "Technosphere® Insulin." *BioDrugs* 24, no. 3(2010): 165–172.

Opanasopit, P., Sakai, M., Nishikawa, M., et al. "Inhibition of Liver Metastasis by Targeting of Immunomodulators Using Mannosylated Liposome Carriers." *J Controlled Release* 80, no. 1-3(2002): 283–294.

Osman, G., Rodriguez, J., Chan, S.Y., et al. "Pegylated Enhanced Cell Penetrating Peptide Nanoparticles for Lung Gene Therapy." *J Controlled Release* 285, (2018): 35–45.

Otterson, G.A., Villalona-Calero, M.A., Hicks, W., et al. "Phase I/Ii Study of Inhaled Doxorubicin Combined with Platinum-Based Therapy for Advanced Non–Small Cell Lung Cancer." *Clin Cancer Res* 16, no. 8(2010): 2466–2473.

Otterson, G.A., Villalona-Calero, M.A., Sharma, S., et al. "Phase I Study of Inhaled Doxorubicin for Patients with Metastatic Tumors to the Lungs." *Clin Cancer Res* 13, no. 4(2007): 1246–1252.

Papagiannaros, A., Dimas, K., Papaioannou, G.T., and Demetzos, C. "Doxorubicin–Pamam Dendrimer Complex Attached to Liposomes: Cytotoxic Studies against Human Cancer Cell Lines." *Int J Pharm* 302, no. 1-2(2005): 29–38.

Patton, J.S., and Byron, P.R. "Inhaling Medicines: Delivering Drugs to the Body through the Lungs." *Nat Rev Drug Discovery* 6, no. 1(2007): 67–74.

Pérez, B., Méndez, G., Lagos, R., and Vargas, M.S. "Mucociliary Clearance System in Lung Defense." *Rev Med Chile* 142, no. 5(2014): 606–615.

Perriello, E.A., and Sobieraj, D.M. "The Respimat Soft Mist Inhaler, a Novel Inhaled Drug Delivery Device." *Conn Med* 80, no. 6(2016): 359–364.

Perumal, O.P., Inapagolla, R., Kannan, S., and Kannan, R.M. "The Effect of Surface Functionality on Cellular Trafficking of Dendrimers." *Biomaterials* 29, no. 24-25(2008): 3469–3476.

Puisney, C., Baeza-Squiban, A., and Boland, S. "Mechanisms of Uptake and Translocation of Nanomaterials in the Lung." In *Cellular and Molecular Toxicology of Nanoparticles.* 21–36: Springer, 2018.

Qu, S., Zhao, L., Zhu, J., et al. "Preparation and Testing of Cefquinome-Loaded Poly Lactic-Co-Glycolic Acid Microspheres for Lung Targeting." *Drug Delivery* 24, no. 1(2017): 745–751.

Rao, R., Markovic, S.N., and Anderson, P. "Aerosol Therapy for Malignancy Involving the Lungs." *Curr Cancer Drug Targets* 3, no. 4(2003): 239–250.

Rodvold, K.A., George, J.M., and Yoo, L. "Penetration of Anti-Infective Agents into Pulmonary Epithelial Lining Fluid." *Clin Pharmacokinet* 50, no. 10(2011): 637–664.

Sahoo, S., Parveen, S., and Panda, J. "The Present and Future of Nanotechnology in Human Health Care." *Nanomedicine* 3, no. 1(2007): 20–31.

Schachter, B. "Slimy Business-the Biotechnology of Biofilms." *Nat Biotechnol* 21, no. 4(2003): 361–365.

Scott, R.W., Wilson, O.M., and Crooks, R.M. "Synthesis, Characterization, and Applications of Dendrimer-Encapsulated Nanoparticles." *J Phys Chem B* 109, no. 2(2005): 692–704.

Servatan, M., Zarrintaj, P., Mahmodi, G., et al. "Zeolites in Drug Delivery: Progress, Challenges and Opportunities." *Drug Discovery Today* 25, no. 4(2020): 642–656.

Shaghaghi, Z., Abedi, S.M., and Hosseinimehr, S.J. "Tricine Co-Ligand Improved the Efficacy of 99mtc-Hynic-(Ser) 3-J18 Peptide for Targeting and Imaging of Non-Small-Cell Lung Cancer." *Biomed Pharmacother* 104 (2018): 325–331.

Sharma, S., White, D., Imondi, A.R., et al. "Development of Inhalational Agents for Oncologic Use." *J Clin Oncol* 19, no. 6(2001): 1839–1847.

Suci, P.A., Mittelman, M., Yu, F.P., and Geesey, G.G. "Investigation of Ciprofloxacin Penetration into *Pseudomonas Aeruginosa* Biofilms." *Antimicrob Agents Chemother* 38, no. 9(1994): 2125–2133.

Tatsumura, T., Koyama, S., Tsujimoto, M., Kitagawa, M., and Kagamimori, S. "Further Study of Nebulisation Chemotherapy, a New Chemotherapeutic Method in the Treatment of Lung Carcinomas: Fundamental and Clinical." *Br J Cancer* 68, no. 6(1993): 1146–1149.

Tatsumura, T., Yamamoto, K., Murakami, A., Tsuda, M., and Sugiyama, S. "New Chemotherapeutic Method for the Treatment of Tracheal and Bronchial Cancers--Nebulization Chemotherapy." *Jpn J Cancer Clin* 29, no. 7(1983): 765–770.

Teicher, B.A., and Chari, R.V. "Antibody Conjugate Therapeutics: Challenges and Potential." *Clin Cancer Res* 17, no. 20(2011): 6389–6397.

Todoroff, J., and Vanbever, R. "Fate of Nanomedicines in the Lungs." *Curr Opin Colloid Interface Sci* 16, no. 3(2011): 246–254.

Tolker-Nielsen, T. "Pseudomonas Aeruginosa Biofilm Infections: From Molecular Biofilm Biology to New Treatment Possibilities." *APMIS* 122, (2014): 1–51.

Tseng, B.S., Zhang, W., Harrison, J.J., et al. "The Extracellular Matrix Protects P Seudomonas Aeruginosa Biofilms by Limiting the Penetration of Tobramycin." *Environ Microbiol* 15, no. 10(2013): 2865–2878.

Tseng, C.-L., Su, W.-Y., Yen, K.-C., Yang, K.-C., and Lin, F.-H. "The Use of Biotinylated-Egf-Modified Gelatin Nanoparticle Carrier to Enhance Cisplatin Accumulation in Cancerous Lungs Via Inhalation." *Biomaterials* 30, no. 20(2009): 3476–3485.

Vallorz, E., Sheth, P., and Myrdal, P. "Pressurized Metered Dose Inhaler Technology: Manufacturing." *AAPS PharmSciTech* 20, no. 5(2019): 1–11.

Vega-Villa, K.R., Takemoto, J.K., Yáñez, J.A., et al. "Clinical Toxicities of Nanocarrier Systems." *Adv Drug Deliv Rev* 60, no. 8(2008): 929–938.

Velkov, T., Rahim, N.A., Zhou, Q.T., Chan, H.-K., and Li, J. "Inhaled Anti-Infective Chemotherapy for Respiratory Tract Infections: Successes, Challenges and the Road Ahead." *Adv Drug Deliv Rev* 85, (2015): 65–82.

Ventura, C.A., Tommasini, S., Crupi, E., et al. "Chitosan Microspheres for Intrapulmonary Administration of Moxifloxacin: Interaction with Biomembrane Models and in Vitro Permeation Studies." *Eur J Pharm Biopharm* 68, no. 2(2008): 235–244.

Vincken, W., Levy, M.L., Scullion, J., et al. "Spacer Devices for Inhaled Therapy: Why Use Them, and How?" *ERJ Open Res* 4, no. 2(2018).

Vishali, D., Monisha, J., Sivakamasundari, S., Moses, J., and Anandharamakrishnan, C. "Spray Freeze Drying: Emerging Applications in Drug Delivery." *J Controlled Release* 300 (2019): 93–101.

Wachtel, H. "Respiratory Drug Delivery." In *Microsystems for Pharmatechnology*. 257–274: Springer, 2016.

Wang, W., Cai, Y., Zhang, G., et al. "Sophoridine-Loaded Plga Microspheres for Lung Targeting: Preparation, in Vitro, and in Vivo Evaluation." *Drug Delivery* 23, no. 9(2016): 3674–3680.

Wauthoz, N., and Amighi, K. "Formulation Strategies for Pulmonary Delivery of Poorly Soluble." *Drugs* (2015).

Wei, Y., Liang, J., Zheng, X., et al. "Lung-Targeting Drug Delivery System of Baicalin-Loaded Nanoliposomes: Development, Biodistribution in Rabbits, and Pharmacodynamics in Nude Mice Bearing Orthotopic Human Lung Cancer." *Int J Nanomed* 12 (2017): 251.

Wenzler, E., Fraidenburg, D.R., Scardina, T., and Danziger, L.H. "Inhaled Antibiotics for Gram-Negative Respiratory Infections." *Clin Microbiol Rev* 29, no. 3(2016): 581–632.

Wicki, A., Witzigmann, D., Balasubramanian, V., and Huwyler, J. "Nanomedicine in Cancer Therapy: Challenges, Opportunities, and Clinical Applications." *J Controlled Release* 200 (2015): 138–157.

Wolff, R. "Safety of Inhaled Proteins for Therapeutic Use." *J Aerosol Med* 11, no. 4(1998): 197–219.

Wolinsky, J.B., and Grinstaff, M.W. "Therapeutic and Diagnostic Applications of Dendrimers for Cancer Treatment." *Adv Drug Deliv Rev* 60, no. 9(2008): 1037–1055.

Xiong, Z., Alves, C.S., Wang, J., et al. "Zwitterion-Functionalized Dendrimer-Entrapped Gold Nanoparticles for Serum-Enhanced Gene Delivery to Inhibit Cancer Cell Metastasis." *Acta Biomater* 99 (2019): 320–329.

Yang, Y., Yuan, L., Li, J., et al. "Preparation and Evaluation of Tilmicosin Microspheres and Lung-Targeting Studies in Rabbits." *Vet J* 246 (2019): 27–34.

Zarogoulidis, P., Darwiche, K., Krauss, L., et al. "Inhaled Cisplatin Deposition and Distribution in Lymph Nodes in Stage Ii Lung Cancer Patients." *Future Oncol* 9, no. 9(2013): 1307–1313.

Zaru, M., Manca, M.-L., Fadda, A.M., and Antimisiaris, S.G. "Chitosan-Coated Liposomes for Delivery to Lungs by Nebulisation." *Colloids Surf, B* 71, no. 1(2009): 88–95.

Zhang, J., and Lu, B. "Studies on Lung Targeted Niosomes of Carboplatin." *Yao Xue Xue Bao* 36, no. 4(2001): 303–306.

Zhang, L., Liu, Z.-h., Cheng, X.-g., et al. "Docetaxel-Loaded Lecithoid Nanoparticles with Enhanced Lung Targeting Efficiency and Reduced Systemic Toxicity: Developed by Solid Dispersion and Effervescent Techniques." *Chem Pharm Bull* (2017): c17–00515.

Zhou, Q.T., Leung, S.S.Y., Tang, P., et al. "Inhaled Formulations and Pulmonary Drug Delivery Systems for Respiratory Infections." *Adv Drug Deliv Rev* 85 (2015): 83–99.

11 Nanostructured Drug Carriers for Nose-to-Brain Drug Delivery
A Novel Approach for Neurological Disorders

Talita Nascimento da Silva[1],
Emanuelle Vasconcellos de Lima[1],
Anna Lecticia Martinez Martinez Toledo[1],
Julia H. Clarke[1], and Thaís Nogueira Barradas[2]
[1]Universidade Federal do Rio de Janeiro
[2]Universidade Federal de Juiz de Fora

CONTENTS

11.1 Introduction 257
11.2 Common Brain Diseases and the Challenge of Treatments 258
11.3 The Nose-to-Brain Route 264
11.4 Limitations from Nose-to-Brain Delivery 266
11.5 Requirements of Nose-to-Brain Formulations 267
11.6 NPs for Nose-to-Brain Targeting 271
    11.6.1 Lipid-Based NP 272
    11.6.2 Polymer NP-Based Systems 276
11.7 Final Remarks 277
References 278

11.1 INTRODUCTION

The increasing occurrence of neurological disorders warns public health organizations and the scientific community to invest in more efficient, noninvasive, and targeted treatments. The main limitation of brain drug delivery lies in the need to overcome the blood-brain barrier (BBB). The nose-to-brain drug delivery, usually via intranasal administration, can be considered an elegant strategy to avoid temporal disruption to BBB. It is also especially interesting to allow the therapy efficacy of nonpermeating drugs since most of the drugs are not able to cross BBB. Thus, to promote a more accurate pharmacological therapy, the nose-to-brain route combined with nanotechnology-based approaches constitute a very promising methodology to enable a more optimize neurological treatment.

According to the World Health Organization (2021), neurological disorders, as dementia, epilepsy, headache, multiple sclerosis, and neuro-infections are presented as greatest threats to public health. Such Central Nervous System (CNS) diseases require the drug to be delivered into the brain. However, drugs with higher molecular weight (above 400 Da) and with hydrophilic character (more than 8 hydrogen bonds) do not freely cross the BBB (Pardridge, 2012). Formulation design plays a crucial role in this context, since, pharmaceutical formulations must allow drugs to cross the BBB and reach the brain. Only few molecules with small lipophilic and hydrophobic regions can achieve those permeation criteria (Poupot, Bergozza, and Fruchon, 2018). In this context, usually almost 98% of drugs are prevented from brain penetration and the successful therapeutic approaches include increased BBB disruption and alternative methods that could lead to CNS infection such as invasive neurosurgical procedures (Furtado et al., 2018).

Intranasal drug administration can provide both local and systemic effects via nasal absorption and nose-to-brain drug delivery with the potential to treat a wide range of diseases. The advantages of nose-to-brain include higher drug brain bioavailability than oral or intravenous routes since it avoids the first-pass metabolism and guarantee direct access to the brain microenvironment. Although the intranasal route was initially used for topical application, it was later considered a potential route for drugs to reach the brain, possibly by taking the olfactory pathway via the nasal mucosa (Hirlekar and Momin, 2018).

Despite the advantages, intranasal administration features some limitations that arise from nasal cavity physiology. Also, drug molecular weight, size, and capacity to resist nasal enzyme activity can impair its absorption and penetration (Ozsoy, Gungor, and Cevher, 2009). Drug encapsulation into nanoparticle carriers is a technological approach to transport the drug from the olfactory epithelium to the CNS. Hence, nanotechnology can improve the performance of such drug delivery systems.

Polymeric nanoparticles, solid lipid nanoparticles (SLNs), nanostructured lipid carriers, microemulsions, nanoemulsions, and liposomes are examples of drug vehicles that can be applied to develop nose-to-brain drug delivery systems (Espinoza et al., 2019). Such nanostructured systems can be optimized with surface modification, increasing drug solubility and vectorization, and avoiding drug degradation by enzymes. Besides, surfactants and mucoadhesive polymers can be added to the formulation to improve drug residence in the nasal cavity and consequently the absorption. Thus, the use of nanoparticles (NP) for brain diseases via the intranasal route is a pharmaceutical strategy to enhance drug pharmacokinetics and facilitates patient compliance.

11.2 COMMON BRAIN DISEASES AND THE CHALLENGE OF TREATMENTS

Neurological disorders have been appointed as the leading cause of disability and second leading cause of death, on a global scale. The number of people who were either affected or died due to brain diseases has been increasing globally for the past

25 years, accounting for 16.5% of total global deaths in 2016 (Feigin et al., 2019). Among them, neurodegenerative diseases and cognitive decline are of great interest due to their impact on life expectancy, life quality, and its common appearance in aged brains (Hou et al., 2019). Even though there are more than 600 diseases that affect the CNS, the most common neurological diseases are dementia and cognitive disorders, Parkinson's disease (PD) and related disorders, stroke, and epilepsy (Dumurgier and Tzourio, 2020; Raggi and Leonardi, 2015). Besides, among the most common neurodegenerative diseases, PD and Alzheimer's diseases (AD) affect millions of people worldwide (Deleidi, Jaggle, and Rubino 2015).

AD and PD are the most prevalent age-associated neurodegenerative diseases, and in the USA alone, these diseases affect 5.7 million (2018) and 930,000 people, respectively (Hou et al., 2019; Marras et al., 2018). Besides, in the USA, people older than 65 years are estimated to increase from 53 million in 2018 to 88 million by 2050; therefore, an increase in age-related diseases can be expected (Hou et al., 2019). AD is marked by neuronal deterioration, inflammation, loss of gray matter, atrophy of the cerebral cortex, loss of cortical neurons, with the presence of amyloid-rich senile plaques, neurofibrillary tangles, and neuronal degeneration. Clinical signs of AD include short-term memory impairment, deterioration of cognitive functions, such as speech, visual-spatial defects, and behavioral changes. Currently, the pharmacological approaches approved by the FDA involve the use of cholinesterase inhibitors, such as tacrine, donepezil, rivastigmine, and galantamine (Samanta et al., 2006). Despite these approved treatments, an effective and satisfactory treatment for AD is still lacking. Due to this, a range of molecular events including amyloid accumulation, neuroinflammation, and tau accumulation are considered potential targets for drug design and development (Lalatsa et al., 2020).

PD is a neurodegenerative disease that leads to the death of dopaminergic neurons in the *substantia nigra pars compacta*. Following dopamine deficiency in the basal, patients can develop a set of movement disorders. Classical motor symptoms include rigidity, resting tremor, and bradykinesia. Its ethiopathogenesis is currently unknown and PD has been linked to genetic, environmental, and immunological factors. Currently available treatments involve the delivery of dopamine precursors (levodopa, l-DOPA, l-3,4 dihidroxifenilalanina), and other symptomatic treatments including dopamine agonists (amantadine, apomorphine, bromocriptine, cabergoline, lisuride, pergolide, pramipexole, ropinirole, rotigotine), monoamine oxidase (MAO) inhibitors (selegiline, rasagiline), and catechol-O-methyltransferase (COMT) inhibitors (entacapone, tolcapone) (Cacabelos, 2017; De Virgilio et al., 2016). The conventional treatments for PD show very small drug bioavailability in the SNC, since only >1% of the drug can overcome the BBB. Clinical trials using intranasal approaches such as intranasal insulin (NCT04251585; NCT04687878), intranasal glutathione (NCT01398748), and intranasal levodopa (NCT03541356) have tried to enhance the drug delivery method, but none so far have reached phase 3.

Apart from PD, brain cancers are a substantial source of morbidity and mortality associated with CNS diseases. In 2016, 330,000 cases of CNS cancer and 227,000 deaths linked to it were globally reported (Ingusci et al., 2019). CNS cancers etiology are mostly unknown, except for some risk factors that have been investigated such as genetic, ionizing radiation exposure, viral infections, asthma, eczema, hormonal contraceptives, hormone replacement therapy, vitamin D level, alcohol, and occupational exposures (Cowppli-Bony 2011; Gonzalez-Rubio et al., 2017; Miranda-Filho et al., 2017; Patel et al., 2019; Spina et al., 2006). The CNS cancer treatment includes several strategies combined, such as surgery, hormonal therapy, chemotherapy, and radiation therapy (Batra, Pawar, and Bahl 2019). Despite those strategies and the aggressive treatment regime, CNS cancers are still associated with high mortality rates, which reveal the need for improvement in the treatments during the early disease stages.

Neuroinflammation induced by viral infections also demands pharmacological approaches with a specific and efficient brain drug delivery. CNS viral infections may cause direct effects that can lead to irreversible disruption of the CNS function, which can be induced by a wide range of viruses. Meningitis is considered the most common unspecific CNS response to viral infections, followed by encephalitis and myelitis (Dahm et al., 2016; Simko et al., 2002). Moreover, some neurological diseases are considered to be long-term and delayed virus-induced disorders such as multiple sclerosis and Guillain-Barré syndrome (Jasti et al., 2016; Palao et al., 2020). Viral CNS infections can also contribute to the development of AD. The neuroinflammation induced by West Nile virus, a mosquito-borne RNA virus belonging to the *flavivirus* family, is a major pathogen of the CNS and causes viral encephalitis. Two other members of the *flavivirus* family that are also linked to viral infections are Zika virus and dengue, such viruses were shown to increase the level of cytokines such as IL-6, TNF-α, and IL-1β and disrupt the BBB (Ben Abid et al., 2018; Dahm et al., 2016; Ingusci et al., 2019). Thus, cytokines can also be targets to novel approaches to treat neuroinflammation using monoclonal antibodies (Figueiredo et al., 2019; Klotz and Wiendl, 2013; Shultz et al., 2013; Tallantyre et al., 2018) or via genetic silencing with iRNA (Ben-David et al., 2020; Chen et al., 2020; Fu et al., 2020; Zhang et al., 2021).

In this context, the BBB constitutes the greatest challenge faced by pharmaceutical formulators, since it is a physiological barrier responsible for protecting the CNS from infections and exogenous substances (Krol, 2012; Wohlfart, Gelperina, and Kreuter, 2012). The BBB is constituted mainly by brain capillary endothelial cells (BCECs), where tight junctions can be found. Along with the endothelial cells, additional cells, such as pericytes, astrocytes end-feet, and a discontinuous basal lamina are present (Krol, 2012). This whole structure proposes a highly controlled environment which can be attributed to: i) tight junctions, which seal the intercellular gap and enable the role as a physical barrier (Luissint et al., 2012), which feature extremely high transendothelial electrical resistance between the blood and the brain, resulting in a restricted passive diffusion to several substances (Gao, 2016); ii) reduced rate of pinocytosis from the luminal side, effectively preventing all the uncontrolled cell arrival; iii) the lack of fenestrations; iv) an enzymatic barrier as second-line protection (Krol, 2012); and v) efflux mechanisms, which are able to prevent drug transport, such as the glycoprotein P, expressed in the lumen of the BBB's endothelium; and, while lipophilic molecules can easily overcome the BBB, this efflux mechanism feature many lipophilic substances as substrate, limiting the influx of those molecules (Gao, 2016). As shown in Figure 11.1, there are several mechanisms through which a drug can overcome the BBB.

Nanostructured Drug Carriers 261

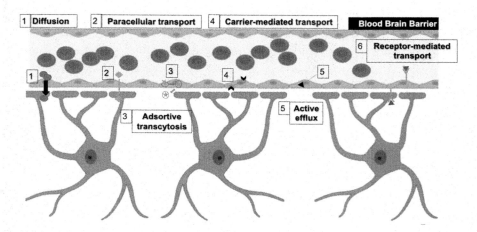

FIGURE 11.1 An overview of different mechanisms through which a drug can cross the BBB. Adapted from Zhou Peng Seven and Leblanc (2018).

Due to the lipophilic character of endothelial cells from BBB, drug targeting to the CNS should consider the hydrophilic environment of the bloodstream and the hydrophobic environment of BBB. Besides, drugs can diffuse through small pores filled with water or cellular channels in the BBB endothelia. Therefore, molecular size, drug partition coefficient, ionization coefficient, and plasma protein binding capacity are characteristics that can influence drug penetration through the BBB. Several authors have shown that lipid-soluble molecules with low molecular weight (around 400–600 Da) are generally transported across the BBB (Bellettato and Scarpa, 2018; Manek and Petroianu, 2021; Poovaiah et al., 2018; Zhang et al., 2016).

One of the main reasons for the clinical failure of CNS pharmacological therapies is the inability of drugs to achieve effective concentrations in the CNS (Oberoi et al., 2016). In addition, some treatments are related to severe metabolic side effects as in the case of Olazanpine, an antipsychotic drug widely used for the treatment of schizophrenia (Townsend et al., 2018). Moreover, antiepileptic oral drugs like carbamazepine, phenobarbital, lamotrigine, gabapentin, retigabine, rufinamide, and stiripentol can present adverse drug reactions even at low concentrations (Jacob and Nair, 2016). Depression and anxiety symptoms associated with psychiatric, physical, and sleep are adverse effects that often lead to the early treatment discontinuation from patients (Dang et al., 2021).

In order to overcome such limitations from these treatments by both oral and intravenous administration, drug delivery via intranasal administration is considered a novel, safer, and alternative route, often called the nose-to-brain route. Several clinical trials are currently being conducted with intranasal drug administration for the treatment of PD and AD, as well as mild cognitive impairment. Table 11.1 shows an overview of clinical trials of intranasal approaches to treat neurological diseases.

TABLE 11.1
An Overview of Clinical Trials with Intranasal Approach for Common Brain Diseases

Disease	Treatment	Stage (Phase)	Status	Outcome	Clinical Trial Identifier
Craniopharyngioma	Oxytocin in Hypothalamic Obesity	2	Recruiting	N/A	NCT03541356
Parkinson	Insulin	2	Recruiting	N/A	NCT04687878/ NCT04251585
Parkinson	Glutathione (in)GSH	1	Completed	Supports the safety and tolerability of (in) GSH in a sample of patients who are within 10 years of PD diagnosis. Needs a larger sample to confirm.	NCT01398748
Parkinson	Reduced Glutathione	1 and 2	Completed	The major limitations of this studies are the lack of verifiable diagnoses reported by respondents, lack of objective symptom improvement, and low survey response rate, all of which were anticipated weaknesses given the study design	NCT02324426 /NCT02424708
Parkinson	Insulin	2	Completed	Even with a small sample the study was able to report an improve in verbal memory, motor	NCT02064166

(Continued)

TABLE 11.1 (Continued)
An Overview of Clinical Trials with Intranasal Approach for Common Brain Diseases

Disease	Treatment	Stage (Phase)	Status	Outcome	Clinical Trial Identifier
Alzheimer	Insulin Glulisine	2	Terminated	performance in comparison with baseline performance The study did not have the power to show difference between groups	NCT02503501
Alzheimer	Insulin	2	Completed	There was no difference between groups regarding hippocampal activation using fMRI	NCT00581867
Alzheimer	Insulin	2	Completed	Both insulin doses preserved cognition and no treatment-related adverse events were observed.	NCT00438568
Mild cognitive impairment and Alzheimer	Insulin	2/3	Completed	In this randomized clinical trial of 289 adults with mild cognitive impairment or Alzheimer disease dementia, no cognitive or functional benefits were observed with intranasal insulin treatment compared with placebo over a 12-month period in the primary analyses.	NCT01767909
Mild cognitive impairment	Davunetide (AL-108)	2	Completed	AL-108 was generally safe and well tolerated. No differences between the efficacy treatment groups on cognitive memory score. Still, AL-108 exhibited merits as a treatment for Alzheimer's disease.	NCT00422981

11.3 THE NOSE-TO-BRAIN ROUTE

Due to the nasal cavity anatomy, the nose-to-brain route allows a direct transport pathway of drugs that enables reaching the brain without an invasive intervention. After intranasal instillation, the drug can be delivered by direct or indirect pathways like olfactory mucosa pathway (direct pathway), trigeminal nerve (direct pathway), and through BBB after reaching systemic circulation (indirect pathway) (Pires and Santos, 2018). As the intranasal route has direct pathways of absorption, it can improve the bioavailability of active ingredients such as therapeutic proteins and peptides (Ghadiri, Young, and Traini, 2019). Furthermore, such direct transport can reduce systemic side effects, improving brain treatment. Figure 11.2 shows an overview of the direct pathways of nose-to-brain administration.

The nasal cavity is divided into three regions: vestibule, respiratory, and olfactory region. The vestibule is more external and is not usually involved in the drug absorption phenomenon. The respiratory region is closely involved with drug absorption, as it provide a wide absorption surface area of almost 160 cm^2 in humans (Selvaraj, Gowthamarajan, and Karri 2018). The olfactory region located at the upper part of the nasal cavity is very close to the olfactory bulb and thus consists of a point of connection between the nose and the brain. This region is composed of

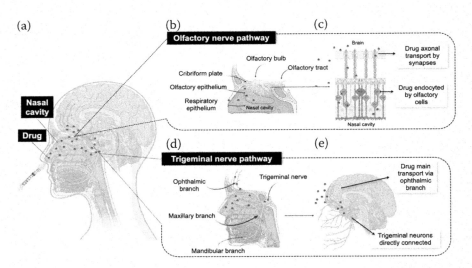

FIGURE 11.2 An overview of the direct pathways of nose to brain administration with the nasal cavity and drug delivery after an intranasal administration (a). The first pathway is the olfactory nerve represented in (b) and once drug is present in nasal cavity, it can be endocyted by olfactory cells (olfactory epithelium) and transported by synapsis (from olfactory nerves) as seen in (c). The second pathway is represented in (d) shows drug transport via trigeminal nerve which is also divided in three branches (ophthalmic, maxillary and mandibular). In (e) the direct connection from trigeminal neurons is highlighted as the main drug transport via the ophthalmic branch. Adapted from Glezer et al. (2020); Hirlekar and Momin (2018); Selvaraj Gowthamarajan and Karri (2018).

basal epithelial cells, sustentacular cells, and olfactory neurons (Samaridou and Alonso, 2018). After intranasal administration, once the bioactive molecules overcome the olfactory mucosa pathway, it is submitted to endocytosis by neurons and supporting olfactory cells (Hong et al., 2019). After this, drug substances are transported by the olfactory cells, followed by the axonal transport to the synapses of the olfactory bulb where it undergoes exocytosis, reaching different brain regions (Bonferoni et al., 2020).

The transport through the trigeminal nerve consists of another route of connection between the nasal cavity and the olfactory bulb. This is an alternative and direct pathway for molecules to bypass the BBB. The trigeminal nerve is composed of three branches: ophthalmic, maxillary, and mandibular branches. The ophthalmic branch is the most important for nose-to-brain drug delivery since the neurons originated from the ophthalmic branch pass directly through the nasal mucosa (Cunha et al., 2017). Drug uptake can also occur through bloodstream circulation, representing an indirect pathway. It is possible, due to the rich vasculature of the olfactory mucosa and the combination of the continuous and fenestrated epithelium that allows the diffusion of small molecules into the bloodstream, reach the BBB (Crowe et al., 2018). Both drug molecular weight and lipophilicity must be considered since large molecules with a high number of hydrophilic bonds cannot overcome the BBB (Ganger and Schindowski, 2018).

One of the key advantages of the nose-to-brain route is the elimination of peripheral side effects since the drugs usually reach minimal plasma levels (Wang et al., 2019). The nasal mucosa is both highly vascularized and permeable. The presence of columnar epithelial cells, which present a large surface area, enables and favors drug systemic absorption. This property, combined with the avoidance of enzymatic degradation and first-pass hepatic metabolism, contributes to a rapid onset of action commonly attributed to the nose-to-brain route (Grassin-Delyle et al., 2012; Shao, 2017).

In this context, nose-to-brain drug delivery emerged as a promising alternative to improve drug targeting to the brain through nanosystems (Costa et al., 2019). The intranasal administration for drug delivery has been considered an emerging alternative for the treatment of CNS disorders like AD and PD as well as depression, anxiety, and others (Shringarpure et al., 2021). To date, it has been an efficient alternative to treat epilepsy in vivo using animal models, as observed by Goncalves et al. (2019). In addition, nose-to-brain drug delivery is an especially trending research topic in order to overcome the limitation of BBB permeability and blood-cerebrospinal fluid barrier (BCSF) (Rassu et al., 2016).

Due to the rapid drug absorption, the ability to avoid the drug degradation in the gastrointestinal tract and the hepatic first-pass metabolism, and better patient compliance, the nose-to-brain should be further explored when designing new drug delivery systems, since it increases drug delivery to the brain (Schwarz and Merkel, 2019). Although low molecular weight drugs often present great bioavailability in the brain region, their absorption can be improved by intranasal formulation, with the addition of surfactants, or using nanostructured drug carriers as a vector for drug delivery.

11.4 LIMITATIONS FROM NOSE-TO-BRAIN DELIVERY

Despite the advantages, it is relevant to mention that intranasal administration presents drawbacks regarding the nasal physiology and formulation common limitations. The BBB represents the major obstacle to brain delivery of therapeutic agents, especially for large molecular weight drugs or hydrophilic molecules, such as bioactive peptides, nucleic acids, or hormones (Schwarz and Merkel, 2019).

The olfactory region plays the most important part in drug transportation to the CNS, since the olfactory neurons reach the olfactory bulb via the cribriform plate of the ethmoid bone, enabling transportation along such neurons (Wu, Hu, and Jiang 2008). The potential intranasal limitations include the small nasal cavity volume combined with the drug's short residence time and the poor paracellular drug transportation through olfactory epithelium (Sonvico et al., 2018).

Despite the noninvasive nature and rapid onset of action, the nasal cavity also feature enzymatic activity and attributes to poor permeability across nasal and olfactory barriers (Mittal et al., 2014). The drug protection from nasal cavity enzymes and degradation combined with nonirritant formulation properties is desirable since an intact olfactory epithelium is important for drug delivery, which involves endocytosis, transcytosis, and paracellular diffusion (Kiparissides et al., 2019). Among the enzymatic activity, it is important to mention the P-glycoproteins responsible for drug efflux, isoforms of CYP450, transferases, epoxide hydroxylase, carboxyl esterase, and aldehyde dehydrogenases that can affect drug bioavailability (Bahadur and Pathak, 2012; Martins, Smyth, and Cui 2019).

Additionally, about 15% of the respiratory cells are covered by long cilia, which produce mucus, a viscous gel with 2–4 μm thickness (Mistry, Stolnik, and Illum 2009). The mucociliary clearance phenomenon occurs every 10–15 minutes and can entrap drugs, microparticles, and other exogenous substances in the cilia and mucus, so they can be eliminated in a short time (Martins, Smyth, and Cui 2019). Despite the noninvasive nature and rapid nose-to-brain action, the nasal cavity also presents enzymatic activity and attributes to poor permeability across nasal and olfactory barriers (Kiparissides et al., 2019).

Along with the olfactory and trigeminal nerve systems that can promote drug transportation via nose-to-brain route, the presence of blood capillaries in the nasal cavity also allows the drug to reach the brain. Despite the abundant bloodstream in nasal cavity, a small number of drugs enter the systemic circulatory system from the respiratory region and subsequently cross the BBB to reach the CNS (Wu, Hu, and Jiang 2008). Concerning the formulation characteristics, the physical-chemical properties of the drug, including its molecular weight and lipophilicity, determine the success of the nose-to-brain treatment. There are also limitations regarding the dose-volume for liquids (100–250 μL) and powders (20–50 mg, depending on their bulk density), which restricts this route to be applied to highly potent drugs. Additionally, to allow frequent administration, the formulation must not cause any harm, irritation, or allergic reactions to the nasal cavity (Ganger and Schindowski, 2018).

Figure 11.3 summarizes the drug, formulation, and nasal physiologic characteristics that promote limitations to the intranasal administration. To avoid these limitations, nose-to-brain formulations can be optimized by using technological

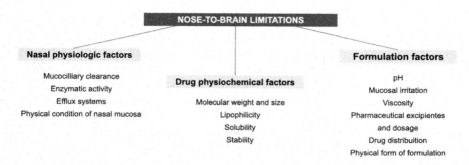

FIGURE 11.3 Examples of nose-to-brain limitations regarding nasal physiological conditions, drug physical-chemical characteristics, and formulation features. Adapted from Pardeshi and Belgamwar (2013).

strategies as nanotechnology, surfactants, and enzymes inhibitors to improve the drug delivery.

11.5 REQUIREMENTS OF NOSE-TO-BRAIN FORMULATIONS

Absorption is a major factor and mandatory performance requirement to be considered during formulation development. Several strategies can be used to optimize and enable the absorption phenomenon, including permeation-enhancing agents, mucolytics agents, mucoadhesive agents, in situ gelling agents, and nanostructured drug carrier technologies. The main goal of nose-to-brain formulations is to enable drug transepithelial absorption. In this context, different formulation strategies as permeation enhancers, mucoadhesive excipients, in situ gelling polymers, and nanostructured carriers can be applied (Martins, Smyth, and Cui, 2019).

Once the formulation is administrated, the first obstacle to achieve biological effect is to cross nasal natural barriers such as the epithelial membranes and mucociliary clearance. Besides, nasal drug absorption depends on drug molecular weight, lipophilicity, and its capacity to resist enzymatic degradation of proteases and peptidases (Poupot, Bergozza, and Fruchon, 2018). According to previous studies, some formulation requirements for the intranasal administration include particle size between 30 and 120 μm in order not to be inhaled and reach the lungs (Trows et al., 2014). Drug molecular weight should be less than 1 kDa to overpass the tight junctions, meanwhile lipophilic and hydrophilic drugs are absorbed through passive diffusion and paracellular pathway, respectively (Costa et al., 2019).

Besides the physical-chemical aspects related to the drug, the formulation can also be optimized, and the use of some excipients may overcome the challenges related to the nasal cavity, as enzymatic activity (Wu, Hu, and Jiang, 2008). Nanotechnology-based formulations can provide drug protection from enzymes. In this way, to improve physical-chemical characteristics from the drugs, NPs are used as a strategy for nose-to-brain delivery. Following this idea, Shobo et al. (2018) developed intranasal oil-in-water nanoemulsion containing pretomanid, an antibiotic used for meningitis treatment. In this work, male rats treated with nanoemulsion via intranasal showed higher concentrations of pretomanid in plasma and brain after 4 hours. Also, brain drug concentration was fourfold higher than what was found after the drug in solution was administered in rats. Thus, nanoemulsions improved drug solubility and pharmacokinetics, meanwhile the intranasal administration facilitated pretomanid to be widespread through different brain sites.

Nanostructured vehicles provide drug protection, time or space-controlled drug release, and enhanced drug absorption (Feng et al., 2018). Thus, NPs can constitute novel formulations, as they can be easily modified to improve drug bioavailability and delivery. In general, the main desired characteristics of an efficient nose-to-brain formulation are odorless, tasteless, mucus solubility and bioadhesivity, suitable dose volume, rapid onset of action, physical-chemical stability, and low irritability (Costa et al., 2019). To meet those requirements, it is common that NPs are often used to facilitate drug incorporation, solubility, absorption, and brain delivery. NPs can be obtained by biocompatible and biodegradable materials like polymers, lipids, and peptides. Furthermore, NP-based formulations can be combined with enzymatic inhibitors, nasal absorption enhancers, and mucoadhesive polymers (Hong et al., 2019).

In order to avoid nasal irritation, the formulation pH should be adjusted to 4.5–6.5, which is close to the pH found in the nasal mucosa, and to ensure drug stability, the pH must favor the drug unionized form (Bahadur and Pathak, 2012). Although the therapeutic effects are dose-dependent, there is a limited administration volume of 200 μL at maximum, due to the nasal cavity anatomy (Pires and Santos, 2018; Wu, Hu, and Jiang 2008).

Besides, the therapeutic effect can be improved with the increasing of nasal drug residence time, consequently favoring drug absorption with mucoadhesive polymers and permeation enhancers (Warnken et al., 2016). Cellulose derivates, alginates, chitosan, polyacrylates, pectin, lectins, and poloxamer are examples of mucoadhesive polymers that can be used in formulations. On the other hand, to enhance permeation, surfactants are commonly used due to their capacity to disrupt the nasal epithelial barrier (Martins, Smyth, and Cui 2019). As surfactants, Cremophor EL, Cremophor RH40, Poloxamer 188, laurate sucrose ester, cyclodextrins, lipids, and even polymers have been already used to enhance drug permeation across the nasal mucosa (Chen et al., 2013). Table 11.2 shows some studies in which polymers or surfactants were used to improve mucoadhesion and drug permeation from nose-to-brain formulations.

The mucoadhesion phenomenon constitutes the result of the interactions between mucoadhesive polymer and mucin chain, which are influenced by polymer-related factors such as average molecular weight, chain flexibility, hydration capacity, hydrogen-bonding capacity, and charge (Russo et al., 2016). There are several theories that can explain mucoadhesion, e.g., electronic theory, adsorption theory, wetting theory, diffusion theory, fraction theory, and mechanical theory (Bappaditya et al., 2017). A well-known mucoadhesive polymer, chitosan, is able to present several mechanisms of mucoadhesion. However, hydrogen bonding with

TABLE 11.2
Recent Studies Using Mucoadhesive and Surfactant Agents for Nose-to-Brain Formulations Improvement

Formulation Additive	Category	Intranasal Formulation	Treatment	Ref
Hydroxypropylmethylcellulose and chitosan	Mucoadhesive polymers	Opiorphin liposomes in gel	Analgesia	Mura et al. (2018)
α-Cyclodextrin	Surfactant	Ribavirin microparticles	Viral encephalitis	Giuliani et al. (2018)
Poloxamer	Mucoadhesive polymer	Naringenin nanoemulsion	Ischemic	Ahmad et al. (2020)
Chitosan, poloxamer and alginate	Mucoadhesive polymers	Levodopa gel	Parkinson	Alipour, Azari, and Ahmadi (2020)
Cremophor RH 40	Surfactant	Resveratrol gold nanoparticles	Neuroprotection and AD	Salem et al. (2019)
β-Cyclodextrin	Surfactant	Quercetin-cyclodextrin complex	Neuroprotection and AD	Manta 2020

glycoprotein of mucin due to presence of –OH and –NH$_2$ groups can correspond to the most important feature to mucoadhesion (Ways, Lau, and Khutoryanskiy 2018). Thus, chitosan is one of the most used mucoadhesive polymers used for mucosal formulations and it has been proposed for nose-to-brain drug delivery systems (Rassu et al., 2016).

Piazzini et al. (2019) developed a formulation of human serum albumin NPs containing sulforhodamine B sodium salt, used as a model drug. In this work, the NPs were coated with chitosan in order to provide the formulation mucoadhesion, stability, drug release modulation, biocompatibility, and permeability. The mucoadhesion evaluation was accessed by the turbidimetric method, which indicated an increase in absorbance difference from NPs. This can be explained by the strong interaction between the positive charges from chitosan and negative charges from mucin chains. Also, ζ-Potential corroborated this interaction, since the positive charges from chitosan-coated NP were neutralized. Besides, once the brain microvascular endothelial cells were exposed to chitosan-coated NPs, they presented less expressive levels of tight junctions, as shown in western blot results. Some previous studies mentioned that chitosan-coated NPs could disrupt the tight junction through deprivation of Ca^{2+} from cadherins at adherents junctions, which could interact with a negative charge from integrin αVβ3 on cell membranes (Hsu et al., 2013; Wang et al., 2017). Therefore, chitosan can open tight junctions, with the capacity to improve drug permeation across nasal mucosa and consequently, olfactory and trigeminal nerves.

Similarly, De Oliveira Junior et al. (2020) produced polycaprolactone (PCL) NPs containing bexarotene, a neuroprotective drug. NPs were covered by a high density of low MW poly(ethylene glycol) (PEG) in order to increase their penetration and stability in mucus. Their results have shown that the presence of PEG did not affect particle size, which was around 100 nm, which was according to the proper size range for this intranasal administration. Also, in vivo results showed that 5% of PEG-coating increased nose-to-brain delivery, once, it allowed better nanoparticle mucus penetration. This was explained by the shield-like protection provided by PEG coating, which neutralized NP surface charge and improved their mucus penetrating ability (Xu et al., 2013). Hence, brain drug concentration was twofold higher by PEG-PCL NPs than PCL NP without PEG-coating.

Besides the improvement of mucoadhesiveness by polymers, the use of nanostructured drug carriers such as nanoemulsions is also a strategy for nose-to-brain due to their small droplet size and ability to enhance the solubility of poorly soluble drugs (Bahadur et al., 2020). Considering those characteristics, Espinoza et al. (2019) aimed at a treatment for AD, combining polymeric nanoemulsion with Pluronic F-127 (PF127)-based thermoreversible gels, to improve bioadhesion, penetration, and delivery of donepezil. As result, the nanoemulsion-based formulation exhibited high viscosity which allowed an enhancement of mucoadhesion in ex vivo assays with porcine nasal mucosa. Furthermore, the nanoformulation enhanced drug solubility in a suitable pH with physical stability and without causing nasal irritation. Therefore, recent advances in nose-to- brain delivery include nanostructured drug vehicles from different materials.

11.6 NPS FOR NOSE-TO-BRAIN TARGETING

In the past decades, researchers have been turning their attention to the use of nanostructured drug carriers to overcome the challenges of drug delivery to the brain. An NP can be defined as a colloidal structure that presents at least one of its dimensions smaller than 100 nm, and further classified as ultradispersion systems (1–50 nm), high dispersion systems (50–100 nm) and sedimentable particles (100–1000 nm) (Strambeanu et al., 2015). NPs for nose-to-brain delivery purposes include solid lipid NPs, liposomes, micelles, dendrimers, nanoemulsions, polymeric micelles, and polymer-based NPs as the most common nanostructured drug vehicles used (Espinoza et al., 2019; Pires and Santos, 2018). These NPs can be tailored to provide improved drug solubility, protection against enzymatic degradation with agents that avoid enzymatic digestion, high encapsulation efficiency, and enhancement cellular internalization (Gao, 2016). Table 11.3 summarizes different types of NPs that can be applied for drug delivery and their advantages for drug delivery. Hence, NPs are a strategy that improves nose-to-brain drug delivery.

Since conventional nasal liquid formulations present extremely low drug penetration and absorption, nanoformulations are applied as a strategy for treatments as

TABLE 11.3

Examples of NPs for Brain Delivery and Their Main Advantages

Nanoparticle or Nanostructured Drug Vehicles	Main Advantages
Micelles	Increase of drug solubility
Liposomes Micro/Nanoemulsions	Drug protection from enzymes
Solid lipid nanoparticles	Increase targeting
Niosomes	Lipophilic/hydrophobic drugs transport
Polymeric micelles	Prevent opsonization
Polymeric nanoparticles	Improve brain pharmacokinetics
Dendrimers	High BBB permeability
Metalic nanoparticles	Active target delivery
Nanotubes	Zero order release
Nanofibers	Drug controlled release
	Suitable for surface ligands
	Resistance of enzymatic degradation
	Accumulation at the target site
	Low production cost
	Good biodegradability
	Long-term stability
	Easy of synthesis
	High target efficiency

Source: Adapted from Martins, Smyth, and Cui (2019); Md et al. (2015), Mendes et al. (2018), and Zhou et al. (2018).

PD, AD, glioblastoma, and epilepsy (Islam et al., 2020; Mignani et al., 2021). Due to the advantages of NPs, already mentioned above, recent studies apply different types of NPs and nanosystems that can improve nose-to-brain therapeutics. Table 11.4 shows some examples of recent studies, displaying the active pharmaceutical ingredient, therapy purpose, and a brief relevant therapeutical outcome.

11.6.1 Lipid-Based NP

Once the nose-to-brain route includes reaching the systemic bloodstream and crossing the BBB, lipid-based nanoparticles based on natural lipids in their composition can be highly compatible with BBB, providing enhanced drug brain delivery. As an example of lipid-based NP, SLNs, which are nanostructured vesicles mainly constituted by a solid lipid core, can entrap lipophilic drugs, genes, DNA, plasmid, and proteins stabilized by a surfactant layer. SLNs feature many advantages, such as the possibility of tailoring their surface properties by modification, increase drug pharmacokinetic, formation of complex structures for delivery, and improve on drug stability. Moreover, aqueous formulations containing SLN can be stored for over 3 years and their gelling tendency can be controlled to withstand modifications (Duan et al., 2020).

Exploring the advantages of SLN, Yasir et al. (2017) developed SLNs of donepezil for intranasal administration, aiming the treatment for mild to moderate AD. In this work, it was observed that the treatment with SLNs provided a higher drug concentration in the mice brain from when in comparison with the drug in solution. This can be explained by the SLN drug protection properties, which avoids rapid clearance in the nasal cavity from mucociliary activity and the protection from enzymatic degradation. Further, the use of Tween 80 as surfactant contributed to the temporary opening of cell junctions, and consequently, NPs endocytosis.

Another example of SLN to treat not only AD but also cerebrovascular disorders were proposed by Wang et al. (2019). In this work, SLN containing *Pueraria flavones* were produced and modified by the incorporation of borneol, a monoterpene known to enhance drug penetration into the brain. Their in vivo results showed that after intranasal administration, the borneol-based SLN improved the drug's pharmacokinetic, reaching a higher maximum concentration in comparison with the plain SLN. Thus, it was confirmed that the borneol-based SNL were capable of cross BBB, reaching the brain quickly, in higher drug concentrations and with a higher distribution in the brain.

More recently, Hady, Sayed, and Akl (2020) also selected SLN to carry levofloxacin and doxycycline (LEVO/DOX) in combination, as an intranasal treatment model for meningitis. By combining different solid lipids, it was observed LEVO/DOX SLNs showed a size range of 14–50 nm and a polydispersity index (PdI) of 0.2 approximately, the size demonstrated that SLNs could be absorbed via nose-to-brain route. These NPs were incorporated into a nasal formulation based on a mucoadhesive gel of hydroxypropyl methylcellulose and then the intranasal administration was performed in rats. The formulation provided a significantly higher maximum concentration and a drug targeting efficiency (DTE) to the brain when compared to intravenous administration. In addition, the nose-to-brain direct

TABLE 11.4
Examples of Nanoparticles for Nose to Brain Delivery with the Active Pharmaceutical Ingredient, Therapy Purpose, and Their Brief Relevant Therapeutical Outcomes

Nanomaterial	Active Pharmaceutical Ingredients	Therapy For	Relevant Therapeutical Outcomes	Reference
Solid lipid nanoparticle	Olanzapine (OLZ)	Psychosis and schizophrenia	Higher drug entrapment and loading capacity. No risk of OLZ hematological and liver toxicities	Gadhave Choudhury and Kokare (2018)
PCL nanoparticles	Curcumin	Brain glioma	Polymeric NPs reach the brain predominately via trigeminal nerve pathway	Li et al. (2019)
Liposomes	Caffeine, hydrocortisone, ibuprofen, ketoprofen, methylprednisolone, and theophylline	Alzheimer's disease	Environment tonicity influences on liposomes size, polydispersity and drug release	Wu et al. (2019)
Nanoemulsion	Letrozole	Convulsion and neuroprotection	Drug solubilization and permeation enhancement. 100% of protection and seizures reduction in mice	Iqbal et al., (2019)
Niosomes	Flibanserin	Premenopause	Low oral bioavailability overcome. Niosomes capacity of increase uptake of brain cells	Fahmy et al. (2020)
Dendrimers	Glutaminase inhibitors (GLS)	Rett syndrome	Microglial GLS inhibition and improve mobility in a mouse mode	Khoury et al. (2020)

transport percentage (DTP) was negative for intranasal free drugs, indicating that the effective delivery from the free drugs to the brain was via an indirect pathway through systemic circulation and BBB permeation. Since the DTE for SLN-gel formulation was higher, it was concluded that the mucoadhesivness of colloidal gel guaranteed NPs deliver to the brain using direct routes as olfactory and trigeminal nerves without crossing BBB.

Besides SLN, liposomes also represent lipid NPs commonly applied to nose-to-brain drug delivery. Liposomes can be defined as spherical vesicles with a lipid bilayer as a shell and an aqueous core. Due to their nature, they can be tailored to mimic cell membranes, incorporate lipophilic, or amphiphilic drugs in their lipid layer or encapsulate hydrophilic compounds in their core. Another advantage to the use of liposomes on brain target delivery systems is that they enable a higher number of drugs to be delivered and can also be considered neutral and inert carriers for encapsulating molecules, and depending on the coupled targeting vector, their pharmacokinetics and tissue distribution can be tailored (Rajpoot, Ganju, and Singh, 2020; Schnyder and Huwyler, 2005). As liposomes can be tailored to target brain and can improve drug pharmacokinetics. Several strategies can be employed to brain vectorization, such as cationization, targeting ligands, and surface functionalization with PEG or peptides, antibodies, aptamers (Lai, Fadda, and Sinico 2013; Vieira and Gamarra, 2016).

Due to such properties, Pashirova et al. (2018) applied cationic liposome as a nose-to-brain drug delivery approach for the treatment of organophosphorus poisoning (OP). In this way, the organophosphate paraoxon was used in a rat model for OP and an acetylcholinesterase (AChE) reactivator pralidoxime chloride (2-PAM) was incorporated into liposomes as a treatment strategy. Their results showed that intranasal administration of free 2-PAM was an ineffective treatment. However, in vivo assays demonstrated that liposomes were capable of reactivating 12% of AChE. Such results proved that modified liposomes are a successful strategy to successfully achieve the nose-to-brain pathway.

Similarly, Dhaliwal et al. (2020) recently produced cationic liposomes containing messenger RNA (mRNA) as an intranasal therapy designed for neurodegenerative disorders. In vitro results showed that different from free mRNA, mRNA-based liposomes were able to reach cytosol of murine macrophages, promoting transfection in 12 and 24 hours. Besides this, in vivo intranasal administration demonstrated that encapsulated mRNA was able to enter different brain regions such as the cortex, striatum, and midbrain. Therefore, liposomes improved the short-term life from mRNA and bioavailability, setting efficient acute and long-term treatments in a noninvasive route.

Polymeric micelles and nanoemulsions are also an example of lipid-based nanostructured vesicles for the nose-to-brain route and their structures are formed by an oily core stabilized by amphiphilic molecules like surfactants or amphiphilic block copolymers (Hanafy, El-Kemary, and Leporatti 2018). They feature the ability to self-assemble when exposed to an aqueous media, resulting in a hydrophobic core surrounded by a hydrophilic corona (Branco and Schneider, 2009). There are several advantages to the use of micellar nanocarriers when applied to nose-to-brain applications. For example, micelles possess a small droplet size, suitable flow properties and low viscosity, ability to incorporate several ingredients, longevity and higher retention effects in the targeted area, enhancing drug permeability across the nasal mucosa (Torchilin, 2007).

As an example of micelles application, Keshari Sonar and Mahajan (2019) developed curcumin-loaded micelles with less than 100 nm. Those micelles presented an in vitro biphasic drug delivery starting with an initial burst release followed by slow and sustained release up to 91.7% of the curcumin from the micelles. The in vitro investigation of sheep nasal mucosa mounted on Franz diffusion cells showed no histopathological changes when compared to the negative control. In vivo studies of brain distribution on Wistar rats showed that after 30 minutes of intranasal administration, the animals which received the micellar nanocarriers had a significantly higher brain concentration after only 30 minutes, with the maximum concentration being reached after 60 minutes. That was due to the micellar nanocarrier being able to be transported via the olfactory pathway of the brain.

Nanoemulsions (NEs) are defined as a heterogeneous system in which one liquid (dispersed phase) is dispersed as nanosized droplets in another liquid (continuous phase) stabilized by emulsifiers or surfactants (Barradas, 2020; Tayeb, Sainsbury, 2018). Their droplet size ranges from around 20–200 nm (Barradas, 2020). Such small droplet size impairs destabilization phenomena like coalescence, creaming, and sedimentation (Bonferoni et al., 2019). Moreover, NE features a superior solubilization capacity of poorly soluble drugs, thermodynamic stability with relatively easy preparation and due to these advantages, they have been widely explored to enhance nose-to-brain drug delivery (Ahmad et al., 2017; Bonferoni et al., 2020).

Rinaldi et al. (2020) developed a NE formulation containing essential oils and antimicrobial drugs for brain infections caused by meningitis and encephalitis. Sorbitane monolaurate (Span 20) is a surfactant accepted by FDA for intranasal administration and was used to obtain a stable NE and also to enhance intranasal permeation. NE was coated in a chitosan solution due to its mucoadhesive properties, hindering the clearance from mucociliary activity. The mucoadhesive assays showed that NE covered by chitosan presented interaction with mucin, the main nasal mucus component. The NE formulation also presented pH values between 3.5 and 6.4, being suitable for nasal route administration. In vitro experiments confirmed the antibacterial action against methicillin-resistant *Staphylococcus aureus*, hence, the proposed formulation would be a treatment for brain infections.

NE formulations to achieve nose-to-brain drug delivery systems as treatments for AD have been recently proposed in order to improve patient compliance and due to NE thermodynamic stability, allowing production with less energy input. Dhaliwal et al. (2020) produced NE containing donepezil, an AChE inhibitor, available in oral formulations. To overcome the side effects from oral formulations and to increase drug brain concentration, their work aimed at an oil/water NE for brain delivery via olfactory nerves. The in vitro results confirmed that the NE formulation did not present toxicity for neurons. Meanwhile, the rat brains showed that after 24 hours of treatment, the amount of donepezil via NE intranasal administration was almost two and sixfold higher than intravenous and oral administrations, respectively. It was concluded that donepezil NE via intranasal was more effective to target the brain than the other routes, being a new approach in the AD treatment.

Another approach for AD treatment with NE was proposed by Kaur et al. (2020), using memantine, an uncompetitive N-methyl-D-Aspartate (NMDA) receptor antagonist. The NE characterization demonstrated that droplet size was 11 nm and PDI 0.080, which indicated a monodispersed formulation. The rats treated via intranasal with the radiolabeled NE formulation presented a maximum brain uptake than the animals treated by oral and intravenous routes. As final observations, drug targeting and transport efficiencies were 207.23% and 51,75%, respectively, showing that memantine NE after intranasal administration could improve the treatment of AD.

11.6.2 Polymer NP-Based Systems

Polymer NP constitute deeply investigated vehicles for drug transportation for brain disorders. Such NP can be defined as solid colloidal particles that are obtained from natural or synthetic polymers. In such systems, the drug can be dissolved, attached, encapsulated in the NP core, or dispersed in the polymer matrix (Sur et al., 2019). Regarding drug delivery, polymer NPs can control release through diffusion, desorption, bulk, or surface erosion (Tan et al., 2020). As advantages, polymeric NP can be easily prepared, and their size distribution can also be controlled by optimizing their technique. Essentially many polymers can be used for NPs, among them polylactic acid (PLA), polyglycolic acid (PGA), and poly lactic-co-glycolic acid (PLGA) are most common due to their biocompatible and biodegradable characteristics (Calzoni et al., 2019).

De Oliveira Junior et al. (2019) applied PCL-based NPs containing melatonin (MLT) as a treatment against glioblastoma. As result, nanoencapsulation increased drug solubility in almost 35-fold and the NP biocompatibility was confirmed by the absence of cytotoxicity from in vitro assays with healthy cells from human glioblastoma and human pulmonary fibroblasts. Also, NPs had their surface modified by Tween 80, which enhanced their penetration in the olfactory tissue of rats. Consequently, with major NP absorption, the intranasal route showed better pharmacokinetics parameters, higher and safe concentration of MLT in the brain than oral administration.

Wang et al. (2020) also produced polymer NPs to increase drug solubility and stability. In this work, polymeric micelles were used as NPs to deliver rotigotine, a dopamine agonist, for PD treatment. The NE was further incorporated into a thermosensitive PF127-based hydrogel, which prolonged the drug residence time and drug absorption. In vivo assays showed that NPs in gels presented higher rotigotine concentration in the following regions: cerebral cortex, cerebellum, and olfactory bulb. Whereas olfactory bulb presented greater drug concentration than the other regions, which indicates that the olfactory bulb is the first target and both, olfactory, and trigeminal nerves, are the primary intranasal routes.

Besides polymer NPs, dendrimers comprise NPs that have been highlighted in recent years their applicability as pharmaceutical excipients due to their composition of highly branched polymers with surfaces that are easily tailored (Santos,

Veiga, and Figueiras 2019). As a definition, dendrimers are nanometric molecules with well-defined branches origined from a central core. Their structure is defined by a fundamental core chemical group from which branches called dendrons rise through diverse chemical reactions (Akbarzadeh et al., 2018). They feature key characteristics such as a globular shape, well defined 3D structures, cavities, and high functionalities. Further, their surface can be tailored with functional groups to enhance vascular permeability and biodistribution, meanwhile, dendrimers size can be controlled with polymerization, resulting in higher solubility, miscibility, and reactivity (Hecht and Fréchet, 2001).

The in vivo drug release from dendrimers can occur in two different forms, being by degradation with enzymes or releasing the drugs due to pH or temperature (Patel and Patel, 2013). Katare et al. (2015) developed dendrimers for haloperidol delivery, an antipsychotic drug. The dendrimers were composed of polyamidoamine (PAMAM), a polymer with amine surface groups, allowing haloperidol binding. In vivo assays with rats compared treatments via intranasal (IN) and intraperitoneal (IP) routes. After the IN treatment, animals had a cataleptic response similar to that achieved when the IP group received a 6.7 times larger dose of haloperidol during a 30–60 minute period post-administration. The IN route was seven times more efficient, when it came to delivering the haloperidol to the striatum when compared to the IP route, demonstrating the efficiency of the dendrimers in delivering that drug.

Recently, Xie et al. (2019) synthesized polyamidoamine (PAMAM) dendrimers associated with in situ gelling systems formed by deacetylated gellan gum. This study aimed to develop an efficient and direct transport to the brain combining an in situ gel with PAMAM dendrimer loaded with paeonol, a neuroprotective agent already applied in rat cerebral ischemic (Zhao et al., 2018) and epileptic (Liu et al., 2019) models. In this work, during in vitro assay, confocal electron microscopy showed that hepatocyte carcinoma cells presented a higher uptake from dendrimers-gel formulation than dendrimer nanocomposites in solution. Thein vitro uptake result was confirmed with in vivo assays in which the animals treated with intranasal in situ gel presented accumulation at 24 hours, indicating that not only the transport was effective, but also NP was retained in the brain. Besides, after de intranasal administration, the gel was capable to change into a gel with a higher viscosity level due to a phase transition in response to Na^+, K^+, and Ca^{2+} presented in the nasal mucus. Therefore, the gel application prolonged the nasal mucosa drug absorption and the drug release time.

11.7 FINAL REMARKS

Pathological conditions that affect the CNS are difficult to treat and they are all serious and important challenges that can greatly impact the quality of life and even lead to death. On the other hand, traditional therapies are often related to mild to severe adverse effects. In this context, the nose-to-brain drug delivery constitutes an elegant, promising, and noninvasive alternative to oral therapies to treat CNS disorders. However, both intranasal administration and brain drug delivery pose several obstacles, which can impair drug absorption and the success of therapeutics.

Strategic formulation design is a crucial step to provide proper nose-to-brain drug delivery systems. Nanostructured drug carriers are promising strategies due to their exceptional characteristics, such as small size, great surface/volume ratio and their capacity to protect drugs from enzymatic degradation, increase both drug solubility and bioavailability, and control drug release. Nanostructured materials, such as polymeric nanoparticles or micelles, lipid nanoparticles, nanoemulsions, and dendrimers, play an important role in the design of the nose-to-brain formulations. Lipid-based formulations are often used due to their capacity to improve the solubilization and encapsulation efficiency of hydrophobic drugs and on the hydrophilic/lipophilic nature of the barriers present in the nose-to-brain pathways.

The presence of surfactants or permeation enhancers can provide a fluidizing effect on endothelial cell membranes, optimizing drug transmucosal absorption. Moreover, mucoadhesive excipients can be incorporated into nanoformulations in order to increase drug residence time and avoid nasal clearance.

As demonstrated in this chapter, there are many reports in the literature with in vivo and in vitro exciting results drug carriers, which point nanostructured represent as an innovative approach with the potential to change the scenario of marketed products and make the delivery of small molecule drugs and high molecular weight bioactive molecules directly to the brain via intranasal administration, a reality in the near future.

REFERENCES

Ahmad, E., Feng, Y., Qi, J., Fan, W. et al. Evidence of nose-to-brain delivery of nanoemulsions: Cargoes but not vehicles. *Nanoscale* 9, no. 3(Jan 19 2017): 1174–1183.

Ahmad, N., Ahmad, R., Ahmad, F.J., Ahmad, W. *et al.* Poloxamer-chitosan-based Naringenin nanoformulation used in brain targeting for the treatment of cerebral ischemia. *Saudi J Biol Sci* 27, no. 1(Jan 2020): 500–517.

Akbarzadeh, A., Khalilov, R., Mostafavi, E., Annabi, N. *et al.* Role of dendrimers in advanced drug delivery and biomedical applications: A review. *Experimental Oncology* 40, no. 3(2018): 178–183.

Alipour, S., Azari, H., Ahmadi, F. In situ thermosensitive gel of levodopa: Potential formulation for nose to brain delivery in Parkinson disease %. *J Trends in Pharmaceutical Sciences* 6, no. 2(2020): 97–104.

Bahadur, S., Pardhi, D.M., Rautio, J., Rosenholm, J.M. *et al.* Intranasal nanoemulsions for direct nose-to-brain delivery of actives for CNS disorders. *Pharmaceutics* 12, no. 12(Dec 18 2020).

Bahadur, S., Pathak, K. Physicochemical and physiological considerations for efficient nose-to-brain targeting. *Expert Opin Drug Deliv* 9, no. 1(Jan 2012): 19–31.

Bappaditya, C., Nursazreen, A., Pinaki, S., Uttam, K.M. Mucoadhesive polymers and their mode of action: A recent update. *J App Pharm Sci* 7(2017): 195–203.

Barradas, T.N., Silva, K.G.D.H.E. Nanoemulsions as optimized vehicles for essential oils. In *Sustainable Agriculture Reviews 44: Pharmaceutical Technology for Natural Products Delivery*, Springer International Publishing, v. 44, Chapter 4. 115–167 (2020). (Sustainable Agriculture Reviews).

Barradas, T.N., Silva, K.G.D.H.E. Nanoemulsions of essential oils to improve solubility, stability and permeability: A review. *Environ Chem Lett* 19, no. 2 (2020).

Batra, H., Pawar, S., Bahl, D. Curcumin in combination with anti-cancer drugs: A nanomedicine review. *Pharmacol Res* 139(Jan 2019): 91–105.

Bellettato, C.M., Scarpa, M. Possible strategies to cross the blood-brain barrier. *Ital J Pediatr* 44, no. Suppl 2(Nov 16 2018): 131.

Ben-David, Y., Kagan, S., Cohen Ben-Ami, H., Rostami, J. et al. RIC3, the cholinergic anti-inflammatory pathway, and neuroinflammation. *Int Immunopharmacol* 83(Jun 2020): 106381.

Ben Abid, F., Abukhattab, M., Ghazouani, H., Khalil, O. et al. Epidemiology and clinical outcomes of viral central nervous system infections. *Int J Infect Dis* 73(Aug 2018): 85–90.

Bonferoni, M.C., Rassu, G., Gavini, E., Sorrenti, M. et al. Nose-to-brain delivery of antioxidants as a potential tool for the therapy of neurological diseases. *Pharmaceutics* 12, no. 12(Dec 21 2020).

Bonferoni, M.C., Rossi, S., Sandri, G., Ferrari, F. et al. Nanoemulsions for "nose-to-brain" drug delivery. *Pharmaceutics* 11, no. 2(Feb 17 2019).

Branco, M.C., and Schneider, J.P. Self-assembling materials for therapeutic delivery. *Acta Biomater* 5, no. 3(Mar 2009): 817–831.

Cacabelos, R. Parkinson's disease: From pathogenesis to pharmacogenomics. *Int J Mol Sci* 18, no. 3(Mar 4 2017).

Calzoni, E., Cesaretti, A., Polchi, A., Di Michele, A. et al. Biocompatible polymer nanoparticles for drug delivery applications in cancer and neurodegenerative disorder therapies. *J Funct Biomater* 10, no. 1(Jan 8 2019).

Chen, X., Zhi, F., Jia, X., Zhang, X. et al. Enhanced brain targeting of curcumin by intranasal administration of a thermosensitive poloxamer hydrogel. *J Pharm Pharmacol* 65, no. 6(Jun 2013): 807–816.

Chen, Y., Zhang, Y., Ye, G., Sheng, C. et al. Knockdown of lncRNA PCAI protects against cognitive decline induced by hippocampal neuroinflammation via regulating SUZ12. *Life Sci* 253(Jul 15 2020): 117626.

Costa, C., Moreira, J.N., Amaral, M.H., Lobo, J.M.S. et al. Nose-to-brain delivery of lipid-based nanosystems for epileptic seizures and anxiety crisis. *J Control Release* 295 (Feb 10 2019): 187–200.

Cowppli-Bony, A., Bouvier, G., Rue, M., Loiseau, H. et al. Brain tumors and hormonal factors: review of the epidemiological literature. *Cancer Causes Control* 22, no. 5 (May 2011): 697–714.

Crowe, T.P., Greenlee, M.H.W., Kanthasamy, A.G., Hsu, W.H. Mechanism of intranasal drug delivery directly to the brain. *Life Sci* 195(Feb 15 2018): 44–52.

Cunha, S., Almeida, H., Amaral, M.H., Lobo, J.M.S. et al. Intranasal lipid nanoparticles for the treatment of neurodegenerative diseases. *Curr Pharm Des* (Nov 27 2017).

Dahm, T., Rudolph, H., Schwerk, C., Schroten, H. et al. Neuroinvasion and inflammation in viral central nervous system infections. *Mediators Inflamm* 2016, (2016): 8562805.

Dang, Y.L., Foster, E., Lloyd, M., Rayner, G. et al. Adverse events related to antiepileptic drugs. *Epilepsy Behav* 115(Feb 2021): 107657.

De Oliveira Junior, E.R., Nascimento, T.L., Salomao, M.A., Da Silva, A.C.G. et al. Increased nose-to-brain delivery of melatonin mediated by polycaprolactone nanoparticles for the treatment of glioblastoma. *Pharm Res* 36, no. 9(Jul 1 2019): 131.

De Oliveira Junior, E.R., Santos, L.C.R., Salomao, M.A., Nascimento, T.L. et al. Nose-to-brain drug delivery mediated by polymeric nanoparticles: Influence of PEG surface coating. *Drug Deliv Transl Res* 10, no. 6(Dec 2020): 1688–1699.

De Virgilio, A., Greco, A., Fabbrini, G., Inghilleri, M. et al. Parkinson's disease: Autoimmunity and neuroinflammation. *Autoimmun Rev* 15, no. 10(Oct 2016): 1005–1011.

Deleidi, M., Jaggle, M., Rubino, G. Immune aging, dysmetabolism, and inflammation in neurological diseases. *Front Neurosci* 9(2015): 172.

Dhaliwal, H.K., Fan, Y., Kim, J., Amiji, M.M. Intranasal delivery and transfection of mRNA therapeutics in the brain using cationic liposomes. *Mol Pharm* 17, no. 6(Jun 1 2020): 1996–2005.

Duan, Y., Dhar, A., Patel, C., Khimani, M. et al. A brief review on solid lipid nanoparticles: Part and parcel of contemporary drug delivery systems. *RSC Advances* 10, no. 45(2020): 26777–26791.

Dumurgier, J., Tzourio, C. Epidemiology of neurological diseases in older adults. *Rev Neurol (Paris)* 176, no. 9(Nov 2020): 642–648.

Espinoza, L.C., Silva-Abreu, M., Clares, B., Rodriguez-Lagunas, M.J. et al. Formulation strategies to improve nose-to-brain delivery of Donepezil. *Pharmaceutic*s 11, no. 2 (Feb 1 2019).

Fahmy, U.A., Badr-Eldin, S.M., Ahmed, O.A.A., Aldawsari, H.M. et al. Intranasal niosomal in situ gel as a promising approach for enhancing flibanserin bioavailability and brain delivery: In vitro optimization and ex vivo/in vivo evaluation. *Pharmaceutics* 12, no. 6(May 27 2020).

Feigin, V.L., Nichols, E., Alam, T., Bannick, M.S. et al. Global, regional, and national burden of neurological disorders, 1990-2016: A systematic analysis for the Global Burden of Disease Study 2016. *Lancet Neurol* 18, no. 5(May 2019): 459–480.

Feng, Y., He, H., Li, F., Lu, Y. et al. An update on the role of nanovehicles in nose-to-brain drug delivery. *Drug Discov Today* 23, no. 5(May 2018): 1079–1088.

Figueiredo, C.P., Barros-Aragao, F.G.Q., Neris, R.L.S., Frost, P.S. et al. Zika virus replicates in adult human brain tissue and impairs synapses and memory in mice. *Nat Commun* 10, no. 1(Sep 5 2019): 3890.

Fu, X., Jiao, J., Qin, T., Yu, J. et al. A New perspective on ameliorating depression-like behaviors: Suppressing neuroinflammation by upregulating PGC-1alpha. *Neurotox Res* (Oct 6 2020).

Furtado, D., Bjornmalm, M., Ayton, S., Bush, A.I. et al. Overcoming the blood-brain barrier: The role of nanomaterials in treating neurological diseases. *Adv Mater* 30, no. 46 (Nov 2018): e1801362.

Gadhave, D., Choudhury, H., Kokare, C. Neutropenia and leukopenia protective intranasal olanzapine-loaded lipid-based nanocarriers engineered for brain delivery. *Appl Nanosci* 9, no. 2(2018): 151–168.

Ganger, S., Schindowski, K. Tailoring formulations for intranasal nose-to-brain delivery: A review on architecture, physico-chemical characteristics and mucociliary clearance of the nasal olfactory mucosa. *Pharmaceutics* 10, no. 3(Aug 3 2018).

Gao, H. Progress and perspectives on targeting nanoparticles for brain drug delivery. *Acta Pharm Sin B* 6, no. 4(Jul 2016): 268–286.

Ghadiri, M., Young, P.M., Traini, D. Strategies to Enhance Drug Absorption via Nasal and Pulmonary Routes. *Pharmaceutics* 11, no. 3(Mar 11 2019).

Giuliani, A., Balducci, A.G., Zironi, E., Colombo, G. et al. In vivo nose-to-brain delivery of the hydrophilic antiviral ribavirin by microparticle agglomerates. *Drug Deliv* 25, no. 1(Nov 2018): 376–387.

Glezer, I., Bruni-Cardoso, A., Schechtman, D., Malnic, B. Viral infection and smell loss: The case of COVID-19. *J Neurochem* (2020).

Goncalves, J., Bicker, J., Gouveia, F., Liberal, J. et al. Nose-to-brain delivery of levetiracetam after intranasal administration to mice. *Int J Pharm* 564(Jun 10 2019): 329–339.

Gonzalez-Rubio, J., Arribas, E., Ramirez-Vazquez, R., Najera, A. Radiofrequency electromagnetic fields and some cancers of unknown etiology: An ecological study. *Sci Total Environ* 599-600(Dec 1 2017): 834–843.

Grassin-Delyle, S., Buenestado, A., Naline, E., Faisy, C. *et al.* Intranasal drug delivery: An efficient and non-invasive route for systemic administration: Focus on opioids. *Pharmacol Ther* 134, no. 3(Jun 2012): 366–379.

Hady, M.A., Sayed, O.M., Akl, M.A. Brain uptake and accumulation of new levofloxacin-doxycycline combination through the use of solid lipid nanoparticles: Formulation; Optimization and in-vivo evaluation. *Colloids Surf B Biointerfaces* 193(Sep 2020): 111076.

Hanafy, N.A.N., El-Kemary, M., Leporatti, S. Micelles structure development as a strategy to improve smart cancer therapy. *Cancers (Basel)* 10, no. 7(Jul 20 2018).

Hecht, S., Fréchet, J.M.J. Dendritic encapsulation of function: Applying nature's site isolation principle from biomimetics to materials science. *Angewandte Chemie International Edition* 40, no. 1(2001): 74–91.

Hirlekar, R.S., Momin, A.M. Advances in drug delivery from nose to brain: An overview. *Curr Drug Ther* 13, no. 1(2018): 4–24.

Hong, S.S., Oh, K.T., Choi, H.G., Lim, S.J. Liposomal formulations for nose-to-brain delivery: Recent advances and future perspectives. *Pharmaceutics* 11, no. 10(Oct 17 2019).

Hou, Y., Dan, X., Babbar, M., Wei, Y. et al. Ageing as a risk factor for neurodegenerative disease. *Nat Rev Neurol* 15, no. 10(Oct 2019): 565–581.

Hsu, L.W., Ho, Y.C., Chuang, E.Y., Chen, C.T. et al. Effects of pH on molecular mechanisms of chitosan-integrin interactions and resulting tight-junction disruptions. *Biomaterials* 34, no. 3(Jan 2013): 784–793.

Ingusci, S., Verlengia, G., Soukupova, M., Zucchini, S. *et al.* Gene therapy tools for brain diseases. *Front Pharmacol* 10(2019): 724.

Iqbal, R., Ahmed, S., Jain, G.K., Vohora, D. Design and development of letrozole nanoemulsion: A comparative evaluation of brain targeted nanoemulsion with free letrozole against status epilepticus and neurodegeneration in mice. *Int J Pharm* 565(Jun 30 2019): 20–32.

Islam, S.U., Shehzad, A., Ahmed, M.B., Lee, Y.S. Intranasal delivery of nanoformulations: A potential way of treatment for neurological disorders. *Molecules* 25, no. 8(Apr 21 2020).

Jacob, S., Nair, A.B. An updated overview on therapeutic drug monitoring of recent antiepileptic drugs. *Drugs R D* 16, no. 4(Dec 2016): 303–316.

Jasti, A.K., Selmi, C., Sarmiento-Monroy, J.C., Vega, D.A. *et al.* Guillain-Barre syndrome: causes, immunopathogenic mechanisms and treatment. *Expert Rev Clin Immunol* 12, no. 11(Nov 2016): 1175–1189.

Katare, Y.K., Daya, R.P., Sookram Gray, C., Luckham, R.E. et al. Brain targeting of a water insoluble antipsychotic drug haloperidol via the intranasal route using PAMAM dendrimer. *Mol Pharm* 12, no. 9(2015): 3380–3388.

Kaur, A., Nigam, K., Srivastava, S., Tyagi, A. *et al.* Memantine nanoemulsion: A new approach to treat Alzheimer's disease. *J Microencapsul* 37, no. 5(Aug 2020): 355–365.

Keshari, P., Sonar, Y., Mahajan, H. Curcumin loaded TPGS micelles for nose to brain drug delivery: In vitro and in vivo studies. *Mater Technol* 34, no. 7(2019): 423–432.

Khoury, E.S., Sharma, A., Ramireddy, R.R., Thomas, A.G. *et al.* Dendrimer-conjugated glutaminase inhibitor selectively targets microglial glutaminase in a mouse model of Rett syndrome. *Theranostics* 10, no. 13(2020): 5736–5748.

Kiparissides, C., Vasileiadou, A., Karageorgos, F., Serpetsi, S. A computational systems approach to rational design of nose-to-brain delivery of biopharmaceutics. *Ind Eng Chem Res* 59, no. 6(2019): 2548–2565.

Klotz, L., Wiendl, H. Monoclonal antibodies in neuroinflammatory diseases. *Expert Opin Biol Ther* 13, no. 6(Jun 2013): 831–846.

Krol, S. Challenges in drug delivery to the brain: nature is against us. *J Control Release* 164, no. 2(Dec 10 2012): 145–155.

Lai, F., Fadda, A.M., Sinico, C. Liposomes for brain delivery. *Expert Opin Drug Deliv* 10, no. 7(Jul 2013): 1003–1022.

Lalatsa, A., Sun, Y., Gamboa, J.I., Knafo, S. Preformulation studies of a stable PTEN-PDZ lipopeptide able to cross an in vitro blood-brain-barrier model as a potential therapy for Alzheimer's disease. *Pharm Res* 37, no. 10(Sep 4 2020): 183.

Li, Y., Wang, C., Zong, S., Qi, J. et al. The trigeminal pathway dominates the nose-to-brain transportation of intact polymeric nanoparticles: Evidence from aggregation-caused quenching probes. *J Biomed Nanotechnol* 15, no. 4(Apr 1 2019): 686–702.

Liu, D.-H., Agbo, E., Zhang, S.-H., Zhu, J.-L. Anticonvulsant and neuroprotective effects of paeonol in epileptic rats. *Neurochem Res* 44, no. 11(2019): 2556–2565.

Luissint, A.C., Artus, C., Glacial, F., Ganeshamoorthy, K. *et al.* Tight junctions at the blood brain barrier: Physiological architecture and disease-associated dysregulation. *Fluids Barriers CNS* 9, no. 1(Nov 9 2012): 23.

Manek, E., Petroianu, G.A. Brain delivery of antidotes by polymeric nanoparticles. *J Appl Toxicol* 41, no. 1(Jan 2021): 20–32.

Manta, K., Papakyriakopoulou, P., Chountoulesi, M., Diamantis, D.A. et al. Preparation and biophysical characterization of quercetin inclusion complexes with beta-cyclodextrin derivatives to be formulated as possible nose-to-brain quercetin delivery systems. *Mol Pharm* 17, no. 11(Nov 2 2020): 4241–4255.

Marras, C., Beck, J.C., Bower, J.H., Roberts, E. *et al.* Prevalence of Parkinson's disease across North America. *NPJ Parkinsons Dis* 4(2018): 21.

Martins, P.P., Smyth, H.D.C., Cui, Z. Strategies to facilitate or block nose-to-brain drug delivery. *Int J Pharm* 570(Oct 30 2019): 118635.

Md, S., Mustafa, G., Baboota, S., Ali, J. Nanoneurotherapeutics approach intended for direct nose to brain delivery. *Drug Dev Ind Pharm* 41, no. 12(2015): 1922–1934.

Mendes, M., Sousa, J.J., Pais, A., Vitorino, C. Targeted theranostic nanoparticles for brain tumor treatment. *Pharmaceutics* 10, no. 4(Oct 9 2018).

Mignani, S., Shi, X., Karpus, A., Majoral, J.P. Non-invasive intranasal administration route directly to the brain using dendrimer nanoplatforms: An opportunity to develop new CNS drugs. *Eur J Med Chem* 209(Jan 1 2021): 112905.

Miranda-Filho, A., Pineros, M., Soerjomataram, I., Deltour, I. *et al.* Cancers of the brain and CNS: Global patterns and trends in incidence. *Neuro Oncol* 19, no. 2(Feb 1 2017): 270–280.

Mistry, A., Stolnik, S., Illum, L. Nanoparticles for direct nose-to-brain delivery of drugs. *Int J Pharm* 379, no. 1(Sep 8 2009): 146–157.

Mittal, D., Ali, A., Md, S., Baboota, S. et al. Insights into direct nose to brain delivery: Current status and future perspective. *Drug Deliv* 21, no. 2(Mar 2014): 75–86.

Mura, P., Mennini, N., Nativi, C., Richichi, B. In situ mucoadhesive-thermosensitive liposomal gel as a novel vehicle for nasal extended delivery of opiorphin. *Eur J Pharm Biopharm* 122(Jan 2018): 54–61.

NCT00422981. Safety, tolerability and efficacy study to evaluate subjects with mild cognitive impairment. *Disponível em*: https://ClinicalTrials.gov/show/NCT00422981. Acesso em: Feb, 13.

NCT00438568. SNIFF 120: Study of nasal insulin to fight forgetfulness (120 days). *Disponível em*: https://ClinicalTrials.gov/show/NCT00438568. Acesso em: Feb, 13.

NCT00581867. Memory and insulin in early Alzheimer's disease. *Disponível em*: https://ClinicalTrials.gov/show/NCT00581867. Acesso em: Feb, 13.

NCT01398748. Intranasal glutathione in Parkinson's disease. *Disponível em*: https://ClinicalTrials.gov/show/NCT01398748. Acesso em: Feb, 13.

NCT01767909. The Study of Nasal Insulin in the Fight Against Forgetfulness (SNIFF). *Disponível em*: https://ClinicalTrials.gov/show/NCT01767909. Acesso em: Feb, 13.

NCT02064166. Treatment of Parkinson disease and multiple system atrophy using intranasal insulin. *Disponível em*: https://ClinicalTrials.gov/show/NCT02064166. Acesso em: Feb, 13.

NCT02324426. CNS uptake of intranasal glutathione. *Disponível em*: https://ClinicalTrials.gov/show/NCT02324426. Acesso em: Feb, 13.

NCT02424708. Phase IIb study of intranasal glutathione in Parkinson's disease. *Disponível em*: https://ClinicalTrials.gov/show/NCT02424708. Acesso em: Feb, 13.

NCT02503501. Intranasal glulisine in amnestic mild cognitive impairment and probable mild Alzheimer's disease. *Disponível em*: https://ClinicalTrials.gov/show/NCT02503501.

NCT03541356. Therapeutic potential for intranasal levodopa in Parkinson's disease -off reversal. *Disponível em*: https://ClinicalTrials.gov/show/NCT03541356. Acesso em: Feb, 13.

NCT04251585. Intranasal insulin in Parkinson's disease. *Disponível em*: https://ClinicalTrials.gov/show/NCT04251585. Acesso em: Feb, 13.

NCT04687878. The effect of intranasal insulin on motor and non-motor symptoms in Parkinson's disease patients. *Disponível em*: https://ClinicalTrials.gov/show/NCT04687878. Acesso em: Feb, 13.

Oberoi, R.K., Parrish, K.E., Sio, T.T., Mittapalli, R.K. et al. Strategies to improve delivery of anticancer drugs across the blood-brain barrier to treat glioblastoma. *Neuro Oncol* 18, no. 1(Jan 2016): 27–36.

Ozsoy, Y., Gungor, S., Cevher, E. Nasal delivery of high molecular weight drugs. *Molecules* 14, no. 9(Sep 23 2009): 3754–3779.

Palao, M., Fernandez-Diaz, E., Gracia-Gil, J., Romero-Sanchez, C.M. et al. Multiple sclerosis following SARS-CoV-2 infection. *Mult Scler Relat Disord* 45(Oct 2020): 102377.

Pardeshi, C.V., Belgamwar, V.S. Direct nose to brain drug delivery via integrated nerve pathways bypassing the blood–brain barrier: An excellent platform for brain targeting. *Expert Opin Drug Deliv* 10, no. 7(2013): 957–972.

Pardridge, W.M. Drug transport across the blood-brain barrier. *J Cereb Blood Flow Metab* 32, no. 11(Nov 2012): 1959–1972.

Pashirova, T.N., Zueva, I.V., Petrov, K.A., Lukashenko, S.S. et al. Mixed cationic liposomes for brain delivery of drugs by the intranasal route: The acetylcholinesterase reactivator 2-PAM as encapsulated drug model. *Colloids Surf B Biointerfaces* 171(Nov 1 2018): 358–367.

Patel, A.P., Fisher, J.L., Nichols, E., Abd-Allah, F. et al. Global, regional, and national burden of brain and other CNS cancer, 1990–2016: A systematic analysis for the Global Burden of Disease Study 2016. *The Lancet Neurology* 18, no. 4(2019): 376–393.

Patel, H.N., Patel, P.M. Dendrimer applications – A review. *Int J Pharma Bio Sci* 4, no. 2(2013): 454–463.

Piazzini, V., Landucci, E., D'ambrosio, M., Tiozzo Fasiolo, L. et al. Chitosan coated human serum albumin nanoparticles: A promising strategy for nose-to-brain drug delivery. *Int J Biol Macromol* 129(May 15 2019): 267–280.

Pires, P.C., Santos, A.O. Nanosystems in nose-to-brain drug delivery: A review of non-clinical brain targeting studies. *J Control Release* 270(Jan 28 2018): 89–100.

Poovaiah, N., Davoudi, Z., Peng, H., Schlichtmann, B. et al. Treatment of neurodegenerative disorders through the blood-brain barrier using nanocarriers. *Nanoscale* 10, no. 36 (Sep 20 2018): 16962–16983.

Poupot, R., Bergozza, D., Fruchon, S. Nanoparticle-Based Strategies to Treat Neuro-Inflammation. *Materials (Basel)* 11, no. 2(Feb 9 2018).

Raggi, A., Leonardi, M. Burden and cost of neurological diseases: a European North-South comparison. *Acta Neurol Scand* 132, no. 1(Jul 2015): 16–22.

Rajpoot, R., Ganju, K., Singh, L.D. Review on Liposome as Novel Approach for Cancer Therapy. *Asian J Pharm Res Dev* 8, no. 3(2020): 122–129.

Rassu, G., Soddu, E., Cossu, M., Gavini, E. et al. Particulate formulations based on chitosan for nose-to-brain delivery of drugs. A review. *J Drug Deliv Sci Technol* 32(2016): 77–87.

Rinaldi, F., Oliva, A., Sabatino, M., Imbriano, A. et al. Antimicrobial essential oil formulation: Chitosan coated nanoemulsions for nose to brain delivery. *Pharmaceutics* 12, no. 7(Jul 17 2020).

Russo, E., Selmin, F., Baldassari, S., Gennari, C.G.M. et al. A focus on mucoadhesive polymers and their application in buccal dosage forms. *J Drug Deliv Sci Technol* 32(2016): 113–125.

Salem, H.F., Kharshoum, R.M., Abou-Taleb, H.A., Naguib, D.M. Brain targeting of resveratrol through intranasal lipid vesicles labelled with gold nanoparticles: in vivo evaluation and bioaccumulation investigation using computed tomography and histopathological examination. *J Drug Target* 27, no. 10(Dec 2019): 1127–1134.

Samanta, M.K., Wilson, B., Santhi, K., Kumar, K.P. et al. Alzheimer disease and its management: A review. *Am J Ther* 13, no. 6(Nov-Dec 2006): 516–526.

Samaridou, E., Alonso, M.J. Nose-to-brain peptide delivery – The potential of nanotechnology. *Bioorg Med Chem* 26, no. 10(Jun 1 2018): 2888–2905.

Santos, A., Veiga, F., Figueiras, A. Dendrimers as pharmaceutical excipients: Synthesis, properties, toxicity and biomedical applications. *Materials (Basel)* 13, no. 1(Dec 21 2019).

Schnyder, A., Huwyler, J. Drug transport to brain with targeted liposomes. *NeuroRx* 2, no. 1(Jan 2005): 99–107.

Schwarz, B., Merkel, O.M. Nose-to-brain delivery of biologics. *Ther Deliv* 10, no. 4 (Apr 2019): 207–210.

Selvaraj, K., Gowthamarajan, K., Karri, V. Nose to brain transport pathways an overview: Potential of nanostructured lipid carriers in nose to brain targeting. *Artif Cells Nanomed Biotechnol* 46, no. 8(Dec 2018): 2088–2095.

Shao, J. Nasal delivery of proteins and peptides. *J. Pharm. Pharm. Sci.* 1, no. 4(2017).

Shobo, A., Pamreddy, A., Kruger, H.G., Makatini, M.M. et al. Enhanced brain penetration of pretomanid by intranasal administration of an oil-in-water nanoemulsion. *Nanomedicine (Lond)* 13, no. 9(May 2018): 997–1008.

Shringarpure, M., Gharat, S., Momin, M., Omri, A. Management of epileptic disorders using nanotechnology-based strategies for nose-to-brain drug delivery. *Expert Opin Drug Deliv* 18, no. 2(Feb 2021): 169–185.

Shultz, S.R., Bao, F., Weaver, L.C., Cain, D.P. et al. Treatment with an anti-CD11d integrin antibody reduces neuroinflammation and improves outcome in a rat model of repeated concussion. *J Neuroinflammation* 10(Feb 15 2013): 26.

Sonvico, F., Clementino, A., Buttini, F., Colombo, G. et al. Surface-modified nanocarriers for nose-to-brain delivery: From bioadhesion to targeting. *Pharmaceutics* 10, no. 1(Mar 15 2018).

Spina, C.S., Tangpricha, V., Uskokovic, M., Adorinic, L. et al. Vitamin D and cancer. *Anticancer Res* 26, no. 4A(Jul-Aug 2006): 2515–2524.

Strambeanu, N., Demetrovici, L., Dragos, D., Lungu, M. Nanoparticles: Definition, classification and general physical properties. *In*: Nanoparticles' Promises and Risks, Springer International Publishing, Chapter 1(2015): 3–8.

Sur, S., Rathore, A., Dave, V., Reddy, K.R. et al. Recent developments in functionalized polymer nanoparticles for efficient drug delivery system. *Nano-Structures & Nano-Objects* 20, no.1 (2019). https://doi.org/10.1016/j.nanoso.2019.100397.

Tallantyre, E.C., Whittam, D.H., Jolles, S., Paling, D. et al. Secondary antibody deficiency: a complication of anti-CD20 therapy for neuroinflammation. *J Neurol* 265, no. 5 (May 2018): 1115–1122.

Tan, M.S.A., Parekh, H.S., Pandey, P., Siskind, D.J. et al. Nose-to-brain delivery of antipsychotics using nanotechnology: a review. *Expert Opin Drug Deliv* 17, no. 6 (Jun 2020): 839–853.
Tayeb, H.H., Sainsbury, F. Nanoemulsions in drug delivery: formulation to medical application. *Nanomedicine (Lond)* 13, no. 19(Oct 2018): 2507–2525.
Torchilin, V.P. Micellar nanocarriers: pharmaceutical perspectives. *Pharm Res* 24, no. 1 (Jan 2007): 1–16.
Townsend, L.K., Peppler, W.T., Bush, N.D., Wright, D.C. Obesity exacerbates the acute metabolic side effects of olanzapine. *Psychoneuroendocrinology* 88(Feb 2018): 121–128.
Trows, S., Wuchner, K., Spycher, R., Steckel, H. Analytical challenges and regulatory requirements for nasal drug products in Europe and the U.S. *Pharmaceutics* 6, no. 2 (Apr 11 2014): 195–219.
Vieira, D.B., Gamarra, L.F. Getting into the brain: liposome-based strategies for effective drug delivery across the blood-brain barrier. *Int J Nanomedicine* 11(2016): 5381–5414.
Wang, F., Yang, Z., Liu, M., Tao, Y. et al. Facile nose-to-brain delivery of rotigotine-loaded polymer micelles thermosensitive hydrogels: In vitro characterization and in vivo behavior study. *Int J Pharm* 577(Mar 15 2020): 119046.
Wang, J., Kong, M., Zhou, Z., Yan, D. et al. Mechanism of surface charge triggered intestinal epithelial tight junction opening upon chitosan nanoparticles for insulin oral delivery. *Carbohydr Polym* 157(Feb 10 2017): 596–602.
Wang, L., Zhao, X., Du, J., Liu, M. et al. Improved brain delivery of pueraria flavones via intranasal administration of borneol-modified solid lipid nanoparticles. *Nanomedicine (Lond)* 14, no. 16(Aug 2019): 2105–2119.
Warnken, Z.N., Smyth, H.D.C., Watts, A.B., Weitman, S. et al. Formulation and device design to increase nose to brain drug delivery. *J Drug Deliv Sci Technol* 35(2016): 213–222.
Ways, T.M., Lau, W.M., Khutoryanskiy, V.V. Chitosan and its derivatives for application in mucoadhesive drug delivery systems. *Polymers (Basel)* 10, no. 3(Mar 5 2018).
Wohlfart, S., Gelperina, S., Kreuter, J. Transport of drugs across the blood-brain barrier by nanoparticles. *J Control Release* 161, no. 2(Jul 20 2012): 264–273.
World Health Organization. Neurological disorders: Public health challenges. https://www.who.int/mental_health/neurology/neurodiso/en/., 2021. *Disponível em*: https://www.who.int/mental_health/neurology/neurodiso/en/. Acesso em: Jan, 15.
Wu, H., Hu, K., Jiang, X. From nose to brain: Understanding transport capacity and transport rate of drugs. *Expert Opin Drug Deliv* 5, no. 10(Oct 2008): 1159–1168.
Wu, I.Y., Nikolaisen, T.E., Skalko-Basnet, N., Di Cagno, M.P. The hypotonic environmental changes affect liposomal formulations for nose-to-brain targeted drug delivery. *J Pharm Sci* 108, no. 8(Aug 2019): 2570–2579.
Xie, H., Li, L., Sun, Y., Wang, Y. et al. An available strategy for nasal brain transport of nanocomposite based on PAMAM dendrimers via in situ gel. *Nanomaterials (Basel)* 9, no. 2(Jan 24 2019).
Xu, Q., Boylan, N.J., Cai, S., Miao, B. et al. Scalable method to produce biodegradable nanoparticles that rapidly penetrate human mucus. *J Control Release* 170, no. 2 (Sep 10 2013): 279–286.
Yasir, M., Sara, U.V.S., Chauhan, I., Gaur, P.K. et al. Solid lipid nanoparticles for nose to brain delivery of donepezil: Formulation, optimization by Box–Behnken design, in vitro and in vivo evaluation. *Artif Cells Nanomed Biotechnol* (2017): 1–14.
Zhang, L., Zhuang, X., Kotitalo, P., Keller, T. et al. Intravenous transplantation of olfactory ensheathing cells reduces neuroinflammation after spinal cord injury via interleukin-1 receptor antagonist. *Theranostics* 11, no. 3(2021): 1147–1161.

Zhang, Y., Walker, J.B., Minic, Z., Liu, F. et al. Transporter protein and drug-conjugated gold nanoparticles capable of bypassing the blood-brain barrier. *Sci Rep* 6 (May 16 2016): 25794.

Zhao, B., Shi, Q.J., Zhang, Z.Z., Wang, S.Y. et al. Protective effects of paeonol on subacute/chronic brain injury during cerebral ischemia in rats. *Exp Ther Med* 2018.

Zhou, Y., Peng, Z., Seven, E.S., Leblanc, R.M. Crossing the blood-brain barrier with nanoparticles. *Control Release* 270(Jan 28 2018): 290–303.

12 Carbon Nanotubes in Cancer Therapy

Renu Sankar, V.K. Ameena Shirin, Chinnu Sabu, and K. Pramod
Government Medical College

CONTENTS

12.1 Introduction 287
12.2 Properties 288
    12.2.1 Cellular Uptake 288
    12.2.2 Biocompatibility 289
    12.2.3 Functionalization 289
    12.2.4 Cellular Toxicity 290
12.3 CNTs in Cancer Therapy 292
12.4 Merits of Cancer Therapy Using CNTs 297
12.5 Photothermal Therapy Using CNT for Cancer 298
12.6 Photodynamic Therapy Using CNT 299
12.7 CNTs for Oral Cancer Therapy 300
12.8 CNT in Cancer Vaccine 300
12.9 Conclusions 303
References 303

12.1 INTRODUCTION

Nanomaterials have gained greater attention in the scientific field due to their size, which is similar to macromolecules like DNA, antibodies, and enzymes (Mohanta et al. 2019). Carbon nanotubes (CNT) are one of the most promising inventions of recent years. CNTs are rolled graphene sheets with sp2 hybridization. The unique properties of CNTs include small size, good conductivity, good tensile strength, lower weight, etc. The recent studies proved the significance of CNTs in the pharmaceutical field also. Their high surface area, strength, and rigid nature make them a potential candidate in the pharmaceutical field. Also, the characteristics of CNTs can be changed by altering their dimensions. The CNTs can be classified into several types based on the number of layers of sheets of graphene. These include single-layered, double-layered, or multi-layered (Jha et al. 2020).

    The reason behind the unique properties of CNTs is the strong bonding pattern found between the atoms and the aspect ratio (Mohanta et al. 2019). The extreme aspect ratio provides excellent properties, which include the ability to penetrate cells, their semi-conductive nature that is responsible for their integration with muscles and nervous system, the high surface area enabling them suitable for integration with biological materials, and hollow nature enabling them to act as cargo for the delivery of drugs or large biomolecules. Moreover, the exterior surface functionalization of CNTs with suitable materials can increase their solubility and biological recognition. The major factor that limits their application is toxicity. The highly reactive nature of the surface of these molecules permits the attachment of molecules like drugs, siRNA, etc., which will improve their biocompatibility (Mahajan et al. 2018).

The synthesis of CNTs, involving high temperature, pressure, and catalyst reactions, is mainly by three processes which include electric arc discharge, laser ablation, and chemical vapor deposition (Ravi Kiran et al., 2020).

Cancer is among the most dangerous diseases which negatively affect the lives of people. Most of the treatment strategies associated with cancer is having dangerous side effects and toxicity to normal cells. Several studies were conducted to reduce the side effects and provide target-specific therapy (Alderton, Gross, and Green 1992). The distinct interaction of CNTs with biological membranes opens a pathway for the use of CNTs in cancer drug delivery. Moreover, the CNTs enter the cells by a needle-like penetration mechanism, which is endocytosis-independent and delivers the drug load directly to the cytoplasm (Kostarelos et al. 2007). Several studies report the successful delivery of anticancer drugs to the target site by using CNTs.

Both single-walled and multi-walled carbon nanotubes (SWCNT and MWCNT respectively) were extensively studied to explore their potential in cancer therapy. They are both classified as a carrier and as a mediator for the targeted tumor therapy. Combined with light energy, the CNTs are also used for photothermal and photodynamic therapy to kill cancer cells (Son, Hong, and Lee 2016).

12.2 PROPERTIES

The CNTs are large molecules having a cylindrical shape in which the carbon atoms are sp2 hybridized. They are formed by the rolling of graphene sheets. Depending on the layers of graphene the CNTs are classified mainly as single-walled, double-walled, or multi-walled, etc. The CNTs formed by the rolling up of a single layer of graphene sheets are known as SWCNTs and by more than one layer are known as MWCNTs. Both these CNTs were wrapped at both ends by fullerenes. The MWCNTs are less stable than the SWCNTs due to the defects in the wall of MWCNTs. In the pharmaceutical field, the attributes which are greatly used include small size, high surface ratio, and high drug loading capacity (Kushwaha et al. 2013).

12.2.1 CELLULAR UPTAKE

The cellular uptake properties of CNTs are more useful for the delivery of drugs and large molecules to the target size. After reaching the cells the CNTs are either accumulated or eliminated from the body rapidly. The excess accumulation of CNTs leads to toxic results whereas the rapid elimination from the body results in reduced therapeutic effect. Hence, to maintain a proper balance between these many studies were conducted (Mu, Broughton, and Yan 2009). The perpendicular positioning of the CNTs on the cell membrane shows that the cellular uptakes of CNTs are similar to nanoneedles (Bianco, Kostarelos, and Prato 2005).

There are two methods by which the CNTs cross the biological membrane. One is passive diffusion and the other is endocytosis. In the passive diffusion method, the CNTs cross the cell by penetration without using any energy (Pantarotto et al. 2004). Whereas in endocytosis, the CNTs gets internalized into the vesicles known as endosome and fused with the lysosome (Kang et al. 2012). Then after reaching the cells the drugs or molecules are released from the CNTs and the CNTs are transported to endosomes, mitochondria, or nucleus. Finally, CNTs are eliminated by the process of biodegradation or exocytosis. This step is crucial in the reduction of toxicity associated with CNTs. After exocytosis, the CNTs again undergo neutrophil-associated enzymatic degradation with help of the enzyme myeloperoxidase (Kagan et al. 2010). The uptake of CNTs by endocytosis was confirmed by the detection of fluoresceinated protein attached to SWNTs-biotin in the endosome (Shi Kam et al. 2004).

12.2.2 BIOCOMPATIBILITY

The CNTs are a class of nanomaterials with diverse applications in the biomedical field due to their unique physicochemical properties (Raphey et al. 2019). The fabrication of CNTs with improved biocompatibility was a greater challenge. The CNT's biocompatibility depends on factors like type, length, surface area, purity, concentration, and functionalization of CNTs (Aoki and Saito 2020). The biocompatibility of CNTs can be enhanced by altering their structure and also their dispensability and solubility can be improved by suitable functionalization (Foldvari and Bagonluri 2008). Among the different factors that affect biocompatibility, functionalization is more prominent. The CNTs functionalized with cations are found to be less toxic than the others (Aboutalebi Anaraki et al. 2015).

12.2.3 FUNCTIONALIZATION

The CNTs are used extensively in various fields including the pharmaceutical and medicinal fields. Their insoluble nature in organic and inorganic solvents extends their applications in various fields (Beg et al. 2011). After reaching the blood, their movement in the blood may create some problems which can be solved by proper functionalization. The functionalization of CNTs with appropriate molecules and methods will enhance the solubility, strength, and absorption characteristics. The functionalization with appropriate polymers will extend the application of CNTs in the pharmacy field. The functionalization depends on mainly two factors, the nature of the polymer used and the type of method (covalent or noncovalent). Usually, functionalization is done in two modes, on the sidewalls of CNT and at the tail end of CNT (Kakade et al. 2008).

The covalent functionalization is mainly performed by two routes, i.e., direct sidewall functionalization or by defect group functionalization. The sidewall covalent functionalization is done by the conversion of sp2 hybridized carbon into sp3 hybridized (Tasis et al. 2006). The defect group functionalization is carried out by forming defects by oxidation to produce carboxylic acid functionalities (Dinesh, Bianco, and Ménard-Moyon 2016).

The noncovalent surface modifications of CNTs are different from covalent functionalization. In noncovalent functionalization, both electrical and optical properties are preserved well. This can be attained by the adsorption of molecules on the surface of CNTs by either π-π interaction or van der walls interaction. Usually, the CNTs are suspended in a polymer-containing solution for wrapping the polymers on the surface of CNTs (Star et al. 2001). Table 12.1 shows the outcomes of several CNT functionalizations.

Anticancer drug delivery system employing doxorubicin was developed using SWCNT modified with chitosan. The noncovalent bounding of chitosan around the SWCNT provides improved water solubility and biocompatibility. To identify the selective destruction of tumor cells, folic acid was also conjugated on the outer side of chitosan. The as-prepared delivery system kills the HCC SMMC 7721 cell lines and also inhibits the tumor growth in liver cancer. Hence the system shows superiority over the free doxorubicin-based therapy. Also, the in vivo toxicity of the system is negligible. So this functionalized CNT-based delivery system is a hopeful pathway for the delivery of cancer drugs (Ji et al. 2012).

The capability of SWCNTs to penetrate the mammalian cell and high drug loading capacity make them a promising candidate for the delivery of cancer drugs to the target cell. The property of these SWCNT to conjugate different molecules will improve their potency to treat cancer and reduce the toxic side effects. For multimodal drug delivery, these SWCNTs were functionalized with three agents that include anticancer drug doxorubicin, a monoclonal antibody for molecular targeting, and finally fluorescein as a fluorescent marker. This study shows that the drug is delivered to the nucleus while the CNT remains within the cytoplasm. Hence the triple functionalization has improved the therapeutic efficacy of the drug (Heister et al. 2009).

The SWCNTs were functionalized with HEMA and NVP monomer. Then they were subjected to free radical polymerization. Studies reveal that the grafting with these polymers improved the solubility of CNTs in both organic solvents and water. This functionalized nanocomposite of CNT serves as a potential carrier for drugs. The functionalization of CNTs with hydrogel polymers will reduce the toxic properties of CNTs thus reducing the damage to normal cells (Abbaszadeh et al. 2014).

12.2.4 Cellular Toxicity

The facts related to the cellular toxicity of CNTs are of prime importance besides their wide range of applications. There are several toxicity problems associated with CNTs, such are tissue accumulation, damage to DNA, oxidative stress, and mitochondrial stress. The CNTs at lower concentrations did not cause toxicity to normal cells directly. But indirectly they cause lymphocyte-mediated cytotoxicity to normal healthy cells (Sun et al. 2011). The smaller size of CNTs comparable to foreign pathogens, induce foreign body response within the body resulting in phagocytosis.

TABLE 12.1
Outcomes of Several CNT Functionalizations

Sl.No.	Anticancer Drug	Functionalization Type	Outcome	Reference
1	Paclitaxel	PEG-functionalized	The improved therapeutic action of paclitaxel	(Arya et al. 2013)
2	Gemcitabine	Folic acid-functionalized	Enhanced cytotoxic action	(Razzazan et al., 2019)
3	Oxaliplatin	PEG600-functionalized	Reduced cytotoxic action	(Wu et al. 2013)
4	Doxorubicin	PEG-functionalized	Increased retention of drug within cancer cells.	(Liu et al. 2012)
5	Doxorubicin	Chitosan and folic acid-functionalized	improved water solubility and biocompatibility	(Ji et al. 2012)
6	Cisplatin	Amide linkage	Improved efficacy	(Bhirde et al. 2009)
7	Dacarbazine	Carboxylated CNT	Spontaneous encapsulation of the drug	(Mirsalari et al. 2020)
8	Platinum drug (IV)	PL-PEG functionalized	Enhanced cytotoxicity	(Feazell et al. 2007)

This will result in oxidative burst causing increased release of ROS thereby promoting oxidative stress (Sharifi et al. 2012). It was found the increase in the length of CNTs enhances their toxic nature. That is shorter CNTs are less toxic than longer CNTs (Ravi Kiran et al., 2020). Several studies have shown that the functionalization of CNTs can reduce their toxic nature.

12.3 CNTs IN CANCER THERAPY

Most of the anticancer drugs available now are nonselective in nature. This nonselectivity results in damage to normal cells causing unwanted side effects to the body. This is the main reason for the increased death rate of cancer patients. The lower bioavailability of these drugs results in increased drug dose requirement leading to toxicities. Additionally, it may result in multidrug resistance. Hence it is necessary to develop a delivery system that targets the cancer cells and produces improved therapeutic effects. Nanomaterials are currently used for the efficient delivery of anticancer drugs to the target site. The CNTs are employed for this purpose. The smaller size of CNTs permits the passive transport of drugs to the target cell (Muller et al. 2009). The properties like high drug loading capacity, large surface area, high aspect ratio, and small size make CNTs an ideal candidate for cancer therapy. The CNTs help for the pH-dependent release of the drug thereby promoting target-specific delivery which will reduce the side effects and frequency of drug administration (Elhissi et al. 2012).

The CD44 targeted alpha-tocopheryl succinate and chondroitin sulfate-modified MWCNTs were developed to treat triple-negative cancer. The study was conducted to detect the potential of the combined effect of anticancer potency of doxorubicin with anticancer and targeting properties of alpha-tocopheryl succinate and chondroitin sulfate. The conjugation with alpha-tocopheryl succinate and chondroitin sulfate enhanced the drug loading and encapsulation capacity. Studies revealed the improved anticancer potential of this system. Further, this system showed only negligible interaction with blood thus ensuring their safety. The stability study confirmed that they are stable under the conditions of refrigerator temperature (Singhai et al. 2020).

The accumulation and distribution of CNTs in glioma cells can be effectively controlled by exposing the tumor cells to the electric field. In a study, the SWCNTs wrapped in DNA were used to treat C6 glioma cells. Then it was exposed to low frequency and low strength electric field. The results revealed that the exposure facilitated the accumulation of CNTs in glioma cells without any damage to the cell membrane and the formation of pore on the plasma membrane. This procedure will reduce the therapeutic concentration of CNTs and also decrease the time required for the whole procedure. This will help the body to recover fast with enhanced therapeutic efficacy (Figure 12.1) (Golubewa et al. 2020). The combination of SWCNTs with low voltage can produce cell electroporation, and thereby molecules of different sizes can be transported to the target cell without affecting the normal cell. High voltage pulses can also produce more pores on the cell membrane and organelles than low voltage pulses. But the mortality rate of cells is more in the case of high voltage pulses. It is found that the combination of SWCNTs with low voltage pulses enhances the EPR effect and thereby increases the accumulation of

Carbon Nanotubes in Cancer Therapy 293

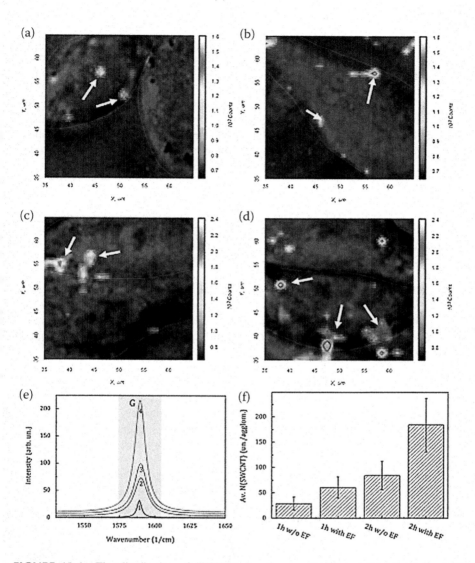

FIGURE 12.1 The distribution of SWCNTs in glioma cells (a) and (b) after 1 hour of exposure to SWCNT suspension. (c) and (d) after 2 hour of exposure to SWCNT suspension. (e) Average Raman spectra of SWCNTs accumulated in glioma cells: 1e for 1 hour, 2e for 1 hour under EF stimulation, 3e for 2 hour, 4e for 2 hour; (f) Calculated average number of SWCNTs per agglomerate for spectra 1–4. Reprinted with permission from (Golubewa et al. 2020)© 2020 Elsevier.

nanoparticles around the cancer cells. Also, the preinjection of SWCNT followed by low voltage pulses enhances the vascular permeability to a greater extent than low voltage alone. Hence this strategy can be definitely employed for cancer therapy (Lee, Peng, and Shieh 2016).

Highly efficient and specific therapy is required for the treatment of triple-negative breast cancer. The doxorubicin-loaded MWCNTs that are functionalized with hyaluronic acid and α-tocopheryl succinate show enhanced antitumor action compared to free MWCNTs. This formulation is a combination of the anticancer potential of Doxorubicin, targeting potential of hyaluronic acid and synergistic action of α-tocopheryl succinate. The antitumor potential was evaluated using the sulforhodamine B assay and apoptotic assay. It shows enhanced cellular uptake and tumor inhibition than the other formulations. The results of the stability study show that this formulation can be safely stored in an amber-colored bottle in a refrigerator for about 3 months (Singhai, Maheshwari, and Ramteke 2020).

The therapeutic efficacy of certain cancer drugs especially drugs acting on the nucleus can be improved by nuclear targeting and cancer cell selectivity. A steroid-MWCNTs delivery system was synthesized for the above purpose. The estradiol-PEG MWCNTs bioconjugate effectively targets the estrogen receptor over-expressing tumor cells through an estrogen receptor-dependent pathway. The loading of anticancer drug doxorubicin to the above system shows the in vivo and in vitro potential of the system in cancer targeting (Das et al. 2013).

MWCNTs delivery system loaded with the platinum-based anticancer drug was synthesized. The abovementioned system shows enhanced tumor inhibition, selective tumor cell accumulation, selective cancer cell targeting, and imaging properties. This system can be used for effective therapy for breast cancer (Fahrenholtz et al. 2016).

It has been already demonstrated that the polo-like kinase (PLK) is a potential target in gene therapy of cancer. The combination of PLK siRNA with CNTs as a delivery system shows more efficacy for cancer therapy. The potential of MWNT-NH^{3+} in delivering siRNA against PLK was studied extensively. This complex can increase animal survival rates. Also, this system shows enhanced accumulation within the tumor cells and cellular uptake by the tumor cells after in vivo local administration (Figure 12.2) (Guo et al. 2015).

The electrostatic bond between the MWCNT and doxorubicin was found to be stronger and shows a slow drug release in cancer cells than the SWCNT. Hence the MWCNTs are more efficacious for the delivery of drugs than the SWCNT (Maleki et al. 2020).

A cancer stem cell has a significant role in cancer therapy. One of the hopeful approaches for cancer therapy is the complete elimination of cancer stem cells. Osteosarcoma is associated with osteosarcoma stem cells. Both modified and unmodified SWCNTs can stop the differentiation of TGFβ1-induced osteosarcoma cells. Also, the treatment with SWCNT downregulates the osteosarcoma stem cells in osteosarcoma and reduces the growth of tumors (Miao et al. 2017).

One of the most frequently employed delivery systems is injectable thermosensitive hydrogels. Using chitosan and β-glycerophosphate a thermosensitive injectable hydrogel was developed into which CNTs were incorporated. To this hydrogel, methotrexate was added. This system remains as a liquid at room temperature and solidifies at body temperature. The cell viability study shows that this system is nontoxic to 3T3 cells. The in vitro drug release study shows that the release of methotrexate from the above-mentioned system is slower than the release

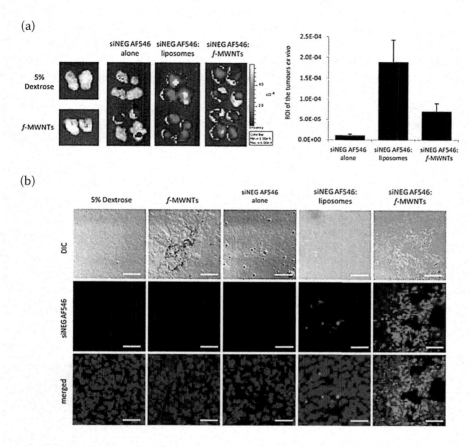

FIGURE 12.2 Uptake of siNEG AF546 in tumor xenografts. (a) *Ex vivo* imaging (left) and quantification of the fluorescence signals (right) detected. (b) Intracellular uptake of siNEG AF546 in the dissected tumors by CLSM. Scale bars, 50 μm. Reprinted with permission from (Guo et al. 2015)© 2015 American Chemical Society.

from the control system without CNT. This investigation reveals that the presence of CNTs in hydrogel enhanced the anticancer efficacy of methotrexate. Hence, this hybrid hydrogel can be used as a promising system for the effective treatment of breast cancer (Saeednia et al. 2019).

A transferrin conjugated MWCNT was fabricated for the effective delivery of docetaxel to treat lung cancer. It shows 74 % encapsulation efficiency. The in vitro drug release study shows a sustained release of docetaxel and enhanced cellular uptake of MWCNT leads to an improved cellular concentration of the drug. This formulation has enhanced cytotoxicity, lower reactive oxygen species generation, lower toxicity, and improved safety (Singh et al. 2016). The MWCNTs conjugated with the chitosan-folate complex can be used as an effective and safe system for the delivery of docetaxel to lung cancer cells. The drug encapsulation efficacy was found to be 79%. The nano-size range of this system was confirmed by TEM analysis. The sustained release of the drug was proved by in vitro release studies. Also, this conjugated system is internalized into the tumor cell through a folate receptor-mediated endocytic pathway. The IC_{50} suggests that this system is about 89-fold more efficacious than free docetaxel for treating human lung cancer (Singh et al. 2017).

The treatment strategies of cancer face several obstacles; the most important among them are multidrug resistance and radioresistance. To hinder the multidrug resistance in hepatocellular carcinoma a CNT-based radiosensitive delivering system was developed. To this system, an anticancer ruthenium complex was loaded and shows enhanced cellular uptake by endocytosis. This nanosystem prolonged the in vivo blood circulation and reduced the toxic effect associated with the loaded drug. Hence can be effectively used for the treatment of multidrug-resistant cancer (Figure 12.3) (Wang et al. 2015).

For the detection of hydrogen peroxide marker from cancer cells, a novel biosensor was developed. The biosensor consists of carbon fiber wrapped by nanoparticles of gold decorated with nitrogen-doped carbon nanotube arrays. This system helps for the detection of hydrogen peroxide secreted by the cancer cells. Here, the carbon nanotube arrays were doped with nitrogen and then grown on the carbon nanofiber. They have a large surface area to volume ratio and several active

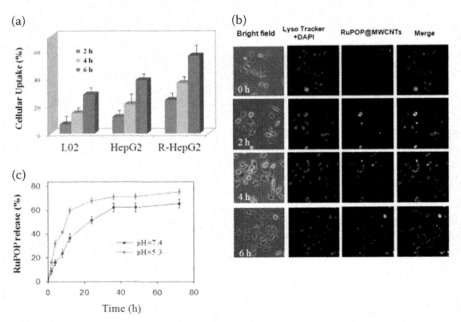

FIGURE 12.3 (a) Cellular uptake efficiency of RuPOP@MWCNTs (160 ng/mL) in several tumors and normal cells after different time periods of incubation. (b) Colocalization of RuPOP@MWCNTs/FITC (green fluorescence) and lysosomes (red fluorescence) in tumor cells. (c) In vitro release profiles of RuPOP from RuPOP@MWCNTs in PBS solution (pH 7.4 and pH 5.3). Reprinted with permission from (N. Wang et al. 2015) © 2015 American Chemical Society.

sites. So the uniformly dispersed high dense nanoparticles can be loaded to it easily. Their excellent performance in biosensing, biocompatibility, flexibility, etc., makes them a promising system for the in situ detection of hydrogen peroxide from breast cancer cells (Y. Zhang et al. 2018). Another potential system was developed for the sensing purpose, which consists of Cu-Au coupled CNT arrays that are grown vertically on carbon spheres. This system has several active sites and to which highly dense molecules can be attached easily. The carcinoembryonic antibody is a biomarker for several cancers, which can be selectively detected by this system. The merits of the fabricated system are a wide linear range of detection, reproducibility, and good stability. Also, it shows excellent accuracy in the detection of carcinoembryonic antibody in samples. The effective attachment of Cu-Au nanocrystals on the 3D structure of the CNTs offers several active sites and a good platform for the immobilization of anticarcinoembryonic antibody for selective detection. This sensing system has the capacity to detect several agents for the effective management and diagnosis of several diseases (Tran et al. 2018).

Fluorescent and magnetic MWCNTs were prepared for the MR and fluorescent imaging of cancer cells. MWCNT-magnetofluorescent carbon quantum dot-doxorubicin nanohybrid was synthesized. It was used as a platform for the delivery of a drug to tumor cells. The advantages of this system include high heat-producing capacity, pH- and near IR-responsive delivery, the temperature-induced release of the drug, etc. The investigations have revealed the potential of this system in releasing drugs upon near IR radiation, delivery to the target tumor cell, efficient elimination of tumor cell by the synergistic action of chemotherapy, and photothermal therapy (M. Zhang et al. 2017).

The SWCNTs loaded with the anticancer drug paclitaxel were synthesized. The biocompatibility of the SWCNTs was enhanced by the noncovalent bonding of chitosan to it. For targeted delivery of the drug, the biocompatible hyaluronan was bonded to the outer side of chitosan. The results showed the pH-stimulated drug release was found to be more at a pH of 5.5. Further investigations proved that this system showed toxic behavior to tumor cells than the normal cells, thus reducing the side effects. More toxicity was shown towards A549 cells. Hence this system which effectively inhibits cell proliferation and kills tumor cells can be used as a potential system for tumor therapy (Yu et al. 2016).

12.4 MERITS OF CANCER THERAPY USING CNTs

The major limitation of conventional therapies of CNTs includes the inability to reach the target tumor locating sites, and risk associated with while operating major organs of the body. The chemotherapy and radiation process results in the occurrence of unwanted toxic effects in the normal cells. Hence these treatment strategies are not efficient enough for the successful killing of cancer cells (Misra, Acharya, and Sahoo 2010). Cancer therapy using CNTs has several advantages. The amount of drug required to produce the therapeutic effect is less than the other methods. Hence, the economic problem associated with the dealing of high-cost drugs can be reduced. Interestingly, there is no need for any solvents for the delivery of a therapeutic agent to the target site. So the risk to normal cells while using additional solvents can be prevented. Also, the target-specific delivery of CNTs prevents us from serious side effects. Further wide range of drugs or molecules can be delivered using them to get therapeutic, diagnostic, and patient-specific therapy (Kushwaha et al. 2013).

12.5 PHOTOTHERMAL THERAPY USING CNT FOR CANCER

Photothermal therapy (PTT) is a therapeutic approach that can be used for the destruction of cancer cells. Here, the photon energy is converted into heat energy by different methods that destroy cancer cells. The CNTs because of their exceptional optical properties are good candidates for the induction of hyperthermia in PTT (Sobhani et al. 2017). This method is less invasive, non-injurious, and highly efficient. The coating of CNTs with ligands will enhance the selectivity and side effects associated (Chen et al. 2017).

In a study, metastatic breast cancer was treated by the combination of PTT with SWCNTs and immunostimulation. The PTT of breast cancer cells using SWCNTs functionalized with annexin A_5 shows an enhanced survival rate in mice. The use of annexin A_5 reduces the amount of SWCNTs to be delivered for the destruction of tumor cells. The histopathological studies reveal that this bioconjugate has no side effects during the period of study (McKernan et al. 2021).

The chitosan-based SWCNTs coated with PEG were developed for the targeting of mitochondria. On reaching the tumor environment the nanoparticles were released which leads to enhanced cell endocytosis and mitochondrial targeting. The PTT using near IR destroys mitochondria leading to cell apoptosis and finally burst of reactive oxygen species within the cell as shown in Figure 12.4 (M. Wang et al. 2020).

The PAMAM-Ag_2S CNTs show enhanced photothermal efficiency than others and attain a temperature of 64.7 °C at 1 g/mL, which is more than the temperature tolerance level of tumor cells. Hence, these CNTs destroy the tumor cell in an effective manner (Neelgund et al. 2019).

Pancreatic cancer is a major threat to life but the available treatment strategies have many disadvantages. A recent study proves the efficiency of PTT in pancreatic

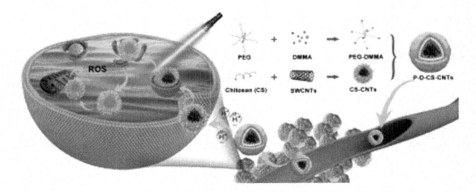

FIGURE 12.4 The illustration of mitochondria-targeted transportation of P-D-CS-CNTs in MB49 cells. Reprinted with permission from (M. Wang et al. 2020) © 2020 Elsevier.

Carbon Nanotubes in Cancer Therapy

cancer therapy. The dye conjugated SWCNT coupled with targeting antibodies will target the pancreatic tumor cells with high selectivity. The CNTs reach the cancer cell through blood circulation by their flexible nature and bind to the tumor which will enhance the retention time. The CNTs will absorb energy and convert to heat which destroys cells (Lu et al. 2019).

Recent studies have shown the effectiveness of MWCNTs in PTT. The MWCNTs are easy to prepare but show increased accumulation leading to cytotoxicity (Eldridge et al. 2016). The oxidation and covering of MWCNTs using PEG will enhance their dispersibility in water and show decreased toxicity against HeLa and HepG2 cell lines. Also, the abovementioned CNTs will reduce the size of tumors sharply in mice than in the mice treated with laser alone (Sobhani et al. 2017).

The combination of Pgp-targeted CNTs and PTT helps for the effective eradication of tumors from the body. A phospholipid-PEG–coated MWCNT was fabricated, which has lower nonspecific cell interactions and high intercellular diffusion. The modification of these MWCNTs with anti-Pgp enhanced cellular uptake. Results of the study show that this MWCNTs on photoirradiation show more cytotoxicity towards multidrug-resistant tumor cell and doesn't show any toxicity towards normal cells that don't express Pgp (Suo et al. 2018).

Using MWCNTs a self-amplified delivery system was developed for tumor PTT. The developed CMWNTs-PEG self-amplified delivery system increases the temperature of cancer cells upon illumination with near IR. The CMWNTs labeled with IR783 shows greater accumulation in tumor cell than the normal cells. The pharmacodynamic studies show that they eradicate the tumor cell after four times of illumination (B. Zhang et al. 2016).

12.6 PHOTODYNAMIC THERAPY USING CNT

Photodynamic therapy consists of the combination of light and photosensitizing agents along with molecular oxygen that kills the cells. Hereafter, the administration of photosensitizing agents to tumor cells the illumination with near IR radiation results in the production of ROS which will cause the death of malignant tumor cells (Dolmans et al., 2003). The efficiency of zinc phthalocyanine-spermine-SWCNT for the photodynamic therapy of MCF-7 breast cancer has been evaluated. This combination improved the photodynamic therapy and showed a cell viability rate of about 95% (Ogbodu and Nyokong 2015).

The near IR illumination of (6,4)-SWCNTs results in the production of OH^- in a sustained manner, which results in the ablation of tumor cells and the disintegration of Aβ-peptide aggregates (Fukuda et al. 2020).

The photodynamic therapy effect of zinc phthalocyanine folic acid conjugate with SWCNT was evaluated and the results suggest improved cell death due to the presence of folic acid (Ogbodu et al., 2015). Similarly, the photodynamic therapy effect of zinc monocarboxy phenoxy phthalocyanine(ZnMPP)-spermine conjugate was evaluated MCF-7 breast cancer cells. The ZnMPP-spermine conjugate was loaded on SWCNT. The results of the study showed improved photophysical nature and also enhanced photodynamic effect of about 97% was observed with this conjugate (Ogbodu et al. 2015).

The conjugation of CNT with hyaluronic acid shows high aqueous solubility, neutral pH, and efficient tumor targeting activity. To this bioconjugate, novel photodynamic agent hematoporphyrin monomethyl ether (HMME) was adsorbed. Then itsin vivo and in vitro tumor inhibition property upon the combination of PTT and photodynamic therapy was evaluated. This combination of PTT and photodynamic therapy shows a synergic effect and enhanced the tumor inhibition than the PTT and photodynamic therapy alone. Also, this combination has shown greater potential in killing tumor cells without affecting the normal cells. Hence this bioconjugate can be used simultaneously with PTT and photodynamic therapy in the future cancer treatment strategy (Shi et al. 2013).

12.7 CNTs FOR ORAL CANCER THERAPY

The CNTs can be effectively used for the delivery of drugs for oral cancer therapy. Because of the unique properties of vertically aligned carbon nanotube arrays, they gained more attention in recent researches. Their promising properties include chemical inertness, absorption of light, and conductivity. These carbon nanotube arrays together with integrated electrode were functionalized with an antibody, which is specific to CIP2A. CIP2A is a tumor protein associated with a variety of diseases like oral cancer. The as-produced immunosensor was efficient in the detection of CIP2A than the ELISA technique. The investigations revealed the potential of this system in the early stage detection of oral cancer (Ding et al. 2018).

Cisplatin and epidermal growth factor (EGF) were attached to SWCNT to deliver the drug to carcinoma cells in the neck and head. The results of the in vivo studies show that these bioconjugates get internalized into the tumor cells rapidly than the control cells without EGF. This study proved that cisplatin–EGF-loaded SWCNT were selectively taken up by the carcinoma cells in the neck and head, whereas the injected SWCNT without EGF were eliminated within 20 minutes of administration. Also, the cisplatin-EGF loaded SWCNT kills the tumor cells selectively than the control group. Moreover, the tumor regression in mice treated with cisplatin-EGF SWCNTs was more rapid than the conjugate without EGF (Bhirde et al. 2009).

12.8 CNT IN CANCER VACCINE

Cancer vaccine consists of antigens that are derived from tumor cells, which will provoke our immune responses (Parish 2003). The anticancer immune responses generated by the cancer vaccine target all tumor cells present in the body and hence these vaccines can be given to metastatic tumors also (Schlom, Arlen, and Gulley 2007). Further, these vaccines will produce a persistent T cell memory against the cancerous cells thereby provides prolonged protection and patient survival (Pagès et al. 2005). The three important components of cancer vaccine are the antigen, the adjuvant, and the delivery vehicle. The success of the cancer vaccine depends on the proper combination of the above three components (Hassan et al. 2019).

The SWCNTs can be used for the delivery of antigen to antigen-presenting cells (APC), which produce a humoral response against cancer cells. The peptide-

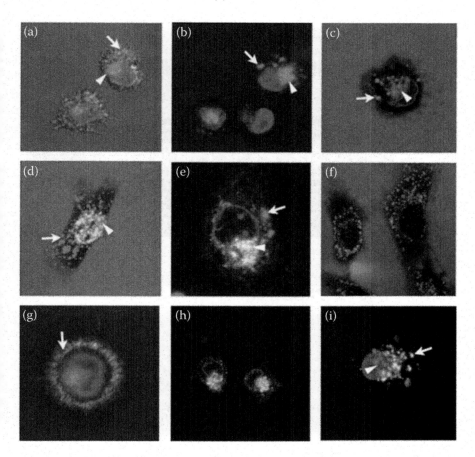

FIGURE 12.5 The internalization of SWNT constructs into DCs by confocal microscopy. Reprinted with permission from (Villa et al. 2011)© 2011 American Chemical Society.

SWCNT combination gets internalized into the APCs easily as shown in Figure 12.5 (Villa et al. 2011).

Similarly in a study, CNT was used as a delivery system for antigen NY-ESO-1 along with synthetic Toll-Like Receptor agonist. This CNT construct gets internalized into the dendritic cell and acts as a depot for antigen. This results in the induction of $CD4^+$ T and $CD8^+$ T cell-mediated immune responses against antigen NY-ESO-1. The investigation has revealed the potential of CNTs as a vaccine delivery system for cancer by providing reduced tumor growth and prolonged survival (Figure 12.6) (Faria et al. 2014). The lentinan–MWCNTs act as an efficient depot for intracellular antigen and improve the accumulation of antigens within the cell. This will enhance cellular and humoral immunity. Thus, the lentinan–MWCNTs can be effectively used for the delivery of vaccines which improves the immunogenicity for therapeutic uses (Xing et al. 2016).

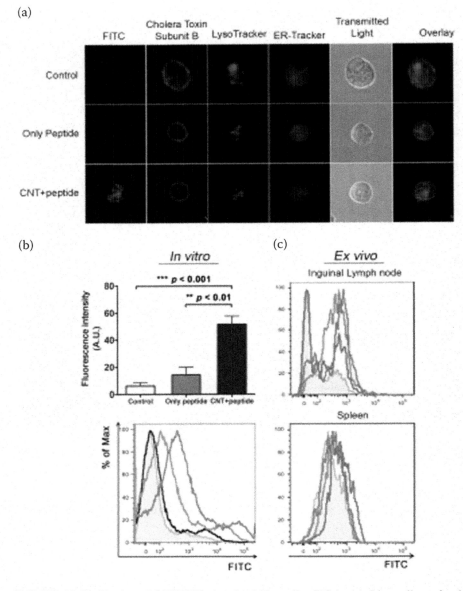

FIGURE 12.6 Uptake of MWCNTs by dendritic cells (DCs). (a) Live-cell confocal fluorescence microscopy images. (b) Optical intensities of randomly selected cells from each sample in panel. (c) MWCNT-peptide-FITC injected by SC route and the presence of intracellular constructs in DCs at 1 hour (green), 20 hour (red), and 3 days (blue) post-injection and evaluated by flow cytometry. Reprinted with permission from (Faria et al. 2014) © 2014 American Chemical Society.

The MWCNTs are efficient as a carrier for the delivery of tumor antigen. The MWCNTs can be used for the delivery of cytosine–phosphate–guanine oligodeoxynucleotide (CpG) and anti-CD40 Ig, together with the antigen ovalbumin, and are found to produce immune responses against tumor cells. Hence, MWCNTs can be used as a delivery system for combinational therapy against cancer (Hassan et al. 2016).

12.9 CONCLUSIONS

Carbon nanotubes are needle-like structures that can deliver the drug, genes, or other biomolecules. Through covalent or noncovalent bonding, the CNTs can be easily functionalized with a variety of molecules. The functionalization improves the aqueous solubility, ability to carry a variety of molecules, and drug targeting properties. The small and needle-like structure of these molecules enables the easy penetration and targeted delivery of the drug. The excellent physicochemical properties of CNTs expanded the applications of CNTs in various fields including pharmaceutical sciences. Several pieces of research were conducted to explore the potential of CNT in cancer therapy. The CNTs have greater potential to enhance the outcomes of cancer therapy. The unique physicochemical properties of CNTs like high drug loading capacity, ability to penetrate the cell membrane, high aspect ratio, and biocompatibility extended their application pharmaceutical field. They can be used as a mediator for photothermal therapy and photodynamic therapy for cancer therapy. Also, CNTs show better cellular uptake properties which lead to the development of a cancer vaccine delivering system using them. But the merits of the CNT cancer vaccine delivering system over the existing vaccine delivering system need to be investigated. Further indepth studies on the physicochemical properties of the CNTs are required to explore the novel applications and clinical practices.

REFERENCES

Abbaszadeh, Fatemeh, Omid Moradi, Mehdi Norouzi, and Omid Sabzevari. "Improvement Single-Wall Carbon Nanotubes (SWCNTs) Based on Functionalizing with Monomers 2-Hydroxyethylmethacryate (HEMA) and N-Vinylpyrrolidone (NVP) for Pharmaceutical Applications as Cancer Therapy." *Journal of Industrial and Engineering Chemistry* 20 (September 2014): 2895–2900. doi:10.1016/j.jiec.2013.11.025.

Aboutalebi Anaraki, Nadia, Leila Roshanfekr Rad, Mohammad Irani, and Ismaeil Haririan. "Fabrication of PLA/PEG/MWCNT Electrospun Nanofibrous Scaffolds for Anticancer Drug Delivery." *Journal of Applied Polymer Science* 132 (3) (2015). John Wiley & Sons, Ltd. doi:10.1002/app.41286.

Alderton, P M, J Gross, and M D Green. "Comparative Study of Doxorubicin, Mitoxantrone, and Epirubicin in Combination with ICRF-187 (ADR-529) in a Chronic Cardiotoxicity Animal Model." *Cancer Research* 52 (1) (1992): 194–201.

Aoki, Kaoru, and Naoto Saito. "Biocompatibility and Carcinogenicity of Carbon Nanotubes as Biomaterials." *Nanomaterials (Basel, Switzerland)* 10 (2) (2020). doi:10.3390/nano10020264.

Arya, Neha, Aditya Arora, K S Vasu, A K Sood, and Dhirendra S Katti. "Combination of Single Walled Carbon Nanotubes/Graphene Oxide with Paclitaxel: A Reactive Oxygen Species Mediated Synergism for Treatment of Lung Cancer." *Nanoscale* 5 (7) (2013): 2818–2829. doi:10.1039/C3NR33190C.

Beg, Sarwar, Mohammad Rizwan, Asif M Sheikh, M Saquib Hasnain, Khalid Anwer, and Kanchan Kohli. "Advancement in Carbon Nanotubes: Basics, Biomedical Applications and Toxicity." *The Journal of Pharmacy and Pharmacology* 63 (2) (2011): 141–163. doi:10.1111/j.2042-7158.2010.01167.x.

Bhirde, Ashwin A, Vyomesh Patel, Julie Gavard, Guofeng Zhang, Alioscka A Sousa, Andrius Masedunskas, Richard D Leapman, Roberto Weigert, J Silvio Gutkind, and James F Rusling. "Targeted Killing of Cancer Cells in Vivo and in Vitro with EGF-Directed Carbon Nanotube-Based Drug Delivery." *ACS Nano* 3 (2) (2009): 307–316. doi:10.1021/nn800551s.

Bianco, Alberto, Kostas Kostarelos, and Maurizio Prato. "Applications of Carbon Nanotubes in Drug Delivery." *Current Opinion in Chemical Biology* 9 (6) (2005): 674–679. doi:10.1016/j.cbpa.2005.10.005.

Chen, Zhou, Aili Zhang, Xiaobing Wang, Jing Zhu, Yamin Fan, Hongmei Yu, and Zhaogang Yang. "The Advances of Carbon Nanotubes in Cancer Diagnostics and Therapeutics." *Journal of Nanomaterials* 2017 (2017). doi:10.1155/2017/3418932.

Das, Manasmita, Raman Preet Singh, Satyajit R Datir, and Sanyog Jain. "Intranuclear Drug Delivery and Effective in Vivo Cancer Therapy via Estradiol–PEG-Appended Multiwalled Carbon Nanotubes." *Molecular Pharmaceutics* 10 (9) (2013): 3404–3416. doi:10.1021/mp4002409.

Dinesh, Bhimareddy, Alberto Bianco, and Cécilia Ménard-Moyon. "Designing Multimodal Carbon Nanotubes by Covalent Multi-Functionalization." *Nanoscale* 8 (44) (2016): 18596–18611. doi:10.1039/C6NR06728J.

Ding, Shaowei, Suprem R Das, Benjamin J Brownlee, Kshama Parate, Taylor M Davis, Loreen R Stromberg, Edward K L Chan, Joseph Katz, Brian D Iverson, and Jonathan C Claussen. "CIP2A Immunosensor Comprised of Vertically-Aligned Carbon Nanotube Interdigitated Electrodes towards Point-of-Care Oral Cancer Screening." *Biosensors and Bioelectronics* 117 (2018): 68–74. doi:10.1016/j.bios.2018.04.016.

Dolmans, Dennis E.J.G.J., Dai Fukumura, and Rakesh K Jain. "Photodynamic Therapy for Cancer." *Nature Reviews Cancer* 3 (5) (2003): 380–387. doi:10.1038/nrc1071.

Eldridge, Brittany N, Brian W Bernish, Cale D Fahrenholtz, and Ravi Singh. "Photothermal Therapy of Glioblastoma Multiforme Using Multiwalled Carbon Nanotubes Optimized for Diffusion in Extracellular Space." *ACS Biomaterials Science & Engineering* 2 (6) (2016): 963–976. doi:10.1021/acsbiomaterials.6b00052.

Elhissi, Abdelbary M A, Waqar Ahmed, Israr Ul Hassan, Vinod. R Dhanak, and Antony D'Emanuele. "Carbon Nanotubes in Cancer Therapy and Drug Delivery." Edited by Rassoul Dinarvand. *Journal of Drug Delivery* (2012). doi:10.1155/2012/837327.

Fahrenholtz, Cale D, Song Ding, Brian W Bernish, Mariah L Wright, Ye Zheng, Mu Yang, Xiyuan Yao, et al. "Design and Cellular Studies of a Carbon Nanotube-Based Delivery System for a Hybrid Platinum-Acridine Anticancer Agent." *Journal of Inorganic Biochemistry* 165 (December 2016): 170–180. doi:10.1016/j.jinorgbio.2016.07.016.

Faria, Paula Cristina Batista de, Luara Isabela dos Santos, João Paulo Coelho, Henrique Bücker Ribeiro, Marcos Assunção Pimenta, Luiz Orlando Ladeira, Dawidson Assis Gomes, Clascídia Aparecida Furtado, and Ricardo Tostes Gazzinelli. "Oxidized Multiwalled Carbon Nanotubes as Antigen Delivery System to Promote Superior CD8+ T Cell Response and Protection against Cancer." *Nano Letters* 14 (9) (2014): 5458–5470. doi:10.1021/nl502911a.

Feazell, Rodney P, Nozomi Nakayama-Ratchford, Hongjie Dai, and Stephen J Lippard. "Soluble Single-Walled Carbon Nanotubes as Longboat Delivery Systems for Platinum (IV) Anticancer Drug Design." *Journal of the American Chemical Society* 129 (27) (2007) 8438–8439. doi:10.1021/ja073231f.

Foldvari, Marianna, and Mukasa Bagonluri. "Carbon Nanotubes as Functional Excipients for Nanomedicines: II. Drug Delivery and Biocompatibility Issues." *Nanomedicine: Nanotechnology, Biology, and Medicine* 4 (3) (2008): 183–200. doi:10.1016/j.nano.2008.04.003.

Fukuda, Ryosuke, Tomokazu Umeyama, Masahiko Tsujimoto, Fumiyoshi Ishidate, Takeshi Tanaka, Hiromichi Kataura, Hiroshi Imahori, and Tatsuya Murakami. "Sustained Photodynamic Effect of Single Chirality-Enriched Single-Walled Carbon Nanotubes." *Carbon* 161 (2020): 718–725. doi:10.1016/j.carbon.2020.02.002.

Golubewa, Lena, Tatsiana Kulahava, Yuliya Kunitskaya, Pavel Bulai, Mikhail Shuba, and Renata Karpicz. "Enhancement of Single-Walled Carbon Nanotube Accumulation in Glioma Cells Exposed to Low-Strength Electric Field: Promising Approach in Cancer Nanotherapy." *Biochemical and Biophysical Research Communications* 529 (3) (2020): 647–651. doi:10.1016/j.bbrc.2020.06.100.

Guo, Chang, Wafa T Al-Jamal, Francesca M Toma, Alberto Bianco, Maurizio Prato, Khuloud T Al-Jamal, and Kostas Kostarelos. "Design of Cationic Multiwalled Carbon Nanotubes as Efficient SiRNA Vectors for Lung Cancer Xenograft Eradication." *Bioconjugate Chemistry* 26 (7) (2015): 1370–1379. doi:10.1021/acs.bioconjchem.5b00249.

Hassan, Hatem A F M, Sandra S Diebold, Lesley A Smyth, Adam A Walters, Giovanna Lombardi, and Khuloud T Al-Jamal. "Application of Carbon Nanotubes in Cancer Vaccines: Achievements, Challenges and Chances." *Journal of Controlled Release: Official Journal of the Controlled Release Society* 297 (March 2019): 79–90. doi:10.1016/j.jconrel.2019.01.017.

Hassan, Hatem A F M, Lesley Smyth, Julie T-W Wang, Pedro M Costa, Kulachelvy Ratnasothy, Sandra S Diebold, Giovanna Lombardi, and Khuloud T Al-Jamal. "Dual Stimulation of Antigen Presenting Cells Using Carbon Nanotube-Based Vaccine Delivery System for Cancer Immunotherapy." *Biomaterials* 104 (October 2016): 310–322. doi:10.1016/j.biomaterials.2016.07.005.

Heister, Elena, Vera Neves, Carmen Tilmaciu, Kamil Lipert, Vanesa Beltrán, Helen Coley, S Ravi Silva, and Johnjoe Mcfadden. "Triple Functionalisation of Single-Walled Carbon Nanotubes with Doxorubicin, a Monoclonal Antibody, and a Fluorescent Marker for Targeted Cancer Therapy." *Carbon* 47 (August 2009): 2152. doi:10.1016/j.carbon.2009.03.057.

Jha, Roopali, Amit Singh, P K Sharma, and Neeraj Kumar Fuloria. "Smart Carbon Nanotubes for Drug Delivery System: A Comprehensive Study." *Journal of Drug Delivery Science and Technology* 58 (2020). doi:10.1016/j.jddst.2020.101811.

Ji, Zongfei, Gaofeng Lin, Qinghua Lu, Lingjie Meng, Xizhong Shen, Ling Dong, Chuanlong Fu, and Xiaoke Zhang. "Targeted Therapy of SMMC-7721 Liver Cancer in Vitro and in Vivo with Carbon Nanotubes Based Drug Delivery System." *Journal of Colloid and Interface Science* 365 (1) (2012): 143–149. doi:10.1016/j.jcis.2011.09.013.

Kagan, Valerian E, Nagarjun V Konduru, Weihong Feng, Brett L Allen, Jennifer Conroy, Yuri Volkov, Irina I Vlasova, et al. "Carbon Nanotubes Degraded by Neutrophil Myeloperoxidase Induce Less Pulmonary Inflammation." *Nature Nanotechnology* 5 (5) (2010): 354–359. doi:10.1038/nnano.2010.44.

Kakade, Bhalchandra, Sanjay Patil, Bhaskar Sathe, Suresh Gokhale, and Vijayamohanan Pillai. "Near-Complete Phase Transfer of Single-Wall Carbon Nanotubes by Covalent Functionalization." *Journal of Chemical Sciences* 120 (6) (2008): 599–606. doi:10.1007/s12039-008-0091-3.

Kang, Bin, Jun Li, Shuquan Chang, Mingzhu Dai, Chao Ren, Yaodong Dai, and Da Chen. "Subcellular Tracking of Drug Release from Carbon Nanotube Vehicles in Living Cells." *Small* 8 (5) (2012): 777–782. doi:10.1002/smll.201101714.

Kostarelos, Kostas, Lara Lacerda, Giorgia Pastorin, Wei Wu, Sébastien Wieckowski, Jacqueline Luangsivilay, Sylvie Godefroy, et al. "Cellular Uptake of Functionalized Carbon Nanotubes Is Independent of Functional Group and Cell Type." *Nature Nanotechnology*. (2007). doi:10.1038/nnano.2006.209.

Kushwaha, Swatantra, Saurav Ghoshal, Awani Rai, and Satyawan Singh. "Carbon Nanotubes as a Novel Drug Delivery System for Anticancer Therapy: A Review." *Brazilian Journal of Pharmaceutical Sciences* 49 (December 2013): 629–643. doi:10.1590/S1984-82502013000400002.

Lee, Pei-Chi, Cheng-Liang Peng, and Ming-Jium Shieh. "Combining the Single-Walled Carbon Nanotubes with Low Voltage Electrical Stimulation to Improve Accumulation of Nanomedicines in Tumor for Effective Cancer Therapy." *Journal of Controlled Release: Official Journal of the Controlled Release Society* 225 (March 2016): 140–151. doi:10.1016/j.jconrel.2016.01.038.

Liu, Hongzhuo, Hui Xu, Yan Wang, Zhonggui He, and Sanming Li. "Effect of Intratumoral Injection on the Biodistribution and Therapeutic Potential of Novel Chemophor EL-Modified Single-Walled Nanotube Loading Doxorubicin." *Drug Development and Industrial Pharmacy* 38 (9) (2012): 1031–1038. doi:10.3109/03639045.2011.637050.

Lu, Guan-Hua, Wen-Ting Shang, Han Deng, Zi-Yu Han, Min Hu, Xiao-Yuan Liang, Chi-Hua Fang, Xin-Hong Zhu, Ying-Fang Fan, and Jie Tian. "Targeting Carbon Nanotubes Based on IGF-1R for Photothermal Therapy of Orthotopic Pancreatic Cancer Guided by Optical Imaging." *Biomaterials* 195 (March 2019): 13–22. doi:10.1016/j.biomaterials.2018.12.025.

Mahajan, Shubhangi, Abhimanyu Patharkar, Kaushik Kuche, Rahul Maheshwari, Pran Kishore Deb, Kiran Kalia, and Rakesh K Tekade. "Functionalized Carbon Nanotubes as Emerging Delivery System for the Treatment of Cancer." *International Journal of Pharmaceutics* 548 (1) (2018): 540–558. doi:10.1016/j.ijpharm.2018.07.027.

Maleki, Reza, Hamid Hassanzadeh Afrouzi, Mirollah Hosseini, Davood Toghraie, Anahita Piranfar, and Sara Rostami. "PH-Sensitive Loading/Releasing of Doxorubicin Using Single-Walled Carbon Nanotube and Multi-Walled Carbon Nanotube: A Molecular Dynamics Study." *Computer Methods and Programs in Biomedicine* 186 (April 2020): 105210. doi:10.1016/j.cmpb.2019.105210.

McKernan, Patrick, Needa A Virani, Gabriela N F Faria, Clément G Karch, Ricardo Prada Silvy, Daniel E Resasco, Linda F Thompson, and Roger G Harrison. "Targeted Single-Walled Carbon Nanotubes for Photothermal Therapy Combined with Immune Checkpoint Inhibition for the Treatment of Metastatic Breast Cancer." *Nanoscale Research Letters* 16 (1) (2021): 9. doi:10.1186/s11671-020-03459-x.

Miao, Yanyan, Haixia Zhang, Yubin Pan, Jian Ren, Miaoman Ye, Fangfang Xia, Rui Huang, et al. "Single-Walled Carbon Nanotube: One Specific Inhibitor of Cancer Stem Cells in Osteosarcoma upon Downregulation of the TGFβ1 Signaling." *Biomaterials* 149 (December 2017): 29–40. doi:10.1016/j.biomaterials.2017.09.032.

Mirsalari, Halimeh, Afsaneh Maleki, Heidar Raissi, and Azim Soltanabadi. "Investigation of the Pristine and Functionalized Carbon Nanotubes as a Delivery System for the Anticancer Drug Dacarbazine: Drug Encapsulation." *Journal of Pharmaceutical Sciences* (November 2020). doi:10.1016/j.xphs.2020.10.062.

Misra, Ranjita, Sarbari Acharya, and Sanjeeb K Sahoo. "Cancer Nanotechnology: Application of Nanotechnology in Cancer Therapy." *Drug Discovery Today* 15 (19–20) (2010) England: 842–850. doi:10.1016/j.drudis.2010.08.006.

Mohanta, Debashish, Soma Patnaik, Sanchit Sood, and Nilanjan Das. "Carbon Nanotubes: Evaluation of Toxicity at Biointerfaces." *Journal of Pharmaceutical Analysis* 9 (5) (2019): 293–300. doi:10.1016/j.jpha.2019.04.003.

Mu, Qingxin, Dana L Broughton, and Bing Yan. "Endosomal Leakage and Nuclear Translocation of Multiwalled Carbon Nanotubes: Developing a Model for Cell Uptake." *Nano Letters* 9 (12) (2009): 4370–4375. doi:10.1021/nl902647x.

Muller, Julie, Monique Delos, Nadtha Panin, Virginie Rabolli, François Huaux, and Dominique Lison. 2009. "Absence of Carcinogenic Response to Multiwall Carbon Nanotubes in a 2-Year Bioassay in the Peritoneal Cavity of the Rat." *Toxicological Sciences: An Official Journal of the Society of Toxicology* 110 (2): 442–448. doi:10.1093/toxsci/kfp100.

Neelgund, Gururaj M, Makobi C Okolie, Folami K Williams, and Aderemi Oki. "Ag2S Nanocrystallites Deposited over Polyamidoamine Grafted Carbon Nanotubes: An Efficient NIR Active Photothermal Agent." *Materials Chemistry and Physics* 234 (2019): 32–37. doi:10.1016/j.matchemphys.2019.05.040.

Ogbodu, Racheal O, Janice L Limson, Earl Prinsloo, and Tebello Nyokong. "Photophysical Properties and Photodynamic Therapy Effect of Zinc Phthalocyanine-Spermine-Single Walled Carbon Nanotube Conjugate on MCF-7 Breast Cancer Cell Line." *Synthetic Metals* 204 (2015): 122–132. doi:10.1016/j.synthmet.2015.03.011.

Ogbodu, Racheal O, and Tebello Nyokong. "The Effect of Ascorbic Acid on the Photophysical Properties and Photodynamic Therapy Activities of Zinc Phthalocyanine-Single Walled Carbon Nanotube Conjugate on MCF-7 Cancer Cells." *Spectrochimica Acta Part A: Molecular and Biomolecular Spectroscopy* 151 (2015): 174–183. doi:10.1016/j.saa.2015.06.063.

Pagès, Franck, Anne Berger, Matthieu Camus, Fatima Sanchez-Cabo, Anne Costes, Robert Molidor, Bernhard Mlecnik, et al. "Effector Memory T Cells, Early Metastasis, and Survival in Colorectal Cancer." *The New England Journal of Medicine* 353 (25) (2005): 2654–2666. doi:10.1056/NEJMoa051424.

Pantarotto, Davide, Ravi Singh, David McCarthy, Mathieu Erhardt, Jean-Paul Briand, Maurizio Prato, Kostas Kostarelos, and Alberto Bianco. "Functionalized Carbon Nanotubes for Plasmid DNA Gene Delivery." *Angewandte Chemie (International Ed. in English)* 43 (39) (2004): 5242–5246. doi:10.1002/anie.200460437.

Parish, Christopher R. "Cancer Immunotherapy: The Past, the Present and the Future." *Immunology and Cell Biology* 81 (2) (2003): 106–113. doi:10.1046/j.0818-9641.2003.01151.x.

Raphey, V R, T K Henna, K P Nivitha, P Mufeedha, Chinnu Sabu, and K Pramod. "Advanced Biomedical Applications of Carbon Nanotube." *Materials Science and Engineering: C* 100 (2019): 616–630. doi:10.1016/j.msec.2019.03.043.

Ravi Kiran, A.V.V., G Kusuma Kumari, and Praveen T Krishnamurthy. "Carbon Nanotubes in Drug Delivery: Focus on Anticancer Therapies." *Journal of Drug Delivery Science and Technology* 59 (2020): 101892. doi:10.1016/j.jddst.2020.101892.

Razzazan, Ali, Fatemeh Atyabi, Bahram Kazemi, and Rassoul Dinarvand. "In Vivo Drug Delivery of Gemcitabine with PEGylated Single-Walled Carbon Nanotubes." *Materials Science & Engineering. C, Materials for Biological Applications* 62 (May) (2019): 614–625. doi:10.1016/j.msec.2016.01.076.

Saeednia, Leyla, Li Yao, Kim Cluff, and Ramazan Asmatulu. "Sustained Releasing of Methotrexate from Injectable and Thermosensitive Chitosan–Carbon Nanotube Hybrid Hydrogels Effectively Controls Tumor Cell Growth." *ACS Omega* 4 (2) (2019): 4040–4048. doi:10.1021/acsomega.8b03212.

Schlom, Jeffrey, Philip M Arlen, and James L Gulley. "Cancer Vaccines: Moving beyond Current Paradigms." *Clinical Cancer Research: An Official Journal of the American Association for Cancer Research* 13 (13) (2007): 3776–3782. doi:10.1158/1078-0432.CCR-07-0588.

Sharifi, Shahriar, Shahed Behzadi, Sophie Laurent, M Laird Forrest, Pieter Stroeve, and Morteza Mahmoudi. "Toxicity of Nanomaterials." *Chemical Society Reviews* 41 (6) (2012): 2323–2343. doi: 10.1039/c1cs15188f.

Shi, Jinjin, Rourou Ma, Lei Wang, Jing Zhang, Ruiyuan Liu, Lulu Li, Yan Liu, et al. "The Application of Hyaluronic Acid-Derivatized Carbon Nanotubes in Hematoporphyrin Monomethyl Ether-Based Photodynamic Therapy for in Vivo and in Vitro Cancer Treatment." *International Journal of Nanomedicine* 8 (2013): 2361–2373. doi: 10.2147/IJN.S45407.

Shi Kam, Nadine Wong, Theodore C Jessop, Paul A Wender, and Hongjie Dai. "Nanotube Molecular Transporters: Internalization of Carbon Nanotube–Protein Conjugates into Mammalian Cells." *Journal of the American Chemical Society* 126 (22) (2004): 6850–6851. doi: 10.1021/ja0486059.

Singh, Rahul Pratap, Gunjan Sharma, Sonali, Sanjay Singh, Shreekant Bharti, Bajarangprasad L Pandey, Biplob Koch, and Madaswamy S Muthu. "Chitosan-Folate Decorated Carbon Nanotubes for Site Specific Lung Cancer Delivery." *Materials Science & Engineering. C, Materials for Biological Applications* 77 (August 2017): 446–458. doi: 10.1016/j.msec.2017.03.225.

Singh, Rahul Pratap, Gunjan Sharma, Sonali, Sanjay Singh, Shashikant C U Patne, Bajarangprasad L Pandey, Biplob Koch, and Madaswamy S Muthu. "Effects of Transferrin Conjugated Multi-Walled Carbon Nanotubes in Lung Cancer Delivery." *Materials Science and Engineering: C* 67 (2016): 313–325. doi: 10.1016/j.msec.2016.05.013.

Singhai, Nidhi Jain, Rahul Maheshwari, Narendra K Jain, and Suman Ramteke. "Chondroitin Sulphate and α-Tocopheryl Succinate Tethered Multiwalled Carbon Nanotubes for Dual-Action Therapy of Triple-Negative Breast Cancer." *Journal of Drug Delivery Science and Technology* 60 (2020): 102080. doi: 10.1016/j.jddst.2020.102080.

Singhai, Nidhi Jain, Rahul Maheshwari, and Suman Ramteke. "CD44 Receptor Targeted 'Smart' Multi-Walled Carbon Nanotubes for Synergistic Therapy of Triple-Negative Breast Cancer." *Colloid and Interface Science Communications* 35 (2020): 100235. doi: 10.1016/j.colcom.2020.100235.

Sobhani, Zahra, Mohammad Ali Behnam, Farzin Emami, Amirreza Dehghanian, and Iman Jamhiri. "Photothermal Therapy of Melanoma Tumor Using Multiwalled Carbon Nanotubes." *International Journal of Nanomedicine* 12 (2017): 4509–4517. doi: 10.2147/IJN.S134661.

Son, Kuk Hui, Jeong Hee Hong, and Jin Woo Lee. "Carbon Nanotubes as Cancer Therapeutic Carriers and Mediators." *International Journal of Nanomedicine* 11 (2016): 5163–5185. doi: 10.2147/IJN.S112660.

Star, Alexander, J Fraser Stoddart, David Steuerman, Mike Diehl, Akram Boukai, Eric W Wong, Xin Yang, Sung-Wook Chung, Hyeon Choi, and James R Heath. "Preparation and Properties of Polymer-Wrapped Single-Walled Carbon Nanotubes." *Angewandte Chemie International Edition* 40 (9) (2001): 1721–1725. doi: 10.1002/1521-3773(20010504)40:9<1721::AID-ANIE17210>3.0.CO;2-F.

Sun, Zhao, Zhe Liu, Jie Meng, Jie Meng, Jinhong Duan, Sishen Xie, Xin Lu, et al. "Carbon Nanotubes Enhance Cytotoxicity Mediated by Human Lymphocytes in Vitro." *PloS One* 6 (6) (2011): e21073–e21073. doi: 10.1371/journal.pone.0021073.

Suo, Xubin, Brittany N Eldridge, Han Zhang, Chengqiong Mao, Yuanzeng Min, Yao Sun, Ravi Singh, and Xin Ming. "P-Glycoprotein-Targeted Photothermal Therapy of Drug-Resistant Cancer Cells Using Antibody-Conjugated Carbon Nanotubes." *ACS Applied Materials & Interfaces* 10 (39) (2018): 33464–33473. doi: 10.1021/acsami.8b11974.

Tasis, Dimitrios, Nikos Tagmatarchis, Alberto Bianco, and Maurizio Prato. "Chemistry of Carbon Nanotubes." *Chemical Reviews* 106 (3) (2006): 1105–1136. doi: 10.1021/cr050569o.

Tran, Duy Thanh, Van Hien Hoa, Le Huu Tuan, Nam Hoon Kim, and Joong Hee Lee. "Cu-Au Nanocrystals Functionalized Carbon Nanotube Arrays Vertically Grown on Carbon Spheres for Highly Sensitive Detecting Cancer Biomarker." *Biosensors and Bioelectronics* 119 (2018): 134–140. doi:10.1016/j.bios.2018.08.022.

Villa, Carlos H, Tao Dao, Ian Ahearn, Nicole Fehrenbacher, Emily Casey, Diego A Rey, Tatyana Korontsvit, et al. "Single-Walled Carbon Nanotubes Deliver Peptide Antigen into Dendritic Cells and Enhance IgG Responses to Tumor-Associated Antigens." *ACS Nano* 5 (7) (2011): 5300–5311. doi:10.1021/nn200182x.

Wang, Miao, Lifo Ruan, Tianyu Zheng, Dongqing Wang, Mengxue Zhou, Huiru Lu, Jimin Gao, Jun Chen, and Yi Hu. "A Surface Convertible Nanoplatform with Enhanced Mitochondrial Targeting for Tumor Photothermal Therapy." *Colloids and Surfaces. B, Biointerfaces* 189 (May 2020): 110854. doi:10.1016/j.colsurfb.2020.110854.

Wang, Ni, Yanxian Feng, Lilan Zeng, Zhennan Zhao, and Tianfeng Chen. "Functionalized Multiwalled Carbon Nanotubes as Carriers of Ruthenium Complexes to Antagonize Cancer Multidrug Resistance and Radioresistance." *ACS Applied Materials & Interfaces* 7 (27) (2015): 14933–14945. doi:10.1021/acsami.5b03739.

Wu, Linlin, Changjun Man, Hong Wang, Xiaohe Lu, Qinghai Ma, Yu Cai, and Wanshan Ma. "PEGylated Multi-Walled Carbon Nanotubes for Encapsulation and Sustained Release of Oxaliplatin." *Pharmaceutical Research* 30 (2) (2013): 412–423. doi:10.1007/s11095-012-0883-5.

Xing, Jie, Zhenguang Liu, Yifan Huang, Tao Qin, Ruonan Bo, Sisi Zheng, Li Luo, Yee Huang, Yale Niu, and Deyun Wang. "Lentinan-Modified Carbon Nanotubes as an Antigen Delivery System Modulate Immune Response in Vitro and in Vivo." *ACS Applied Materials & Interfaces* 8 (30) (2016): 19276–19283. doi:10.1021/acsami.6b04591.

Yu, Baodan, Li Tan, Runhui Zheng, Huo Tan, and Lixia Zheng. "Targeted Delivery and Controlled Release of Paclitaxel for the Treatment of Lung Cancer Using Single-Walled Carbon Nanotubes." *Materials Science & Engineering. C, Materials for Biological Applications* 68 (November 2016): 579–584. doi:10.1016/j.msec.2016.06.025.

Zhang, Bo, Huafang Wang, Shun Shen, Xiaojian She, Wei Shi, Jun Chen, Qizhi Zhang, Yu Hu, Zhiqing Pang, and Xinguo Jiang. "Fibrin-Targeting Peptide CREKA-Conjugated Multi-Walled Carbon Nanotubes for Self-Amplified Photothermal Therapy of Tumor." *Biomaterials* 79 (February 2016): 46–55. doi:10.1016/j.biomaterials.2015.11.061.

Zhang, Ming, Wentao Wang, Fan Wu, Ping Yuan, Cheng Chi, and Ninglin Zhou. "Magnetic and Fluorescent Carbon Nanotubes for Dual Modal Imaging and Photothermal and Chemo-Therapy of Cancer Cells in Living Mice." *Carbon* 123 (2017): 70–83. doi:10.1016/j.carbon.2017.07.032.

Zhang, Yan, Jian Xiao, Yimin Sun, Lu Wang, Xulin Dong, Jinghua Ren, Wenshan He, and Fei Xiao. "Flexible Nanohybrid Microelectrode Based on Carbon Fiber Wrapped by Gold Nanoparticles Decorated Nitrogen Doped Carbon Nanotube Arrays: In Situ Electrochemical Detection in Live Cancer Cells." *Biosensors and Bioelectronics* 100 (2018): 453–461. doi:10.1016/j.bios.2017.09.038.

13 Applications of Silica-Based Nanomaterials for Combinatorial Drug Delivery in Breast Cancer Treatment

Mubin Tarannum and Juan L. Vivero-Escoto
The University of North Carolina at Charlotte

CONTENTS

13.1 Introduction 311
13.2 Silica-Based Nanomaterials 312
13.3 Synthesis and Functionalization of Silica-Based Nanoparticles 313
13.4 Breast Cancer 314
13.5 Nanoparticles for Combination Therapy 315
    13.5.1 Chemotherapy Agents Combined with Multidrug Resistance Inhibitors 316
    13.5.2 Delivery of Nucleic Acids and Chemotherapy Agents Combined with Nucleic Acids 317
    13.5.3 Combination Based on Photothermal/Photodynamic Therapy ...318
    13.5.4 Hybrid Silica Nanoparticles 319
13.6 Conclusions 320
References 320

13.1 INTRODUCTION

Nanotechnology has evolved to be at the forefront of diagnostics, imaging, and therapeutic drug delivery. Nanomaterials are applied in a plethora of scientific areas by virtue of their unique properties associated with the nanoscale, such as surface area and quantum effect (Vieira and Gamarra 2016). Nanomedicine brings nanotechnology and medicine together to improve current diagnosis and treatment as an effective solution to address some of the critical issues associated with cancer. Nanomaterials such as liposomes, albumin-based nanoparticles, polymeric nanoparticles, micelles, gold, and silica nanoparticles have been extensively used to develop novel imaging probes and therapies to improve cancer treatment (Sailor and Park 2012; Zhu et al. 2017). The applications of nanomedicine have made their way to the benefit of cancer patients with products like Doxil, Abraxane, and MM-398; additionally, many other platforms are under clinical investigation (van der Meel, Lammers, and Hennink 2017).

The rationale design of nanocarriers for the anticancer drug delivery has resulted in increased plasma half-life, improved pharmacodynamic and pharmacokinetic properties of drugs, target-specific drug delivery via passive or active targeting, and decreased off-target toxicity (Hu and Zhang 2012; Parhi, Mohanty, and Sahoo 2012). The anticancer application of the nanomedicine field is based on the leaky vasculature and poor lymphatic drainage in tumors. These features of tumor results in the accumulation of nanoparticles and is referred to as enhanced permeability and retention effect (EPR effect) (Stylianopoulos 2013). The nanoparticles are large enough to escape filtration by the kidneys yet small enough to evade phagocytic removal by Kupffer cells and splenocytes in the mononuclear phagocyte system (MPS). The nanoparticles between 1 and 500 nm provide the inherent property to interact with the cell surface and organelles. In conjunction, nanoparticles can actively target tumor tissue with the use of targeting moieties, referred to as active targeting which is advantageous, specifically in tumors where the EPR effect is diminished (Rosenblum et al. 2018). Nanomaterials provide an additional advantage for combination therapies, involving the addition of multiple chemo agents, nucleic acid agents, therapeutic proteins, and imaging agents (Ma, Kohli, and Smith 2013; Shrestha, Tang, and Romero 2019).

13.2 SILICA-BASED NANOMATERIALS

Among the plethora of nanomaterials, silica nanoparticles have attracted significant interest in various fields of science and engineering due to their unique properties. They offer several unique and advantageous properties such as chemical and thermal stability, high surface area, tunable particle size, shape and porosity, chemically modifiable surfaces, and facile functionalization (Pasqua et al. 2016; Yang and Yu 2016). There are various types of silica nanoparticles, solid, mesoporous, hollow, and hybrid silica nanoparticles (Selvarajan, Obuobi, and Ee 2020). Furthermore, mesoporous silica nanoparticles (MSNs) have been highly explored for catalysis, food manufacturing, biosensing, delivery of drugs, and contrast agents (Colilla, González, and Vallet-Regí 2013; Vivero-Escoto et al. 2010).

Silica-based materials exhibit several advantages, which make them an excellent candidate for biomedical applications (Pasqua et al. 2016; Rosenholm et al. 2012; Zhou et al. 2018): **(i)** The silanol-containing silica surfaces can be functionalized with high selectivity to achieve control over drug loading and release. In addition, the external surface of nanoparticles can be conjugated with various targeting moieties, imaging agents, and stimuli-responsive molecules for efficient target-specific drug delivery (Song et al. 2016). **(ii)** Silica is "Generally Recognized as Safe" material by the US FDA and Cornell dots (C dots), another class of silica nanoparticles were approved for human clinical trial for targeted molecular imaging (Phillips et al. 2014). Various in vivo preclinical evaluations have established that silica NPs are biocompatible. In addition, the biodistribution, passive targeting, and clearance can be tuned by modifying the structural parameters like particle size, morphology, porosity, surface properties, functionalization, and administration routes. (**iii**) Their surface can be modified with macrocycle molecules, polymers, and proteins, which act as capping agents or pore gating agents. Another approach for controlled drug release is to use stimuli-responsive chemical handles to conjugate drugs to nanoparticles. These chemical linkages can prevent the premature release of drugs in the systemic circulation, whereas once in the target tissue (tumor), the chemical linkages break in response to the stimuli. (**iv**) Porous silica nanoparticles exhibit additional advantages for drug delivery. MSNs' high surface area and ordered porous structure allow for high drug loading. Drugs can be encapsulated in the internal surface and/or conjugated to the external MSN surface (Pasqua et al. 2016; Rosenholm et al. 2012; Zhou et al. 2018).

13.3 SYNTHESIS AND FUNCTIONALIZATION OF SILICA-BASED NANOPARTICLES

The Stöber process is one of the widely employed methods for the fabrication of silica nanoparticles. This method enables a high control over the polymerization and reaction kinetics (Selvarajan, Obuobi, and Ee 2020; S.-H. Wu, Mou, and Lin 2013). The Stöber process is applicable for a variety of silica nanomaterials like solid, mesoporous, or hollow (Stöber, Fink, and Bohn 1968). The surfactant-template approach is a variation of the Stöber method used for the fabrication of MSNs (Trewyn et al. 2007; S.-H. Wu, Mou, and Lin 2013). In this approach, a surfactant like cetyltrimethylammonium bromide (CTAB) is utilized, where the surfactant assembles into micelles at a concentration greater than the critical micellar concentration (CMC). Upon the reaction conditions, this process leads to the formation of a liquid crystal, which acts as a template for the formation of a porous structure in MSNs. Once the surfactant template is formed, the addition of silica precursor (tetraethyl orthosilicate, TEOS) condenses around the micellar template. The pH of the reaction solution controls the hydrolysis/condensation rates, which plays a major role in the polymerization of silica (Trewyn et al. 2007; S.-H. Wu, Mou, and Lin 2013). In order to generate the porous structure, the template surfactant is removed by calcination or by solvent extraction.

Apart from the Stöber process, the reverse microemulsion is another method that uses water-in-oil reverse microemulsion (Han et al. 2008). The microemulsion method is based on selective distribution of the hydrophobic ligand in the oil phase. This fabrication process is an effective method used to synthesize uniform nanoparticles and reduce the polydispersity (López-Quintela 2003). In addition, hydrophobic therapeutic agents can be solubilized in the nanodroplet forming the monomeric units of the nanoparticle (Lyles et al. 2020). Surfactant-stabilized water droplets are dispersed in an oil phase to create micelles and these micelles act as nanoreactors for controlled nucleation and nanoparticle growth. In general, this process is used for smaller and monodispersed nanoparticles. Nevertheless, it is limited by the excessive use of organic solvent and excessive purification. The reverse microemulsion approach is very desirable for creating a layer of silica shells on other nanoparticles (Han et al. 2008). Silica shell on other nanoparticles adds biocompatibility and facile functionalization. Furthermore, direct micelle-assisted or dual-phase approaches are also used for the synthesis of silica nanoparticles.

Unmodified silica nanoparticles are inherently negatively charged due to the surface hydroxyl groups. Nevertheless, the surface can be modified post-synthesis using alkoxysilane derivatives. Functionalization of silica NPs can be carried out through two main approaches: post-synthesis grating or co-condensation (Kuthati et al. 2013; Trewyn et al. 2007). In the post-synthesis grafting method, the organosilanes (example: aminopropyl triethoxysilane, mercaptopropyl trimethoxysilane) react with the silanol groups on the surface of the nanoparticles. Post-synthesis grafting can be carried out before or after surfactant template extraction. This reaction process is specifically advantageous to functionalize the exterior surface of the nanoparticles. In the co-condensation method, the organosilanes are introduced at the time of silica framework formation. This approach is mainly advantageous to functionalize the interior surface of porous silica nanoparticles, MSNs (Kuthati et al. 2013; Trewyn et al. 2007). One of the outstanding features of MSNs compared to other nanomaterials is the ability of multi-functionalization with high precision. The interior or exterior surface of the MSNs can be selectively functionalized with different functional groups. We have used this approach to functionalize the interior surface with pro-drugs and/or imaging agents and the external surface with other types of drugs and/or targeting moieties (Alvarez-Berríos et al. 2016; Vivero-Escoto, Huxford-Phillips, and Lin 2012). Based on the superior features and versatility of silica-based nanomaterials, in this chapter, we report the recent approaches for cancer therapy with a focus on breast cancer.

13.4 BREAST CANCER

Breast cancer (BC) is one of the most frequently diagnosed cancer in women, with 30% of all new cancer diagnoses and is the second leading cause of cancer-related deaths among women in the USA (Kucharczyk et al. 2017). Currently, BC is divided into subtypes based on the molecular heterogeneity: estrogen/progesterone receptor, HER2 receptor, luminal A, luminal B and, triple-negative breast cancer (TNBC). These receptors serve as markers for diagnosis and targeted hormonal therapies. The molecular subtype of BC is an important prognostic variable along with tumor size and nodal size which impact the prognosis and influence the decision making in BC treatment (Prat et al. 2015). For example, the HER2-positive and basal-like subtypes are typically aggressive and suffer from poor outcomes, whereas the Luminal A tumors showed better outcomes. Traditionally, BC therapy involves a multimodal strategy including neoadjuvant chemotherapy, surgery, and radiotherapy accompanied with adjuvant chemotherapy and/or endocrine therapy (Sachdev and Jahanzeb 2016). Systemic neoadjuvant therapy, mostly chemotherapy, may decrease the tumor burden to increase the possibility of surgery usually provides greater chances for breast-conserving surgery. Adjuvant therapy involves local radiation, systemic chemotherapy, molecular targeted therapies, or their combination (Kucharczyk et al. 2017). Adjuvant systemic chemotherapy is a mainstay in the clinic for controlling the disease and improving survival as well as chemotherapy remains the core treatment for metastatic breast cancer (Sachdev and Jahanzeb 2016).

Combination therapies for BC are performed by combining alkylating agents, cyclophosphamide, and antimetabolites (methotrexate and 5-FU). Combining multiple chemotherapy agents results in a higher response rate, better efficacy, and dose reduction, reduce drug resistance. One of the first few combination trials includes a phase II neoadjuvant trial, docetaxel plus epirubicin combination was evaluated in advanced stage BC and the trial showed a response rate of 76.7% with the participating drugs (de Matteis et al. 2002). In the study reported by National Surgical Adjuvant Breast and Bowel Project Protocol B-27, docetaxel after doxorubicin (DOX) plus cyclophosphamide resulted in increased pathological response rates for operable BC (Bear et al. 2003). Further, sequencing chemo taxanes prior to anthracyclines could improve the complete response (Earl et al. 2014; von Minckwitz et al. 2012). Docetaxel and carboplatin combination was investigated in stage II or III of BC and showed 16% pCR including patients with TNBC. The study concluded four cycles of neoadjuvant carboplatin and docetaxel given 2 weeks apart, followed by pegfilgrastim (Roy et al. 2013). Similarly, other drugs like DOX, platinum-based drugs (cisplatin and carboplatin) are also included in combinations. These trials support the usage of multidrug chemo agents for improved BC therapy. Taxanes form another important class of chemo agents in BC which include paclitaxel and docetaxel. Paclitaxel has been combined with DOX (L Gianni et al. 1995), cisplatin (Gelmon et al. 1996; Wasserheit et al. 1996), cyclophosphamide, 5-FU, and mitomycin. Chemo agents are also combined with hormonal therapy and molecular agents including panitumumab and lapatinib. EGFR inhibitors including panitumumab were added to various chemo agents with improved outcomes. In addition, heat shock protein (Hsp90) inhibitors were also combined with chemo agents and molecular therapies, specifically, 17-AAG plus trastuzumab showed potent anticancer activity in metastatic BC (Lu et al. 2012; Modi et al. 2011).

13.5 NANOPARTICLES FOR COMBINATION THERAPY

Traditional combination therapy suffers from limitations of low bioavailability, selectivity, multidrug resistance, nonspecific drug accumulation, and increased dosage. Critically, exploiting the full potential of combination therapies requires the development of versatile delivery platforms that can simultaneously transport and release multiple drugs that exhibit different physicochemical properties (Dicko, Mayer, and Tardi 2010; Mayer et al. 2006; F. Meng, Han, and Yeo 2017). Nanocarriers provide a great opportunity in this direction for the improvement of combination therapies. Use of multidrug regimen where the participating drugs have different targets with additive effects, overcome resistance and low availability in tumors. Nanoparticles can increase the stability of therapeutic agents, increase their bioavailability in the tumor, decrease drug degradation (Dicko, Mayer, and Tardi 2010). Nanoparticles can carry multiple classes of anticancer agents and provide co-delivery of drugs concerning time/space and ratio, without increasing the frequency of administration. As participating drugs usually have different physiochemical properties, NPs can normalize pharmacokinetics and biodistribution of drugs as they exhibit different physicochemical properties (Ma, Kohli, and Smith 2013).

Nanoparticles have been a huge focus for the multidrug combinations for BC therapy with lipid, micelles, inorganic nanomaterials been investigated for combining chemo agents, hormonal therapy, and molecular agents, which have been reviewed recently. The nanoparticle formulation of paclitaxel (nab-paclitaxel) is already being investigated in clinical trials, combined with other chemo agents and/or molecular agents (Adams et al. 2019; Gianni et al. 2018). Silica nanoparticles are a versatile delivery system that provides several unique advantages for the co-delivery of drugs. The nanotechnology application for diagnosis and treatment of BC is detailed in the various review in the past decade (Du et al. 2019; Grobmyer et al. 2012; Hussain, Khan, and Murtaza 2018; Islamian, Hatamian, and Rashidi 2015; Olov, Bagheri-Khoulenjani, and Mirzadeh 2018; Prados et al. 2012; Saadeh et al. 2014; Singh et al. 2017; Tharkar et al. 2015). The vast applications of MSNs for breast cancer therapy have been recently reviewed (Poonia, Lather, and Pandita 2018). MSNs have been used for the delivery of single chemotherapy agents including DOX, curcumin, or cisplatin (Alvarez-Berríos and Vivero-Escoto 2016; Ma'mani et al. 2014; J. Wang et al. 2016). Silica nanoparticles have also been extensively used for imaging and diagnosis of BC (Dréau et al. 2016; Hanafi-Bojd et al. 2018; Milgroom et al. 2014). The silica layer is coated on various other nanoparticles for diagnosis and therapy (Fathy et al. 2019). In this chapter, we focus on the silica-based nanoparticles for the combination therapy of BC (Figure 13.1).

13.5.1 CHEMOTHERAPY AGENTS COMBINED WITH MULTIDRUG RESISTANCE INHIBITORS

The major limitation and factor of failure in chemotherapy for BC being drug resistance. Multi-drug resistance (MDR) can be a result of various drug pumps including P-glycoprotein (P-gp) and Ca^{2+} channel. Hence, chemo agents have been combined with MDR inhibitors including to improve chemotherapy in BC. Wang et al. used

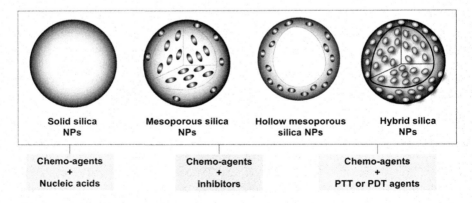

FIGURE 13.1 Various types of silica nanoparticles including solid, mesoporous, hollow mesoporous, and hybrid silica nanoparticles for combinatorial applications in breast cancer therapy.

MSNs for combinatorial delivery of DOX with Ca^{2+} channel siRNA which showed high drug loading of DOX and siRNA based on the DOX loaded in the nanoparticle pores and the siRNA. The investigation of DOX-siRNA-MSNs showed a synergistic response in drug-resistant BC cells by decreasing the cytosolic Ca^{2+} concentration resulting in G_0/G_1 phase cell-cycle arrest (S. Wang et al. 2019). Similarly, Liu et al. combined PTX and P-gp inhibitor (quercetin) using chondroitin sulfate-coated MSNs. The combination resulted in higher apoptosis, G2M arrest, and stronger microtubule destruction in drug-resistant BC cells. Quercetin improved the paclitaxel sensitivity in breast cancer cells as well as can affect carcinogenic pathways as quercetin has been reported to exhibit multiple anticancer benefits (Liu et al. 2020).

Li et al. used MSNs for combining DOX and proapoptotic peptide (KLAK), which disrupts mitochondrial membranes. MSN pores were encapsulated with DOX and the surface was conjugated to KLAK peptides with high loading of both molecules as well as pH-sensitive and redox-responsive release of therapeutics. The material showed increased apoptosis in BC cells and high tumor efficiency and safety profile in vivo (X. Li et al. 2019). In another study, MSNs were coated with a synthetic dual-functional polymer-lipid material P123-DPE which acted as the barrier against drug leakage. The polymer also acted as an inhibitor against breast cancer resistance protein (BCRP). These nanoparticles lead to the increased concentration of irinotecan in resistant BC cells. The system showed a high therapeutic response in xenograft mice (Xinxin Zhang et al. 2014).

13.5.2 Delivery of Nucleic Acids and Chemotherapy Agents Combined with Nucleic Acids

Nucleic acid-based therapeutics including DNA, small interfering RNA (siRNA), microRNA (miRNA), and antisense oligonucleotides exhibit huge therapeutic potential. These agents can selectively knock down genes, decrease the expression of gene products, alter mRNA splicing, and thereby regulate cellular processes dependent on the specific genes or their products. Nucleic acids can selectively target various genes, alter mRNA splicing and hence regulate the progression of cancer (Alvarez-Salas 2008; F. Li and Mahato 2015). siRNA therapy involves a short double-stranded RNA to cause RNA interference leading to the inhibition of protein transcription. In contrast, miRNA regulates a network of genes by degrading transcripts that share a partially complementary motif on a 3' untranslated region of the mRNA (F. Li and Mahato 2015). Drawbacks of both siRNA and miRNA, as well as other nucleic acid therapeutics, share the same limitations of systemic stability, degradation by nucleases, tissue penetration, off-target effects, and ineffective cellular internalization (Haussecker 2014; Kim, Park, and Sailor 2019). In this direction, nanoparticles protect the premature degradation, cellular uptake at the target site. Solid silica nanoparticles were modified to deliver the combination of trastuzumab and DOX as well as target the HER2 positive BC cells. The material exerted enhanced anticancer therapy in a synergistic effect of hormone and chemotherapy (L. Meng et al. 2018). Meng et al. used PEG-PEI copolymer coated

MSNs for the co-delivery of DOX and siRNA to overcome drug resistance by silencing the P-glycoprotein (P-gp) drug exporter. The intravenous administration of DOX/siP-gp loaded MSNs, allowed P-gp knockdown and synergistic inhibition of tumor growth (H. Meng et al. 2013). Similarly, Wang et al. combined T-type Ca^{2+} channel siRNA and DOX using novel hollow MSNs, where the DOX was loaded in the hollow region and the siRNA was loaded onto the amines on the MSN surface. The combination therapy induced G0/G1 arrest in resistant BC cells in vitro and exhibited a remarkable tumor inhibition rate (S. Wang et al. 2019). Other studies have shown the benefit of suing MSNs for combining DOX or epirubicin with Bcl2-siRNA or P-gp (Chen et al. 2009; M. Wu et al. 2016).

MSNs have also been used to combine multiple nucleic acid agents. In particular, Wang et al. reported MSNs for the co-delivery of siRNA against Polo-like kinase1 (siPlk1) and miR-200c. Plk1 is a cell cycle regulator overexpressed in tumors, whereas miR-200c plays a vital role in the epithelial mesenchymal transition (EMT) essential for cancer metastasis. Photochemical internalization (PCI) can improve endosomal processing and the escape of RNA. Thus, MSNs were also loaded with ICG to achieve the PCI effect. The Plk1 plus miR-200c loaded MSNs showed superior regression of the primary tumor as well as exhibited antimetastatic ability in an aggressive metastatic triple-negative BC model (Y. Wang et al. 2020).

A recent breakthrough for NA therapeutics against cancer is the development of programmable nucleic acid nanoparticles (NANPs). These NANPs can be designed with different morphologies, such as globular, planar, or fibrous. Our group engineered MSNs for the efficient delivery of fibrous RNA NANPs. The internalization of RNA fibers using MSNs and subsequent GFP silencing was observed in a triple-negative breast cancer (TNBC) cell line. We further used this system to co-deliver DOX with anti-Bcl2 siRNA where the combination increased apoptosis with efficient Bcl2 gene silencing (Juneja et al. 2020).

13.5.3 Combination Based on Photothermal/Photodynamic Therapy

The clinical need to develop new therapeutic alternatives in BC has led to the investigation into photo-based therapies like photothermal and photodynamic therapy. PDT is a minimally invasive and selective technique which uses nontoxic photosensitizers and oxygen and upon activation with light of a specific wavelength generates reactive oxygen species (ROS) to eliminate cancer cells. PDT combined with chemotherapy is advantageous due to high selectivity, minimal invasiveness, and effectiveness in overcoming drug resistance. Wang et al. designed a magnetic MSN to co-deliver DOX and photosensitizer chlorin e6 (Ce6). The nanocomposites effectively reversed the resistance and induced apoptosis in resistant BC cells. This material showed responsiveness to the magnetic field as well as pH-controlled release contributing to the highly efficient antitumor characteristics against DOX-resistant MCF-7/ADR cells. The nanoparticles showed tumor-targeting under magnetic field resulting in the suppressed tumor growth in vivo (D. Wang et al. 2019).

Photothermal therapy is an extension to PDT where photosensitizers or light absorbents use light energy for heat production. This causes increase in temperature of the target tissue to trigger apoptosis. Chemo agents are combined with PTT, for instance, Milan et al. used topotecan-loaded hollow MSNs coated with graphene oxide (GO)-nanoflakes. The system showed excellent heat-absorption for cell ablation leading to cell killing and tumor inhibition in the MDA-MB-231 xenograft model (Gautam et al. 2019). Similarly, DOX and NIR dye-loaded MSNs were synthesized with Fe_3O_4 core. The surface of the nanocomposite was conjugated with hyaluronic acid (HA) to release the dual therapeutics in response to HAase. The material showed T2 MR and heat generation when irradiated with NIR (Fang et al. 2019).

Apart from chemotherapy agents, PTT agents have also been combined with antibodies and molecular inhibitors. For instance, silica NPs were conjugated to cetuximab which targets EGFR on BC cells, and indocyanine green (ICG) for photothermal effect. The material was designed to co-deliver the antibody and ICG and showed combination benefits in vitro and in vivo (Xiaoxue Zhang et al. 2019). Similarly, Kayuan et al. used HB5 aptamer functionalized mesoporous silica-carbon-based DOX loaded NPs. HER2 positive breast cancer cells showed increased uptake of these NPs which exhibited increased cell killing effect due to combined chemo plus PTT (K. Wang et al. 2015). In summary, these targeted particles rapidly invaded cancer cells, delivering the drug intracellularly and improving the anticancer activity of the free drug. Photosensitizers like Ce6 can also act as sonosensitizers, where MSNs loaded with DOX and Ce6 combined with US treatment. Dox-Ce6-MSNs could effectively inhibit cell proliferation under ultrasound and showed combined benefit in xenograft tumor-bearing mice. Sonodynamic therapy (Xu et al. 2020).

MSNs have been used for magnetic hyperthermia, where heat is generated from alternating magnetic fields. MSNs containing superparamagnetic ferroferric oxide and PTX were used and the surface was coated with the cancer cell membrane to provide homotypic targeting ability. PTX and superparamagnetic ferroferric oxide were both loaded in MSNs (SiFeNPs) and the material generated heat under the alternating magnetic field to increase apoptosis in tumor cells (Cai et al. 2019). In another approach, Thomas et al. incorporated zinc-doped iron oxide nanocrystals in the MSN framework and the surface has been modified with pseudorotaxanes. The nanocrystals generate heat upon exposure to the magnetic field which causes the release of cargo. When breast cancer cells were treated with DOX-loaded nanoparticles and exposed to AC field (Thomas et al. 2010).

13.5.4 Hybrid Silica Nanoparticles

There are various silica-based hybrid NPs that are designed to incorporate the benefit of silica materials and the degradability of organic material. Our group and others investigate polysilsesquioxane (PSilQ) NPs are a promising platform which have benefits such as biocompatibility and benefit of chemical tunability as other silica platforms. They have the additional advantage of increased drug loading as the drug forms the monomeric units of the material. They have increased degradability when compared to other silica-based materials discussed above. We have developed PSilQ platform for the delivery of photosensitizers including porphyrin and are currently expanded to Ce6. The material is designed to exhibit redox-responsive degradability, as the PSilQ nanoplatform consists of redox-responsive protoporphyrin silane derivative as the monomeric unit of the network. The biodegradability of the material was demonstrated in vivo and the PDT therapy was shown in orthotopic TNBC mice (Lyles et al. 2020). We recently explored this material for combination therapy for TNBC by loading hydrophobic drug curcumin in the hydrophobic core. The material was further modified with PEI that allowed for NA delivery via electrostatic interaction. The material generated singlet oxygen and ROS after irradiation. In addition, the NA delivery was enhanced after the irradiation owing to the photochemical internalization (PCI). Altogether, we observed a synergistic effect in TNBC cells as well as efficient GFP silencing (Juneja et al. 2019).

13.6 CONCLUSIONS

Silica nanoparticles have emerged as a promising nanocarriers for drug delivery, importantly for the combination of therapeutic agents. Various types of silica nanoparticles, solid, porous, hollow porous, and hybrid materials, have been investigated for the combination therapy. The unique properties of silica nanomaterials include modifiable particle size, pore diameter, degradability, and functionalization. Breast cancer is one of the highly diagnosed cancer in women and requires multimodal therapies for improving the prognosis. Silica nanoparticles have been used for diagnosis and drug delivery with success in breast cancer treatment. The preclinical applications of silica nanoparticles for the combination of chemo agents with inhibitors, nucleic acid therapeutics, and photosensitizers were discussed in this chapter. The silica material needs urther investigation based on its safety and biocompatibility. Compared to organic nanoparticles, inorganic materials like silica nanoparticles are accumulated in organs and decreased biodegradability. Hence, the long-term effect of the silica nanoparticles on the tissues needs to be further investigated. In addition, there is intense research towards increasing the degradability of silica nanomaterials to improve their clearance. The promising features of silica nanoparticles will future improve the translation of this versatile platform to clinical use.

REFERENCES

Adams, Sylvia, Jennifer R Diamond, Erika Hamilton, Paula R Pohlmann, Sara M Tolaney, Ching-Wei Chang, Wei Zhang, et al. "Atezolizumab Plus Nab-Paclitaxel in the Treatment of Metastatic Triple-Negative Breast Cancer With 2-Year Survival Follow-up: A Phase 1b Clinical Trial." *JAMA Oncology* 5 (3) (2019): 334–342. https://doi.org//10.1001/jamaoncol.2018.5152.

Alvarez-Berríos, Merlis P, Naisha Sosa-Cintron, Mariel Rodriguez-Lugo, Ridhima Juneja, and Juan L Vivero-Escoto. "Hybrid Nanomaterials Based on Iron Oxide Nanoparticles and Mesoporous Silica Nanoparticles: Overcoming Challenges in Current Cancer Treatments." Edited by Dennis Douroumis. *Journal of Chemistry* 2016 (2016): 2672740. https://doi.org//10.1155/2016/2672740.

Alvarez-Berríos, Merlis P, and Juan L Vivero-Escoto. "In Vitro Evaluation of Folic Acid-Conjugated Redox-Responsive Mesoporous Silica Nanoparticles for the Delivery of Cisplatin." *International Journal of Nanomedicine* 11 (November 2016): 6251–6265. https://doi.org//10.2147/IJN.S118196.

Alvarez-Salas, Luis M. "Nucleic Acids as Therapeutic Agents." *Current Topics in Medicinal Chemistry* 8 (15) (2008): 1379–1404. https://doi.org//10.2174/156802608786141133.

Bear, Harry D, Stewart Anderson, Ann Brown, Roy Smith, Eleftherios P Mamounas, Bernard Fisher, Richard Margolese, et al. "The Effect on Tumor Response of Adding Sequential Preoperative Docetaxel to Preoperative Doxorubicin and Cyclophosphamide: Preliminary Results from National Surgical Adjuvant Breast and Bowel Project Protocol B-27." *Journal of Clinical Oncology: Official Journal of the American Society of Clinical Oncology* 21 (22) (2003): 4165–4174. https://doi.org//10.1200/JCO.2003.12.005.

Cai, D., Liu, L., Han, C. et al. Cancer Cell Membrane-Coated Mesoporous Silica Loaded with Superparamagnetic Ferroferric Oxide and Paclitaxel for the Combination of Chemo/Magnetocaloric Therapy on MDA-MB-231 Cells. Sci Rep 9, 14475 (2019). https://doi.org/10.1038/s41598-019-51029-8.

Chen, Alex M, Min Zhang, Dongguang Wei, Dirk Stueber, Oleh Taratula, Tamara Minko, and Huixin He. "Co-Delivery of Doxorubicin and Bcl-2 SiRNA by Mesoporous Silica Nanoparticles Enhances the Efficacy of Chemotherapy in Multidrug-Resistant Cancer Cells." *Small (Weinheim an Der Bergstrasse, Germany)* 5 (23) (2009): 2673–2677. https://doi.org//10.1002/smll.200900621.

Colilla, Montserrat, Blanca González, and María Vallet-Regí. "Mesoporous Silica Nanoparticles for the Design of Smart Delivery Nanodevices." *Biomaterials Science* 1 (2): (2013): 114–134. https://doi.org//10.1039/C2BM00085G.

Dicko, Awa, Lawrence D Mayer, and Paul G Tardi. "Use of Nanoscale Delivery Systems to Maintain Synergistic Drug Ratios in Vivo." *Expert Opinion on Drug Delivery* 7 (12) (2010): 1329–1341. https://doi.org//10.1517/17425247.2010.538678.

Dréau, D., L.J. Moore, M.P. Alvarez-Berrios, M. Tarannum, P. Mukherjee, and J.L. Vivero-Escoto. "Mucin-1-Antibody-Conjugated Mesoporous Silica Nanoparticles for Selective Breast Cancer Detection in a Mucin-1 Transgenic Murine Mouse Model." *Journal of Biomedical Nanotechnology* 12 (12) (2016): https://doi.org//10.1166/jbn.2016.2318.

Du, Manling, Yong Ouyang, Fansu Meng, Qianqian Ma, Hui Liu, Yong Zhuang, Mujuan Pang, Tiange Cai, and Yu Cai. "Nanotargeted Agents: An Emerging Therapeutic Strategy for Breast Cancer." *Nanomedicine (London, England)* 14 (13) (2019): 1771–1786. https://doi.org//10.2217/nnm-2018-0481.

Earl, Helena M, Anne-Laure Vallier, Louise Hiller, Nicola Fenwick, Jennie Young, Mahesh Iddawela, Jean Abraham, et al. "Effects of the Addition of Gemcitabine, and Paclitaxel-First Sequencing, in Neoadjuvant Sequential Epirubicin, Cyclophosphamide, and Paclitaxel for Women with High-Risk Early Breast Cancer (Neo-TAnGo): An Open-Label, 2 × 2 Factorial Randomised Phase 3 Tria." *The Lancet. Oncology* 15 (2) (2014): 201–212. https://doi.org//10.1016/S1470-2045(13)70554-0.

Fang, Zhengzou, Xinyuan Li, Zeyan Xu, Fengyi Du, Wendi Wang, Ruihua Shi, and Daqing Gao. "Hyaluronic Acid-Modified Mesoporous Silica-Coated Superparamagnetic Fe(3)O (4) Nanoparticles for Targeted Drug Delivery." *International Journal of Nanomedicine* 14 (July 2019): 5785–5797. https://doi.org//10.2147/IJN.S213974.

Fathy, Mohamed M, Heba M Fahmy, Omnia A Saad, and Wael M Elshemey. "Silica-Coated Iron Oxide Nanoparticles as a Novel Nano-Radiosensitizer for Electron Therapy." *Life Sciences* 234 (October 2019): 116756. https://doi.org//10.1016/j.lfs.2019.116756.

Gautam, Milan, Raj Kumar Thapa, Bijay Kumar Poudel, Biki Gupta, Hima Bindu Ruttala, Hanh Thuy Nguyen, Zar Chi Soe, et al. "Aerosol Technique-Based Carbon-Encapsulated Hollow Mesoporous Silica Nanoparticles for Synergistic Chemo-Photothermal Therapy." *Acta Biomaterialia* 88 (2019): 448–461. https://doi.org//10.1016/j.actbio.2019.02.029.

Gelmon, K A, S E O'Reilly, A W Tolcher, C Campbell, C Bryce, J Ragaz, C Coppin, et al. "Phase I/II Trial of Biweekly Paclitaxel and Cisplatin in the Treatment of Metastatic Breast Cancer." *Journal of Clinical Oncology: Official Journal of the American Society of Clinical Oncology* 14 (4) (1996): 1185–1191. https://doi.org//10.1200/JCO.1996.14.4.1185.

Gianni, L, E Munzone, G Capri, F Fulfaro, E Tarenzi, F Villani, C Spreafico, A Laffranchi, A Caraceni, and C Martini. "Paclitaxel by 3-Hour Infusion in Combination with Bolus Doxorubicin in Women with Untreated Metastatic Breast Cancer: High Antitumor Efficacy and Cardiac Effects in a Dose-Finding and Sequence-Finding Study." *Journal of Clinical Oncology: Official Journal of the American Society of Clinical Oncology* 13 (11) (1995): 2688–2699. https://doi.org//10.1200/JCO.1995.13.11.2688.

Gianni, Luca, Mauro Mansutti, Antonio Anton, Lourdes Calvo, Giancarlo Bisagni, Begoña Bermejo, Vladimir Semiglazov, et al. "Comparing Neoadjuvant Nab-Paclitaxel vs Paclitaxel Both Followed by Anthracycline Regimens in Women With ERBB2/HER2-Negative Breast Cancer-The Evaluating Treatment With Neoadjuvant Abraxane (ETNA) Trial: A Randomized Phase 3 Clinical Trial." *JAMA Oncology* 4 (3) (2018): 302–308. https://doi.org//10.1001/jamaoncol.2017.4612.

Grobmyer, Stephen R, Guangyin Zhou, Luke G Gutwein, Nobutaka Iwakuma, Parvesh Sharma, and Steven N Hochwald. "Nanoparticle Delivery for Metastatic Breast Cancer." *Maturitas* 73 (1) (2012): 19–26. https://doi.org//10.1016/j.maturitas.2012.02.003.

Han, Yu, Jiang Jiang, Su Seong Lee, and Jackie Y Ying. "Reverse Microemulsion-Mediated Synthesis of Silica-Coated Gold and Silver Nanoparticles." *Langmuir: The ACS Journal of Surfaces and Colloids* 24 (11) (2008): 5842–5848. https://doi.org//10.1021/la703440p.

Hanafi-Bojd, Mohammad Yahya, Seyedeh Alia Moosavian Kalat, Seyed Mohammad Taghdisi, Legha Ansari, Khalil Abnous, and Bizhan Malaekeh-Nikouei. "MUC1 Aptamer-Conjugated Mesoporous Silica Nanoparticles Effectively Target Breast Cancer Cells." *Drug Development and Industrial Pharmacy* 44 (1) (2018): 13–18. https://doi.org//10.1080/03639045.2017.1371734.

Haussecker, D. "Current Issues of RNAi Therapeutics Delivery and Development." *Journal of Controlled Release: Official Journal of the Controlled Release Society* 195 (December 2014): 49–54. https://doi.org//10.1016/j.jconrel.2014.07.056.

Hu, Che-Ming Jack, and Liangfang Zhang. "Nanoparticle-Based Combination Therapy toward Overcoming Drug Resistance in Cancer." *Biochemical Pharmacology* 83 (8) (2012): 1104–1111. https://doi.org//10.1016/j.bcp.2012.01.008.

Hussain, Zulfia, Junaid Ali Khan, and Sehrish Murtaza. "Nanotechnology: An Emerging Therapeutic Option for Breast Cancer." *Critical Reviews in Eukaryotic Gene Expression* 28 (2) (2018): 163–175. https://doi.org//10.1615/CritRevEukaryotGeneExpr.2018022771.

Islamian, Jalil Pirayesh, Milad Hatamian, and Mohammad Reza Rashidi. "Nanoparticles Promise New Methods to Boost Oncology Outcomes in Breast Cancer." *Asian Pacific Journal of Cancer Prevention: APJCP* 16 (5) (2015): 1683–1686. https://doi.org//10.7314/apjcp.2015.16.5.1683.

Juneja, Ridhima, Zachary Lyles, Hemapriyadarshini Vadarevu, Kirill A Afonin, and Juan L Vivero-Escoto. "Multimodal Polysilsesquioxane Nanoparticles for Combinatorial Therapy and Gene Delivery in Triple-Negative Breast Cancer." *ACS Applied Materials & Interfaces* 11 (13) (2019): 12308–12320. https://doi.org//10.1021/acsami.9b00704.

Juneja, Ridhima, Hemapriyadarshini Vadarevu, Justin Halman, Mubin Tarannum, Lauren Rackley, Jacob Dobbs, Jose Marquez, Morgan Chandler, Kirill Afonin, and Juan L Vivero-Escoto. "Combination of Nucleic Acid and Mesoporous Silica Nanoparticles: Optimization and Therapeutic Performance In Vitro." *ACS Applied Materials & Interfaces* 12 (35) (2020): 38873–38886. https://doi.org//10.1021/acsami.0c07106.

Kim, Byungji, Ji-Ho Park, and Michael J Sailor. "Rekindling RNAi Therapy: Materials Design Requirements for In Vivo SiRNA Delivery." *Advanced Materials* 31 (49) (2019): 1903637. https://doi.org//10.1002/adma.201903637.

Kucharczyk, Michael Jonathan, Sameer Parpia, Cindy Walker-Dilks, Laura Banfield, and Anand Swaminath. "Ablative Therapies in Metastatic Breast Cancer: A Systematic Review." *Breast Cancer Research and Treatment* 164 (1) (2017): 13–25. https://doi.org//10.1007/s10549-017-4228-2.

Kuthati, Yaswanth, Ping-Jyun Sung, Ching-Feng Weng, Chung-Yuan Mou, and Chia-Hung Lee. "Functionalization of Mesoporous Silica Nanoparticles for Targeting, Biocompatibility, Combined Cancer Therapies and Theragnosis." *Journal of Nanoscience and Nanotechnology* 13 (4) (2013): 2399–2430. https://doi.org//10.1166/jnn.2013.7363.

Li, Feng, and Ram I Mahato. "MiRNAs as Targets for Cancer Treatment: Therapeutics Design and Delivery. Preface." *Advanced Drug Delivery Reviews* 81 (January 2015): v–vi. https://doi.org//10.1016/j.addr.2014.11.005.

Li, Xiang, Gao He, Hui Jin, Jing Tao, Xinping Li, Changyuan Zhai, Yu Luo, and Xiaoan Liu. "Dual-Therapeutics-Loaded Mesoporous Silica Nanoparticles Applied for Breast Tumor Therapy." *ACS Applied Materials & Interfaces* 11 (50) (2019): 46497–46503. https://doi.org//10.1021/acsami.9b16270.

Liu, Mengyao, Manfei Fu, Xiaoye Yang, Guoyong Jia, Xiaoqun Shi, Jianbo Ji, Xianghong Liu, and Guangxi Zhai. "Paclitaxel and Quercetin Co-Loaded Functional Mesoporous Silica Nanoparticles Overcoming Multidrug Resistance in Breast Cancer." *Colloids and Surfaces. B, Biointerfaces* 196 (December 2020): 111284. https://doi.org//10.1016/j.colsurfb.2020.111284.

López-Quintela, M.Arturo. "Synthesis of Nanomaterials in Microemulsions: Formation Mechanisms and Growth Control." *Current Opinion in Colloid & Interface Science* 8 (2) (2003): 137–144. https://doi.org//10.1016/S1359-0294(03)00019-0.

Lu, Xiangyi, Li Xiao, Luan Wang, and Douglas M Ruden. "Hsp90 Inhibitors and Drug Resistance in Cancer: The Potential Benefits of Combination Therapies of Hsp90 Inhibitors and Other Anti-Cancer Drugs." *Biochemical Pharmacology* 83 (8) (2012): 995–1004. https://doi.org//10.1016/j.bcp.2011.11.011.

Lyles, Zachary K, Mubin Tarannum, Cayli Mena, Natalia M Inada, Vanderlei S Bagnato, and Juan L Vivero-Escoto. "Biodegradable Silica-Based Nanoparticles with Improved and Safe Delivery of Protoporphyrin IX for the In Vivo Photodynamic Therapy of Breast Cancer." *Advanced Therapeutics* 3 (7) (2020): 2000022. https://doi.org//10.1002/adtp.202000022.

Ma'mani, Leila, Safoora Nikzad, Hamidreza Kheiri-Manjili, Sharafaldin Al-Musawi, Mina Saeedi, Sonia Askarlou, Alireza Foroumadi, and Abbas Shafiee. "Curcumin-Loaded Guanidine Functionalized PEGylated I3ad Mesoporous Silica Nanoparticles KIT-6: Practical Strategy for the Breast Cancer Therapy." *European Journal of Medicinal Chemistry* 83 (August 2014): 646–654. https://doi.org//10.1016/j.ejmech.2014.06.069.

Ma, Liang, Manish Kohli, and Andrew Smith. "Nanoparticles for Combination Drug Therapy." *ACS Nano* 7 (11) (2013): 9518–9525. https://doi.org//10.1021/nn405674m.

Matteis, Andrea de, Francesco Nuzzo, Giuseppe D'Aiuto, Vincenzo Labonia, Gabriella Landi, Emanuela Rossi, Angelo Antonio Mastro, Gerardo Botti, Ermelinda De Maio, and Francesco Perrone. "Docetaxel plus Epidoxorubicin as Neoadjuvant Treatment in Patients with Large Operable or Locally Advanced Carcinoma of the Breast: A Single-Center, Phase II Study." *Cancer* 94 (4) (2002): 895–901. https://doi.org//10.1002/cncr.20335.abs.

Mayer, Lawrence D, Troy O Harasym, Paul G Tardi, Natashia L Harasym, Clifford R Shew, Sharon A Johnstone, Euan C Ramsay, Marcel B Bally, and Andrew S Janoff.

"Ratiometric Dosing of Anticancer Drug Combinations: Controlling Drug Ratios after Systemic Administration Regulates Therapeutic Activity in Tumor-Bearing Mice." *Molecular Cancer Therapeutics* 5 (7) (2006): 1854–1863. https://doi.org//10.1158/1535-7163.MCT-06-0118.

Meel, Roy van der, Twan Lammers, and Wim E Hennink. "Cancer Nanomedicines: Oversold or Underappreciated?" *Expert Opinion on Drug Delivery* 14 (1) (2017): 1–5. https://doi.org//10.1080/17425247.2017.1262346.

Meng, Fanfei, Ning Han, and Yoon Yeo. "Organic Nanoparticle Systems for Spatiotemporal Control of Multimodal Chemotherapy." *Expert Opinion on Drug Delivery* 14 (3) (2017): 427–446. https://doi.org//10.1080/17425247.2016.1218464.

Meng, Huan, Wilson X Mai, Haiyuan Zhang, Min Xue, Tian Xia, Sijie Lin, Xiang Wang, et al. "Codelivery of an Optimal Drug/SiRNA Combination Using Mesoporous Silica Nanoparticles To Overcome Drug Resistance in Breast Cancer in Vitro and in Vivo." *ACS Nano* 7 (2) (2013): 994–1005. https://doi.org//10.1021/nn3044066.

Meng, Ling-xin, Qiang Ren, Qin Meng, Yu-xiu Zheng, Mao-lei He, Shu-yan Sun, Zhao-jun Ding, Bing-cheng Li, and Hui-yun Wang. "Trastuzumab Modified Silica Nanoparticles Loaded with Doxorubicin for Targeted and Synergic Therapy of Breast Cancer." *Artificial Cells, Nanomedicine, and Biotechnology* 46 (sup3 2018): S556–S563. https://doi.org//10.1080/21691401.2018.1501380.

Milgroom, Andrew, Miranda Intrator, Krishna Madhavan, Luciano Mazzaro, Robin Shandas, Bolin Liu, and Daewon Park. "Mesoporous Silica Nanoparticles as a Breast-Cancer Targeting Ultrasound Contrast Agent." *Colloids and Surfaces. B, Biointerfaces* 116 (April 2014): 652–657. https://doi.org//10.1016/j.colsurfb.2013.10.038.

Minckwitz, Gunter von, Michael Untch, Jens-Uwe Blohmer, Serban D Costa, Holger Eidtmann, Peter A Fasching, Bernd Gerber, et al. "Definition and Impact of Pathologic Complete Response on Prognosis after Neoadjuvant Chemotherapy in Various Intrinsic Breast Cancer Subtypes." *Journal of Clinical Oncology: Official Journal of the American Society of Clinical Oncology* 30 (15) (2012): 1796–1804. https://doi.org//10.1200/JCO.2011.38.8595.

Modi, Shanu, Alison Stopeck, Hannah Linden, David Solit, Sarat Chandarlapaty, Neal Rosen, Gabriella D'Andrea, et al. "HSP90 Inhibition Is Effective in Breast Cancer: A Phase II Trial of Tanespimycin (17-AAG) plus Trastuzumab in Patients with HER2-Positive Metastatic Breast Cancer Progressing on Trastuzumab." *Clinical Cancer Research: An Official Journal of the American Association for Cancer Research* 17 (15) (2011): 5132–5139. https://doi.org//10.1158/1078-0432.CCR-11-0072.

Olov, Nafise, Shadab Bagheri-Khoulenjani, and Hamid Mirzadeh. "Combinational Drug Delivery Using Nanocarriers for Breast Cancer Treatments: A Review." *Journal of Biomedical Materials Research. Part A* 106 (8) (2018): 2272–2283. https://doi.org//10.1002/jbm.a.36410.

Parhi, Priyambada, Chandana Mohanty, and Sanjeeb Kumar Sahoo. "Nanotechnology-Based Combinational Drug Delivery: An Emerging Approach for Cancer Therapy." *Drug Discovery Today* 17 (17–18) (2012): 1044–1052. https://doi.org//10.1016/j.drudis.2012.05.010.

Pasqua, Luigi, Antonella Leggio, Diego Sisci, Sebastiano Andò, and Catia Morelli. "Mesoporous Silica Nanoparticles in Cancer Therapy: Relevance of the Targeting Function." *Mini Reviews in Medicinal Chemistry* 16 (9) (2016): 743–753. https://doi.org//10.2174/1389557516666160321113620.

Phillips, Evan, Oula Penate-Medina, Pat B Zanzonico, Richard D Carvajal, Pauliah Mohan, Yunpeng Ye, John Humm, et al. "Clinical Translation of an Ultrasmall Inorganic Optical-PET Imaging Nanoparticle Probe." *Science Translational Medicine* 6 (260) (2014): 260ra149–260ra149. https://doi.org//10.1126/scitranslmed.3009524.

Poonia, Neelam, Viney Lather, and Deepti Pandita. "Mesoporous Silica Nanoparticles: A Smart Nanosystem for Management of Breast Cancer." *Drug Discovery Today* 23 (2) (2018): 315–332. https://doi.org//10.1016/j.drudis.2017.10.022.

Prados, Jose, Consolación Melguizo, Raul Ortiz, Celia Vélez, Pablo J Alvarez, Jose L Arias, Maria A Ruíz, Visitacion Gallardo, and Antonia Aranega. "Doxorubicin-Loaded Nanoparticles: New Advances in Breast Cancer Therapy." *Anti-Cancer Agents in Medicinal Chemistry* 12 (9) (2012): 1058–1070. https://doi.org//10.2174/187152012803529646.

Prat, Aleix, Estela Pineda, Barbara Adamo, Patricia Galván, Aranzazu Fernández, Lydia Gaba, Marc Díez, Margarita Viladot, Ana Arance, and Montserrat Muñoz. "Clinical Implications of the Intrinsic Molecular Subtypes of Breast Cancer." *Breast (Edinburgh, Scotland)* 24 Suppl 2 (November 2015): S26–S35. https://doi.org//10.1016/j.breast.2015.07.008.

Rosenblum, Daniel, Nitin Joshi, Wei Tao, Jeffrey M Karp, and Dan Peer. "Progress and Challenges towards Targeted Delivery of Cancer Therapeutics." *Nature Communications* 9 (1) (2018): 1410. https://doi.org//10.1038/s41467-018-03705-y.

Rosenholm, Jessica M, Veronika Mamaeva, Cecilia Sahlgren, and Mika Lindén. "Nanoparticles in Targeted Cancer Therapy: Mesoporous Silica Nanoparticles Entering Preclinical Development Stage." *Nanomedicine (London, England)* 7 (1) (2012): 111–120. https://doi.org//10.2217/nnm.11.166.

Roy, Vivek, Barbara A Pockaj, Jacob B Allred, Heidi Apsey, Donald W Northfelt, Daniel Nikcevich, Bassam Mattar, and Edith A Perez. "A Phase II Trial of Docetaxel and Carboplatin Administered Every 2 Weeks as Preoperative Therapy for Stage II or III Breast Cancer: NCCTG Study N0338." *American Journal of Clinical Oncology* 36 (6) (2013): 540–544. https://doi.org//10.1097/COC.0b013e318256f619.

Saadeh, Yamaan, Tiffany Leung, Arpita Vyas, Lakshmi Shankar Chaturvedi, Omathanu Perumal, and Dinesh Vyas. "Applications of Nanomedicine in Breast Cancer Detection, Imaging, and Therapy." *Journal of Nanoscience and Nanotechnology* 14 (1) (2014): 913–923. https://doi.org//10.1166/jnn.2014.8755.

Sachdev, Jasgit C, and Mohammad Jahanzeb. "Use of Cytotoxic Chemotherapy in Metastatic Breast Cancer: Putting Taxanes in Perspective." *Clinical Breast Cancer* 16 (2) (2016): 73–81. https://doi.org//10.1016/j.clbc.2015.09.007.

Sailor, Michael J, and Ji-Ho Park. "Hybrid Nanoparticles for Detection and Treatment of Cancer." *Advanced Materials* 24 (28) (2012): 3779–3802. https://doi.org//10.1002/adma.201200653.

Selvarajan, Vanitha, Sybil Obuobi, and Pui Lai Rachel Ee. "Silica Nanoparticles—A Versatile Tool for the Treatment of Bacterial Infections." *Frontiers in Chemistry* 8 (2020): 602. https://doi.org//10.3389/fchem.2020.00602.

Shrestha, Binita, Liang Tang, and Gabriela Romero. "Nanoparticles-Mediated Combination Therapies for Cancer Treatment." *Advanced Therapeutics* 2 (11) (2019): 1900076. https://doi.org//10.1002/adtp.201900076.

Singh, Santosh Kumar, Shriti Singh, James W Jr Lillard, and Rajesh Singh. "Drug Delivery Approaches for Breast Cancer." *International Journal of Nanomedicine* 12 (2017): 6205–6218. https://doi.org//10.2147/IJN.S140325.

Song, Yuanhui, Yihong Li, Qien Xu, and Zhe Liu. "Mesoporous Silica Nanoparticles for Stimuli-Responsive Controlled Drug Delivery: Advances, Challenges, and Outlook." *International Journal of Nanomedicine* 12 (December 2016): 87–110. https://doi.org//10.2147/IJN.S117495.

Stöber, Werner, Arthur Fink, and Ernst Bohn. "Controlled Growth of Monodisperse Silica Spheres in the Micron Size Range." *Journal of Colloid and Interface Science* 26 (1) (1968): 62–69. https://doi.org//10.1016/0021-9797(68)90272-5.

Stylianopoulos, Triantafyllos. "EPR-Effect: Utilizing Size-Dependent Nanoparticle Delivery to Solid Tumors." *Therapeutic Delivery* 4 (4) (2013): 421–423. https://doi.org//10.4155/tde.13.8.

Tharkar, Priyanka, Asad Ullah Madani, Annette Lasham, Andrew N Shelling, and Raida Al-Kassas. "Nanoparticulate Carriers: An Emerging Tool for Breast Cancer Therapy." *Journal of Drug Targeting* 23 (2) (2015): 97–108. https://doi.org//10.3109/1061186X.2014.958844.

Thomas, Courtney R, Daniel P Ferris, Jae-Hyun Lee, Eunjoo Choi, Mi Hyeon Cho, Eun Sook Kim, J Fraser Stoddart, Jeon-Soo Shin, Jinwoo Cheon, and Jeffrey I Zink. "Noninvasive Remote-Controlled Release of Drug Molecules in Vitro Using Magnetic Actuation of Mechanized Nanoparticles." *Journal of the American Chemical Society* 132 (31) (2010): 10623–10625. https://doi.org//10.1021/ja1022267.

Trewyn, Brian G, Igor I Slowing, Supratim Giri, Hung-Ting Chen, and Victor S.-Y. Lin. "Synthesis and Functionalization of a Mesoporous Silica Nanoparticle Based on the Sol–Gel Process and Applications in Controlled Release." *Accounts of Chemical Research* 40 (9) (2007): 846–853. https://doi.org//10.1021/ar600032u.

Vieira, Débora Braga, and Lionel Fernel Gamarra. "Advances in the Use of Nanocarriers for Cancer Diagnosis and Treatment." *Einstein (Sao Paulo, Brazil)* 14 (1) (2016): 99–103. https://doi.org//10.1590/S1679-45082016RB3475.

Vivero-Escoto, Juan L, Rachel C Huxford-Phillips, and Wenbin Lin. "Silica-Based Nanoprobes for Biomedical Imaging and Theranostic Applications." *Chemical Society Reviews* 41 (7) (2012): 2673–2685. https://doi.org//10.1039/c2cs15229k.

Vivero-Escoto, Juan L, Igor I Slowing, Brian G Trewyn, and Victor S.-Y. Lin. "Mesoporous Silica Nanoparticles for Intracellular Controlled Drug Delivery." *Small* 6 (18) (2010): 1952–1967. https://doi.org//10.1002/smll.200901789.

Wang, Dan, Xuefen Li, Xinfang Li, Anfeng Kang, Linhong Sun, Miao Sun, Feng Yang, and Congjian Xu. "Magnetic And PH Dual-Responsive Nanoparticles For Synergistic Drug-Resistant Breast Cancer Chemo/Photodynamic Therapy." *International Journal of Nanomedicine* 14 (2019): 7665–7679. https://doi.org//10.2147/IJN.S214377.

Wang, Jiao, Yue Wang, Qiang Liu, Linnan Yang, Rongrong Zhu, Chengzhong Yu, and Shilong Wang. "Rational Design of Multifunctional Dendritic Mesoporous Silica Nanoparticles to Load Curcumin and Enhance Efficacy for Breast Cancer Therapy." *ACS Applied Materials & Interfaces* 8 (40) (2016): 26511–26523. https://doi.org//10.1021/acsami.6b08400.

Wang, Kaiyuan, Hui Yao, Ying Meng, Yi Wang, Xueying Yan, and Rongqin Huang. "Specific Aptamer-Conjugated Mesoporous Silica-Carbon Nanoparticles for HER2-Targeted Chemo-Photothermal Combined Therapy." *Acta Biomaterialia* 16 (April 2015): 196–205. https://doi.org//10.1016/j.actbio.2015.01.002.

Wang, Shu, Xi Liu, Shizhu Chen, Zhirong Liu, Xiaodi Zhang, Xing-Jie Liang, and Linlin Li. "Regulation of Ca2+ Signaling for Drug-Resistant Breast Cancer Therapy with Mesoporous Silica Nanocapsule Encapsulated Doxorubicin/SiRNA Cocktail." *ACS Nano* 13 (1) (2019): 274–283. https://doi.org//10.1021/acsnano.8b05639.

Wang, Yazhe, Ying Xie, Kameron V Kilchrist, Jing Li, Craig L Duvall, and David Oupický. "Endosomolytic and Tumor-Penetrating Mesoporous Silica Nanoparticles for SiRNA/MiRNA Combination Cancer Therapy." *ACS Applied Materials & Interfaces* 12 (4) (2020): 4308–4322. https://doi.org//10.1021/acsami.9b21214.

Wasserheit, C, A Frazein, R Oratz, J Sorich, A Downey, H Hochster, A Chachoua, et al. "Phase II Trial of Paclitaxel and Cisplatin in Women with Advanced Breast Cancer: An Active Regimen with Limiting Neurotoxicity." *Journal of Clinical Oncology: Official Journal of the American Society of Clinical Oncology* 14 (7) (1996): 1993–1999. https://doi.org//10.1200/JCO.1996.14.7.1993.

Wu, Meiying, Qingshuo Meng, Yu Chen, Lingxia Zhang, Mengli Li, Xiaojun Cai, Yaping Li, Pengcheng Yu, Linlin Zhang, and Jianlin Shi. "Large Pore-Sized Hollow Mesoporous Organosilica for Redox-Responsive Gene Delivery and Synergistic Cancer Chemotherapy." *Advanced Materials (Deerfield Beach, Fla.)* 28 (10) (2016): 1963–1969. https://doi.org//10.1002/adma.201505524.

Wu, Si-Han, Chung-Yuan Mou, and Hong-Ping Lin. "Synthesis of Mesoporous Silica Nanoparticles." *Chem. Soc. Rev.* 42 (9) (2013): 3862–3875. https://doi.org//10.1039/C3CS35405A.

Xu, Peng, Jia Yao, Zhen Li, Meng Wang, Linghui Zhou, Guansheng Zhong, Yi Zheng, et al. "Therapeutic Effect of Doxorubicin-Chlorin E6-Loaded Mesoporous Silica Nanoparticles Combined with Ultrasound on Triple-Negative Breast Cancer." *International Journal of Nanomedicine* 15 (2020): 2659–2668. https://doi.org//10.2147/IJN.S243037.

Yang, Yannan, and Chengzhong Yu. "Advances in Silica Based Nanoparticles for Targeted Cancer Therapy." *Nanomedicine: Nanotechnology, Biology and Medicine* 12 (2) (2016): 317–332. https://doi.org//10.1016/j.nano.2015.10.018.

Zhang, Xiaoxue, Yinyan Li, Minjie Wei, Chang Liu, Tao Yu, and Jun Yang. "Cetuximab-Modified Silica Nanoparticle Loaded with ICG for Tumor-Targeted Combinational Therapy of Breast Cancer." *Drug Delivery* 26 (1) (2019): 129–136. https://doi.org//10.1080/10717544.2018.1564403.

Zhang, Xinxin, Feifei Li, Shiyan Guo, Xi Chen, Xiaoli Wang, Juan Li, and Yong Gan. "Biofunctionalized Polymer-Lipid Supported Mesoporous Silica Nanoparticles for Release of Chemotherapeutics in Multidrug Resistant Cancer Cells." *Biomaterials* 35 (11) (2014): 3650–3665. https://doi.org//10.1016/j.biomaterials.2014.01.013.

Zhou, Yixian, Guilan Quan, Qiaoli Wu, Xiaoxu Zhang, Boyi Niu, Biyuan Wu, Ying Huang, Xin Pan, and Chuanbin Wu. "Mesoporous Silica Nanoparticles for Drug and Gene Delivery." *Acta Pharmaceutica Sinica B* 8 (2) (2018): 165–177. https://doi.org//10.1016/j.apsb.2018.01.007.

Zhu, Lei, Charles Staley, David Kooby, Bassel El-Rays, Hui Mao, and Lily Yang. "Current Status of Biomarker and Targeted Nanoparticle Development: The Precision Oncology Approach for Pancreatic Cancer Therapy." *Cancer Letters* 388 (March 2017): 139–148. https://doi.org//10.1016/j.canlet.2016.11.030.

14 Cationic Gemini Surfactants as Genes and Drug Carriers

Mays Al-Dulaymi[1], Anas El-Aneed[2], and Ildiko Badea[2]
[1]Avro Life Science and University of Waterloo
[2]University of Saskatchewan

CONTENTS

14.1 Introduction 329
14.2 Overview of the Physicochemical Properties of Gemini Surfactants 331
    14.2.1 Critical Micelle Concentration 331
    14.2.2 Aggregation Number 333
    14.2.3 Aggregate Shape 333
14.3 Application of Gemini Surfactants in the Pharmaceutical and Biomedical Field 336
    14.3.1 Delivery of Small Drug Molecules 336
    14.3.2 Delivery of Nucleic Acids 338
        14.3.2.1 Effect of Head Region (Head Group and Spacer) 338
        14.3.2.2 Effect of the Hydrophobic Tail 345
14.4 Conclusion 347
References 347

14.1 INTRODUCTION

Surfactants are amphiphilic molecules composed of spatially distinctive hydrophilic and hydrophobic regions. They are best known for their ability to reduce surface tension and unique self-assembly characteristics, which resulted in their wide applications as emulsifiers, solubilizers, wetting agents, foaming agents, and disinfectants (Anestopoulos et al. 2020; Hargreaves 2007). In 1991, Fredric M. Menger coined the term "gemini surfactant" to describe an emerging family of amphiphiles composed of two polar head groups and two hydrophobic tails covalently linked by a spacer (Figure 14.1) (Menger and Littau 1991). The definition was later expanded to include all dimeric surfactants with various head groups (i.e., cationic, anionic, or neutral) and spacer regions.

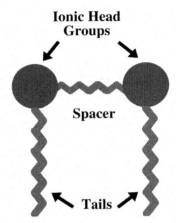

FIGURE 14.1 Schematic representation of the general structure of gemini surfactants displaying the ionic head groups, hydrocarbon tails, and the spacer region.

Gemini surfactants possess a number of superior characteristics compared to their monomeric counterparts, including (i) one or two orders of magnitude lower critical micelle concentrations (CMC), (ii) lower Krafft temperature, (iii) higher efficiency in reducing surface tension, and (iv) enhanced wetting properties (Menger and Littau 1991). Such enhanced properties resulted in their application in various fields, such as agriculture, cosmetics, textiles, and the petroleum industry (Kumar and Tyagi 2014; Mpelwa et al. 2020; Sanchez-Martin et al. 2006; Tehrani-Bagha et al. 2012). In particular, gemini surfactants has attracted attention as an important class of compounds in pharmaceutical preparations, especially in colloidal drug and gene delivery systems.

Gemini surfactants are commonly classified according to their head groups into (i) cationic, such as ammonium and imidazolium, (ii) anionic such as carboxylate and sulfate, (iii) nonionic as glucosides, and (iv) zwitterionic containing one cationic and one anionic headgroup (Sharma et al. 2017). They also vary depending on the nature and structure of the spacer that connects the two head groups, including degree of hydrophobicity, rigidity, and length. The majority of the investigated gemini surfactants contain two hydrophobic tails, but molecules with a greater number of tails have also been reported (Sharma et al. 2017). Hydrophobic regions also greatly vary in their length, flexibility, and geometry. The nomenclature *m-s-m* is generally used for gemini surfactants, where *m* is the number of carbon atoms in the alkyl tails and *s* is the number of carbon atoms in the spacer region (Menger and Littau 1993). Modifications to the head group areas, spacer, and hydrophobic regions can be strategically conducted to tailor the characteristics of the compounds to best suite specific applications. In this chapter, we focus on cationic gemini surfactants since they are the most extensively studied subclass of gemini surfactants. We briefly outline their physicochemical characteristics and discuss their application in the pharmaceutical and biomedical field, specifically as drug and gene carriers.

14.2 OVERVIEW OF THE PHYSICOCHEMICAL PROPERTIES OF GEMINI SURFACTANTS

14.2.1 CRITICAL MICELLE CONCENTRATION

Critical micelle concentration (CMC) is defined as the minimum concentration of surfactants in water (usually) or other solvents above which, aggregates of surfactant molecules, termed micelles, start to form spontaneously. It is a quantitative representation of the propensity of the amphiphilic surfactants to assemble in a specific solvent, where lower CMC value indicates a higher tendency of micellization. The self-assembly process is determined by all the structural elements of the surfactants, such as the size and charge of the head group(s), hydrophobicity and steric hindrance of the hydrophobic tail(s), hydrophobicity and rigidity of the spacer, and the type of the counterions. It reflects the complex interplay between the coulombic repulsive forces as well as the attractive forces resulting from the hydrophobic interactions and hydrogen bonding.

Gemini surfactants display one or two orders of magnitude lower CMC compared to their monomeric counterparts. For example, gemini surfactant 12-3-12 has a CMC of merely 1 mM, whereas its monomer congener dodecyltrimethylammonium bromide (DTAB) demonstrate a value of 14 mM (Basílio et al. 2015; Wettig et al. 2008). While the addition of a second ionic head group increases the hydrophilicity and contributes to the coulombic repulsive forces, its impact is counterbalanced by the hydrophobicity of the second tail (Silva et al. 2013). Furthermore, the presence of the spacer that links the two surfactants restricts their individual movement, causing the precipitous drop in CMC value (Silva et al. 2013).

The effect of modifying the head groups on the CMC values relies mainly on the charge density of the groups, where softer charges caused lower CMC values as result of the reduced repulsion between the head groups. This can be observed by comparing quaternary ammonium-based compounds to pyridinium and imidazolium congeners (Table 14.1, group 1) (Alami et al. 1993; Ao et al. 2008; Zhou et al. 2007). Increasing the length of the hydrophobic tails results in a decrease in the CMC values in a trend similar to those of monomeric counterparts (Zana et al. 1991; Zana 2002). Such an observation is noted for different classes of gemini surfactants, such as quaternary ammonium salts, serine-based bis-quats, and heterocyclic-based compounds (Ao et al. 2008; Silva et al. 2012; Zana et al. 1991).

The length, hydrophobicity, and fixability of the spacer collectively drive the micellization process and subsequently influence CMC values. Unlike the hydrophobic tail, the impact of increasing the number of carbon atoms in the spacer on the CMC is not linear. It was found that increasing the length of the spacer up to 4–5 carbon atoms led to an increase in the CMC values, while further elongation resulted in lower values (Table 14.1, group 2) (Zana et al. 1991). This initial rise in CMC was justified by the increased rigidity of the spacer that restricted the distance between the head groups, whereas further elongation resulted in a progressive folding of the spacer into the micelle hydrophobic core, which enhanced the tail hydrophobic effect, and brought the head groups closer to each other. Such trend was also maintained in different classes of gemini surfactants such as the ones with

TABLE 14.1
Chemical Structure of Selective Examples of Gemini Surfactants and Their CMC Values

Group Number	Chemical Structure	CMC (mM)	Reference
Group 1	[structure]	1.17	(Alami et al. 1993)
	[structure]	0.70	(Ao et al. 2008)
	[structure]	0.56	(Zhou et al. 2007)
Group 2	[structure]	1.1	(Alami et al. 1993)
	[structure]	1.17	(Alami et al. 1993)
	[structure]	0.83	(Alami et al. 1993)
	[structure]	0.12	(Alami et al. 1993)
Group 3	[structure]	0.52	(Fisicaro et al. 2008)
	[structure]	1.15	(Fisicaro et al. 2008)
	[structure]	0.29	(Fisicaro et al. 2008)
	[structure]	0.11	(Fisicaro et al. 2008)
Group 4	[structure]	1.35	(Wang et al. 2004)
	[structure]	1.40	(Wang et al. 2004)
	[structure]	1.52	(Wang et al. 2004)

heterocyclic head groups (Table 14.1, group 3) (Fisicaro et al. 2008; Quagliotto et al. 2003).

Modifications to the hydrophobicity and the flexibility of spacer region was investigated by Wang et al. by assessing the aggregation properties of three gemini surfactants with different types of spacers of similar overall length (Wang et al. 2004). The tested compounds included (i) hydrophobic, flexible spacer of hexyl group; (ii) hydrophilic, flexible spacer of diethyl ether; and hydrophobic, rigid spacer of p-xylyl (Table 14.1, group 4). Flexible spacer prompts micelle formation as it allows for a more closely packed micelles than hydrophobic, rigid spacers, and lead to lower CMC values (Wang et al. 2004).

14.2.2 Aggregation Number

Aggregation number is defined as the number of surfactant molecules forming a stable micelle upon reaching the CMC. Similar to CMC, aggregation number is impacted by all the structural elements of the gemini surfactants. In particular, increasing the head group polarity enhances hydration and results in an increased aggregation number. This was apparent in Borse et al. work in which the polarity of ammonium salt headgroups was increased by incorporating either one or two hydroxyl groups, resulting in an increase in aggregation number from 60 to 69 then to 75 (Table 14.2, group 1). (Devínsky et al. 1990).

The nature, hydrophobicity, and length of the spacer impact the aggregation number of gemini surfactants. Spacers that are more hydrophilic and flexible in nature prompts micelle growth as the hydration of the spacer reduces the coulombic repulsion between the head groups. This was observed in the class of gemini surfactants investigated by Wang et al. (Table 14.2, group 2) where the compound with diethyl ether exhibited the highest aggregation number (21) compared to the ones with p-xylyl and hexyl group (16 and 19, respectively) (Wang et al. 2004). Concerning the length of the spacer, increasing the length of the spacer decreased the aggregation number (Table 14.2, group 3). It is believed that the formation of loop toward the hydrophobic core of the micelles is the reason for such behavior (Borse et al. 2005).

14.2.3 Aggregate Shape

Gemini surfactants have the ability to self-assemble into a wide variety of aggregate structures, such as spherical or spheroidal micelles, bilayers vesicles, and threadlike micelles at higher concentrations, above CMC. The shape of the supramolecular assemblies controls the properties of the bulk phases, which impact their rheological properties, ability to solubilize drugs and interact with the nucleic acids. The shape of the supramolecular assembly is mainly governed by the chemical structure and the geometry of the gemini surfactants. It can be predicted depending on the proportion between the tail and head regions (head group and spacer), depicted by the lipid packing parameter, $P,$ a concept developed by Israelachvili (Cullis et al. 1986).

TABLE 14.2
Chemical Structure of Selective Examples of Gemini Surfactants and Their Aggregation Number Values

Group Number	Chemical Structure	CMC (mM)	Reference
Group 1	$H_3C-N^+(CH_3)(C_{12}H_{25})-(CH_2)_4-N^+(CH_3)(C_{12}H_{25})-CH_3$	60	(Devínsky et al. 1990)
	$HO-CH_2CH_2-N^+(CH_3)(C_{12}H_{25})-(CH_2)_4-N^+(CH_3)(C_{12}H_{25})-CH_2CH_2-OH$	69	(Devínsky et al. 1990)
	$(HOCH_2CH_2)_2N^+(C_{12}H_{25})-(CH_2)_4-N^+(C_{12}H_{25})(CH_2CH_2OH)_2$	75	(Devínsky et al. 1990)
Group 2	$H_3C-N^+(CH_3)(C_{12}H_{25})-CH_2CH_2-O-CH_2CH_2-N^+(CH_3)(C_{12}H_{25})-CH_3$	21	(Wang et al. 2004)
	$H_3C-N^+(CH_3)(C_{12}H_{25})-(CH_2)_6-N^+(CH_3)(C_{12}H_{25})-CH_3$	19	(Wang et al. 2004)
	$H_3C-N^+(CH_3)(C_{12}H_{25})-CH_2-C_6H_4-CH_2-N^+(CH_3)(C_{12}H_{25})-CH_3$	16	(Wang et al. 2004)
Group 3	$H_3C-N^+(CH_3)(C_{12}H_{25})-(CH_2)_4-N^+(CH_3)(C_{12}H_{25})-CH_2CH_2-OH$	108	(Borse et al. 2005)
	$H_3C-N^+(CH_3)(C_{12}H_{25})-(CH_2)_6-N^+(CH_3)(C_{12}H_{25})-CH_2CH_2-OH$	56	(Borse et al. 2005)
	$H_3C-N^+(CH_3)(C_{12}H_{25})-(CH_2)_{10}-N^+(CH_3)(C_{12}H_{25})-CH_2CH_2-OH$	52	(Borse et al. 2005)

Cationic Gemini Surfactants

$$P = V/a_0 l \qquad (14.1)$$

Where V is the volume of the hydrophobic chain, l the length of the hydrophobic chain, and a_0 is the surface area occupied by the head regions (head group and spacer). The P value gives an insight about the preferred curvature of the structure. For example, a P value of 0.3 is typical for spherical micellar structure while a bilayer structure forms when $P = 1$. On the other hand, an inverted micellar organization is formed when $P > 1$ (Karlsson et al. 2002). Therefore, P values can be rationally modified to form various aggregate structures by modifying one or more of the following: 1) the head group size and valence, 2) the spacer nature and length, and 3) the tail length and its degree of unsaturation.

The head region in gemini surfactants is determined by both the head groups and the spacer that connects the two monomers. Larger head group region translates into larger a_0 and smaller P, favoring the formation of more positive interfacial curvature such as spherical shape micelles. While the headgroup size and charge density impact the shape of the supramolecular complex, variation in the linker length and structure has a more profound effect in gemini surfactants. For example, increasing the length of the spacer in gemini surfactants with quaternary amine headgroups above two or three carbon atoms results in transition from elongated threadlike micelles into spheroidal ones (Danino et al. 1995). However, further increase in the length of the spacer (≥16 carbon atom) lead to the formation of bilayer vesicles. This transition is prompted by the high hydrophobicity of the lengthy spacer and its inability to maintain contact with water, which forces the spacer to fold into the micelles core (Danino et al. 1995).

Furthermore, chemical modification to the spacer by inserting functional groups or organic molecules vastly affect the headgroup area and lipid packing parameter depending on the polarity, bulkiness, and geometry of the inserted moiety. We evaluated systematically the impact of branching the spacer with amino acids (Al-Dulaymi et al. 2018). Increasing the number of lysine moieties from one, in gemini surfactants 16-7N(G-K)-16, to three, in compound 16-7N(G-K_3)-16, then seven, in 16-7N(G-K_7)-16 led to significant enlargements in the headgroup areas from 40 to 123 to 190, respectively, as a result of the increased bulkiness of the incorporated motifs (Figure 14.2). This gave rise to P values of 1.26, 0.41, and 0.27 that supported the formation of inverted micelles, cylindrical and spherical micelles, respectively (Figure 14.2).

Variation to the hydrophobic tails can affect the lipid packing parameter, by directly contributing to V and l values in equation (14.1). In fact, there has been a common notion in the literature which suggest that a_0 value is the main determinant of the packing parameter since V/l ratio is fixed for common surfactants regardless of the tail length (Nagarajan 2002; Tanford 1980). However, Nagarajan showed that the hydrophobic tails could have an explicit impact on the P values by affecting the a_0 in the equilibrium (Nagarajan 2002). For example, increasing the length of the tails in a homologous gemini surfactants series of from 10 to 12 then to 16 carbon groups changed the shape of the micelles from spherical to filomicelle, and then to disklike micelles (Aswal et al. 1998; Danino et al. 1995; Hirata et al. 1995).

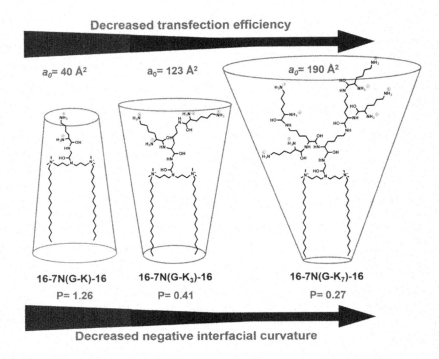

FIGURE 14.2 Schematic representation demonstrating the impact of increasing the number of terminal lysine moieties on the molecular packing parameter (P), head group are a_0, preferred curvature, and biological activity. Adapted with permission from Molecular Engineering as an Approach to Modulate Gene Delivery Efficiency of Peptide-modified Gemini Surfactants. Mays Al-Dulaymi, Deborah Michel, Jackson M. Chitanda, Anas El-Aneed, Ronald E. Verrall, Pawel Grochulski, and Ildiko Badea. Bioconjugate Chemistry 2018 29 (10), 3293–3308. Copyright 2018 American Chemical Society.

14.3 APPLICATION OF GEMINI SURFACTANTS IN THE PHARMACEUTICAL AND BIOMEDICAL FIELD

14.3.1 Delivery of Small Drug Molecules

The tendency of gemini surfactants to interact with different kinds of active pharmaceutical ingredients along with their improved tissue permeability and enhanced residence time at the therapeutic target site make gemini surfactants favorable drug delivery candidates (Sanan et al. 2014; Sharma, Nandni, et al. 2014; Mahajan and Mahajan 2013). Casal-Dujat et al. highlighted the potential of bis-imidazolium gemini surfactants as nano carriers by investigating their ability to solubilize an anionic model drug, valproate (Casal-Dujat et al. 2013). Assessment of the interaction between the micelles and the drug indicated a progressive decrease in the self-diffusion coefficient of micelles as a function of an elevated drug concentration, providing insights about the solubility of the valproate (Casal-Dujat et al. 2013). The interaction of promethazine hydrochloride, an amphiphilic

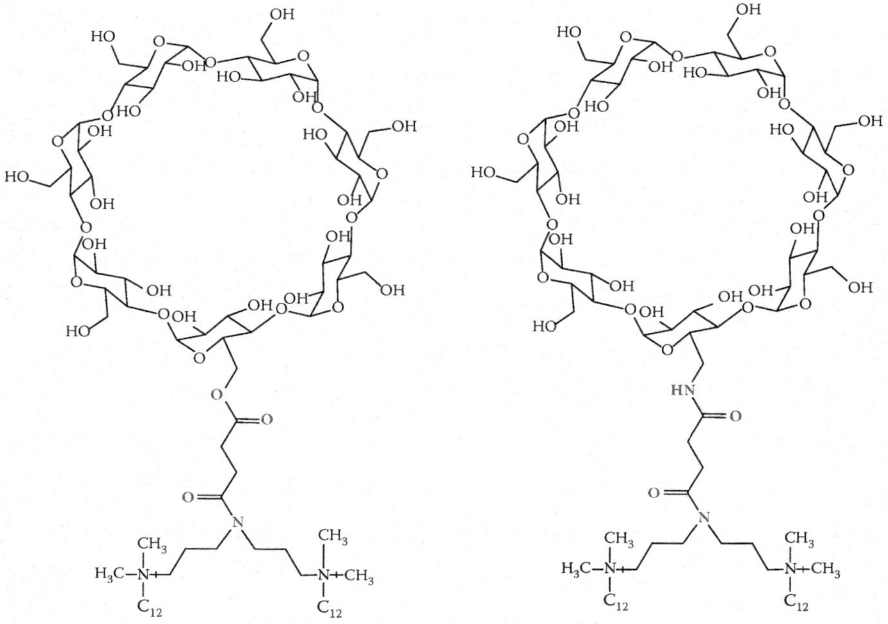

FIGURE 14.3 Chemical structure of β-Cyclodextrin-gemini surfactant conjugates.

phenothiazine tranquilizer, with the pyridinium gemini surfactants was thoroughly evaluated (Mahajan et al. 2012). Results shown a cation-π interactions and hydrophobic interactions between the drug and gemini surfactants (Mahajan et al. 2012). Furthermore, it revealed a synergism in the mixed micelles system and spontaneity of the adsorption process (Mahajan et al. 2012).

The structure of gemini surfactants offers vast possibilities of structural modifications, allowing for the emergence of compounds that are specifically tailored to address delivery hurdles of many active pharmaceutical ingredients. The conjugation of cyclodextrin, a naturally occurring cyclic oligosaccharides, to gemini surfactants is an example of such modifications (Figure 14.3) (Badea et al. 2015). It aimed at leveraging the unique characteristics for each component to improve the delivery of poorly soluble drugs and enhance intracellular penetration. Cyclodextrin provides a hydrophilic outer surface and lipophilic inner cavity that support the noncovalent inclusion of a wide variety of active pharmaceutical ingredients (Davis and Brewster 2004; Loftsson et al. 2005). Cyclodextrin-based gemini surfactants were utilized to improve the solubility of a poorly soluble curcumin analog (NC 2067) developed to treat in-transit melanoma metastasis. The drug/cyclodextrin-gemini surfactant carrier at 1:2 mole ratio formed complexes in the range of 100–200 nm (Michel et al. 2012). The complex demonstrated higher efficiency than the cyclodextrin-drug complexes and selectivity in inhibiting the growth of melanoma cells (A375) with no cytotoxicity to the healthy human epidermal keratinocytes (Michel et al. 2012).

Cyclodextrin-based gemini surfactants were further evaluated in the delivery of melphalan, a poorly soluble drug currently used as an adjunctive therapy for patient with regional metastatic melanoma (Michel et al. 2016; Testori et al. 2011). Melphalan poor solubility requires the addition of an organic co-solvent, namely propylene glycol, which lead to poor stability and significant toxicity (Zar et al. 2007). Cyclodextrin-based gemini surfactants enhanced the chemotherapeutic efficiency of melphalan manifested by the decrease in the 50% inhibitory concentration (IC_{50}) from 82 μM for un-complexed melphalan to 32 μM for the complexed drug in melanoma cell line model (Michel et al. 2016). The cyclodextrin-based gemini surfactants did not exhibited any intrinsic toxicity or change the pathway of the cellular death triggered by melphalan (Michel et al. 2016). This suggests the high versatility and potential of cyclodextrin-based gemini surfactants as drug carriers poorly soluble anticancer agents.

14.3.2 Delivery of Nucleic Acids

Cationic gemini surfactants have emerged as promising class of nonviral gene delivery systems. They possess essential characteristics that make them of particular interest for gene delivery. First, the positively charged head groups of the gemini surfactants allow for an electrostatic interaction with the negatively charged phosphate groups on the nucleic acids backbone (DNA or RNA); which results in shielding, neutralizing, condensing, and encapsulating the nucleic acids into nano-sized particles (Rosenzweig et al. 2001). Second, the hydrophobic domains of the gemini surfactants engage in cooperative hydrophobic interactions, hydrogen bonding and van der Waals interactions that stabilize the complexes during the delivery process (Ahmed et al. 2016a). Furthermore, the hydrocarbon tails act as a fusogenic group that can facilitate penetration of the nanoparticle into the cell. Finally, the low immunogenicity, unlimited packaging capacity and ease of production make gemini surfactants an attractive class of gene vectors.

The structure of the gemini surfactants dictate the nature of the interaction with the nucleic acids, controlling critical steps such as strength of compaction and subsequent intracellular release (Uhríková et al. 2005). It also determines the shape of supramolecular arrangement of the formed lipoplexes. Both the strength of the interactions and morphology of the nanoparticles influence the stability of the formulation and ability to deliver the genetic cargo inside the cells (transfection efficiency) (Ma et al. 2007; Wang et al. 2007; Wasungu et al. 2006). Over the past 30 years, a large variety of gemini surfactants have been synthesized by combination of various head groups, spacer regions and hydrophobic domains to improve transfection efficiency and biocompatibility. The following sections review the main structural variations to the gemini surfactants and highlight the impact of each modification.

14.3.2.1 Effect of Head Region (Head Group and Spacer)

The positively charged head group of the gemini surfactants is considered the main driving force for the electrostatic interaction with the negatively charged genetic material (Rosenzweig et al. 2001). As such, the nature, size, and charge density of the gemini surfactant head group greatly impacts the physicochemical characteristics, toxicity, and transfection efficiency of the delivery system (Wettig et al. 2008; Karlsson et al. 2002; Ahmed et al. 2016a; 2016b).

The spacer of the gemini surfactants also plays a crucial role in determining the efficiency of the delivery system. It impacts the shape and size of the gemini surfactants affecting the self-assembly process and the CMC value (Ahmed et al. 2016a; 2016b; Karlsson et al. 2002; Wettig et al. 2008). In addition, the length and composition of the spacer greatly influence the binding of the gemini surfactant with the DNA (Ahmed et al. 2016a; Ahmed et al. 2016b; Karlsson et al. 2002; Wettig et al. 2008). The following section briefly highlights the most commonly used cationic gemini surfactants' head groups and spacer modifications, discussing their impact on the gene delivery process.

A Quaternary Ammonium Head Groups

Cationic gemini surfactants bearing quaternary ammonium head groups are, by far, the most extensively studied class of gemini surfactants, mainly due to their ease of preparation and efficiency in compacting the genetic material (Elsabahy et al. 2014). They were first introduced in Menger's seminal paper in 1991 (Menger and Littau, 1991), after which a series of compounds have been synthesised and utilized in delivering genetic materials both in vitro and in vivo.

The first generation of the gemini surfactants is the cationic N, N-bis(dimethylalkyl)-α,ω-alkane-diammonium, which is the simplest and the most frequently encountered family of gemini surfactants with the general formula [$C_mH_{2m+1}(CH_3)_2N^+$ $(CH_2)_s$ $N^+(CH_3)_2$ $C_mH_{2m+1}.2X^-$], abbreviated as *m-s-m*, where m = the number of carbon atoms in the alkyl tails, s = the number of carbon atoms in the spacer, and X = the counter ion (Table 14.3 group 1). In the *m-s-m* family, varying the length of the spacer (s = 2-16) induced significant differences in the level of the transfection efficiency. An in vitro transfection study in murine keratinocytes revealed that compounds with s = 3 exhibited the highest level of gene expression and had a lower toxicity profile than the commercially available transfection agent, Lipofectamine Plus (Badea et al. 2005). This was attributed to the distance between the two amine groups (0.49 nm) which is suitable for optimal electrostatic interactions with the two adjacent phosphate groups in the DNA backbone that have a distance of 0.34 nm (Badea et al. 2005). Further elongation in the spacer was accompanied by a decreasing trend in the transfection up to s = 8, after which the transfection efficiency increased gradually (Badea et al. 2005). This was explained by the folding of the spacer into a U-shape as a function of spacer elongation resulting in a decrease in the distance between the two amine head groups especially with s = 12-16 (Badea et al. 2005).

The potential of the first-generation quaternary ammoniums gemini surfactants as a gene carrier for the treatment of localized scleroderma was evaluated in normal, knock out and diseased animal models (Badea et al. 2005; 2007; 2011). Application of the lead compound-based lipoplexes, 16-3-16, revealed a significant level of gene expression in animals treated with gemini surfactant-based lipoplexes (Badea et al. 2005; 2007). Moreover, collagen production in diseased animal model dropped by

TABLE 14.3
Chemical Structure of Gemini Surfactants Used as Gene Delivery Vectors

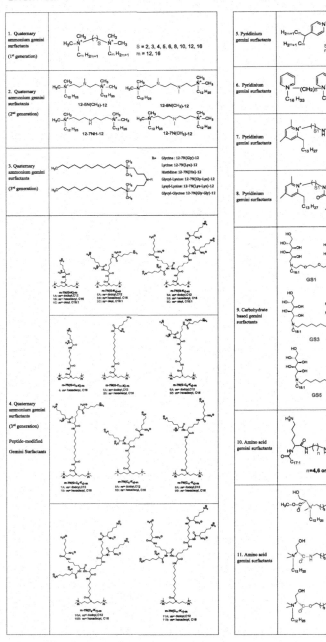

70%, demonstrating the promises of this family as effective nonviral gene delivery vectors (Badea et al. 2011).

In light of the apparent importance of the distance between the two amine head groups in the interaction with the DNA, a second generation of gemini surfactants was introduced by inserting secondary or tertiary amine functional groups in the spacer region in an attempt to improve transfection. The gemini surfactants 12–7NH–12, 12–7N(CH_3)–12, 12–5N(CH_3)–12 and 12-8N(CH_3)-12 (Table 14.3, group 2) were evaluated in monkey kidney-derived fibroblast (COS-7) cell-line. Wettig et al. (2007) Experimental data showed that compounds with a three methylene (CH_2- CH_2- CH_2) separation between adjacent nitrogen centers, such as 12–7NH–12 and 12–7N(CH_3)–12 exhibited higher level of transfection compared to compounds separated with merely two methylene (CH_2- CH_2) moieties, 12–5N(CH_3)–12 and 12-8N(CH_3)-12 (Wettig et al. 2007). The findings demonstrate the significance of optimal spacing between the nitrogen atoms within the spacer region (Wettig et al. 2007). In addition, 12–7NH–12 exhibited higher level of gene expression with 9-fold increase compared to an unsubstituted gemini surfactant 12-3-12 and a threefold increase compared to the aza analog, 12–7N(CH_3)–12. This was explained by the pH-dependent change in morphology arising from the protonated secondary amine functional group (Wettig et al. 2007).

In an attempt to enhance the transfection efficiency and reduce the cytotoxicity of the second generation gemini surfactants, a third generation of structures was designed by coupling amino acids onto the N position of the spacer region of the 12-7NH-12 gemini surfactant (Yang et al. 2010). Novel compounds with the general chemical formula $C_{12}H_{25}(CH_3)_2N^+(CH_2)_3-N(AA)-(CH_2)_3-N^+(CH_3)_2-C_{12}H_{25}$ where AA= glycine, lysine, glycyl–lysine or, lysyl-lysine (Table 14.3, group 3) were evaluated in three different cell-lines, namely monkey kidney fibroblasts (COS-7), rabbit epithelial cells, and murine keratinocyte cells (Yang et al. 2010). Results showed that the compounds substituted with either glycine or glycyl–lysine moieties had higher transfection efficiency compared to the unsubstituted compound, 12-7NH-12, in all the tested cell lines (Yang et al. 2010). The insertion of amino acids provided additional terminal amino groups that contributed to: (i) better cellular membrane binding due to the added hydrogen bonding, (ii) enhanced electrostatic interactions between the gemini surfactant and nucleic acid as a result of the high pK_a value of the terminal amines, and (iii) induced flip-flop liposomal fusion mediated by the strong electrostatic interaction between the nanoparticles and the negatively charged cell membrane (Singh et al. 2011; Yang et al. 2010). Additionally, cytotoxicity of the third-generation compounds was as low as the parent unsubstituted compounds' and significantly lower than commercial Lipofectamine Plus demonstrating the intrinsic biocompatibility of amino acids Singh et al. (2011). In order to assess their in vivo behavior, the glycyl-lysine substituted gemini surfactant-based lipoplexes was topically applied onto rabbit vaginal cavities, exhibiting higher transfection efficiency compared to the unsubstituted compound without visible toxicity (Singh et al. 2015).

In order to capitalize on the potential of amino acids-modified gemini surfactants as effective and safe gene delivery vectors, the third generation series was later optimized and expanded by systematically evaluating the transgene efficiency of 22 peptide-modified gemini surfactants in the murine keratinocyte cell line originally researched by our group, and COS-7 cell line, which was added as the most common in vitro tool for gene delivery studies (Table 14.3, group 4) (Al-Dulaymi et al. 2018). The structural modifications focused on increasing the number of conjugated lysine moieties, motivated by the previously demonstrated role of terminal amino groups in compacting the nucleic acid and production of pH responsive nanoparticles (Singh et al. 2012; Yang et al. 2010). Hence, compounds with tri- and heptalysine motifs were synthesized, generating for the first time a dendrimer-like gemini surfactants (Table 14.3, group 4) (Al-Dulaymi et al. 2018). Results showed that increasing the number of terminal lysine moieties from mono- to tri-, then to hepta led to a significant decrease in the level of expressed protein in both cells (2.5–3.5-folds in murine keratinocytes and 7–10-folds in COS-7). This reduction was attributed to (i) the production of compounds that support the formation of positive interfacial curvature due to the alteration to the micelles' geometry (Figure 14.2), (ii) growth in particle size due to the added steric demand, (iii) increase in surface charge as result of the higher charge density, and (iv) disruption to the compaction and the subsequent release of the nucleic acid due to the higher CMC values and higher charge density associated with tri- and to hepta-lysine compounds (Al-Dulaymi et al. 2018).

The impact of introducing a space between the quaternary ammonium headgroups and the amino acid substituent was also evaluated by incorporating a hexyl or undecyl chains (Table 14.3, group 4). A drop in the level of expressed protein was reported in compounds with mono-lysine regardless of the length of the added hydrocarbon linker, whereas a significant increase in gene expression was reported in compounds with tri- and hepta-lysine residues. Such a trend can be explained by the impact of the linker on the overall "balanced" interaction of the gemini surfactant with the DNA. While the added linker in compounds with a mono-lysine moiety gave rise to more hydrophobic compounds that could strongly interact with the nucleic acid and hamper the subcellular release, it balanced the high charge density in compounds with increased number of lysine building blocks (Al-Dulaymi et al. 2018).

Among the 22 compounds, lipoplexes prepared with 16-7N(G-K)-16, shown an increase in the level of protein secretion by eightfolds in murine keratinocytes and 20% enhancement in cell viability compared to first-generation unsubstituted gemini surfactants (Al-Dulaymi et al. 2018; Badea et al. 2005). It could be concluded that balanced hydrophilic and hydrophobic characteristics are an essential in defining the aggregation behavior, efficiency and safety profile of the peptide-modified gemini surfactants.

B Heterocyclic Head Group

Heterocyclic chemical groups such as pyridinium, pyrrolidinium, and imidazolium were incorporated into the gemini surfactants as "softer" charged systems than quaternary amines. This was attributed to the delocalization of the positive charge on the atoms of the heterocyclic group (Sharma and Ilies 2014). As such, heterocyclic head groups can mediate for a balanced interaction with the nucleic acid accommodating for both processes of compaction and subsequent release (Sharma and Ilies 2014; Sharma, Lees, et al. 2014). In addition, the lower charge density of the heterocyclic-based gemini surfactants minimized the repulsion between the neighboring gemini surfactants, resulting in lower CMC compared to the ammonium-based gemini surfactants (Sharma and Ilies 2014). A low CMC value in gemini surfactants usually translates into higher transfection efficiency owing to the enhanced stability of the lipoplexes during the delivery process (Dauty et al. 2001).

Several generations of gemini surfactants with heterocyclic head-groups were introduced over the past two decades showing promising results (Fisicaro et al. 2014; Ilies et al. 2005; Meekel et al. 2000; Sharma and Ilies 2014). The impact of varying the spacer length on the transfection efficiency was more pronounced compared to the gemini surfactants bearing quaternary ammonium head groups. This was linked to the larger steric demand of the heterocyclic head groups relative to quaternary ammonium head groups. Engberts et al. were the first to synthesize pyridinium gemini surfactants with four hydrophobic chains varying the length of the aliphatic spacer ($s = 3$, 4, and 5) (Table 14.3, group 5) (Meekel et al. 2000). Transfection studies conducted on the COS-7 cell-line revealed that the compound with spacing s = 4 demonstrated the highest transfection efficiency relative to the shorter s = 3 or longer s = 5 spacers, however, no clear explanation was presented (Meekel et al. 2000). In another study, pyridinium cationic gemini surfactants having two aliphatic chains and various spacer lengths ($s = 3$, 4, 8, and 12) (Table 14.3, group 6) were evaluated in a human rhabdomyosarcoma cell line (Fisicaro et al. 2014). Similar to past observations (Meekel et al. 2000), gemini surfactants with s = 4 showed the highest transfection (Fisicaro et al. 2014). It was suggested that a spacer of 4 carbons allowed the compound to act like molecular tongs gripping the basic groups near each other, resulting in efficient compaction of the DNA (Fisicaro et al. 2014).

Unlike quaternary ammonium head group gemini surfactants (Wettig et al. 2007) the insertion of secondary amine functional groups in the spacer region did not translate into an increase in the transfection efficiency (Table 14.3, group 7) (Ilies et al. 2006). This was explained by the strong interaction with the DNA that impedes its release upon cellular entry. The authors tested this theory by assessing the transfection activity of the gemini surfactants' Boc-protected synthetic precursors (Table 14.3, group 8) (Ilies et al. 2006). Results supported the hypothesis revealing a fivefold elevation in the transfection efficiency in the Boc-protected precursor-based lipoplexes (Ilies et al. 2006).

C Glycosylated Head Groups

The replacement of the head group region with acyclic carbohydrate moieties such as glucose and mannose resulted in the production of the sugar-based gemini surfactants (Johnsson and Engberts 2004). Since sugars are nontoxic, biodegradable materials, less cytotoxicity is expected for sugar-based gemini surfactants. In addition, some mono-carbohydrates, such as mannose and galactose have been used as a targeting ligands for specific tissues and cells. For example, mannose-containing lipoplexes were efficiently recognized by the mannose receptors expressed in the liver and macrophages exhibiting higher gene expression both in vitro and in vivo (Kawakami et al. 2000).

Engberts et al. synthesized sugar-based gemini surfactants and explore their structure activity relationship by examining the effects of the head group size, carbohydrate stereochemistry as well as the length and nature of the spacer region (Table 14.3, group 9) (Fielden et al. 2001; Wasungu et al. 2006). Results suggested that neither variations in the head group nor the spacer significantly impacted the level of transfection efficiency (Wasungu et al. 2006). However, the use of an aliphatic chain spacer was remarkably more toxic than the use of a more hydrophilic spacer such as ethylene oxide (Wasungu et al. 2006).

Sugar-based surfactants exhibit a pH-dependent aggregation in an endosome simulated environment (Bell et al. 2003; Johnsson et al. 2003; Wasungu et al. 2006). Upon gradual acidification, sugar-based gemini surfactant lipoplexes exhibited different phase behaviors, namely lamellar phase and inverted hexagonal phase (Bell et al. 2003; Wasungu et al. 2006). Adopting such a phase transition can contribute to the escape of the nucleic acid-carrier from the late endosomal compartment and the subsequent release of the cargo.

D Amino Acids Head Groups

The use of amino acids as the gemini surfactant head groups was aimed at producing lipoamino acids mimicking compounds with biocompatible and biodegradable features (Morán et al. 2004; Pérez et al. 2014). Both basic (e.g., lysine) and aromatic (e.g., serine) amino acids were successfully utilized in delivering genetic materials showing a reduced cytotoxicity (Cardoso et al. 2015; Castro et al. 2004; Pérez et al. 2014; Morán et al. 2004). In addition to the fact that amino acids are naturally occurring nontoxic compounds, they possess multifunctional properties, such as the presence of a chiral center in the amino acid. The chirality induces changes in the molecular orientation, allowing the construction of various aggregate structures (Brito et al. 2013). Furthermore, amino acids are pH-sensitive motifs that respond to the pH drop in the endosomal compartment to release the encapsulated genetic material, resulting in higher transfection efficiency (Xu et al. 2010).

Varying the spacer length and the nature of the amino acid head group-gemini surfactants did not translate into major differences in transfection efficiency. For example, experimental data of gemini surfactants composed of two N-acyl-lysines linked through a di-amine spacer (Table 14.3, group 10) revealed no changes in the transfection efficiency upon changing the spacer length (Pérez et al. 2014). This was attributed to the distance between the spacer and cationic charges which led to a negligible role for the spacer (Table 14.3, group 10) (Pérez et al. 2014).

The nature of the linkage (amine, amide, and ester) between the spacer and the head groups was also evaluated in serine-based gemini surfactants (Table 14.3, group 11) (Cardoso et al. 2015). Transfection studies exhibited contradicting results between the tested cell-lines: human epithelial cervical carcinoma and human embryonic kidney cells, which prevent drawing a solid conclusion (Cardoso et al. 2015). However, it was suggested that the C–O bond in the ester series made the spacer more flexible, which resulted in optimized interaction with the DNA and lowered cytotoxicity (Cardoso et al. 2015).

While the head region is recognized as an important factor in the efficiency and safety of gemini surfactants-based gene delivery systems, it is not possible to find common trend among various classes of gemini surfactants. That could be due to the lack of common structural thread, gaps in the availability of physicochemical parameters (CMC, Krafft temperature, packing paramenter, etc) of all compounds in various studies, diversity of cell lines and experimental conditions, and diversity of analytical and bioanalytical methodologies.

14.3.2.2 Effect of the Hydrophobic Tail

The hydrophobic domain of gemini surfactants impacts their interactions with the hydrophobic domain of the DNA. It also affects the aggregation behavior, supramolecular assembly shape, and membrane fluidity. The following section briefly explores the most common structural variation of the hydrophobic domain of gemini surfactants and their influence on biological activity.

A Aliphatic Tail Length

An aliphatic tail is, by far, the most widely used hydrophobic domain in gemini surfactant-based gene carriers. The length of the alkyl tail was found to significantly alter the physicochemical characteristics of gemini surfactants, modulating their gene delivery efficacy (Al-Dulaymi et al. 2016; Badea et al. 2005; Castro et al. 2004; McGregor et al. 2001; Nagarajan 2002). Several studies have suggested that an increase in the tail length could be translated into superior transfection efficiency (Al-Dulaymi et al. 2016; Castro et al. 2004; McGregor 2001) For example, increasing the tail length from 12 to 16 in m-7N(G-K)-m resulted in 5-fold increase in the level of protein expression. Since the cooperative hydrophobic interaction between the tail groups plays a key role in the interaction with the nucleic acid, increasing the alkyl tail length (higher hydrophobicity) produced a stronger DNA interaction (Hayakawa et al. 1983; Rudiuk et al. 2012). In fact, Matulis et al. revealed that the addition of a methylene group in the alkyl chain led to fourfold increase in the lipid-DNA binding constant (Matulis et al. 2002). In addition, alkyl tail elongation is associated with decrease in the aggregation properties and the CMC (Donkuru et al. 2012), resulting in a higher tendency of gemini surfactants to self-assemble with enhanced stability (Menger and Keiper 2000).

It should be noted that increasing the alkyl tail length may have negative consequences. Longer chains will elevate the micelles' critical temperature (Krafft temperature), which cause the supramolecular assemblies to become stiffer at physiological temperature (Chu and Feng 2011). This will eventually interfere with the lipid mixing at the cellular membrane affecting both cellular uptake and endosomal escape. In addition, further elongation in the alkyl tail might produce a larger particle size which could affect critical steps, such as the route of cellular entry, rate of cellular uptake, and intracellular fate (Prabha et al. 2002, 2014; Sharma and Ilies 2014).

Although several studies have attempted to identify the optimal alkyl tail length, a definitive conclusion is rarely obtained due to the complex interplay between the head group, the spacer region and the nature of the tail that is specific to each series of gemini surfactants. Nevertheless, high transfection efficacy was usually achieved with tail length between C12 to C18 (Ahmed et al. 2016).

B Degree of Unsaturation and Stereochemistry

The use of unsaturated hydrocarbon tails is a common structural modification that has been extensively assessed to optimize the transfection efficiency of gemini surfactants. The introduction of unsaturation (double and triple bond) in the surfactants tails resulted in an increased CMC compared to the saturated equivalents (Kuiper et al. 2001). In addition, it lowered the Krafft temperature below the physiological temperature resulting in enhanced susceptibility of the nano-assemblies to morphological changes at physiological temperature, which, in turn, will ease the process of lipid mixing (membrane fusion) during endocytosis (Kirby et al. 2003). Thus, unsaturation of the alkyl tail can be used as a strategy to overcome the stiffness of the supramolecular assemblies that occurs at physiological temperature with tail elongation.

Membrane fluidity is an essential requirement for successful gene transfer as it greatly impacts critical steps in the transfection process such as membrane fusion and endosomal escape (Ramezani et al. 2009). Unsaturated hydrophobic tails are known to have higher membrane fluidity than saturated ones (Deinum et al. 1988; Róg et al. 2004). This is mainly linked to the geometrical characteristics of the unsaturated bond that hamper the packing of lipids at the molecular level inducing a looser packing arrangement.

The conformational arrangement of the unsaturated hydrophobic tails is another important element in influencing the transfection efficiency of the lipid-based gene delivery systems (Kudsiova et al. 2011). There is an apparent debate in the literature on whether trans or cis configuration is more favorable for higher transfection, however, the overwhelming majority of evidence showed that cis orientation is a better choice for higher transfection (Fielden et al. 2001; McGregor et al. 2001). A cis-configured double bond is projected to hamper the lipid packing more than a trans-configured bond, resulting in a higher CMC value (Kuiper et al. 2001).

The position of the double or triple bond in the alkyl chain is an additional parameter that should be considered when the effect of unsaturation is evaluated. It can affect the CMC and the lipid packing arrangement within the supramolecular assembly (Kuiper et al. 2001). For example, unsaturation adjacent to the head group or at the end of the hydrocarbon chain has less impact on the aggregation properties than unsaturated bond in the middle of the chain. This could be attributed to the lesser impact in hampering the packing of the lipids when the unsaturation is positioned around the extremities (Kuiper et al. 2001). Despite the added advantages of unsaturation via decreasing the compounds' melting point and increasing solubility, the existence of unsaturation makes the compound more susceptible to oxidation reducing stability during storage (Fujiwara et al. 2000).

C Asymmetry

Previous sections focused on the most commonly used gemini surfactants, namely symmetrical compounds. The past decade, however, has witnessed a growing interest in the use of asymmetrical gemini surfactants, m-s-n where m ≠ n (Bai et al. 2002; Jiang et al. 2005; Wang et al. 2003). Thermodynamic studies reported no major differences in the aggregation behavior when the number of $m + n$ is equal to the total carbons number in the tails of the corresponding symmetrical analogs (Bai et al. 2002; Wang et al. 2003). However, in compounds where m was fixed but n varied, resulting in different degrees of asymmetry, the CMC was decreased and the micellization process was significantly altered (Fan et al. 2007). In order to assess how such results could affect the DNA complexation, gemini surfactants with varying degrees of asymmetry have been synthesized. A weaker interaction with the nucleic acid and gemini surfactants was attained with higher degree of asymmetry due to disruption in the intermolecular hydrophobic interactions (Jiang et al. 2005).

14.4 CONCLUSION

Gemini surfactants offer a great technological potential to the field of pharmaceutical and biomedical sciences. They possess unique solution properties, reduced toxicity, and enhanced complexing ability. Their physicochemical properties can be easily tuned by modulating the hydrophobic tails, head groups, and/or spacer region. In this chapter, we reviewed prominent structural modification to enhance efficiency and reduce cytotoxicity. Furthermore, we outlined the unique physicochemical properties of gemini surfactants that is relevant to their use as gene and drug carriers. Despite the significant advancement in the design and production of gemini surfactants over the past three decades, new generations of compounds should continue to be produced to further augment their efficiency and biocompatibility. We envision a more robust and prevalent application of the gemini surfactants in the field of biomedical and pharmaceutical sciences. Their clinical translation in areas such as genetic immunization and theranostic nanomedicine could happen in the near future.

REFERENCES

Ahmed, T.; Kamel, A. O.; Wettig, S. D., Interactions between DNA and Gemini surfactant: Impact on gene therapy: part I. *Nanomedicine 11* (3) (2016a) 289–306.

Ahmed, T.; Kamel, A. O.; Wettig, S. D., Interactions between DNA and gemini surfactant: impact on gene therapy: part II. *Nanomedicine 11* (4) (2016b) 403–420.

Alami, E.; Beinert, G.; Marie, P.; Zana, R., Alkanediyl-. alpha., omega.-bis (dimethylalkylammonium bromide) surfactants. 3. Behavior at the air-water interface. *Langmuir9* (6) (1993) 1465–1467.

Al-Dulaymi, M. A.; Chitanda, J. M.; Mohammed-Saeid, W.; Araghi, H. Y.; Verrall, R. E.; Grochulski, P.; Badea, I., Di-Peptide-Modified Gemini Surfactants as Gene Delivery Vectors: Exploring the Role of the Alkyl Tail in Their Physicochemical Behavior and Biological Activity. *The AAPS Journal 18* (5) (2016) 1168–1181.

Al-Dulaymi, M.; Michel, D.; Chitanda, J. M.; El-Aneed, A.; Verrall, R. E.; Grochulski, P.; Badea, I., Molecular Engineering as an Approach To Modulate Gene Delivery Efficiency of Peptide-Modified Gemini Surfactants. *Bioconjugate Chemistry 29* (10) (2018) 3293–3308.

Anestopoulos, I.; Kiousi, D. E.; Klavaris, A.; Galanis, A.; Salek, K.; Euston, S. R.; Pappa, A.; Panayiotidis, M. I., Surface active agents and their health-promoting properties: Molecules of multifunctional significance. *Pharmaceutics 12* (7) (2020) 688.

Ao, M.; Xu, G.; Zhu, Y.; Bai, Y., Synthesis and properties of ionic liquid-type Gemini imidazolium surfactants. *Journal of Colloid and Interface Science* **326** (2) (2008) 490–495.

Aswal, V.; Goyal, P.; Heenan, R., Transition from disc to rod-like shape of 16-3-16 dimeric micelles in aqueous solutions. *Journal of the Chemical Society, Faraday Transactions* **94** (19) (1998) 2965–2967.

Badea, I.; Wettig, S.; Verrall, R.; Foldvari, M., Topical non-invasive gene delivery using gemini nanoparticles in interferon-γ-deficient mice. *European Journal of Pharmaceutics and Biopharmaceutics* **65** (3) (2007) 414–422.

Badea, I.; Virtanen, C.; Verrall, R.; Rosenberg, A.; Foldvari, M., Effect of topical interferon-γ gene therapy using gemini nanoparticles on pathophysiological markers of cutaneous scleroderma in Tsk/+ mice. *Gene Therapy* **19** (10) (2011) 978–987.

Badea, I.; Verrall, R.; Yang, P.; Foldvari, M.; Chitanda, J.; Michel, D., Drug delivery agents comprising cyclodextrin covalently linked to a gemini surfactant, and pharmaceutical compositions comprising the same. Google Patents: 2015.

Badea, I.; Verrall, R.; Baca-Estrada, M.; Tikoo, S.; Rosenberg, A.; Kumar, P.; Foldvari, M., In vivo cutaneous interferon-γ gene delivery using novel dicationic (gemini) surfactant–plasmid complexes. *The Journal of Gene Medicine* **7** (9) (2005) 1200–1214.

Bai, G.; Wang, J.; Wang, Y.; Yan, H.; Thomas, R. K., Thermodynamics of hydrophobic interaction of dissymmetric gemini surfactants in aqueous solutions. *The Journal of Physical Chemistry B* **106** (26) (2002) 6614–6616.

Basílio, N.; Spudeit, D. A.; Bastos, J.; Scorsin, L.; Fiedler, H. D.; Nome, F.; García-Río, L., Exploring the charged nature of supramolecular micelles based on p-sulfonatocalix [6] arene and dodecyltrimethylammonium bromide. *Physical Chemistry Chemical Physics* **17** (39) (2015) 26378–26385.

Bell, P. C.; Bergsma, M.; Dolbnya, I. P.; Bras, W.; Stuart, M. C.; Rowan, A. E.; Feiters, M. C.; Engberts, J. B., Transfection mediated by gemini surfactants: engineered escape from the endosomal compartment. *Journal of the American Chemical Society* **125** (6) (2003) 1551–1558.

Borse, M.; Sharma, V.; Aswal, V.; Goyal, P.; Devi, S., Effect of head group polarity and spacer chain length on the aggregation properties of Gemini surfactants in an aquatic environment. *Journal of Colloid and Interface Science* **284** (1) (2005) 282–288.

Brito, R. O.; Oliveira, I. S.; Araújo, M. J.; Marques, E. F., Morphology, thermal behavior, and stability of self-assembled supramolecular tubules from lysine-based surfactants. *The Journal of Physical Chemistry B* **117** (32) (2013) 9400–9411.

Cardoso, A. M.; Morais, C. M.; Cruz, A. R.; Silva, S. G.; do Vale, M. L.; Marques, E. F.; de Lima, M. C. P.; Jurado, A. S., New serine-derived gemini surfactants as gene delivery systems. *European Journal of Pharmaceutics and Biopharmaceutics* **89** (2015) 347–356.

Casal-Dujat, L.; Griffiths, P. C.; Rodríguez-Abreu, C.; Solans, C.; Rogers, S.; Pérez-García, L., Nanocarriers from dicationic bis-imidazolium amphiphiles and their interaction with anionic drugs. *Journal of Materials Chemistry B* **1** (38) (2013) 4963–4971.

Castro, M.; Griffiths, D.; Patel, A.; Pattrick, N.; Kitson, C.; Ladlow, M., Effect of chain length on transfection properties of spermine-based gemini surfactants. *Organic & Biomolecular Chemistry* **2** (19) (2004) 2814–2820.

Chu, Z.; Feng, Y., Empirical correlations between Krafft temperature and tail length for amidosulfobetaine surfactants in the presence of inorganic salt. *Langmuir* **28** (2) (2011) 1175–1181.

Cullis, P. R.; Hope, M. J.; Tilcock, C. P., Lipid polymorphism and the roles of lipids in membranes. *Chemistry and Physics of Lipids* **40** (2) (1986) 127–144.

Danino, D.; Talmon, Y.; Zana, R., Alkanediyl-. alpha.,. omega.-bis (dimethylalkylammonium bromide) surfactants (dimeric surfactants). 5. aggregation and microstructure in aqueous solutions. *Langmuir* **11** (5) (1995) 1448–1456.

Dauty, E.; Remy, J.-S.; Blessing, T.; Behr, J.-P., Dimerizable cationic detergents with a low CMC condense plasmid DNA into nanometric particles and transfect cells in culture. *Journal of the American Chemical Society 123* (38) (2001) 9227–9234.
Davis, M. E.; Brewster, M. E., Cyclodextrin-based pharmaceutics: Past, present and future. *Nature Reviews Drug Discovery 3* (12) (2004) 1023–1035.
Deinum, G.; Van Langen, H.; Van Ginkel, G.; Levine, Y. K., Molecular order and dynamics in planar lipid bilayers: Effects of unsaturation and sterols. *Biochemistry 27* (3) (1988) 852–860.
Devínsky, F.; Lacko, I.; Mlynarčík, D.; Švajdlenka, E.; Borovská, V., Quaternary ammonium-salts. 33. QSAR of antimicrobially active Niketamide derivatives. *Chemical Papers 44* (1990) 159–170.
Donkuru, M.; Wettig, S. D.; Verrall, R. E.; Badea, I.; Foldvari, M., Designing pH-sensitive gemini nanoparticles for non-viral gene delivery into keratinocytes. *Journal of Materials Chemistry 22* (13) (2012) 6232–6244.
Elsabahy, M.; Badea, I.; Verrall, R.; Donkuru, M.; Foldvari, M., Dicationic gemini nanoparticle design for gene therapy. *Organic Nanomaterials: Synthesis, Characterization, and Device Applications*. Wiley Online, 2014, 509–528. https://doi.org/10.1002/9781118354377.ch23
Fan, Y.; Li, Y.; Cao, M.; Wang, J.; Wang, Y.; Thomas, R. K., Micellization of dissymmetric cationic gemini surfactants and their interaction with dimyristoylphosphatidylcholine vesicles. *Langmuir 23* (23) (2007) 11458–11464.
Fielden, M. L.; Perrin, C.; Kremer, A.; Bergsma, M.; Stuart, M. C.; Camilleri, P.; Engberts, J. B., Sugar-based tertiary amino gemini surfactants with a vesicle-to-micelle transition in the endosomal pH range mediate efficient transfection in vitro. *European Journal of Biochemistry 268* (5) (2001) 1269–1279.
Fisicaro, E.; Compari, C.; Biemmi, M.; Duce, E.; Peroni, M.; Barbero, N.; Viscardi, G.; Quagliotto, P., Unusual behavior of the aqueous solutions of gemini bispyridinium surfactants: Apparent and partial molar enthalpies of the dimethanesulfonates. *The Journal of Physical Chemistry B 112* (39) (2008) 12312–12317.
Fisicaro, E.; Compari, C.; Bacciottini, F.; Contardi, L.; Barbero, N.; Viscardi, G.; Quagliotto, P.; Donofrio, G.; Różycka-Roszak, B. e.; Misiak, P., Nonviral gene delivery: gemini bispyridinium surfactant-based DNA nanoparticles. *The Journal of Physical Chemistry B 118* (46) (2014) 13183–13191.
Fujiwara, T.; Hasegawa, S.; Hirashima, N.; Nakanishi, M.; Ohwada, T., Gene transfection activities of amphiphilic steroid–polyamine conjugates. *Biochimica et Biophysica Acta (BBA)-Biomembranes 1468* (1) (2000) 396–402.
Hargreaves, A. E., *Chemical formulation: An overview of surfactant based chemical preparations used in everyday life*. Royal Society of Chemistry. 2007.
Hayakawa, K.; Santerre, J. P.; Kwak, J. C., The binding of cationic surfactants by DNA. *Biophysical Chemistry 17* (3) (1983) 175–181.
Hirata, H.; Hattori, N.; Ishida, M.; Okabayashi, H.; Frusaka, M.; Zana, R., Small-angle neutron-scattering study of bis (quaternary ammonium bromide) surfactant micelles in water. Effect of the spacer chain length on micellar structure. *The Journal of Physical Chemistry 99* (50) (1995) 17778–17784.
Ilies, M. A.; Johnson, B. H.; Makori, F.; Miller, A.; Seitz, W. A.; Thompson, E. B.; Balaban, A. T., Pyridinium cationic lipids in gene delivery: An in vitro and in vivo comparison of transfection efficiency versus a tetraalkylammonium congener. *Archives of Biochemistry and Biophysics 435* (1) (2005) 217–226.
Ilies, M. A.; Seitz, W. A.; Johnson, B. H.; Ezell, E. L.; Miller, A. L.; Thompson, E. B.; Balaban, A. T., Lipophilic pyrylium salts in the synthesis of efficient pyridinium-based cationic lipids, gemini surfactants, and lipophilic oligomers for gene delivery. *Journal of Medicinal Chemistry 49* (13) (2006) 3872–3887.

Jiang, N.; Wang, J.; Wang, Y.; Yan, H.; Thomas, R. K., Microcalorimetric study on the interaction of dissymmetric gemini surfactants with DNA. *Journal of Colloid and Interface Science* 284 (2) (2005) 759–764.

Johnsson, M.; Engberts, J. B., Novel sugar-based gemini surfactants: aggregation properties in aqueous solution. *Journal of Physical Organic Chemistry* 17 (11) (2004) 934–944.

Johnsson, M.; Wagenaar, A.; Stuart, M. C.; Engberts, J. B., Sugar-based gemini surfactants with pH-dependent aggregation behavior: vesicle-to-micelle transition, critical micelle concentration, and vesicle surface charge reversal. *Langmuir* 19 (11) (2003) 4609–4618.

Karlsson, L.; van Eijk, M. C.; Söderman, O., Compaction of DNA by gemini surfactants: effects of surfactant architecture. *Journal of Colloid and Interface Science* 252 (2) (2002) 290–296.

Kawakami, S.; Sato, A.; Nishikawa, M.; Yamashita, F.; Hashida, M., Mannose receptor-mediated gene transfer into macrophages using novel mannosylated cationic liposomes. *Gene Therapy* 7 (4) (2000) 292–299.

Kirby, A. J.; Camilleri, P.; Engberts, J. B.; Feiters, M. C.; Nolte, R. J.; Söderman, O.; Bergsma, M.; Bell, P. C.; Fielden, M. L.; García Rodríguez, C. L., Gemini surfactants: new synthetic vectors for gene transfection. *Angewandte Chemie International Edition* 42 (13) (2003) 1448–1457.

Kudsiova, L.; Ho, J.; Fridrich, B.; Harvey, R.; Keppler, M.; Ng, T.; Hart, S. L.; Tabor, A. B.; Hailes, H. C.; Lawrence, M. J., Lipid chain geometry of C14 glycerol-based lipids: effect on lipoplex structure and transfection. *Molecular BioSystems* 7 (2) (2011) 422–436.

Kuiper, J. M.; Buwalda, R. T.; Hulst, R.; Engberts, J. B., Novel pyridinium surfactants with unsaturated alkyl chains: aggregation behavior and interactions with methyl orange in aqueous solution. *Langmuir* 17 (17) (2001) 5216–5224.

Kumar, N.; Tyagi, R., Industrial applications of dimeric surfactants: A review. *Journal of Dispersion Science and Technology* 35 (2) (2014) 205–214.

Loftsson, T.; Jarho, P.; Másson, M.; Järvinen, T., Cyclodextrins in drug delivery. *Expert Opinion on Drug Delivery* 2 (2) (2005) 335–351.

Ma, B.; Zhang, S.; Jiang, H.; Zhao, B.; Lv, H., Lipoplex morphologies and their influences on transfection efficiency in gene delivery. *Journal of Controlled Release* 123 (3) (2007) 184–194.

Mahajan, S.; Mahajan, R. K., Interactions of phenothiazine drugs with surfactants: A detailed physicochemical overview. *Advances in Colloid and Interface Science* 199 (2013) 1–14.

Mahajan, R. K.; Mahajan, S.; Bhadani, A.; Singh, S., Physicochemical studies of pyridinium gemini surfactants with promethazine hydrochloride in aqueous solution. *Physical Chemistry Chemical Physics* 14 (2) (2012) 887–898.

Matulis, D.; Rouzina, I.; Bloomfield, V. A., Thermodynamics of cationic lipid binding to DNA and DNA condensation: Roles of electrostatics and hydrophobicity. *Journal of the American Chemical Society* 124 (25) (2002) 7331–7342.

McGregor, C.; Perrin, C.; Monck, M.; Camilleri, P.; Kirby, A. J., Rational approaches to the design of cationic gemini surfactants for gene delivery. *Journal of the American Chemical Society* 123 (26) (2001) 6215–6220.

Meekel, A. A.; Wagenaar, A.; Šmisterová, J.; Kroeze, J. E.; Haadsma, P.; Bosgraaf, B.; Stuart, M.; Brisson, A.; Ruiters, M.; Hoekstra, D., Synthesis of pyridinium amphiphiles used for transfection and some characteristics of amphiphile/DNA complex formation. *European Journal of Organic Chemistry* 2000 (4) (2000) 665–673.

Menger, F. M.; Littau, C., Gemini-surfactants: Synthesis and properties. *Journal of the American Chemical Society* 113 (4) (1991) 1451–1452.

Menger, F.; Littau, C., Gemini surfactants: A new class of self-assembling molecules. *Journal of the American Chemical Society* 115 (22) (1993) 10083–10090.

Menger, F. M.; Keiper, J. S., Gemini surfactants. *Angewandte Chemie International Edition* 39 (11) (2000) 1906–1920.

Michel, D.; Mohammed-Saeid, W.; Getson, H.; Roy, C.; Poorghorban, M.; Chitanda, J. M.; Verrall, R.; Badea, I., Evaluation of β-cyclodextrin-modified gemini surfactant-based delivery systems in melanoma models. *International Journal of Nanomedicine 11* (2016) 6703.

Michel, D.; Chitanda, J. M.; Balogh, R.; Yang, P.; Singh, J.; Das, U.; El-Aneed, A.; Dimmock, J.; Verrall, R.; Badea, I., Design and evaluation of cyclodextrin-based delivery systems to incorporate poorly soluble curcumin analogs for the treatment of melanoma. *European Journal of Pharmaceutics and Biopharmaceutics 81* (3) (2012) 548–556.

Morán, M. C.; Pinazo, A.; Pérez, L.; Clapés, P.; Angelet, M.; García, M. T.; Vinardell, M. P.; Infante, M. R., "Green" amino acid-based surfactants. *Green Chemistry 6* (5) (2004) 233–240.

Mpelwa, M.; Tang, S.; Jin, L.; Hu, R.; Wang, C.; Hu, Y., The study on the properties of the newly extended Gemini surfactants and their application potentials in the petroleum industry. *Journal of Petroleum Science and Engineering 186* (2020) 106799.

Nagarajan, R., Molecular packing parameter and surfactant self-assembly: the neglected role of the surfactant tail. *Langmuir 18* (1) (2002) 31–38.

Pérez, L.; Pinazo, A.; Pons, R.; Infante, M., Gemini surfactants from natural amino acids. *Advances in Colloid and Interface Science 205* (2014) 134–155.

Prabha, S.; Zhou, W.-Z.; Panyam, J.; Labhasetwar, V., Size-dependency of nanoparticle-mediated gene transfection: studies with fractionated nanoparticles. *International Journal of Pharmaceutics 244* (1) (2002) 105–115.

Prabha, S.; Arya, G.; Chandra, R.; Ahmed, B.; Nimesh, S., Effect of size on biological properties of nanoparticles employed in gene delivery. *Artificial Cells, Nanomedicine, and Biotechnology 44* (1) (2014) 1–9.

Quagliotto, P.; Viscardi, G.; Barolo, C.; Barni, E.; Bellinvia, S.; Fisicaro, E.; Compari, C., Gemini pyridinium surfactants: Synthesis and conductometric study of a novel class of amphiphiles1. *The Journal of Organic Chemistry 68* (20) (2003) 7651–7660.

Ramezani, M.; Khoshhamdam, M.; Dehshahri, A.; Malaekeh-Nikouei, B., The influence of size, lipid composition and bilayer fluidity of cationic liposomes on the transfection efficiency of nanolipoplexes. *Colloids and Surfaces B: Biointerfaces 72* (1) (2009) 1–5.

Róg, T.; Murzyn, K.; Gurbiel, R.; Takaoka, Y.; Kusumi, A.; Pasenkiewicz-Gierula, M., Effects of phospholipid unsaturation on the bilayer nonpolar region a molecular simulation study. *Journal of Lipid Research 45* (2) (2004) 326–336.

Rosenzweig, H. S.; Rakhmanova, V. A.; MacDonald, R. C., Diquaternary ammonium compounds as transfection agents. *Bioconjugate Chemistry 12* (2) (2001) 258–263.

Rudiuk, S.; Yoshikawa, K.; Baigl, D., Enhancement of DNA compaction by negatively charged nanoparticles: Effect of nanoparticle size and surfactant chain length. *Journal of Colloid and Interface Science 368* (1) (2012) 372–377.

Sanan, R.; Kaur, R.; Mahajan, R. K., Micellar transitions in catanionic ionic liquid–ibuprofen aqueous mixtures: Effects of composition and dilution. *RSC Advances 4* (110) (2014) 64877–64889.

Sanchez-Martin, M.; Rodriguez-Cruz, M.; Andrades, M.; Sanchez-Camazano, M., Efficiency of different clay minerals modified with a cationic surfactant in the adsorption of pesticides: Influence of clay type and pesticide hydrophobicity. *Applied Clay Science 31* (3-4) (2006) 216–228.

Sharma, V. D.; Ilies, M. A., Heterocyclic cationic gemini surfactants: a comparative overview of their synthesis, self-assembling, physicochemical, and biological properties. *Medicinal Research Reviews 34* (1) (2014) 1–44.

Sharma, R.; Nandni, D.; Mahajan, R. K., Interfacial and micellar properties of mixed systems of tricyclic antidepressant drugs with polyoxyethylene alkyl ether surfactants. *Colloids and Surfaces A: Physicochemical and Engineering Aspects 451* (2014) 107–116.

Sharma, R.; Kamal, A.; Abdinejad, M.; Mahajan, R. K.; Kraatz, H.-B., Advances in the synthesis, molecular architectures and potential applications of gemini surfactants. *Advances in Colloid and Interface Science 248* (2017) 35–68.

Sharma, V. D.; Lees, J.; Hoffman, N. E.; Brailoiu, E.; Madesh, M.; Wunder, S. L.; Ilies, M. A., Modulation of pyridinium cationic lipid–DNA complex properties by pyridinium gemini surfactants and its impact on lipoplex transfection properties. *Molecular Pharmaceutics 11* (2) (2014) 545–559.

Silva, S. G.; Fernandes, R. F.; Marques, E. F.; Vale, M., Serine-based bis-quat gemini surfactants: Synthesis and micellization properties. *European Journal of Organic Chemistry 2012* (2) (2012) 345–352.

Silva, S. G.; Alves, C.; Cardoso, A. M.; Jurado, A. S.; Pedroso de Lima, M. C.; Vale, M. L. C.; Marques, E. F., Synthesis of gemini surfactants and evaluation of their interfacial and cytotoxic properties: Exploring the multifunctionality of serine as headgroup. *European Journal of Organic Chemistry 2013* (9) (2013) 1758–1769.

Singh, J.; Michel, D.; Chitanda, J. M.; Verrall, R. E.; Badea, I., Evaluation of cellular uptake and intracellular trafficking as determining factors of gene expression for amino acid-substituted gemini surfactant-based DNA nanoparticles. *Journal of Nanobiotechnology 10* (1) (2012) 7.

Singh, J.; Michel, D.; Getson, H. M.; Chitanda, J. M.; Verrall, R. E.; Badea, I., Development of amino acid substituted gemini surfactant-based mucoadhesive gene delivery systems for potential use as noninvasive vaginal genetic vaccination. *Nanomedicine 10* (3) (2015) 405–417.

Singh, J.; Yang, P.; Michel, D.; E Verrall, R.; Foldvari, M.; Badea, I., Amino acid-substituted gemini surfactant-based nanoparticles as safe and versatile gene delivery agents. *Current Drug Delivery 8* (3) (2011) 299–306.

Tanford, C., *The Hydrophobic Effect: Formation of Micelles and Biological Membranes* 2d Ed. J. Wiley: 1980.

Tehrani-Bagha, A. R.; Singh, R.; Holmberg, K., Solubilization of two organic dyes by cationic ester-containing gemini surfactants. *Journal of Colloid and Interface Science 376* (1) (2012) 112–118.

Testori, A.; Verhoef, C.; Kroon, H. M.; Pennacchioli, E.; Faries, M. B.; Eggermont, A. M.; Thompson, J. F., Treatment of melanoma metastases in a limb by isolated limb perfusion and isolated limb infusion. *Journal of Surgical Oncology 104* (4) (2011) 397–404.

Uhríková, D.; Zajac, I.; Dubničková, M.; Pisárčik, M.; Funari, S. S.; Rapp, G.; Balgavý, P. J. C. A. S. B. B., Interaction of gemini surfactants butane-1, 4-diyl-bis (alkyldimethylammonium bromide) with DNA. *Colloids and Surfaces B: Biointerfaces 42* (1) (2005) 59–68.

Wang, X.; Wang, J.; Wang, Y.; Ye, J.; Yan, H.; Thomas, R. K., Micellization of a series of dissymmetric gemini surfactants in aqueous solution. *The Journal of Physical Chemistry B 107* (41) (2003) 11428–11432.

Wang, X.; Wang, J.; Wang, Y.; Yan, H.; Li, P.; Thomas, R. K., Effect of the nature of the spacer on the aggregation properties of Gemini surfactants in an aqueous solution. *Langmuir 20* (1) (2004) 53–56.

Wang, C.; Li, X.; Wettig, S. D.; Badea, I.; Foldvari, M.; Verrall, R. E. J. P. C. C. P., Investigation of complexes formed by interaction of cationic gemini surfactants with deoxyribonucleic acid. *Physical Chemistry Chemical Physics 9* (13) (2007) 1616–1628.

Wasungu, L.; Stuart, M. C.; Scarzello, M.; Engberts, J. B.; Hoekstra, D., Lipoplexes formed from sugar-based gemini surfactants undergo a lamellar-to-micellar phase transition at acidic pH. Evidence for a non-inverted membrane-destabilizing hexagonal phase of lipoplexes. *Biochimica et Biophysica Acta (BBA)-Biomembranes 1758* (10) (2006) 1677–1684.

Wasungu, L.; Scarzello, M.; van Dam, G.; Molema, G.; Wagenaar, A.; Engberts, J. B.; Hoekstra, D., Transfection mediated by pH-sensitive sugar-based gemini surfactants; potential for in vivo gene therapy applications. *Journal of Molecular Medicine 84* (9) (2006) 774–784.

Wettig, S. D.; Verrall, R. E.; Foldvari, M., Gemini surfactants: A new family of building blocks for non-viral gene delivery systems. *Current Gene Therapy 8* (1) (2008) 9–23.

Wettig, S. D.; Badea, I.; Donkuru, M.; Verrall, R. E.; Foldvari, M., Structural and transfection properties of amine-substituted gemini surfactant-based nanoparticles. *The Journal of Gene Medicine 9* (8) (2007) 649–658.

Xu, R.; Wang, X.-L.; Lu, Z.-R., New amphiphilic carriers forming pH-sensitive nanoparticles for nucleic acid delivery. *Langmuir 26* (17) (2010) 13874–13882.

Yang, P.; Singh, J.; Wettig, S.; Foldvari, M.; Verrall, R. E.; Badea, I., Enhanced gene expression in epithelial cells transfected with amino acid-substituted gemini nanoparticles. *European Journal of Pharmaceutics and Biopharmaceutics 75* (3) (2010) 311–320.

Zana, R., Dimeric and oligomeric surfactants. Behavior at interfaces and in aqueous solution: a review. *Advances in Colloid and Interface Science 97* (1-3) (2002) 205–253.

Zana, R.; Benrraou, M.; Rueff, R., Alkanediyl-. alpha.,. omega.-bis (dimethylalkylammonium bromide) surfactants. 1. Effect of the spacer chain length on the critical micelle concentration and micelle ionization degree. *Langmuir 7* (6) (1991) 1072–1075.

Zar, T.; Graeber, C.; Perazella, M. A. In *Reviews: recognition, treatment, and prevention of propylene glycol toxicity*, Seminars in dialysis, Wiley Online Library: 2007; pp 217–219.

Zhou, L.; Jiang, X.; Li, Y.; Chen, Z.; Hu, X., Synthesis and properties of a novel class of gemini pyridinium surfactants. *Langmuir 23* (23) (2007) 11404–11408.

15 Stability of Nanomaterials

Mulham Alfatama[1], Ahmed R. Gardouh[2], and Abd Almonem Doolaane[3]
[1]Universiti Sultan Zainal Abidin, Besut Campus
[2]Suez Canal University and Jadara University
[3]International Islamic University Malaysia (IIUM)

CONTENTS

15.1	Introduction	355
15.2	Transformation Processes	356
	15.2.1 Physical Transformations	357
	15.2.2 Chemical Transformation	357
	15.2.3 Biological Transformations	358
15.3	Factors Influencing the Stability of Nanoparticles	359
	15.3.1 Aggregation	359
	15.3.2 Sedimentation	359
15.4	Stabilization Mechanisms	360
	15.4.1 Charge or Electrostatic Stabilization	360
	15.4.2 Steric Stabilization	361
15.5	Nanoparticles Surface Chemistry	361
	15.5.1 Stabilizers	361
	15.5.1.1 PEGylation	362
	15.5.1.2 Surfactants	366
15.6	Lyophilization-Based Stability	371
15.7	Stability Tests	371
	15.7.1 Shelf-Life (Aging)	371
	15.7.2 Stability in Relevant Media	373
	15.7.3 pH	373
	15.7.4 Temperature	373
	15.7.5 Chemical Resistance	373
15.8	Stability Measurements of Nanoparticles	373
15.9	Conclusion	374
References		374

15.1 INTRODUCTION

Nanotechnology is a term endowed by Professor Norio Taniguchi in 1974 to indicate the technology, engineering, and science performed at the nano-level (1–100 nm).

Nanoparticles referring to the nanoscale materials are of high importance owing to the large surface area-to-volume ratio, optical attributes, small size, and multifunctional nature. Bionanotechnology on the other hand, is the application of nanotechnology in the biological sciences (Sapsford et al. 2011). Nanoparticles can be porous core/hollow such as polymeric nanoparticles of synthetic or natural origin (Kozielski et al. 2014), liposomes (Xing et al. 2018), and nanocages (Li et al. 2019) or solid core nanocarriers that comprise inorganic metals like gold, platinum, silicon, iron oxide, quantum dots, gadolinium, copper oxide, selenium, zinc oxide, or organic carbon or hybrid nanoparticles that resembles metals. The inorganic nanoparticle surfaces are typically modified with natural or synthetic monomer and/or polymers that can be of biological sources such as proteins, peptides, lipids, carbohydrates, RNA, DNA, PNA, hybrid biosynthetic molecules, aptamers, and others. Appropriate selection of capping ligand can impart the biocompatibility, functionality, and stability to the nanoparticles for further modification or applications.

Drugs possess poor pharmacodynamics can be delivered via nanoparticle that may conquer these obstacles by improving bioavailabilities, half-lives, and stabilities (Patra et al. 2018). Nevertheless, the applications of nanoparticles are not limited to the drug delivery systems (DDS), but also their great applications in analytical biosensors/probes and imaging render them unique theranostics agents (Salunkhe et al. 2020). A total of 51 nanodrugs have been approved by FDA since 1995, and about 77 products were in clinical trials till 2016 (Bobo et al. 2016). Recently, nanomaterials are being employed in the fight against SARS-CoV-2 (Chan 2020). Immunoassays-based gold nanoparticles have been used in rapid detection of SARS-CoV-2 in asymptomatic individuals or patient with mild symptoms (Luan et al. 2020). A vaccine of mRNA was processed to clinical trial phase I in March 2000, encapsulated into lipid nanoparticles as effective delivery system (Natarajan and Tomich 2020). This chapter aims to highlight the most common nanoparticle-related stability issues, stabilizers, stabilization tests, and measurements.

15.2 TRANSFORMATION PROCESSES

Nanoparticles are prone to various physical, chemical, or biological transformation processes based on their intrinsic attributes and the suspending fluid (in case of nanosuspension) (Li et al. 2019). Physical transition including sedimentation, deposition, homo/hetero aggregation, and agglomeration is mediated by the tendency of these nanoparticles to revert back to their thermodynamically stable state causing them to lose their desired space and size (Cuenya and Behafarid 2015). On the other hand, chemical processes that could take place are redox reactions, photochemical reaction, and dissolution. When nanoparticles are introduced into a biological system, their physicochemical properties are altered, causing changes in stability caused by particle-particle interactions (Mohapatra et al. 2018). Whereas, biodegradation-mediated microbial contamination and biomodification events are counted for biological transformation (Lead et al. 2018). Figure 15.1 represents the main processes that nanoparticles undergo, which has been influenced by environmental conditions such as pH, existence of natural organic matter (NOM) and

Stability of Nanomaterials

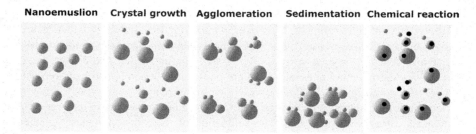

FIGURE 15.1 Possible transformation phenomenon of synthesized nanoparticles.

ionic forces and their composition. The modification of surface chemistry when nanoparticles and NOM interact will influence their behavior in the suspending fluid (Biswas and Sarkar 2019).

15.2.1 PHYSICAL TRANSFORMATIONS

The possible physical transformation may occur at any stage of nanoparticles' processing in a form of agglomeration or aggregation. These terms are typically used reciprocally owing to the difficulty to set them apart. In theory, aggregation is an irremediable phenomenon that link nanosurfaces together via strong chemical bonds or electrostatic forces, resulting in a reduction in surface area. There are two kinds of aggregation: homo-aggregation which involves the same kind of particles and hetero-aggregation that occurs between nanoparticles and other surrounding particles. Agglomeration causes nanoparticle size enlargement and hence modifying the reactivity, transport, and toxicity attributes. On the contrary to aggregation, agglomeration is a reversible process of connecting nanoparticles together as individual entities without affecting surface area via Van der Waals forces (Aitken et al. 2010).

Agglomeration takes place as a result of the attractive forces of Van der Waals become greater than the repulsive ones. It is developed by high ionic strength of divalent ions where pH is close to a value that negates the photocatalytic reactions generated by the source of sunlight (Yin et al. 2015). The decrease in coulombic repulsion forces between the nanoparticles encourages agglomeration through negating electrostatic stabilization (Rameshkumar and Ramaraj 2013). Agglomerated nanoparticles tend to sediment by means of gravity strength that controlled by Stoke's law, which can be used to extract the nanoparticles from suspension. Random motion or Brownian movement of nanoparticles is the prevailing force promoting particles agglomeration, while the gravity attractions facilitate their tendency to sediment (Petosa et al. 2010).

15.2.2 CHEMICAL TRANSFORMATION

Electron transfer among chemical moieties determines the reduction and oxidation reactions in the natural system. This mainly occurs in elemental metals such as iron and silver. There are three different types of environment in redox state: the oxygen-rich oxidizing environment, the depleted oxygen reductive environment, and redox dynamic environment. Moreover, photoreduction and photooxidation are sunlight-induced redox processes that influence oxidation state of the nanoparticles, persistence, reactive oxygen species (ROS), and coating. Chen and Jafvert have reported that carbon nanotubes and TiO2 are photoactive innately and can give rise to ROS (Chen and Jafvert 2011).

Nanoparticles of class B metal ions like Cu, Zn, and Ag can easily undergo sulfidation and dissolution processes. The sulfidation of these nanoparticles typically occurs due to the great affinity to inorganic and organic sulfide binding agents, while they are able to produce slightly soluble metal oxide followed their dissolution. Eventually, sulfidation and dissolution impact the nanoparticles' surface characteristics, toxicity, and persistence. For example, insoluble metal-oxide can be deposited on the surfaces of the nanoparticles causing alteration of charges and aggregation as a consequence of sulfidation of class B metals. Whereas the toxicity of these metal nanoparticles is reported to be driven by toxic cations and poor persistence after dissolution (Levard et al. 2011). Adsorption process of organic or inorganic macromolecules and ligands on nanosurfaces can highly modify the surface chemistry and behavior. For example, the dissolution, charge, and stability of nanoparticles are highly influenced by incorporation of thiol groups as organic ligands.

15.2.3 BIOLOGICAL TRANSFORMATIONS

This kind of transformation is inevitable when the nanoparticles prepared or stored in sterile-free conditions. The nanoparticles uptake into living cells or organisms is mediated by various processes including changes related to coating/core, interactions with other molecules and redox reactions. These transformation will directly impact the surface charge, aggregation, reactivity, and toxicity characteristics of the nanoparticles (Lowry et al. 2012). Low et al. have fundamentally established an approach of oxidation-reduction in bacteria like *Shewanella* and *Geobacter* that enabled Ag^+ reduction and synthesis of silver nanoparticles (Lowry et al. 2012). Another study has observed the nanoparticles biotransformation after coating with poly(ethylene glycol) (PEG) that led to initiation of their aggregation (Lowry et al. 2012). Silver nanoparticles (AgNPs) have been widely produced in the form of different products used by human such as food, medicine, fabrics, and washing machines (Shimabuku et al. 2017). However, the extensive application of AgNPs is associated with negative impacts on the environment, specifically on the aquatic ecosystem including aquatic organism, plants, and potentially to human (Begum et al. 2016).

Various interactions between AgNPs and environmental factors such as organic compounds, inorganic anions, and metal cations could take place to change the composition and surface of AgNPs. The nanoparticles can undergo oxidation and emit silver ions that are toxic to environment and human health. Natural organic matter (NOM) can be adsorbed on AgNPs to facilitate their deagglomeration. Additionally, environmental ionic strength may also affect AgNPs stability by improving their agglomeration, particularly in acid pH (de Souza, Souza, and Franchi 2019). Polymeric nanosystems and mesoporous silica nanomaterials that commonly used in biological systems due to their ability to interact with cells, inferring from their cellular uptake enhancement (Hao et al. 2016). To conclude, the toxicity, behavior, and biological transformation processes of engineered nanoparticles are mainly depending on the chemical environmental properties where the nanoparticles are placed.

15.3 FACTORS INFLUENCING THE STABILITY OF NANOPARTICLES

15.3.1 AGGREGATION

Nanoparticles are highly reactive entities due to the high density of dangling bonds scattered at their large surfaces. The small size generates unstable high surface energy and attempts to enhance the stability is achieved by the reduction of surface area. Thus, agglomeration or aggregation can take place to form weak or strong bonds, respectively, where the latter resists breaking up into discrete nanoparticles by mechanical forces (Kamiya et al. 2008).

The functionality of the nanoparticles can be altered by uncontrolled aggregation. The surface energy of nanoparticles suspended in a media is reducible by either dissolution into smaller species or via aggregation (Casals, Gonzalez, and Puntes 2012). Hence, in order to smartly utilize the unique features of the nano-sized particles, both these processes should be minimized. Ostwald ripening reaction is referred to the dissolution and aggregation of the suspended remnant particles in a media. Aggregation can be retarded by modulating the surface charge or by steric approach. Deep understanding of individual nanoparticle that comprising material and surface group is mandatory to enable formulation of nanoparticles without aggregation. The encapsulant and the required size of the nanoparticle determine the strategies to avoid aggregation.

Alteration of property caused by aggregation is evidenced in many ways including bioavailability, toxicity, reactivity, and photoreactivity. Reactivity is affected by reduction of the surfaces that are able to be catalyzed. The environmental application of nanoparticles such as pollutant treatment by iron particles is influenced by aggregation that retards mobility, thereby reducing the functionality to reach to the pollutant. Another study also demonstrated the impact of particle size to the reactivity of ZnO and TiO_2 nanoparticles when they form aggregates that negate the formation of hydroxyl radicals (Jassby et al., 2012). Incorporation of appropriate ligand on the surface of nanoparticles can prevent their aggregation as an attempt to stabilize the surface reactive sites (Schmudde et al. 2016).

15.3.2 SEDIMENTATION

The sedimentation tendency of nanosuspensions has been investigated intensively. It is widespread in numerous applications including gravity settling, fluid-solid segregation, and refining processes of particulate sludge and slurries. Moreover, sedimentation is widely examined for the industrial applications for improving the shelf-life of many products, such as pharmaceuticals, cosmetics, foods, consumer products, coatings, paints, and nanomaterials. The concentration reliance of the sedimentation velocity on the fluid–particle and particle–particle interactivity is of great focus in establishing the quality and stability of the suspensions (Vesaratchanon, Nikolov, and Wasan 2007).

15.4 STABILIZATION MECHANISMS

As the attractive forces between similar colloidal particles cause aggregation that reduces system stability, repulsion forces that able to resist strong attraction can confer stability to the system (Hierrezuelo et al. 2010). Stability can be imparted via two mechanisms including charge and steric stabilization, depending on the type of repulsive forces.

15.4.1 Charge or Electrostatic Stabilization

Electric double layer (EDL) is formed on the nanosufraces to impart stabilization in the polar liquids via ions adsorption to equate the attraction strength of Van der Waals between colloidal particles (Figure 15.2). EDL possesses a natural net of electrical charges with double-aligned charged layers of nanoparticle. The outer sheet is recognized as stern that constructed by deposition of negative or positive ions onto particle surfaces. Whereas the second layer (diffuse layer) is formed by ion attractions through Coulomb forces towards deposited surface charges that form a repellent blockade. The charges present within the diffuse layer are attached poorly to the surface ions that give rise to move freely. System stability is attained by mutual resistance of the double strata surrounding the nanoparticles (Hierrezuelo et al. 2010).

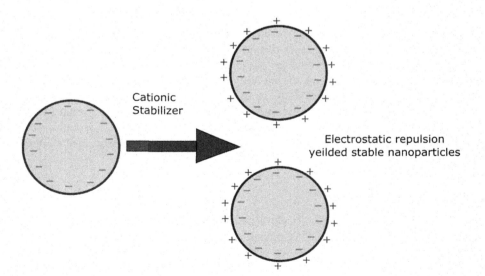

FIGURE 15.2 Schematic representation of the electrostatic stabilization of the nanoparticles.

15.4.2 Steric Stabilization

Steric stabilization is a process where adsorbed polymers or nonionic surfactants give rise to strong repulsion between particles and surrounding droplets in a dispersion. The layer formed of polymer of nonionic surfactant on the particle surface or droplets provides an adsorbed barrier with thickness d and solvation property by the molecules of the solvent, giving that the latter is great solvent for the polymer or surfactant layer. Upon approaching of two particles of radius R to a distance of separation h that is lesser than 2d, compressing or overlapping of the adsorbed layer takes place, leading to a great repulsion due to two main factors: i) improper mixing of the adsorbed layer while these are in suitable mixing state (this is known as mixing interaction or osmotic repulsion, G_{mix}) and ii) decreasing of adsorbed layers configurational entropy with considerable overlap (this is known as elastic repulsion or entropic, G_{el}). The steric repulsion G_s is the sum of G_{el} and G_{mix}, where a plot of G_s and separation distance h exhibits intense incline when $h < 2d$, the origin of steric stabilization.

There are several requirements for an efficient steric stabilization. Firstly, a sufficient amount of the polymer should be used to completely cover all the particles. Uncoated patches can lead to flocculation either by bridging where polymer molecules are adsorbed on more than one particle simultaneously or by Van der Waals force attractions. Moreover, the polymer should be firmly attached to the surfaces of the particles to avoid any detachment during particle approach, especially in concentrated suspensions. In addition, the stabilizing agent should be freely soluble in the medium and highly solvated by its component molecules (Tadros 2013).

15.5 NANOPARTICLES SURFACE CHEMISTRY

Modification of the surface chemistry is usually achieved by incorporation of molecules such as ligands to impart stability to the nanoparticles. Despite the type of the ligand used, anchoring it onto the surface of the nanoparticles is a general challenge that destabilase the system. This ligand incorporation process has been investigated widely and is highly related to the composition of the nanoformulation (Alvarez-Puebla et al. 2016). Therefore, different approaches are proposed depending on the surface affinity to various chemical groups. Generally, there are three main classes: i) oxides, commonly utilized as with magnetic nanoparticles such as iron oxides via oxygen bonding with hydroxyl and acidic groups; ii) binary compounds that typically comprise elements from groups 12–16 including nanoparticles of fluorescent semiconductor like quantum dots. This group exhibits great affinity towards hydroxyl, thiol and amino groups and iii) noble metals (Figure 15.3).

15.5.1 Stabilizers

Prevention of nanoparticles agglomeration and particle growth to confer stability to the system is mediated by various stabilizer agents. Stabilizers are functioning via

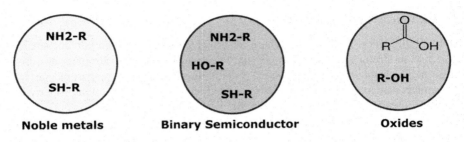

FIGURE 15.3 Schematic representation of the most widely used surface ligands functional groups.

adsorbing at growth sites of nanoparticles to reduce the rate of growth and coalescence. Adsorption propensity of stabilizing agent is proportional to the ability to interact with the nanoparticle surfaces. Therefore, careful selection of stabilizer is paramount to endow stability to the nanoparticles.

15.5.1.1 PEGylation

Despite the numerous advantages of nanoparticles, some substantial limitations hamper their clinical applications. These include reticuloendothelial system (RES) uptake in which the nanoparticles are promptly cleared from the blood circulation to the liver, bone marrow, or spleen, where nonspecific binding of nanoparticles to nontargeting area takes place. This rises concern about the stability and toxicity of the nanoparticles due to RES accumulation. Nanoparticles can be aggregated and entrapped to the live, lungs, or other organs as a consequence of capillary occlusion.

RES is a component of the immune system that responsible to eliminate foreign materials like viruses and bacteria through circulating monocytes and macrophages, spleen, and liver Kupffer cells and other lymphatic vessels. Since viruses and bacteria carry the same negative charge as phagocytes, opsonin proteins are essential in reducing the repulsion charges between them (Figure 15.4). Phagocytic cells then transport the foreign bodies to the liver and spleen following engulfing to be degraded and excreted. Addition phagocytic macrophages (Kupffer cells) are constantly located in the liver to filter vast range of nanoparticles, thus, significantly modulate the half-life of the encapsulants. Incorporation of PEG onto the nanoparticle surfaces enables opsonization reduction that leads to avoiding monocytes and macrophages recognition, permitting the nanoparticles to remain in the blood circulation for extended period of time, and hence prolonging the half-life. Hydrophobic nanoparticles are likewise highly susceptible to the RES and incorporation of hydrophilic PEG can overcome these limitations.

Various strategies have been proposed to overcome the stability-related issues of nanoparticles (Pillai et al. 2013). Polymeric stabilizing agent has been demonstrated to be effective approach to provide a physical barrier onto the surface of the nanoparticles and preventing reciprocal contact (Chairam and Somsook 2008). The technique of poly(ethylene glycol) (PEG) binding to a given system is known as PEGylation, a well-established approach to develop systematic administration drug delivery system (Kong, Campbell, and Kros 2019). PEGylation forms a hydrated cloud with a

Stability of Nanomaterials

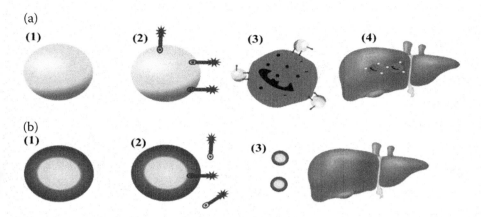

FIGURE 15.4 Schematic representation of (A): Nanoparticle (A1) is attached with opsonin (A2) then associated with macrophages (A3) to be transported to the liver (A4). (B): PEGylated nanoparticles (B1) prevent opsonization (B2), yielding reduced liver deposition and improve bioavailability of encapsulants (B3).

relatively massive excluded volume around the nanoparticles to prevent the interaction sterically between the component of the blood and the nanoparticles (Yang and Samuel 2015). Moreover, the flexibility of PEG offers a great conformational freedom that unfavored the penetration of foreign substances into the PEG corona thermodynamically (Suk et al. 2016). Therefore, PEGylation confers the nanoparticles with extended blood circulation time as a result of reduced protein adsorption. This will consequently promote drug permeability, retention, and accumulation in the target site (C. Yin et al. 2019; G. Yu et al. 2019). Table 15.1 represents examples of different types of PEGylated nanoparticles for stability enhancement.

15.5.1.1.1 PEG Applications

PEG (Carboxwax) as a nontoxic polymer, has been widely used as water soluble dispersant/stabilizer prior the revolutionize applications in nanotechnology. It can be used in tablet formulations, beauty and health products, laxative preparations, toothpastes, eye drops, and even as a stabilizer in organs and blood donations. Early applications of PEGylated nanoparticles emerged mostly in drug delivery. Davis and Abuchowski were among the first researchers to report PEGylation of bovine serum albumin and liver catalase via attaching methoxy-PEGs (1900 and 5000 Da) covalently (Abuchowski et al. 1977). Acrylic microspheres were functionalized with PEG-modified human serum albumin improved half-life in vivo. Li et al. have reported that latex nanoparticles enabled 40-fold higher circulation time in rat for PEGylated particles compared to PEG-free ones (Tan et al. 1993). In other study, introducing PEG to the liposomes conferred significant increase in half-life while avoiding the leakage of liposome interior (Klibanov et al. 1990). This widened the applications of nanoparticles to reach the market in the mid of 1990s with the first FDA-approved nanoparticle therapeutic anticancer product (Doxil), comprising liposomal delivery carrier and PEG-1 asparaginase for doxorubicin delivery (Petros

TABLE 15.1
Examples of Various PEGylated Nanoparticles

Nanoparticle Composition	Method of Preparation	Particle Size (nm)	Zeta Potential (mV)	Stability/Advantages Remarks	Ref.
Zein, PEG-35000	Desolvation and, then, coating by incubation with PEG.	200	Negative	PEG-coated zein nanoparticles, prepared without incorporating reactive reagents, represent adequate carriers for promoting bioavailability of oral biomacromolecules with low permeability properties.	(Reboredo et al. 2021)
Silver	NPs formulated in an ethanol solution applying thiol chemistry.	1–20	-3	Silver NPs coated with PEG and attached covalently to their surface have yielded a stable colloid. The stability of this system, even at high ionic forces, was maintained (0.1 M NaCl).	(Shkilnyy et al. 2009)
Silver, PEG-200 g/mol	Silver nitrate reduction with PEG 200 at basic pH in polar solvent.	51–170	-	Degradation of PEG by thermal oxidation took place at 100 °C. This has risen the number of molecules and functional groups that can decrease silver species and confer stabilization to the silver NPs.	(Fleitas-Salazar et al. 2017)
Ferrous chloride tetrahydrate, ammonium hydroxide, ferric chloride hexahydrate, PEG-600 Da	In situ coprecipitation technique.	32–43	-	PEG coated NPs displayed improved colloidal stability in an alkaline medium (pH = 10) and nitrile (NBR) latex for up to 21 days in comparison to the PEG-free formulation in the course of the sedimentation experiment.	(Tai, Lai, and Hamid 2016)
Poly(D,L-lactic acid) (PLA), PEG	Double emulsion solvent evaporation method	300		PEGylated NPs endowed stable in acidic and alkaline pH of dissolution medium, which efficiently protected the cargo from enzymatic degradation in the GIT.	(Singh, Mandal, and Khan 2017)

(Continued)

TABLE 15.1 (Continued)
Examples of Various PEGylated Nanoparticles

Nanoparticle Composition	Method of Preparation	Particle Size (nm)	Zeta Potential (mV)	Stability/Advantages Remarks	Ref.
Au, PEG 5000	Heating of chloroauric acid and then reaction with sodium citrate	17.3–6.4		NPs capping density and stability exhibited a linear relationship in DCM and water, while the stability was reduced in PBS and PBS/BSA.	(Manson et al. 2011)
Silver, chitosan, PEG (Mw 5 KDa), dextran sulfate	Layer-by-layer (LbL) assembly	34.3	−37.5	Copolymers grafting permits colloids encapsulation that exhibited relative stability over a vast range of NaCl strengths and enhance targetability biocompatibility of the nanoparticles prepared for imaging or therapeutic applications.	(Shutava and Livanovich 2020)
Silver nitrate, PEG-400	Turkevich's method to prepare silver nanoparticles (AgNPs)	19–50	−40 AgNPs-37 PEG-AgNPs	The fabricated AgNPs and PEG-AgNPs NPs possessed spherical shape, narrow size distributions and high stability. Coated silver NPs could be used as targeted nanovehicle for dermal pathologies and diagnostic.	(Pinzaru et al. 2018)
PEG-PLGA	Nanoprecipitation technique	43		Long-circulating nanoparticles were achieved to overcome the drug drawbacks including short elimination half-life and low oral bioavailability.	(Ahmed and Badr-Eldin 2020)
Copper (II) nitrate trihydrate Cu (NO3)2. PEG 4000	Chemical reduction method	30–50		Particle size and size distribution were significantly reduced when using PEG as the stabilizing polymer.	(Olad, Alipour, and Nosrati 2017)
Chitosan-PEG 8000	Ionotropic gelation	135–229	+20 to +28	Horseradish peroxidase was immobilized into chitosan and chitosan/PEGylated NPs, yielding improved immobilization efficiency in the PEG-free formulations.	(Melo et al. 2020)

and DeSimone 2010). At one week of injection, Doxil has improved doxorubicin bioavailability nearly 90-times of PEGylated liposomes compared to free drug. The incorporation of PEG in the formulations of doxorubicin could modulate the half-life up to 72 hours and circulation half-life up to 36 hours.

15.5.1.2 Surfactants

The morphology of the nanoparticles can be modified during the synthesis process by means of surfactant incorporation (Bakshi 2016). Surface active molecules can be utilized as adjuvant to tailor the shape of nanomaterials. It was reported that the quality of the nanoparticles dispersion has been enhanced via coating with various species of surfactants including anionic (SDS), cationic (CTAB, PEI), and charge-free (PEG-6000, Triton X 100). Anchoring mechanism of surfactant to the nanosurfaces can be either noncovalent adsorption or covalent assembly. Noncovalent adsorption approach that comprises hydrogen bonds, ion pairing, and ion exchange enables zeta potential shifting, conferring improved stability (Heinz et al. 2017). Covalent assembly, on the other hand, is based on attachment of the end groups of the nanoparticle with the surfactant. Monolayer of surfactant growth on the surface of the nanoparticles typically produces more stable and smaller size nanoparticles compared to bilayers surfactant (Astete et al. 2011). Anand and Varghese 2017., have investigated the stability of zinc oxide nanosuspension by means of application of various surfactant species (Anand and Varghese 2017). They have reported superior stability upon applying SDS as a surfactant compared to other species (STAB, PEG 6000, Triton X 100). This might be attributed to the high adsorption levels of SDS on the nanosurfaces which was translated by zeta potential value of 28.9 mV. Other researcher has also reported a significant reduction in nanoparticles agglomeration when using SLS as a surfactant (Hwang et al. 2008).

Thermal conductivity in water dispersion was 1.4-fold higher in iron oxide rod-shape nanoparticles prepared via co-precipitation through sodium dodecylbenzenesulfonate (SDBS) as a surfactant (Gayadhthri et al. 2014). Carbon nanotube-based nanofluids showed better stability when surfactant was incorporated compared to surfactant-free formulation. The presence of hexadecyltrimethylammonium bromide has associated with the highest zeta potential value, whereas highest thermal conductivity (25.7%) was attained with gum Arabic as a surfactant (Leong et al. 2016). Similarly, Cu-H2O nanofluids synthesized by two-step approach exhibited enhanced thermal conductivity and improved dispersion behavior via optimization the SDBS levels (Zhu et al. 2009). Also, the stability of TiO2 nanosuspensions was preserved even after a period of 7 days in the base fluid with the aid of surfactant (Aziz, Khalid, and Khalid 2018). Moreover, the stability of Ag-silicon oil nanofluid and carbon black-water was enabled by utilizing different types of surfactants, such as SDS or oleic acid (Aziz, Khalid, and Khalid 2018). Introducing of the surfactant in various nanofluid preparations facilitates homogenous dispersion of the nanoparticles in the base fluid as a result of electrostatic repulsion (F. Yu et al. 2017).

The effects of surfactant on the stability of ammonia-water-nanofluids was investigated. Yang et al. have reported an initial proportional increase of the absorbances of nanofluids comprising 0.1% carbon black nanoparticles upon introducing emulsifier, followed by a decrease during static storage period due to insufficient

Stability of Nanomaterials

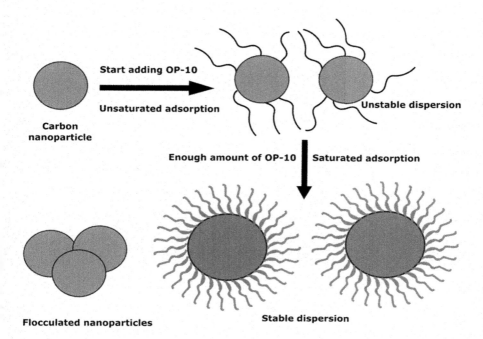

FIGURE 15.5 Mechanism of stabilization of carbon nanoparticle by PO-10 surfactant.

adsorption process which necessitate finite time (L. Yang et al. 2011). After completion of adsorption process, a reduction of absorbency has been recorded at specific time intervals. This process can be demonstrated by the formation of single layer adsorption concept for carbon nanofluid stabilization as represented in Figure 15.5. Prior adding a surfactant, nanoparticles generate high surface energy, Van der Waals forces, and Brownian motion that gave rise to aggregation. Then, when inadequate amount of dispersant is introduced, the surfactant molecules tilted or flat on the solid surface of the nanoparticles as a result of unsaturated adsorption. Adding a greater amount of surfactant will render efficient adsorption process on the nanosurfaces, where hydrophilic sides are directed perpendicularly into the aqueous phase to form electrostatic hooking that allows the nanoparticulate to be suspended and separated for extended period of time (Jagannathan and Irvin, 2005). However, excessive amount of surfactant causes oversaturation adsorption on the nanosurfaces that weakens the stability of carbon black nanofluid (L. Yang et al. 2011).

Diverse categories of surfactants like Triton, polyvinylpyrrolidone (PVP), SDS, NaDDBs, Pluronic, and polyethylene glycol have been demonstrated to enhance dispersion stability of nanoparticles (Sultana et al. 2020). Binary mixture of SDS and dodecyltrimethylammonium bromide (DTAB) conferred carbon nanotubes of multi-wall dispersion stable at reduced total surfactant concentration compared to solo use even at high concentration (Javadian et al. 2017). Similarly, combination of ethylenediaminetetraacetic acid (EDTA) and SDS surfactants enabled formation of energy barrier that hinders agglomeration of the nanoparticles. Table 15.2 represents examples of surfactants used in nanoformulations for stability endowment.

TABLE 15.2
Examples of Surfactants Used for Nanostabilization

Nanoparticle Composition	Preparation Method	Surfactant	Particles Size (nm)	Zeta Potential (mV)	Stability Remark	Ref.
ZnOAgCu	Commercial nanoparticles	CTABSDS	195.2295.5	>70b>70	Ionic surfactants reduced the agglomeration tendency of NPs and enhanced their colloidal stability. However, the presence of ionic surfactant with ZnO nanoparticles exhibited a better long-term stability than AgCu ones.	(Kowalczyk and Kaminska 2020)
Silver	Chemical reduction method.	Cationic phosphonium	40–109	27–50	Stabilizing effect was availed by increasing alkyl chain length of the surfactant.	(Pisárčik et al. 2020)
Fe, Fe3O4, Fe2O3	Chemical reduction method	PVP	12–200	NA	High amount of dispersant confers particles with improved stability and homogeneity.	(Ruíz-Baltazar et al. 2015)
CrFe2O4	Green synthesis via coprecipitation	*Cucumis sativus* peel extract	NA	NA	Natural surface-active agent extracted from Cucumis sativus peel confers a novel approach to formulate large-scale magnetic NPs.	(Lestari, Taufiq, and Hidayat 2020)
Metal of palladium (Pd)	Chemical reduction of palladium acetylacetonate (Pd (acac)2)	Brij series surfactants, TX-100	NA	NA	Control of the size of Pd NPs was achieved by setting the temperature of the reaction to 433 K and utilization of surfactant. Brij-58 and Brij-35 were the most adequate for the formation of a stable surfactant/IL hybrid system of Pd NPs.	(Harada, Yamamoto, and Sakata 2020)

(Continued)

TABLE 15.2 (Continued)
Examples of Surfactants Used for Nanostabilization

Nanoparticle Composition	Preparation Method	Surfactant	Particles Size (nm)	Zeta Potential (mV)	Stability Remark	Ref.
Iron oxide	Hydrothermal method	SDSTX100 CTAB	12–205	−25.9-25.1+15.2	A nanofluid formulated in 2 CMC SDS solution possessed the highest colloidal stability and thermal conductivity of 0.4787 W/m.K.	(Kongsat et al. 2021)
Fe3O4	Electro-oxidation of iron by a chrono-amperometric technique	Tetramethy-lammoniumchloride	NA	NA	Introducing high amounts of surfactant yielded particles with reduced size and narrower size distribution, and also the shape was transformed to quasi-spherical from tetragonal, cubic, hexagonal, and triangular.	(Kazeminezhad and Mosivand 2012)
Amylose	Precipitation with absolute ethanol	Tween80, Span80, and mixtures of Tween80 and Span80 with ratios of 25/75, 50/50, and 75/25	155.2–183.1	NA	The size of prepared NPs was significantly influenced by surfactant species, strength and hydrophilic– lipophilic balance (HLB) value. The smallest size of ANPs was attainable by incorporation of 0.5% of Tween80/Span80 mixture (HLB = 12.33).	(Dong et al. 2016)
ZnO	Mechao-chemical approach	TX-100, PEG- 6000 CTABSDS	186–716	−28.9	Better stability of ZnO NPs through particle size analysis and zeta potential measurement was obtained when anionic surfactant (SDS) was introduced.	(Anand and Varghese 2017)

(Continued)

TABLE 15.2 (Continued)
Examples of Surfactants Used for Nanostabilization

Nanoparticle Composition	Preparation Method	Surfactant	Particles Size (nm)	Zeta Potential (mV)	Stability Remark	Ref.
Curcumin nano-emulsions	Emulsification	Lecithin Tween 80	44.4–193.9	-48 to -62.9	There is a proportional relationship between surfactant concentrations and encapsulation efficacy and stability for 15 days with reduced particle size of the nanoemulsion.	(Chuacharoen, Prasongsuk, and Sabliov 2019)

Nanoparticles (NPs), Not available (NA)

15.6 LYOPHILIZATION-BASED STABILITY

Freeze-drying or lyophilization is a dehydration process utilized to confer stability to nanosuspension, improving the shelf-life and facilitating handling and storage. The low temperature allows transition from a solid to a gas state, avoiding the liquid state. This process comprises three steps: freezing, primary, and secondary drying. Sublimation of ice takes place to remove the water from the sample, while unfrozen water is desorbed under vacuum at temperature and pressure below the triple point of water. The processing conditions of freezing and desiccation stresses may introduce instabilities to the nanoparticles. Various adjuvant materials such as lyoprotectants and cryoprotectants can be used to reduce the lyophilization stresses and conserve the physio-chemical attributes of the nanoparticles (Fonte et al. 2015). For instance, sugar can be used as cryo- and lyoprotectants to protect proteins nanoparticles from inactivation during lyophilization and storage owing to the innocuous property and the ability to vitrify during the freezing step. Lyophilization technique is a routinely used industrial process to promote the long-term stability of pharmaceutical formulations including nanoparticles on account of its unique advantages (Mohammady and Yousefi 2020). In particular, its ability of application for small- and large-scale production. Alike solid dosage form pharmaceuticals or supplements, such as capsules or pill, the nanosuspension is transformed to dry powder prior to conversion to required solid dosage forms (Wong, Yu, and Hadinoto 2018).

Lyophilization is an optimizable process where the most influential factors including concentration and type of lyoprotectant/cryoprotectant, stabilizer concentration and structure, nanoparticles concentration, freezing, annealing, drying rate, solvent type and solvent to antisolvent ratio, and the interaction between stabilizer and protectants. Careful optimization of specific variables in relation to the type of nanomaterials should take place in order to achieve the best results of drying. For example, incorporation of 5%–20% of saccharides as cryoprotectants for emulsified nanoparticles has yielded the stabilized product, and for nanocrystals, the freezing rate, and solvent/antisolvent ratio are considered highly important variables to be optimized. Therefore, optimization of the effective factors related to lyophilization process and formulation perspective ensures minimized aggregation and improved stability of the nanoparticles (Mohammady and Yousefi 2020).

15.7 STABILITY TESTS

There are several key parameters for test conditions discussed in the following sections. Table 15.3 shows examples of stability tests employed for nanoparticles.

15.7.1 Shelf-Life (Aging)

Typically, nanoparticles stability of 6 months to 1 year under ambient laboratory conditions is considered sufficient for most applications (Cho and Hackley 2018). Shelf-life is an essential characteristic in food and medicine. Each product possesses specific shelf-life that should be precisely determined and labelled. Theoretically, shelf-life can be defined as a finite length of time post production, during which the

TABLE 15.3
Examples of Stability Tests Used in Nanoformulation

Type	Tasted Variables	Protocol	Results	Reference
Solid lipid nanoparticles (SLNPs)	EE, particle size, ZP and PDI	SLNPs were stored in sealed glass vials at 25 ± 2 °C/60 ± 5% RH in a stability chamber (Remi, Mumbai, India) for 3 months. Control samples were stored at 2–8 °C at corresponding time intervals.	Under low temperature (2–8 °C), no significant variations in any of the examined parameters had taken place. With more time passed, drug expulsion from solid-lipid matrix (due to crystallization of lipid), resulting in lowering of the EE.	(Ravi et al. 2014)
Tacrolimus loaded poly(ε-caprolactone)	pH, particle size, PDI, ZP	The nanoparticles were stored at room temperature for a period of 120 days	Significant decrease of pH occurred after 60 days. Particle size and PDI were unchanged. Zeta potential was increased after day 120 of the experiment.	(dos Anjos Camargo et al. 2020)
Chitosan-DNA-FAP-B NPs	Size and ZP of polyplexes, DNA binding affinity of chitosan and in-vitro transfection efficiency of NPs.	Chitosan-DNA-FAP-B NPs were placed into a clear glass-vials and kept at 5 and 25 °C for a period of 6 months and at –20 °C for 12 months. Samples were withdrawn at specific intervals (0, 1, 3, 6 months and 12 months for frozen sample).	Stability studies demonstrated that chitosan-DNA-FAP-B NPs were stable after 1 month when placed at –20 °C and preserved their initial size, ZP and transfection efficiency. However, their stability was not preferable at 5 and 24 °C.	(Mohammadi et al. 2012)
Au NPs (AuNPs)	Surface plasmon resonance (SPR).	The stability of AuNps were investigated over 30 to 180 days at 4 °C and the surface plasmon resonance (SPR) was measured accordingly. With reference to the stability under physiological body conditions, the samples were stored in the solution (pH 7.2) comprising several inorganic salts such as NaCl, KCl, NaHCO3, MgCl2.6H2O, CaCl2, and KH2PO4 over a period of 3 hours and accordingly measured the SPR.	Results suggest desirable compatibility and stability of protein functionalized NPs when compared with the blank NPs.	(Thilagam and Gnanamani 2020)

Nanoparticles (NPs), zeta potential (ZP), encapsulation efficiency (EE), polydispersity index (PDI) product (nanoparticles) conserves a required level of quality under well-defined storage conditions (Nicoli 2012). The concept of shelf-life assessment displays an exciting challenge for scientists even with the great efforts by many researchers over few decades. Parallelly, it is highly considered as a vital process to comply with the current regulations to improve market share and preserve company brand reputation.

15.7.2 Stability in Relevant Media

Stability of pharmaceutical nanoformulations in physiologically relevant media is crucial to ensure the appropriateness for biomedical applications. Media can be salt solutions, deionized water, cell culture solution embedded, or serum protein-free, since serum comprises thiolated proteins that are able to replace nucleic acid segments/DNA, proteins, and native surface-bound ligand.

15.7.3 pH

The pH stability should be carefully considered for pharmaceutical products like nanoparticles. For instance, orally administered dosage form of nanoparticles will undergo a wide range of pH, starting from acidic in the stomach to alkaline in the intestine prior absorption into blood circulation from near natural pH. In addition, lysosomes in mammalian cells possess slightly acidic pH levels (4–5). Thus, nanoparticles may experience a broad range of conditions within a biological system from highly acidic to mildly alkaline.

15.7.4 Temperature

Thermal stability of nanoparticles should be assessed over a range of ambient, physiological and higher temperatures in the range of 20–60 °C to include the relevant environment for storage, transport, biomedical applications, and most biological assays.

15.7.5 Chemical Resistance

Nanomaterial formulations designed for therapeutic applications are probably transiting the blood circulation prior distributing to the target sites. Suitable hydrophilic and stable functionalized nanoparticles can reduce protein binding, minimize immune system recognition and clearance, prolong retention in blood circulation, yielding a summative effect of improved bioavailability. An ideal nanoparticulate platform should possess an efficient coating that resist ligand displacement (surface functionalization disruption) and chemical attack.

15.8 STABILITY MEASUREMENTS OF NANOPARTICLES

The molecular, atomic structure, and behavior of the nanomaterials can be evaluated via several techniques such as X-ray diffraction and nuclear magnetic resonance (NMR) spectroscopy. The NMR spectroscopy is a useful technique to investigate molecular structures in solution and able to produce tertiary structural information, making it a powerful tool for protein nanoparticles. On the other hand, X-ray diffraction generates electromagnetic radiation as a result of electron collide with metal atom like copper. It is also widely used to evaluate the 3-D structure of proteins in crystallography due to the wide wavelength range of X-ray. However, large amount of samples are required by both techniques, imposing constraint in many biological settings (Rohiwal et al. 2015).

Physical characterization techniques of the nanoparticles can be achieved by Dynamic Light Scattering (DLS) to measure the particle size and to determine the surface charge. Moreover, the spectroscopic method is widely employed to demonstrate the particle aggregation and coagulation and to investigate the conformational changes of nanoparticles. The fluorescence spectroscopy offers an orthogonal comparison as compared to the physical techniques. In addition, the structural morphology of the nanoparticles can be observed by atomic force microscopy (AFM) and transmission electron microscopy (TEM).

15.9 CONCLUSION

The emerging of nanomaterials introduces significant benefits associated with vast applications. In order to make full use of these nanoparticles, reliable synthesis and storage are prerequisites to ensure high degree of stability of the products. This can be achieved by employing different approaches to convey the product development including stabilizing agents of ligands, surfactants, and sterically substances. This enables characteristics, such as solubility, shape, size, functionality, and surface chemistry of the produced nanoparticles, to be efficiently controlled.

REFERENCES

Abuchowski, Abraham et al. "Effect of Covalent Attachment of Polyethylene Glycol on Immunogenicity and Circulating Life of Bovine Liver Catalase." *Journal of Biological Chemistry* 252(11) (1977) 3582–3586.

Ahmed, Osama A A, and Shaimaa M Badr-Eldin. "Biodegradable Self-Assembled Nanoparticles of PEG-PLGA Amphiphilic Diblock Copolymer as a Promising Stealth System for Augmented Vinpocetine Brain Delivery." *International Journal of Pharmaceutics* 588 (2020) 119778.

Aitken, Robert J, Alan D Jones, S. Peters, and Vicki Stone. "Regulation of Carbon Nanotubes and Other High Aspect Ratio Nanoparticles: Approaching." In *International Handbook on Regulation Nanotechnologies*. Edited by Graeme A Hodge, Diana M Bowman, Andrew D Maynard, Edward Elgar Publishing Ltd (2010) 205–237.

Alvarez-Puebla, Ramon A et al. "Ultrasensitive Multiplex Optical Quantification of Bacteria in Large Samples of Biofluids." *Scientific Reports* 6(1) (2016). https://doi.org/10.1038/srep29014.

Anand, K., and Siby Varghese. "Role of Surfactants on the Stability of Nano-Zinc Oxide Dispersions." *Particulate Science and Technology* 35(1) (2017) 67–70.

dos Anjos Camargo, Guilherme et al. "Stability Testing of Tacrolimus-Loaded Poly (ε-Caprolactone) Nanoparticles by Physicochemical Assays and Raman Spectroscopy." *Vibrational Spectroscopy* 110 (2020) 103139.

Astete, Carlos E et al. "Antioxidant Poly (Lactic-Co-Glycolic) Acid Nanoparticles Made with α-Tocopherol–Ascorbic Acid Surfactant." *ACS Nano* 5(12) (2011) 9313–9325.

Aziz, Saba, Shahid Khalid, and Hina Khalid. "Influence of Surfactant and Volume Fraction on the Dispersion Stability of TiO2/Deionized Water Based Nanofluids for Heat Transfer Applications." *Materials Research Express* 6(1) (2018) 15031.

Bakshi, Mandeep Singh. "How Surfactants Control Crystal Growth of Nanomaterials." *Crystal Growth & Design* 16(2) (2016) 1104–1133.

Begum, Aynun N, Jose S Aguilar, Lourdes Elias, and Yiling Hong. "Silver Nanoparticles Exhibit Coating and Dose-Dependent Neurotoxicity in Glutamatergic Neurons Derived from Human Embryonic Stem Cells." *Neurotoxicology* 57 (2016) 45–53.

Biswas, Jayanta Kumar, and Dibyendu Sarkar. "Nanopollution in the Aquatic Environment and Ecotoxicity: No Nano Issue!" *Current Pollution Reports* 5(1) (2019) 4–7.

Bobo, Daniel et al. "Nanoparticle-Based Medicines: A Review of FDA-Approved Materials and Clinical Trials to Date." *Pharmaceutical Research* 33(10) (2016) 2373–2387.

Casals, Eudald, E. Gonzalez, and Victor Franco Puntes. "Reactivity of Inorganic Nanoparticles in Biological Environments: Insights into Nanotoxicity Mechanisms." *Journal of Physics D: Applied Physics* 45(44) (2012) 443001.

Chairam, Sanoe, and Ekasith Somsook. "Starch Vermicelli Template for Synthesis of Magnetic Iron Oxide Nanoclusters." *Journal of Magnetism and Magnetic Materials* 320(15) (2008) 2039–2043.

Chan, Warren C W. "Nano Research for COVID-19." *ACS Nano* 14(4) (2020) 3719–3720.

Chen, Chia-Ying, and Chad T Jafvert. "The Role of Surface Functionalization in the Solar Light-Induced Production of Reactive Oxygen Species by Single-Walled Carbon Nanotubes in Water." *Carbon* 49(15) (2011) 5099–5106.

Cho, Tae Joon, and Vincent A Hackley. "Assessing the Chemical and Colloidal Stability of Functionalized Gold Nanoparticles." *NIST Special Publication*(1200) (2018) 26.

Chuacharoen, Thanida, Sehanat Prasongsuk, and Cristina M Sabliov. "Effect of Surfactant Concentrations on Physicochemical Properties and Functionality of Curcumin Nanoemulsions under Conditions Relevant to Commercial Utilization." *Molecules* 24(15) (2019) 2744.

Cuenya, Beatriz Roldan, and Farzad Behafarid. "Nanocatalysis: Size-and Shape-Dependent Chemisorption and Catalytic Reactivity." *Surface Science Reports* 70(2) (2015) 135–187.

de Souza Tiago Alves Jorge, Lilian Rodrigues Rosa Souza, and Leonardo Pereira Franchi. "Silver Nanoparticles: An Integrated View of Green Synthesis Methods, Transformation in the Environment, and Toxicity." *Ecotoxicology and Environmental Safety* 171 (2019) 691–700.

Dong, Yan et al. "Effects of Surfactants on Size and Structure of Amylose Nanoparticles Prepared by Precipitation." *Bulletin of Materials Science* 39(1) (2016) 35–39.

Fleitas-Salazar, Noralvis et al. "Effect of Temperature on the Synthesis of Silver Nanoparticles with Polyethylene Glycol: New Insights into the Reduction Mechanism." *Journal of Nanoparticle Research* 19(3) (2017) 1–12.

Fonte, Pedro et al. "Polymer-Based Nanoparticles for Oral Insulin Delivery: Revisited Approaches." *Biotechnology Advances* 33(6, Part 3) (2015) 1342–1354. http://www.sciencedirect.com/science/article/pii/S073497501500035X.

Gayadhthri, V., K. S. Suganthi, S. Manikandan, and K. S. Rajan. "Role of Surfactants in Colloidal Stability and Properties of α-Fe2O3 Based Nanofluid." *Asian Journal of Scientific Research* 7(3) (2014) 320.

Hao, N. et al. "Roles of Particle Size, Shape and Surface Chemistry of Mesoporous Silica Nanomaterials on Biological Systems." *International Materials Reviews* 62(2) 2016) 1–21.

Harada, Masafumi, Miho Yamamoto, and Maharu Sakata. "Temperature Dependence on the Size Control of Palladium Nanoparticles by Chemical Reduction in Nonionic Surfactant/Ionic Liquid Hybrid Systems." *Journal of Molecular Liquids* 311 (2020) 113255.

Heinz, Hendrik et al. "Nanoparticle Decoration with Surfactants: Molecular Interactions, Assembly, and Applications." *Surface Science Reports* 72(1) (2017) 1–58.

Hierrezuelo, José et al. "Electrostatic Stabilization of Charged Colloidal Particles with Adsorbed Polyelectrolytes of Opposite Charge." *Langmuir* 26(19) (2010) 15109–15111.

Hwang, Yujin et al. "Production and Dispersion Stability of Nanoparticles in Nanofluids." *Powder Technology* 186(2) (2008) 145–153.

Jagannathan, Ramesh, and Glen C Irvin Jr. "Nanofluids: A New Class of Materials Produced from Nanoparticle Assemblies." *Advanced Functional Materials* 15(9) (2005) 1501–1510.

Jassby, David, Jeffrey Farner Budarz, and Mark Wiesner. "Impact of Aggregate Size and Structure on the Photocatalytic Properties of TiO2 and ZnO Nanoparticles." *Environmental Science & Technology* 46(13) (2012) 6934–6941.

Javadian, Soheila et al. "Dispersion Stability of Multi-Walled Carbon Nanotubes in Catanionic Surfactant Mixtures." *Colloids and Surfaces A: Physicochemical and Engineering Aspects* 531 (2017) 141–149.

Kamiya, Hidehiro et al. "Characteristics and Behavior of Nanoparticles and Its Dispersion Systems." In *Nanoparticle Technology Handbook*, Elsevier (2008) 113–176.

Kazeminezhad, Iraj, and Saba Mosivand. "Effect of Surfactant Concentration on Size and Morphology of Sonoelectrooxidized Fe3O4 Nanoparticles." *Current Nanoscience* 8(4) (2012) 623–627.

Klibanov, A. L., K. Maruyama, V. P. Torchilin, and L. Huang. "Amphipathic Polyethyleneglycols Effectively Prolong the Circulation Time of Liposomes." *FEBS Letters* 268(1) (1990) 235–237. http://www.ncbi.nlm.nih.gov/pubmed/2384160 (March 28, 2016).

Kong, Li, Frederick Campbell, and Alexander Kros. "DePEGylation Strategies to Increase Cancer Nanomedicine Efficacy." *Nanoscale Horizons* 4(2) (2019) 378–387.

Kongsat, Pantharee et al. "Synthesis of Structure-Controlled Hematite Nanoparticles by a Surfactant-Assisted Hydrothermal Method and Property Analysis." *Journal of Physics and Chemistry of Solids* 148 (2021) 109685.

Kowalczyk, Dorota, and Irena Kaminska. "Effect of PH and Surfactants on the Electrokinetic Properties of Nanoparticles Dispersions and Their Application to the PET Fibres Modification." *Journal of Molecular Liquids* 320 (2020) 114426.

Kozielski, Kristen L, Stephany Y Tzeng, Bolivia A Hurtado De Mendoza, and Jordan J Green. "Bioreducible Cationic Polymer-Based Nanoparticles for Efficient and Environmentally Triggered Cytoplasmic SiRNA Delivery to Primary Human Brain Cancer Cells." *ACS Nano* 8(4) (2014) 3232–3241.

Lead, Jamie R et al. "Nanomaterials in the Environment: Behavior, Fate, Bioavailability, and Effects—an Updated Review." *Environmental Toxicology and Chemistry* 37(8) (2018) 2029–2063.

Leong, Kin Yuen, Hanafi Nurfadhillah Mohd, Sohaimi Risby Mohd, and Noor Hafizah Amer. "The Effect of Surfactant on Stability and Thermal Conductivity of Carbon Nanotube Based Nanofluids." *Thermal Science* 20(2) (2016) 429–436.

Lestari, Merinda, Ahmad Taufiq, and Arif Hidayat. "Green Synthesis of CrFe2O4 Nanoparticles Using Cucumis Sativus as a Natural Surfactant." *Materials Today: Proceedings* 44 (2020): 3221–3224.

Levard, Clément et al. "Sulfidation Processes of PVP-Coated Silver Nanoparticles in Aqueous Solution: Impact on Dissolution Rate." *Environmental Science & Technology* 45(12) (2011) 5260–5266.

Li, Zhenxing et al. "Mesoporous Hollow Cu–Ni Alloy Nanocage from Core–Shell Cu@ Ni Nanocube for Efficient Hydrogen Evolution Reaction." *ACS Catalysis* 9(6) (2019) 5084–5095.

Lowry, Gregory V, Kelvin B Gregory, Simon C Apte, and Jamie R Lead. "Transformations of Nanomaterials in the Environment." *Environmental Science & Technology* 46 (13) (2012) 6893–6899. https://doi.org/10.1021/es300839e.

Luan, Jingyi et al. "Ultrabright Fluorescent Nanoscale Labels for the Femtomolar Detection of Analytes with Standard Bioassays." *Nature Biomedical Engineering* 4(5) (2020) 518–530.

Manson, Joanne, Dhiraj Kumar, Brian J Meenan, and Dorian Dixon. "Polyethylene Glycol Functionalized Gold Nanoparticles: The Influence of Capping Density on Stability in Various Media." *Gold Bulletin* 44(2) (2011) 99–105.

Melo, Micael Nunes et al. "Immobilization and Characterization of Horseradish Peroxidase into Chitosan and Chitosan/PEG Nanoparticles: A Comparative Study." *Process Biochemistry* 98 (2020) 160–171.

Mohammadi, Zohreh et al. "Stability Studies of Chitosan-DNA-FAP-B Nanoparticles for Gene Delivery to Lung Epithelial Cells." *Acta Pharmaceutica* 62(1) (2012) 83–92.

Mohammady, Mohsen, and Gholamhossein Yousefi. "Freeze-Drying of Pharmaceutical and Nutraceutical Nanoparticles: The Effects of Formulation and Technique Parameters on Nanoparticles Characteristics." *Journal of Pharmaceutical Sciences* 109(11) 2020.

Mohapatra, Shyam et al. *Characterization and Biology of Nanomaterials for Drug Delivery: Nanoscience and Nanotechnology in Drug Delivery*, Elsevier. (2018)

Natarajan, Pavithra, and John M Tomich. "Understanding the Influence of Experimental Factors on Bio-Interactions of Nanoparticles: Towards Improving Correlation between in Vitro and in Vivo Studies." *Archives of Biochemistry and Biophysics* 694 (2020) 108592.

Nicoli, Maria Cristina. "The Shelf Life Assessment Process." In *Shelf Life Assessment of Food, Nicoli*, edited by Nicoli, M. C., CRC PRESS, Taylor & Francis Group: Boca Raton (2012) 17–36.

Olad, Ali, Mahnaz Alipour, and Rahimeh Nosrati. "The Use of Biodegradable Polymers for the Stabilization of Copper Nanoparticles Synthesized by Chemical Reduction Method." *Bulletin of Materials Science* 40(5) (2017) 1013–1020.

Patra, Jayanta Kumar et al. "Nano Based Drug Delivery Systems: Recent Developments and Future Prospects 10 Technology 1007 Nanotechnology 03 Chemical Sciences 0306 Physical Chemistry (Incl. Structural) 03 Chemical Sciences 0303 Macromolecular and Materials Chemistry 11 Medical and He." *Journal of Nanobiotechnology* 16(1) (2018) 1–33.

Petosa, Adamo R et al. "Aggregation and Deposition of Engineered Nanomaterials in Aquatic Environments: Role of Physicochemical Interactions." *Environmental Science & Technology* 44(17) (2010) 6532–6549.

Petros, Robby A, and Joseph M DeSimone. "Strategies in the Design of Nanoparticles for Therapeutic Applications." *Nature Reviews. Drug discovery* 9(8) (2010) 615–627. http://www.ncbi.nlm.nih.gov/pubmed/20616808 (July 14, 2014).

Pillai, Pramod P, Sabil Huda, Bartlomiej Kowalczyk, and Bartosz A Grzybowski. "Controlled PH Stability and Adjustable Cellular Uptake of Mixed-Charge Nanoparticles." *Journal of the American Chemical Society* 135(17) (2013) 6392–6395.

Pinzaru, Iulia et al. "Stable PEG-Coated Silver Nanoparticles–A Comprehensive Toxicological Profile." *Food and Chemical Toxicology* 111 (2018) 546–556.

Pisárčik, Martin et al. "Phosphonium Surfactant Stabilised Silver Nanoparticles. Correlation of Surfactant Structure with Physical Properties and Biological Activity of Silver Nanoparticles." *Journal of Molecular Liquids* 314 (2020) 113683.

Rameshkumar, Perumal, and Ramasamy Ramaraj. "Gold Nanoparticles Deposited on Amine Functionalized Silica Sphere and Its Modified Electrode for Hydrogen Peroxide Sensing." *Journal of Applied Electrochemistry* 43(10) (2013) 1005–1010.

Ravi, Punna Rao et al. "Lipid Nanoparticles for Oral Delivery of Raloxifene: Optimization, Stability, in Vivo Evaluation and Uptake Mechanism." *European Journal of Pharmaceutics and Biopharmaceutics* 87(1) (2014) 114–124.

Reboredo, C. et al. "Preparation and Evaluation of PEG-Coated Zein Nanoparticles for Oral Drug Delivery Purposes." *International Journal of Pharmaceutics* 597 (2021) 120287.

Rohiwal, S. S., A. P. Tiwari, G. Verma, and S. H. Pawar. "Preparation and Evaluation of Bovine Serum Albumin Nanoparticles for Ex Vivo Colloidal Stability in Biological Media." *Colloids and Surfaces A: Physicochemical and Engineering Aspects* 480 (2015) 28–37.

Ruíz-Baltazar, Alvaro, Rodrigo Esparza, Gerardo Rosas, and Ramiro Pérez. "Effect of the Surfactant on the Growth and Oxidation of Iron Nanoparticles." *Journal of Nanomaterials* 2015.(2015). https://doi.org/10.1155/2015/240948.

Salunkhe, Ashwini et al. "MRI Guided Magneto-Chemotherapy with High-Magnetic-Moment Iron Oxide Nanoparticles for Cancer Theranostics." *ACS Applied Bio Materials* 3(4) (2020) 2305–2313.

Sapsford, Kim E et al. "Analyzing Nanomaterial Bioconjugates: A Review of Current and Emerging Purification and Characterization Techniques." *Analytical Chemistry* 83(12) (2011) 4453–4488.

Schmudde, Madlen et al. "Controlling the Interaction and Non-Close-Packed Arrangement of Nanoparticles on Large Areas." *ACS Nano* 10(3) (2016) 3525–3535.

Shimabuku, Quelen Letícia et al. "Water Treatment with Exceptional Virus Inactivation Using Activated Carbon Modified with Silver (Ag) and Copper Oxide (CuO) Nanoparticles." *Environmental Technology* 38(16) (2017) 2058–2069.

Shkilnyy, Andriy et al. "Poly (Ethylene Glycol)-Stabilized Silver Nanoparticles for Bioanalytical Applications of SERS Spectroscopy." *Analyst* 134(9) (2009) 1868–1872.

Shutava, Tatsiana G, and Kanstantsin S Livanovich. "Colloidal Stability of Silver Nanoparticles with Layer-by-Layer Shell of Chitosan Copolymers." *Journal of Nanoparticle Research* 22 (2020) 1–14.

Singh, Neha Atulkumar, Abul Kalam Azad Mandal, and Zaved Ahmed Khan. "Fabrication of PLA-PEG Nanoparticles as Delivery Systems for Improved Stability and Controlled Release of Catechin." *Journal of Nanomaterials* 2017 (2017).

Suk, Jung Soo et al. "PEGylation as a Strategy for Improving Nanoparticle-Based Drug and Gene Delivery." *Advanced Drug Delivery Reviews* 99 (2016) 28–51.

Sultana, Shaheen et al. "Stability Issues and Approaches to Stabilised Nanoparticles Based Drug Delivery System." *Journal of Drug Targeting* 28(5) (2020) 468–486.

Tadros, Tharwat. "Steric Stabilization." In *Encyclopedia of Colloid and Interface Science*, edited by Tharwat Tadroser, Springer Verlag (2013) 1048–1049. https://doi.org/10.1007/978-3-642-20665-8_146.

Tai, Mun Foong, Chin Wei Lai, and Sharifah Bee Abdul Hamid. "Facile Synthesis Polyethylene Glycol Coated Magnetite Nanoparticles for High Colloidal Stability." *Journal of Nanomaterials* 2016 (2016) 1–2.

Tan, J. S. et al. "Surface Modification of Nanoparticles by PEO/PPO Block Copolymers to Minimize Interactions with Blood Components and Prolong Blood Circulation in Rats." *Biomaterials* 14(11) (1993) 823–833. http://www.ncbi.nlm.nih.gov/pubmed/8218736 (April 25, 2016).

Thilagam, R., and A. Gnanamani. "Preparation, Characterization and Stability Assessment of Keratin and Albumin Functionalized Gold Nanoparticles for Biomedical Applications." *Applied Nanoscience* 10 (6) (2020) 1–14.

Vesaratchanon, Sudaporn, Alex Nikolov, and Darsh T Wasan. "Sedimentation in Nano-Colloidal Dispersions: Effects of Collective Interactions and Particle Charge." *Advances in Colloid and Interface Science* 134 (2007) 268–278.

Wong, Jerome Jie Long, Hong Yu, and Kunn Hadinoto. "Examining Practical Feasibility of Amorphous Curcumin-Chitosan Nanoparticle Complex as Solubility Enhancement Strategy of Curcumin: Scaled-up Production, Dry Powder Transformation, and Long-Term Physical Stability." *Colloids and Surfaces A: Physicochemical and Engineering Aspects* 537 (2018) 36–43.

Xing, Shanshan et al. "Doxorubicin/Gold Nanoparticles Coated with Liposomes for Chemo-Photothermal Synergetic Antitumor Therapy." *Nanotechnology* 29(40) (2018) 405101.

Yang, Liu et al. "An Experimental and Theoretical Study of the Influence of Surfactant on the Preparation and Stability of Ammonia-Water Nanofluids." *International Journal of Refrigeration* 34(8) (2011) 1741–1748.

Yang, Qi, and Samuel K Lai. "Anti-PEG Immunity: Emergence, Characteristics, and Unaddressed Questions." *Wiley Interdisciplinary Reviews: Nanomedicine and Nanobiotechnology* 7(5) (2015) 655–677.

Yin, Chunyang et al. "Radial Extracorporeal Shock Wave Promotes the Enhanced Permeability and Retention Effect to Reinforce Cancer Nanothermotherapeutics." *Science Bulletin* 64(10) (2019) 679–689.

Yin, Yongguang et al. "Water Chemistry Controlled Aggregation and Photo-Transformation of Silver Nanoparticles in Environmental Waters." *Journal of Environmental Sciences* 34 (2015) 116–125.

Yu, Fan et al. "Dispersion Stability of Thermal Nanofluids." *Progress in natural science: Materials International* 27(5) (2017) 531–542.

Yu, Guocan et al. "Porphyrin Nanocage-Embedded Single-Molecular Nanoparticles for Cancer Nanotheranostics." *Angewandte Chemie International Edition* 58(26) (2019) 8799–8803.

Zhu, Dongsheng et al. "Dispersion Behavior and Thermal Conductivity Characteristics of Al2O3–H2O Nanofluids." *Current Applied Physics* 9(1) (2009) 131–139.

16 Nanotoxicology and Regulatory Aspects of Nanomaterials and Nanomedicines

Nashwa Osman[1] and Imran Saleem[2]
[1]Liverpool John Moores University and Sohag University
[2]Liverpool John Moores University

CONTENTS

- 16.1 Nanoparticles and Nanotoxicology 382
- 16.2 Regulatory Aspects of Nanoparticles 382
- 16.3 Nanoparticle Interactions at Cellular and Molecular Levels 383
  - 16.3.1 Proposed Mechanisms of NPs Toxicity 387
    - 16.3.1.1 Cell Membrane Disruption 387
    - 16.3.1.2 Oxidative Stress 389
    - 16.3.1.3 Cell Organelles Damage and Mitochondrial Shutdown 389
    - 16.3.1.4 DNA Damage and Mutagenicity 389
    - 16.3.1.5 Inflammation 390
    - 16.3.1.6 Cell Death Mechanisms 390
  - 16.3.2 NP Toxicity Determinants 391
    - 16.3.2.1 Nanoparticle Size and Surface Area 391
    - 16.3.2.2 Nanoparticle Surface Chemistry 391
    - 16.3.2.3 Nanoparticle Chemistry 393
    - 16.3.2.4 Nanoparticle Aggregation 393
    - 16.3.2.5 Repeated Dose Exposure 393
    - 16.3.2.6 The Dose-Dependent Toxicity 394
- 16.4 Nanotoxicology Assessment Methods 394
  - 16.4.1 In Vitro Methods 394
  - 16.4.2 Ex Vivo Methods 395
  - 16.4.3 In Vivo Methods 395
  - 16.4.4 In Silico Methods 395
- 16.5 Conclusions 397
- References 397

16.1 NANOPARTICLES AND NANOTOXICOLOGY

Nanoparticles (NPs) are particles or materials with nanoscale size, i.e., 1–1000 nm. NPs have widespread applications in many fields, such as many industrial products, cosmetics, medicines, and environmental applications (Mohapatra et al., 2018).

Lowering the size to nanoscale (1–1000 nm) is associated with increasing the surface area for the same mass or volume. The NPs or nanomaterials exhibit exceptional physicochemical properties that render them different in some or all aspects from their parent bulk materials. This explains their high reactivity and strong adsorption properties that plays both roles in facilitating their biomolecular interaction and NP–NP interactions (aggregating/agglomerating) as well as their toxicity (Garcia-Mouton et al., 2019). Hence, the importance of evaluating the safety of these NPs, understanding their biological and environmental interactions, and improve the methods of their assessments. This has been covered by a new branch of science, known as *Nanotoxicology*, to recognize, determine, and regulate the main factors underlying NPs toxicology (Donaldson et al., 2004; Lombardo et al., 2019; Pontes and Grenha, 2020).

Human exposure to NPs occurs via entry routes such as through ingestion, inhalation, dermal penetration, and parenteral injection such as IV. Engineered NPs (ENP) are intentionally engineered to meet the desired applications, such as NPs fabricated for nanomedicines, cosmetics, and other consumer-based products. Based on the type of the materials, NPs can stem from organic materials, i.e., carbon-based, lipid-based, polymeric, or inorganic/metallic NPs, i.e., silver, gold, titanium NPs. This book chapter focuses on the toxicological aspects of NPs for biomedical applications (Lamon et al., 2020; Mohapatra et al., 2018).

16.2 REGULATORY ASPECTS OF NANOPARTICLES

Given the exceptional properties of NPs, they behave in a completely different way to their bulk materials. Hence, they might pose certain nano-specific risks to the humans requiring more data on their safety assessments at the NP-biological interface (Osman et al., 2020). Moreover, conventional chemical testing are not suitable for NPs testing with increasing reports of NPs interferences with many assays (Joris et al., 2013). Therefore, evaluations of NPs have to be regulated and adapted to ensure the accurate conclusions of their hazard assessments. Many government and non-government bodies are working closely to address the regulatory challenges, to facilitate and bridge the international harmonization, assessments, guidelines, and practise standardizations (Lamon et al., 2020). This will help broaden the understanding of NPs, their biological behavior, hazards, and safety evaluation and draw fruitful knowledge and conclusions among scientific community (Gordon et al., 2015).

In the USA, the Organisation for Economic Co-operation and Development (OECD) that regulates the safety of chemical materials has launched an active OECD's Working Party on Manufactured Nanomaterials (WPMN) to address the safety aspects of ENPs, their suitable testing measures, and implementing strategies to evaluate their safety in intended applications (Ventola, 2017). WPMN liaises with many organizations, industries, and committees to ensure the efficiency of the testing measures, and acceptability and updating the methods and the results (where

Nanotoxicology and Regulatory Aspects

it is covered by OECD agreement of Mutual Acceptance of Data (MAD)) with the international widespread harmonization. Moreover, the FDA is regulating the nano-based products for medicinal, cosmetics, food applications, and animal food products. However, there is not yet such a safe category of nanoproducts or harmful list that exists (Lamon et al., 2020). The FDA approval process is based on case-by-case scenario to determine the safety of the final product. Post-marketing review by FDA is maintained to ensure the safety of the products. While the main safety evaluation remains the responsibility of the manufacturer. Up till now, FDA has not adapted its regulatory guidelines to cover nanoproducts. The US Nanotechnology Characterization Laboratory (USNCL) is another US institute that contribute to tackle the issues and contribute to regulate the nanomedicines (Miernicki et al., 2019; Rodríguez, 2018; Sainz et al., 2015).

In Europe, the safety of nanomaterial-containing chemicals is being regulated by Registration, Evaluation, Authorisation, Characterisation, and Hazard-restriction of chemicals (REACH) and its executive agency European Chemical Agency (ECHA) (Sainz et al., 2015). ECHA has two nanomaterials related groups (nanomaterials working group, NMWG, and group assessing already registered nanomaterials, GAARN) that work closely with stakeholders and many organizations to provide expert scientific and technical advice on Classification, Labelling and Packaging (CLP) legislation for nanoproducts, and provide "the best practise" guidelines for testing and safety evaluation for nanomaterials. However, their updates are considered as advisory and under development (Miernicki et al., 2019).

For medicinal purposes, there are many regulatory organizations that regulate the preclinical characterization and safety of nanomedicines such as European Medicines Agency (EMA, similar to US FDA), European Nanomedicine Characterization Laboratory (EU-NCL). These are still treating nanomaterials and nanomedicines on case by case scenario (Miernicki et al., 2019).

For regulatory purposes of food and cosmetics that contain nanomaterials, these are under the regulatory sector of European Food Safety Authority (EFSA) that regulate the safety assessments of such applications such as nanomaterials as food additives, flavoring, or preservatives, and provides development of the methods of their assessments with harmonizing efforts to exchange any scientific expertise (Rauscher et al., 2017; Sainz et al., 2015).

As a consequence, there is an existing challenge for standardization of methods used to assess the nanomaterials and their applications. This is due to many organizations adopting the nanomaterials, the lack of harmonizing the efforts of worldwide expertise with recommended guidelines and technical development, and the compliance within the scientific/industrial communities with any updates from the regulatory bodies. Further readings can be found in the following sources (Lamon et al., 2020; Miernicki et al., 2019; Rauscher et al., 2017; Rodríguez, 2018; Yildirim et al., 2018).

16.3 NANOPARTICLE INTERACTIONS AT CELLULAR AND MOLECULAR LEVELS

NPs have inherent abilities to overcome the biological barriers and interact with many biological components in the cell in physicochemical-dependent manner.

Cellular uptake can be mediated by various mechanisms; passive or carrier-mediated or endocytosis. Endocytosis pathways involve many processes – pinocytosis and phagocytosis. Pinocytosis is the cellular uptake of small-sized particles and fluids, and includes micropinocytosis and clathrin- and caveolin-dependant and independent mechanisms (Figure 16.1). While the phagocytosis is the uptake of larger debris, particles, and bacteria and is only carried by professional immune cells, such as macrophages and neutrophils. Further details of these different mechanisms can be reviewed in these sources (Aderem and Underhill, 1999; Conner and Schmid, 2003; Donahue et al., 2019; Elkin et al., 2016; Jameson et al., 2019). NP uptake mechanisms are physicochemical dependent processes and cell-type dependent (Mohapatra et al., 2018).

Phagocytosis or "cell eating" is the process of engulfing large-sized particles/debris, dead cells, bacteria, or viruses found in the extracellular environment, with the aim to dispose unwanted materials. It is executed by professional phagocytes, such as macrophages or neutrophils or eosinophils. It is largely triggered by opsonization of the unwanted materials by antibodies or complement proteins (i.e., antibodies: IgG, IgA, IgM, Fcγ, complement proteins [C3, C4, C5], serum proteins that can be found in protein corona; fibronectin, C-reactive protein, type-I collagen) prior to binding to surface receptors (such as Fc, complement) initiating cellular recognition. Engulfing nonopsonized materials by identifying certain molecular patterns (i.e., mannose, C-type lectins such as Dectin-1 or -2, scavenger receptors or fructose receptors, apoptotic receptors: TIM-1, TIM-4, stabilin-2, and BAI-1) also exist (Uribe-Querol and Rosales, 2020). Hence, shielding NPs with stealth polymers, such as PEG, reduces the opsonization process, evades the RES recognition, and increases the lifespan of NPs.

Many studies conducted in vitro and in vivo have demonstrated that MC is size-dependent where NP with a size range below 0.5 μm showed limited clearance but those with size above 1 μm had higher clearance, more than those of size above 6 μm (Champion et al., 2007; Hillaireau and Couvreur, 2009; Nelemans and Gurevich, 2020). In addition to size, the shape and surface properties of the NPs have influence on the MC. Particle shape showed an effect on the initial contact and subsequent progress into phagocytosis (Champion and Mitragotri, 2006). For example, rod or fiber-like NP are very challenging for phagocytosis resulting in frustrated macrophages (Bakand and Hayes, 2016).

After triggering, signalling cascade Rho family GTPases stimulate cell membrane actin filaments condensation into pseudopodia that zipper up enclosing the foreign material forming the characteristic cupping. The vacuole closes around the target by the actin constrictions and very recently it has been discovered with the activation of session protein GTPase dynamin 2, it is released into the cytoplasm (Marie-Anaïs et al., 2016; Susanne and Thomas, 2018). Once enclosed in a phagosome, it is then fused with the lysosomes that starts digesting with acidic hydrolase (such as lipases and esterases, proteases, phosphatases, and nucleases) under acidic pH (Joseph and Liu, 2020; Manzanares and Ceña, 2020; Uribe-Querol and Rosales, 2020).

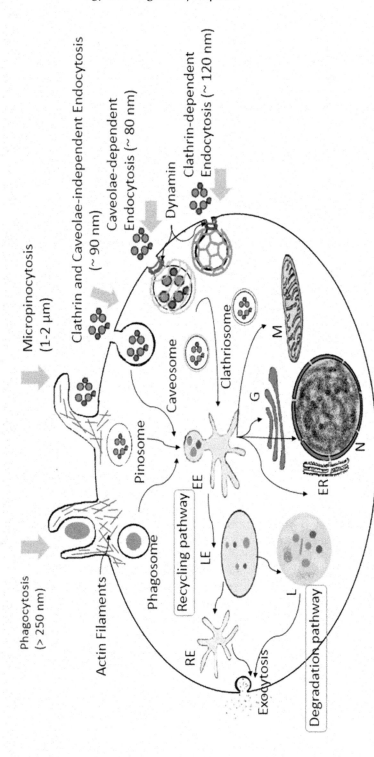

FIGURE 16.1 Main uptake mechanism of NPs and the endocytic pathways. (EE: Early Endosome-low pH, LE: Late endosome-low pH, ER: Endoplasmic reticulum, L: Lysosome-very acidic, N: Nucleus, M: Mitochondria, G: Golgi, RE: Recycle endosome) (A permission has been requested (52)).

Pinocytosis or "cell drinking" is the cellular uptake of small-sized particles and fluids, and subcategorized into macropinocytosis, clathrin-dependent, caveolin-dependant, and clathrin- and caveolin-independent mechanisms.

Macropinocytosis process uptakes a large volume of extracellular fluids (0.5–10 μm) upon activation by growth factors initiating signalling cascade of Rho family GTPase (Kumar et al., 2020). Unlike phagocytosis, the actin filaments condense into membranous protrusions or rufflings around the target and collapse to fuse with the cell membrane generating large macropinosomes that can be partly dependent on dynamin2 or its variants for scission (Park et al., 2013). It is involved in diverse functions, such as immune recognition, cell migration, disposing unwanted materials, and sampling fluids and nutrients from the extracellular environment and ending with lysosomal degradations (Canton, 2018; Foroozandeh and Aziz, 2018).

Clathrin-dependent endocytosis is the most characterized endocytosis process as far it's the most active and constitutively present in all types of cells (Wang et al., 2020). Its role is essential for the cellular hemostasis for nutrients and macromolecule transport such as cholesterol via low density lipoproteins (LDL receptors), and iron via transferrin (Tfn receptors) that are considered as markers for this route of uptake, serum proteins, membranous ion pumps. It has a crucial role in cellular communication during organogenesis, cell signalling regulation by controlling and downregulation of receptor levels, synaptic neuronal transmission (Ca+2-gated channels regulation, recycling of neurotransmitter vesicles), and reabsorption of serum proteins after filtration in kidney tubules (Kaksonen and Roux, 2018; López-Hernández et al., 2020). It is membranous invaginations mediated through clathrin triskelia that coat or cage the incoming vesicles (Conner and Schmid, 2003). It is initiated by a binding of certain ligands to surface receptors at clathrin-nucleation sites/pits with subsequent clathrin-1 protein assembly with the adaptor/assembly protein complexes into clathrin-coated lattices. Dynamin GTPase scission mechanism proceeds around the vesicle neck. Clathrisome disassemble its coatings prior to fusion with the endolysosmes. The molecules/particles size ranges from ~100 to 200 nm (Wang et al., 2020).

Caveolin-dependent endocytosis is a process of internalization of small particles and fluids through membrane invaginations as flask-shaped vesicles. It is commonly abundant in endothelial linings facilitating the extravasation of serum proteins and nutrients to the surrounding tissues. Caveolae are triggered through certain receptors, i.e., serum albumin gp60. Subsequent signalling cascade through Src tyrosine kinase activates phosphorylation and formation of Caveolin-1; a dimeric protein that forms a coating of caveolin striations around certain cholesterol/sphingolipids-rich bindings on the inner membrane layer. Caving-in or invaginating membranous vesicles ensues with the contraction of dynamin and dynamin arrangement by the actin cytoskeleton forming caveosomes (Conner and Schmid, 2003). Caveosomes have neutral pH and bypass the lysosomes protecting its package from degradation with subcellular smooth ER and cytosolic delivery. This route has been used by some pathogens and bacteria to evade lysosomal degradation (Pelkmans et al., 2001), and has been under investigations to enhance NPs internalization (Dauty et al., 2002). Caveolae have many biological roles, such as lipid and cholesterol hemostasis, regulating some cellular cascades, regulating endothelial NO synthase, and transcellular transport of serum albumin and nutrients

(Cohen et al., 2004; Parton and Del Pozo, 2013). Caveolae involve transport small fluid vesicles of ~80 nm (Foroozandeh and Aziz, 2018).

Clathrin- and caveolin-independent endocytosis are many processes and yet they are negatively described as being independent to the aforementioned cornerstone molecules in other endocytosis processes. They involve lipid rafts that are highly organized lipid clusters and cholesterol-rich that exist on the cell membrane exerting endocytosis uptake in absence of clathrin, caveolin, and/or dynamin but mostly requires actin polymerization. Examples are Flotillin, endophilins, clathrin-independent carriers (CLIC/GEEC) pathways have been recently discovered and full understanding is under research (Joseph and Liu, 2020; Sezgin et al., 2017).

The size of NPs is suitable for many uptake and endocytosis mechanisms where NP size around 100–200 nm are endocytosed via clathrin- or caveolin-mediated pathway while NP size > 250 nm up to 3 μm occur via macropinocytosis and phagocytosis (Manzanares and Ceña, 2020; Shin et al., 2015). However, there are many examples of larger or smaller NPs uptaken by various endocytic mechanisms (Manzanares and Ceña, 2020; Shin et al., 2015).

Post uptake of NPs, vesicles will be fused with the early endosome (EE) where a low pH digestive activity can take place degrading the NPs. Endosomes have a complex machinery that allows for NP vesicular sorting, digestion, and degradation, and waste-exocytosis and recycling as well as initiating cellular death in case of toxic NP overload. Full endosomal maturation cycle and their role in NP degradation can be reviewed in these sources (Elkin 2016; Rothen-Rutishauser 2019; Scott et al., 2014). Some NPs can be found without vesicles in the cytoplasm (Bourquin et al., 2018). NPs may translocate to cytoplasm, mitochondria, nucleus, or other cellular organelles and molecules, and may evoke a cytotoxic response (Donahue et al., 2019; Jiang et al., 2010; Shi et al., 2011).

16.3.1 Proposed Mechanisms of NPs Toxicity

NP toxicity is still one of the hot topic areas due to the novelty of their materials that rendered the NP with unique physicochemical properties (Saifi et al., 2018). Cellular injury might vary from trivial reversible injuries recovered by the efficient cellular repair mechanisms to severe or irreversible injuries inducing cell death or long-term adverse effects (Donahue et al., 2019). Various intersecting toxic mechanisms were reported upon exposure to various types of NPs, or even to the same chemical structure NP with variable physiochemical properties (Figure 16.2). Up till now, there is no agreed solid background as to which is the single and the most critical parameter for NP toxicity, for example, is it the size only, or the chemical composition, or the mass? Unlike the same bulk counterparts, the mass dose is not such critical for the toxicity (apart from being overtly overdosed that would generate toxicity anyway) and the full identity of physicochemical properties of the NPs is critical (Xia et al., 2016).

16.3.1.1 Cell Membrane Disruption

Membrane disruption due to NP interactions could be mediated via various mechanisms (Donahue et al., 2019; Farnoud and Nazemidashtarjandi, 2019). Positively

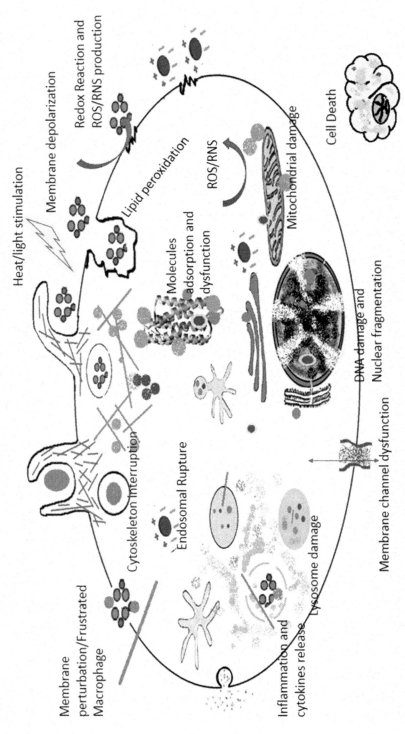

FIGURE 16.2 NP cellular uptake and interactions with different mechanisms of cytotoxicity (A permission has been requested (52)).

charged NPs interact with the negatively charged cell membrane altering, depolarizing or damaging the membrane, i.e., thinning, pore formations, and erosions (Jameson et al., 2019; Yeh et al., 2013). The negative charge NPs could also affect the membrane potential through the electrostatic interactions with the lipids causing lipid leakage (Mu et al., 2014). Cytoskeleton alteration is another mechanism where NPs can functionally block the F-actin and α- or β-tubulins, which are the major functional proteins. This cytoskeleton plays an important role in preserving the cellular shape, motility, adhesion, transport, and cellular division and proliferation. PEGylated NPs were found to reduce this cytoskeleton affection (Tarantola et al., 2009). Polystyrene NPs showed inverse correlation between the size and the membrane interaction in vitro cell membrane model (Accomasso et al., 2016; Peetla et al., 2009).

16.3.1.2 Oxidative Stress

When ROS production exceeds the cellular ability to inhibit, this induces what is called oxidative stress where free radicals will react with the cellular components, proteins, membranes inducing cellular dysfunction, inflammation, lipid peroxidation, mitochondrial shutdown, molecular and DNA damage, and eventually cell death, either through the apoptosis if its less severe insult or through necrosis with severe toxic insults (Donahue et al., 2019; Evans et al., 2019; Mu et al., 2014). Furthermore, almost all types of NPs had showed toxicity that is linked to the production of ROS and this could be measured by measuring free radicals and GSH levels (Anttila et al., 1997). NPs ability to induce the ROS is dependent on its physicochemical properties.

16.3.1.3 Cell Organelles Damage and Mitochondrial Shutdown

NPs' ability to interact with the cellular components and induce membrane damage and ROS will suggest potential toxicity on the mitochondria. Direct NP-mitochondrial interaction was noticed causing lipid membrane damage and leakage in which 3 nm gold NPs were localized into the mitochondrial membranes causing its dysfunction and leakage (Salnikov et al., 2007). Lysosomal damage either by chemical sponge theory or physical needle-like shaped NPs rupturing the lysosome and releasing Ca^{2+} from the endoplasmic reticulum with eventually loss of control of mitochondrial permeability. Lysosomal rupture and enzymatic spillage might activate apoptosis pathways. Other carbon-based and fullerenes are reported to cause mitochondrial damage. The mitochondrial damage with loss of energy production and antioxidant properties will results in high ROS, nuclear and DNA damage, and membrane disruption (Donahue et al., 2019; Saifi et al., 2018).

16.3.1.4 DNA Damage and Mutagenicity

This could be direct NP genotoxic effect or indirect effect through the induction of inflammation, inflammatory mediators or ROS, molecular, and organelle damage (Doak and Dusinska, 2016; Singh et al., 2017). Due to the high surface energy of the NPs, DNA as any other biological molecule is subjected to surface adsorption. This binding will induce both conformational and functional deformity of these biomolecules. Carbon-based NPs were found to induce DNA-double stranded to cleave under light or with the presence of copper ions (Mu et al., 2014). Gold NPs were found to induce conformational changes as relaxation of the DNA strand coils, cleavage of the double strands (Railsback et al., 2012) and functional changes as inhibition of the transcription (McIntosh et al., 2001). Long-term studies are required to investigate NPs-genotoxicity with the development of cancer.

16.3.1.5 Inflammation

Upon macrophage recognition and phagocytosis, inflammatory mediators are released, which initiate an inflammatory response. This is a physio-pathological reaction which in excess or persistence of inflammation could predispose to autoimmune disease, long-term diseases, and cancer (Donahue et al., 2019). When the macrophages are overwhelmed by NP size, shape, chemistry, particle number, and agglomerates or surface group with prolonged pulmonary retention due to slow or incomplete degradation and metabolism, inflammatory response will be evoked. It has been known from the pathogenesis of lung silicosis that long fibers were trapped longer due to defective clearance with subsequent inflammatory induction. Hence, the multiple parameters of NPs physicochemical characteristics play a role in inflammation induction (Braakhuis et al., 2014). Amorphous silica NP induce inflammation and cell death with generation of ROS (Fu et al., 2014). A new potential mechanism involved in the inflammation cascades is the NLRP3 inflammasome activation with rising IL-1β. Certain pathogens, toxins or particles might induce this mechanism with a rise of the IL-1β through activation of caspase 1 and initiation of inflammatory induced apoptosis or known as pyroptosis. Several studies has showed NPs can induce the NLRP3 activation, e.g., some liposomes and polymer-based NPs. A detailed source for this mechanism and its assessment can be found in these following sources (Sharma et al., 2018; Shirasuna et al., 2019).

16.3.1.6 Cell Death Mechanisms

The cell death is commonly the final fate after any irreversible cytotoxic or stress insult. Cell death was known to be either programmed cell death known as apoptosis, and unprogrammed cell death known as necrosis. Very recently, the Nomenclature Committee on Cell Death had adopted new considerations to replace the old cell death classification to the new adopted and emerging mechanisms proved by many studies that have discovered variable mechanisms of the programmed cell death, such as apoptosis, autophagy, ferroptosis, pyroptosis, necroptosis, paraptosis, lysosomal-induced, autoimmune-induced, and many others. These processes have different fundamental mechanisms with different underlying biological and histological picture of the cell death (Mohammadinejad et al., 2019; Tang et al., 2019). The uncontrolled cell death or known as necrosis is commonly a passive accidental cell death due to energy failure or simply bursting in response to severe acute insult. Detailed information for these mechanisms is found in these sources (Rothen-Rutishauser et al., 2019; Tang et al., 2019). NPs can induce complex death mechanism with many intersecting activation of different death pathways with a physicochemical dependant manner (Mohammadinejad et al., 2019). NPs can trigger many intrinsic and extrinsic pathways and induce apoptosis, autophagy, lysosomal-induced cell death. Although some inconsistencies are found in the literature regarding the necrosis where some researchers considered the loss of cell or the decreased viability as necrosis.

16.3.2 NP TOXICITY DETERMINANTS

16.3.2.1 Nanoparticle Size and Surface Area

NPs have intrinsic properties that play a role in their beneficial advantages as well as their toxicity (Figure 16.3 and Table 16.1). Lowering the size to nanoscale (1-1000 nm) is associated with increasing the surface area for the same mass or volume. This explains their high reactivity and strong adsorption properties that play both roles in facilitating their biomolecular interaction and NP–NP interactions (aggregating/agglomerating) as well as their toxicity (Garcia-Mouton et al., 2019).

16.3.2.2 Nanoparticle Surface Chemistry

NPs with specific charge, surface coatings, or modifications are intentionally designed to achieve some pharmaceutical or therapeutic targets, such as increasing the stability of the formulation, increasing the efficiency of protein adsorption with cationic NPs (Kunda et al., 2015; Mohamed et al., 2019), reducing the immune system recognition, prolonging the circulatory half-life (Figure 16.3 and Table 16.1) (Mu et al., 2014). This surface charge is designed using different surface coatings or modifications and measured by determining their zeta potential. For example, positively charged NP are coated with amine groups using cationic emulsifiers, such as 1,2-dioleoyl-3-trimethylammoniumpropane (DOTAP), chitosan (CS), CTAB, DMAB), while negatively charged NP are coated with acidic groups, such as polyvinyl alcohol (PVA), poloxamer188 (PF68), or neutral NP charge (near the neutral or slightly negative charged NPs) (Mura et al., 2011). The NP charge plays a major role in NP biomolecular interactions. The cationic NPs are commonly reported to be cytotoxic than the negatively and neutrally charged NP by interacting

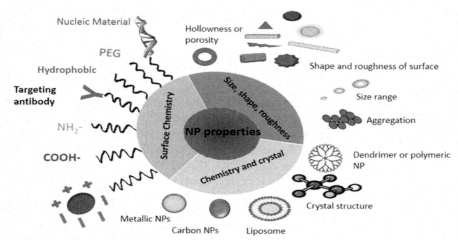

FIGURE 16.3 Nanoparticle physiochemical properties (A permission has been requested (52)).

TABLE 16.1
NP Properties and Possible Cytotoxic Effects

NP Properties	Physicochemical Dependant Cytotoxicity
NP size	• Cellular penetration and crossing tissue barriers • Cellular injury and membrane damage • Escape defence mechanisms and longer retention • Translocation to other organs • Inflammation • Subcellular localization and organelle impairment • DNA damage
NP agglomeration	• Phagocytosis and inflammatory induction • Reduces the uptake
Surface area	• Increased reactivity and toxicity • Increase ROS production
NP shape	• Rod or fibre-like more toxic than rounded shaped NP • Inflammatory induction, membrane, or organelle damage
NP charge	• Positive NP: more cellular interaction and cytotoxicity • Negative NP: might increase or decrease cellular uptake, prone to macrophages clearance
Chemical structure, composition, and purity	• Increased toxicity • Membrane depolarization • ROS generation • Mitochondrial damage • Inflammation and immune modulation • Cellular injury and metabolic impairment • DNA damage • Transporting/adsorbing contaminants and toxins
Poor solubility/ biodurability	• Bioresistant NP induces long term effects as chronic inflammation • Long term effects • Cancer

Source: Osman et al., 2020.

with cell membrane by different mechanisms (Jameson et al., 2019; Leroueil et al., 2008; Mecke et al., 2005; Pang et al., 2020; Yeh et al., 2013) and higher bimolecular interaction and protein adsorption (Mohamed et al., 2019; Kunda et al., 2015) that makes them suitable for delivery of negatively charged molecules, such as RNAs (Haque et al., 2018; Jensen et al., 2012; Mohamed et al., 2019) and different types of macromolecules (Kaye et al., 2009; Kunda et al., 2015; Mohamed et al., 2019; Osman et al., 2018; Rodrigues et al., 2018). However, the surface charge and the high adsorption energy and interaction of NPs were involved in the development of protein corona once in contact with a biological medium (Hu et al., 2013; Monopoli et al., 2011).

NP shape can vary from rounded or oval spheres, ellipsoids, wires or rods, cubes, sheets, cylinders, and many others (Figure 16.3 and Table 16.1). Aspect ratio (AR) is a term to describe nonspherical particles that have a length and a width rather than a diameter. AR is the ratio of length to width (Zhang 2017).

Generally, the spherical NPs are safer and faster to be endocytosed than rod- or fiber-like NPs. Needle-shaped PLGA-PEG NPs showed more cytotoxicity than the spherical particles (similar volume, surface chemistry, and negatively charged, prepared by film-stretching from the spherical NPs) when assessed in vitro (in HepG2 and Hela cell lines): more cytotoxicity evaluated by MTT, loss of LDH, and DNA fragmentation by (TUNEL assay). The cytotoxicity of needle-shaped PLGA NPs were due to damage of the cell membranes and the lysosomal vesicles disruption with upregulating caspase 3/7 and DNA damage and cell apoptosis (Zhang et al., 2017).

16.3.2.3 Nanoparticle Chemistry

Generally, NP chemistry is being referred to parent bulk materials of NP core. It usually used to classify NPs such as organic NP, i.e., carbon-based, lipid-based, polymeric, or metallic NPs, i.e., silver, gold, titanium NPs. However, the exceptional physicochemical properties gained at the nanoscale must be assessed prior to their intended application (Figure 16.3 and Table 16.1). Chemical properties of NPs have an important role in both the pharmaceutical and clinical performances – chemical structure, composition, crystal structure, chirality and isomerism, molecular weight, monomeric ratio, degradability and degradation products, hydrophobicity, drug-loading capacity, drug release kinetics, surface properties, or functionalization.

16.3.2.4 Nanoparticle Aggregation

NP aggregation is another property that depends upon other NPs characteristics and the dispersing media (Figure 16.3 and Table 16.1). It needs to be considered to understand the poor correlation between different toxicity studies, i.e., inhalation or instillation or in vitro studies. NP aggregation could play a double effect; it could increase NPs uptake as larger amount of particles in contact with cellular surface (Limbach et al., 2005), or it could reduce their uptake if NP aggregates are bigger than cellular size to permit the uptake (Drescher et al., 2011). In vivo, it may play an effective role in reducing their toxicity due to easier macrophages clearance (Takenaka et al., 2001). NP aggregates might increase the retention of NPs either in the lung or RES exerting inflammatory potentials (Gatoo et al., 2014).

16.3.2.5 Repeated Dose Exposure

Here, small doses deposited may accumulate inducing chronic effects. Hence, NPs formulations designed for chronic diseases must be evaluated to ensure the safety and suitability for long term treatments. However, it is difficult to study the chronic exposures in vitro and usually preferred to be evaluated in vivo. Some studies tried to study chronic exposure by repeated short-term exposures.

16.3.2.6 The Dose-Dependent Toxicity

The mass dose of the NPs alone is not solely responsible for the toxicity but also the conventional parent bulk molecules. However, interpretation of NP physicochemical induced toxicity should be considered. In vitro to in vivo (IVIV) nanotoxicology studies have shown conflicting results regarding the dose-dependent NP toxicity. It is commonly due to high doses used in conducting these studies to obtain detectable signals to uncover the mechanistic interactions under In vitro conditions. Therefore, discrimination between test conditions with high doses or within the range of the aimed therapeutic doses should be considered (Takenaka et al., 2001).

16.4 NANOTOXICOLOGY ASSESSMENT METHODS

For nanotoxicological evaluation, NPs biocompatibility must be confirmed by two methods; in vitro then progress into the in vivo methods (Table 6) prior to clinical trials. There are different methods that are increasingly getting more popular in the research communities, i.e., ex vivo and in silico methods. Prior to NP testing, thorough physicochemical characterization of NPs is a must. This allows better understanding, easy comparability of different studies, and correlating the NP physicochemical characters with their toxicity or biological profiles that helps better optimization for nanomedical applications (Saifi et al., 2018).

16.4.1 IN VITRO METHODS

In vitro methods using cell-based assays are the initial first step for nanotoxicological evaluation. These provide a faster screen, of lower cost, better experimental control, wider range of materials can be tested in a short time, and more ethically accepted.

These methods usually use enzyme linked assays with a final endpoint such as cytotoxicity (Alamar Blue, Tetrazolium/MTT, Neutral Red, Trypan blue-based assays), internalization, membrane potential (LDH assay), mitochondrial membrane potential (JC1 dye), genotoxicity (micronucleus, COMET, and chromosomal aberrations tests), ROS (DCF Fluorescence assay), and cell death, that discriminates between the NPs compatibility versus toxicity. The cells are usually exposed to NP dispersions within a variable range of time (few hours up to 2 or 3 days) then processing the cells for the endpoint results (Joris et al., 2013).

The choice of the cell is usually dependant on the aimed target tissue, the route of administration, and the intended application. Types of cells used are either primary or secondary cell lines. The secondary cell lines are genetically transformed immortal cell lines, hence called continuous or cancerous cell line. They are more widely used and commercially available from many vendors, i.e., American Tissue Type Culture Collection (ATCC). They provide easier experimental handling and maintenance, faster growth, and shorter experiment time. The primary cell lines are a harvest of a target tissue on special plates with nutritional media (Joris et al., 2013). They are more challenging in isolation, differentiation and phenotyping, maintenance, and special experimental and ethical requirements. As a result, their use is very expensive and onerous testing but more closely representative for the in

Nanotoxicology and Regulatory Aspects

vivo counterparts. The 2D monoculture is a single and classic cell culture model and commonly blamed to lack the multiple cell interactions and cross communications. The co-cultures of different cell lines could provide more accurate results by the resemblance of the different cell mixture within the target tissue (Sayes and Warheit, 2009). 3D models have closer representation of a mixture of different cell types within the tissue and provide the cell-cell interaction and communication with enhanced in vivo correlated results. 3D co-culture models known as spheroids or organoids are grown in a 3D scaffold made of inert materials as collagen, matrigel, or others, supplemented with stem cells or tissue cell mixtures and stimulated by different stimuli to differentiate the cells into target organ structures. "Organ on chip" and microfluidic systems are another 3D co-culture models where the target tissue organoid is subjected to mechanical and physical factors mimicking the biological environment (Rothen-Rutishauser et al., 2019).

16.4.2 Ex Vivo Methods

These methods use recently cut or removed tissues from recently sacrificed animals (for any other cause excluding experimentation)/or human donors. These tissues are maintained under simulated conditions such as excised human skin, e.g., in vitro skin absorption test (OECD TG 428), ex vivo rat lung or human lung slices. They offer better biological representation, better control, and of lower cost. However, still challenges regarding their standardizations are remaining to be overcome. Despite their novelty, these promising studies can provide very complex and realistic information about the NP uptake, transport and mechanism of action (Joris et al., 2013).

16.4.3 In Vivo Methods

The in vivo methods will follow the in vitro/ex vivo methods to investigate more realistic animal exposures and understand a multiorgan body response, i.e., pharmacokinetics and dynamics. The main challenges lie in the interspecies variability, the dose range to humans, and the complex translation to human.

REACH regulations were adopted for NPs evaluation even though they are different from bulk chemicals (Sayes and Warheit, 2009). "When assessing nanotoxicity in vivo, the following aspects ought to be evaluated according to REACH guidelines: acute, subchronic and chronic toxicity, skin and eye irritation or corrosion and skin sensitization, genotoxicity, reproductive toxicity, carcinogenicity and the NPs toxicokinetic". Full detailed reports of the animal conditions, sampling procedures, and toxicity are to be documented (Sayes and Warheit, 2009). (Table 16.2).

16.4.4 In Silico Methods

Recently, in silico models are getting popular. Quantitative structure–activity relationships (QSAR) models are successfully developed for many chemicals employing pre-setted standard of parameters that foresees NPs physicochemical

TABLE 16.2
Advantages and Limitations of Nanotoxicological Assessment Methods

	In Vitro	Ex Vivo	In Vivo	In Silico
Advantages	• Initial Faster screen • High throughput screen • Easy to perform/control • Easier dosing • Immortal continuous cell lines • Mechanistic and toxicity studies • Permeability and uptake studies • Non-animal alternative	• Fast screen • Relatively controlled and easy dosing • Better multicellular and organ response for mechanistic and toxicity study	• Whole body exposure • Biodistribution data • Single or repeated exposure • Acute or chronic toxicity • Short-term and long-term studies	• Predictive ability for mechanistic and toxicity • No animal cruelty • Computer-based studies
Limitations	• Lack of multicellular interactions • Difficult to translate in-vivo • Short term exposure • Lack of standardizations • Difficult to compare between different studies • Variations between primary and immortalized cell lines	• Lack of biodistribution data • Difficult to maintain and handling • Short term exposure	• Training and handling • Expensive and technically demanding • Animal discomfort and cruelty • Labor demanding • Interspecies variability • Sometimes poor human translation	• The availability of enough information that enables the study • Lack of experiments standardizations makes it difficult to compare different results

Source: Osman et al., 2020.

characteristics and their possible biological and toxicological effects. However, successful QSAR models require data availability of accurate and consistent NP grouping, a standardized set of physicochemical characters, mechanisms of action, exposure scenarios, and standardized experimental conditions. This represents a huge challenge giving the novelty of NPs as well as their widespread and fast-growing applications (Joris et al., 2013; Sayes et al., 2007).

16.5 CONCLUSIONS

NPs unique properties have made them very appealing for many applications ranging from nanomedicines to industrial products. These exceptional properties are also involved in their cytotoxicity. The exposure to NPs can be through any relevant route of entry due to the side applications of NPs such as oral, inhalation, dermal, or parenteral routes. Nanotoxicity is commonly reported due to different mechanisms such as cell membrane damage, organelle damage, ROS, inflammation, and genotoxicity. Nanotoxicology is assessed by variety of methods such as in vitro/ex vivo, in silico, and then in vivo methods to uncover NP's computability and the mechanisms of toxicity.

Meanwhile, a lot of efforts are offered by the growing research and awareness of scientific communities and different organizations to standardize methods for characterization, testing conditions, and best laboratory practises to handle NPs and their products prior their availability for customer use, building up reliable data bases for NPs safety and exposure.

REFERENCES

Accomasso L., Gallina C., Turinetto V., *et al.* Stem cell tracking with nanoparticles for regenerative medicine purposes: An overview. *Stem Cells International*. 2016 (2016) 7920358.

Aderem A., Underhill D. M. Mechanisms of phagocytosis in macrophages. *Annual Review Immunology*. 17 (1999) 593–623.

Anttila S., Hukkanen J., Hakkola J., *et al.* Expression and localization of CYP3A4 and CYP3A5 in human lung. *American Journal of Respiratory Cell and Molecular Biology*. 16(3) (1997) 242–249.

Bakand S., Hayes A. Toxicological considerations, toxicity assessment, and risk management of inhaled nanoparticles. *International Journal Molecular Sciences*. 17(6) (2016).

Bourquin J., Milosevic A., Hauser D., *et al.* Biodistribution, clearance, and long-term fate of clinically relevant nanomaterials. *Advanced Materials*. 30(19) (2018) e1704307.

Braakhuis H. M., Park M. V., Gosens I., *et al.* Physicochemical characteristics of nanomaterials that affect pulmonary inflammation. *Particle and Fibre Toxicology*. 11 (2014) 18.

Canton J. Macropinocytosis: New insights into its underappreciated role in innate immune cell surveillance. *Frontiers in Immunology*. 9(2286) (2018) 22–86.

Champion J. A., Katare Y. K., Mitragotri S. Making polymeric micro- and nanoparticles of complex shapes. *Proceedings of the National Academy of Sciences*. 104(29) (2007) 11901–11904.

Champion J. A., Mitragotri S. Role of target geometry in phagocytosis. *Proceedings of the National Academy of Sciences of the United States of America.* 103(13) (2006) 4930–4934.

Cohen A. W., Hnasko R., Schubert W., et al. Role of caveolae and caveolins in health and disease. *Physiological Reviews.* 84(4) (2004) 1341–1379.

Conner S. D., Schmid S. L. Regulated portals of entry into the cell. *Nature.* 422(6927) (2003) 37–44.

Dauty E., Remy J. S., Zuber G., et al. Intracellular delivery of nanometric DNA particles via the folate receptor. *Bioconjugated Chemistry.* 13(4) (2002) 831–839.

Doak S. H., Dusinska M. NanoGenotoxicology: Present and the future. *Mutagenesis.* 32(1) (2016) 1–4.

Donahue N. D., Acar H., Wilhelm S. Concepts of nanoparticle cellular uptake, intracellular trafficking, and kinetics in nanomedicine. *Advanced Drug Delivery Reviews* 143. 2019. 68–96.

Donaldson K., Stone V., Tran C. L., et al. Nanotoxicology. *Occupational and Environmental Medicine.* 61(9) (2004) 727–728.

Drescher D., Orts-Gil G., Laube G., et al. Toxicity of amorphous silica nanoparticles on eukaryotic cell model is determined by particle agglomeration and serum protein adsorption effects. *Analytical and Bioanalytical Chemistry.* 400(5) (2011) 1367–1373.

Elkin S. R., Lakoduk A. M., Schmid S. L. Endocytic pathways and endosomal trafficking: A primer. *Wiener medizinische Wochenschriftien.* 166(7-8) (2016) 196–204.

Evans S. J., Jenkins G. J., Doak S. H., et al. Cellular defense mechanisms following nanomaterial exposure: A focus on oxidative stress and cytotoxicity. *Biological Responses to Nanoscale Particles*, Springer, 2019. 243–254.

Farnoud A. M., Nazemidashtarjandi S. Emerging investigator series: Interactions of engineered nanomaterials with the cell plasma membrane; what have we learned from membrane models? *Environmental Science-Nano.* 6(1) (2019) 13–40.

Foroozandeh P., Aziz A. A. Insight into cellular uptake and intracellular trafficking of nanoparticles. *Nanoscale Research Letters.* 13(1) (2018) 339.

Fu P. P., Xia Q., Hwang H. M., et al. Mechanisms of nanotoxicity: Generation of reactive oxygen species. *Journal of Food and Drug Analysis.* 22(1) (2014) 64–75.

Garcia-Mouton C., Hidalgo A., Cruz A., et al. The lord of the lungs: The essential role of pulmonary surfactant upon inhalation of nanoparticles. *European Journal of Pharmaceutics and Biopharmaceutics.* 144 (2019) 230–243.

Gatoo M. A., Naseem S., Arfat M. Y., et al. Physicochemical properties of nanomaterials: Implication in associated toxic manifestations. *BioMed Research International.* 2014:8 (2014).

Gordon S., Daneshian M., Bouwstra J., et al. Non-animal models of epithelial barriers (skin, intestine and lung) in research, industrial applications and regulatory toxicology. *Altex.* 32(4) (2015) 327–378.

Haque A. K. M. A., Dewerth A., Antony J. S., et al. Chemically modified hCFTR mRNAs recuperate lung function in a mouse model of cystic fibrosis. *Scientific Reports.* 8(1) (2018) 16776.

Hillaireau H., Couvreur P. Nanocarriers' entry into the cell: Relevance to drug delivery. *Cellular and Molecular Life Sciences.* 66(17) (2009) 2873–2896.

Hu G., Jiao B., Shi X., et al. Physicochemical properties of nanoparticles regulate translocation across pulmonary surfactant monolayer and formation of lipoprotein corona. *ACS Nano.* 7(12) (2013) 10525–10533.

Jameson C. J., Oroskar P., Song B., et al. Molecular dynamics studies of nanoparticle transport through model lipid membranes. *Biomimetic Lipid Membranes: Fundamentals, Applications, and Commercialization*, Springer, 2019. p. 109–165.

Jensen D. K., Jensen L. B., Koocheki S., et al. Design of an inhalable dry powder formulation of DOTAP-modified PLGA nanoparticles loaded with siRNA. *Journal of Controlled Release*. 157(1) (2012) 141–148.
Jiang X., Rocker C., Hafner M., et al. Endo- and exocytosis of zwitterionic quantum dot nanoparticles by live HeLa cells. *ACS Nano*. 4(11) (2010) 6787–6797.
Joris F., Manshian B. B., Peynshaert K., et al. Assessing nanoparticle toxicity in cell-based assays: influence of cell culture parameters and optimized models for bridging the in vitro-in vivo gap. *Chemistry Society Reviews*. 42(21) (2013) 8339–8359.
Joseph J. G., Liu A. P. Mechanical regulation of endocytosis: New insights and recent advances. *Advanced Biosystems*. 4 (2020) 190–278.
Kaksonen M., Roux A. Mechanisms of clathrin-mediated endocytosis. *Nature Reviews Molecular Cell Biology*. 19(5) (2018) 313–326.
Kaye R. S., Purewal T. S., Alpar H. O. Simultaneously manufactured nano-in-micro (SIMANIM) particles for dry-powder modified-release delivery of antibodies. *Journal of Pharmaceutical Sciences*. 98(11) (2009) 4055–4068.
Kumar A., Ahmad A., Vyawahare A., et al. Membrane trafficking and subcellular drug targeting pathways. *Frontiers in Pharmacology*. 11 (2020) 629.
Kunda N. K., Alfagih I. M., Dennison S. R., et al. Bovine serum albumin adsorbed PGA-co-PDL nanocarriers for vaccine delivery via dry powder inhalation. *Pharmaceutical Research*. 32(4) (2015) 1341–1353.
Lamon L., Asturiol D., Aschberger K., et al. Modeling of nanomaterials for safety assessment: From regulatory requirements to supporting scientific theories. In *Computational Nanotoxicology: Challenges and Perspectives*. Jenny Stanford Publishing, 2020, 1–97.
Leroueil P. R., Berry S. A., Duthie K., et al. Wide varieties of cationic nanoparticles induce defects in supported lipid bilayers. *Nano Letters*. 8(2) (2008) 420–424.
Limbach L. K., Li Y., Grass R. N., et al. Oxide nanoparticle uptake in human lung fibroblasts: Effects of particle size, agglomeration, and diffusion at low concentrations. *Environmental Science & Technology*. 39(23) (2005) 9370–9376.
Lombardo D., Kiselev M. A., Caccamo M. T. Smart nanoparticles for drug delivery application: Development of versatile nanocarrier platforms in biotechnology and nanomedicine. *Journal of Nanomaterials*. 2019 (2019) 26.
López-Hernández T., Puchkov D., Krause E., et al. Endocytic regulation of cellular ion homeostasis controls lysosome biogenesis. *Nature Cell Biology* 22(7) (2020) 815–827.
Manzanares D., Ceña V. Endocytosis: The nanoparticle and submicron nanocompounds gateway into the cell. *Pharmaceutics*. 12(4) (2020) 371.
Marie-Anaïs F., Mazzolini J., Bourdoncle P., et al. "Phagosome closure assay" to visualize phagosome formation in three dimensions using total internal reflection fluorescent microscopy (TIRFM). *Journal of Visualized Experiments: JoVE*. (114) (2016) 54470.
McIntosh C. M., Esposito E. A., 3rd, Boal A. K., et al. Inhibition of DNA transcription using cationic mixed monolayer protected gold clusters. *Journal American Chemistry Society*. 123(31) (2001) 7626–7629.
Mecke A., Majoros I. J., Patri A. K., et al. Lipid Bilayer disruption by polycationic polymers: The roles of size and chemical functional group. *Langmuir*. 21(23) (2005) 10348–10354.
Miernicki M., Hofmann T., Eisenberger I., et al. Legal and practical challenges in classifying nanomaterials according to regulatory definitions. *Nature Nanotechnology*. 14(3) (2019) 208–216.
Mohamed A., Pekoz A. Y., Ross K., et al. Pulmonary delivery of Nanocomposite Microparticles (NCMPs) incorporating miR-146a for treatment of COPD. *International Journal of Pharmaceutics*. 569 (2019) 18–24.
Mohammadinejad R., Moosavi M. A., Tavakol S., et al. Necrotic, apoptotic and autophagic cell fates triggered by nanoparticles. *Autophagy*. 15(1) (2019) 4–33.

Mohapatra S., Ranjan S., Dasgupta N., et al. *Characterization and Biology of Nanomaterials for Drug Delivery: Nanoscience and Nanotechnology in Drug Delivery*. Elsevier, 2018.

Monopoli M. P., Walczyk D., Campbell A., et al. Physical-chemical aspects of protein corona: relevance to in vitro and in vivo biological impacts of nanoparticles. *Journal American Chemistry Society*. 133(8) (2011) 2525–2534.

Mu Q., Jiang G., Chen L., et al. Chemical basis of interactions between engineered nanoparticles and biological systems. *Chemistry Reviews*. 114(15) (2014) 7740–7781.

Mura S., Hillaireau H., Nicolas J., et al. Influence of surface charge on the potential toxicity of PLGA nanoparticles towards Calu-3 cells. *International Journal of Nanomedicine*. 6 (2011) 2591–2605.

Nelemans L. C., Gurevich L. Drug delivery with polymeric nanocarriers—cellular uptake mechanisms. *Materials*. 13(2) (2020) 366.

Osman N., Kaneko K., Carini V., et al. Carriers for the targeted delivery of aerosolized macromolecules for pulmonary pathologies. *Expert Opinion on Drug Delivery*. 15(8) (2018) 821–834.

Osman N. M., Sexton D. W., Saleem I. Y. Toxicological assessment of nanoparticle interactions with the pulmonary system. *Nanotoxicology*. 14(1) (2020) 21–58.

Pang Y. T., Ge Z., Zhang B., et al. Pore formation induced by nanoparticles binding to a lipid membrane. *Nanoscale*. 12(14) (2020) 7902–7913.

Park R. J., Shen H., Liu L., et al. Dynamin triple knockout cells reveal off target effects of commonly used dynamin inhibitors. *Journal of Cell Science*. 126(22) (2013) 5305–5312.

Parton R. G., Del Pozo M. A. Caveolae as plasma membrane sensors, protectors and organizers. *Nature Reviews Molecular Cell Biology*. 14(2) (2013) 98–112.

Peetla C., Stine A., Labhasetwar V. Biophysical interactions with model lipid membranes: Applications in drug discovery and drug delivery. *Molecular Pharmaceutics*. 6(5) (2009) 264–276.

Pelkmans L., Kartenbeck J., Helenius A. Caveolar endocytosis of simian virus 40 reveals a new two-step vesicular-transport pathway to the ER. *Nature Cell Biology*. 3(5) (2001) 473–483.

Pontes J. F., Grenha A. Multifunctional nanocarriers for lung drug delivery. *Nanomaterials*. 10(2) (2020) 183.

Railsback J. G., Singh A., Pearce R. C., et al. Weakly charged cationic nanoparticles induce DNA bending and strand separation. *Advanced Materials*. 24(31) (2012) 4261–4265.

Rauscher H., Rasmussen K., Sokull-Klüttgen B. Regulatory aspects of nanomaterials in the EU. *Chemie Ingenieur Technik*. 89(3) (2017) 224–231.

Rodríguez H. Nanotechnology and risk governance in the European union: The constitution of safety in highly promoted and contested innovation areas. *NanoEthics*. 12(1) (2018) 5–26.

Rodrigues T. C., Oliveira M. L. S., Soares-Schanoski A., et al. Mucosal immunization with PspA (Pneumococcal surface protein A)-adsorbed nanoparticles targeting the lungs for protection against pneumococcal infection. *PLOS ONE*. 13(1) (2018) e0191692.

Rothen-Rutishauser B., Bourquin J., Petri-Fink A. Nanoparticle-cell interactions: Overview of uptake, intracellular fate and induction of cell responses. *Biological Responses to Nanoscale Particles*, Springer; 2019. p. 153–170.

Saifi M. A., Khan W., Godugu C. Cytotoxicity of nanomaterials: Using nanotoxicology to address the safety concerns of nanoparticles. *Pharmaceutical Nanotechnology*. 6(1) (2018) 3–16.

Sainz V., Conniot J., Matos A. I., et al. Regulatory aspects on nanomedicines. *Biochemical and Biophysical Research Communications*. 468(3) (2015) 504–510.

Salnikov V., Lukyanenko Y. O., Frederick C. A., et al. Probing the outer mitochondrial membrane in cardiac mitochondria with nanoparticles. *Biophysical Journal.* 92(3) (2007) 1058–1071.

Sayes C. M., Warheit D. B. Characterization of nanomaterials for toxicity assessment. *Wiley Interdisciplinary Reviews. Nanomedicine and Nanobiotechnology.* 1(6) (2009) 660–670.

Sayes C. M., Reed K. L., Warheit D. B. Assessing toxicity of fine and nanoparticles: Comparing in vitro measurements to in vivo pulmonary toxicity profiles. *Toxicological Sciences.* 97(1) (2007) 163–180.

Scott C. C., Vacca F., Gruenberg J. Endosome maturation, transport and functions. *Seminars in Cell and Developmental Biology.* 31 (2014) 2–10.

Sezgin E., Levental I., Mayor S., et al. The mystery of membrane organization: Composition, regulation and roles of lipid rafts. *Nature Reviews Molecular Cell Biology.* 18(6) (2017) 361–374.

Sharma B., McLeland, C.B., Potter, T.M., Stern, S.T., Adiseshaiah, P.P. *Assessing NLRP3 Inflammasome Activation by Nanoparticles.* Humana Press, 2018.

Shi X., von dem Bussche A., Hurt R. H., et al. Cell entry of one-dimensional nanomaterials occurs by tip recognition and rotation. *Nature Nanotechnology.* 6(11) (2011) 714–719.

Shin S. W., Song I. H., Um S. H. Role of physicochemical properties in nanoparticle toxicity. *Nanomaterials.* 5(3) (2015) 1351–1365.

Shirasuna K., Karasawa T., Takahashi M. Exogenous nanoparticles and endogenous crystalline molecules as danger signals for the NLRP3 inflammasomes. *Journal of Cellular Physiology.* 234(5) (2019) 5436–5450.

Singh N., Nelson B. C., Scanlan L. D., et al. Exposure to engineered nanomaterials: Impact on DNA repair pathways. *International Journal of Molecular Sciences.* 18(7) (2017) 1515.

Susanne E., Thomas F. R. Modulation of dynamin function by small molecules. *Biological Chemistry.* 399(12) (2018) 1421–1432.

Takenaka S., Karg E., Roth C., et al. Pulmonary and systemic distribution of inhaled ultrafine silver particles in rats. *Environmental Health Perspectives.* 109 Suppl 4 (2001) 547–551.

Tang D., Kang R., Berghe T. V., et al. The molecular machinery of regulated cell death. *Cell Research.* 29(5) (2019) 347–364.

Tarantola M., Schneider D., Sunnick E., et al. Cytotoxicity of metal and semiconductor nanoparticles indicated by cellular micromotility. *ACS Nano.* 3(1) (2009) 213–222.

Uribe-Querol E., Rosales C. Phagocytosis: Our current understanding of a universal biological process. *Frontiers in Immunology.* 11 (2020) 1066.

Ventola C. L. Progress in nanomedicine: Approved and investigational nanodrugs. *P & T: A Peer-reviewed Journal for Formulary Management.* 42(12) (2017) 742–755.

Wang X., Chen Z., Mettlen M., et al. DASC, a sensitive classifier for measuring discrete early stages in clathrin-mediated endocytosis. *eLife.* 9 (2020) e53686.

Xia T., Zhu Y., Mu L., et al. Pulmonary diseases induced by ambient ultrafine and engineered nanoparticles in twenty-first century. *National Science Review.* 3(4) (2016) 416–429.

Yeh Y. C., Saha K., Yan B., et al. The role of ligand coordination on the cytotoxicity of cationic quantum dots in HeLa cells. *Nanoscale.* 5(24) (2013) 12140–12143.

Yildirim S., Röcker B., Pettersen M. K., et al. Active packaging applications for food. *Comprehensive Reviews in Food Science and Food Safety.* 17(1) (2018) 165–199.

Zhang B., Lung P. S., Zhao S., et al. Shape dependent cytotoxicity of PLGA-PEG nanoparticles on human cells. *Scientific Reports.* 7(1) (2017) 1–8.

Index

"3d print drugs" and "3d print medicine," number of published papers, 118f
3D printing, 98–99
3D printing in the pharmaceutical industry, 117–121
3D-printed nanocrystals for oral administration of drugs: 3D printing in the pharmaceutical industry, 117–121; challenges of oral drug delivery, 110–111; conclusions and future prospects, 131; electromagnetic radiation-based printing systems, 128–131; incorporation of nanocrystals in oral solid dosage forms, 116–117; oral solid formulations of nanocrystals by #DP, 121–128; pharmaceutical nanocrystals, 111–113; techniques for obtaining and post-processing nanocrystals, 113–116
3DP printing methods, classification of, 120f
3DP stages of, 118–120, 119f
3DP types of, 120–121
3DP usage, advantages of, 117–118
5-fluorouracil (5-FU), 233

A

Abellan-Llobregat, A., 140–141
Abraxene, 312
acrylic acid (AA), 88
acrylonitrile (AN), 88
active ingredients (AIs), 182
active targeting, 157–158
acute lung injury (ALI), 242
adsorptive mediated transcytosis (AMT), 63–64
agglomeration, 356–357, 359
aggregate shape, 333–336
aggregation, 359; see also NP toxicity determinants
aggregation number, 333
albumin, 52
albumin-based bioinspired nano-formulations, 100–101
albumin-based nanoparticles, 311
alginate, 52
allotropes, carbon, 142–143
alpha-1 antitrypsin deficiency, 14
anticancer drugs, 13, 88, 95–96, 182, 201, 288, 292

anticancer therapy see cancer therapy
antiepileptic oral drugs, 261
antigen-presenting cells (APC), 300
aqueous solubility, 110
armchair (electrical conductivity > copper) (SWCNTS), 143–144
artificial microneedle (MN), 102
atomic force microscopy (AFM), 374

B

baicalin liposomes in various rabbit tissues after intravenous administration, distribution of, 246f
baicalin-loaded nanoliposomes, 246
B-cell lymphoma, 219
ß-cyclodextrin-gemini surfactant conjugates, 337f
bead milling, 43
bioactive silicate, bioinspired nano-formulations, 96–97
bioavailability, defined, 110
biocompatibility, CNTS, 289
bioinspiration in drug delivery, approaches of, 87–92
bioinspired materials, 93–94
bioinspired materials, viral origin, 95–96
bioinspired nano-formulations: approaches to in drug delivery, 87–92; background, 86–87; of bacterial origin, 94–96; bioinspired materials, 93–94; bioinspired nano-formulations, development of, 99–101; bioinspired nano-formulations, fabrication methods of, 97–99; bioinspired nano-formulations, types of, 96–97; bioinspired-based nano-formulation for transdermal delivery, 101–102; of microbial origin, 94
bioinspired nano-formulations, flow diagrams of 87s
bioinspired nano-formulations, types of, 96–97
bioinspired nano-formulations, typical examples to prepared, 98f
bioinspired nano-formulations (BioIns-NFs), 102–103
bioinspired nano-formulations for drug delivery, development of, 99–101
bioinspired nanomaterials, microbial origin, 94–95

403

bioinspired shielding strategies, representative examples, 92f
bioinspired shielding strategies for nano-formulations, 9–92
bioinspired stimulus responsive release, 88f
bioinspired stimulus-responsive release, 88–90
biomimetic movements of bioinspired nano-formulations, 92
biomimetic vesicles, 100
biomimicking of inner architecture, 92
biomolecule responsiveness as endogenous stimuli, 89
bionanotechnology, 356
biotemplates and biomimetics, bioinspired nanomaterials, 94
biotransformation, 358
blood brain barrier (BBB), 15, 56–64; see also nose-to-brain drug delivery, nanostructured drug carriers for
blood brain barrier (BBB), schematics of, 63f
blood brain drug delivery nanosystems, pharmacokinetics summary, 2, 65–69t
blood cells, 100
Bohr radius, 142
bottom-up technique, NCs preparation, 41, 113–115
bovine serum albumin (BSA), 100–101
brain, drug delivery to, 15, 56–69
brain cancers, 259–260
brain diseases, common, 258–263
brain drug delivery of nanoparticles, pharmacokinetics of, 63–64
breast cancer, 294, 298, 314–315
breast cancer treatment, silica-based nanomaterials for combinatorial drug delivery: breast cancer, 314–315; introduction to, 311–312; nanoparticles for combination therapy, 315–320; silica-based nanomaterials, 312–313; synthesis and functionalization of silica-based nanoparticles, 313–314
"bricks and mortar," stratum corneum and keratinocytes, 54–56
Brodie method, 211, 213
brome mosaic virus (BMV)-like particles, 95
Brust-Schiffrin scheme, 8
Buckminsterfullerene (C60), 144–145
buckyballs, 144–145

C

cancer diagnosis, 221
cancer stem cell, 294
cancer therapy see carbon nanotubes in cancer therapy; carbon nanotubes in cancer therapy; lung targeting; nose-to-brain drug delivery, nanostructured drug carriers for; amphiphilic-formed liposomes and polymeric nanosystems, 56; breast cancer, 314–315; CNT in cancer vaccine, 300–303; CNTs for oral cancer therapy, 300; graphene-based nanomaterials, 218–221; lipid-based nano-formulations, 100; MOF as drug carrier, 192; nanoscale uses, 197; semiconductor quantum dots (QDs), 142; smart systems in, 162
cancer therapy using CNTs, merits of, 297–298
cancer vaccines, 300–303
cancer-causing cells, 13–14
carbon allotropes, 143–144
carbon nanodiamonds (CNDs), 147
carbon nanofibers, 136
carbon nanoonions, 147–148
carbon nanotubes (CNTs), 13, 136, 143–144
carbon nanotubes (CNTs), defined, 287
carbon nanotubes (CNTs), properties: biocompatibility, 289; cellular toxicity, 290–292; cellular uptake, 288–289; functionalization, 289–290
carbon nanotubes in cancer therapy: CNT in cancer vaccine, 300–303; CNTS in cancer therapy, 292–297; CNTs or oral cancer therapy, 300; introduction to, 287–288; merits of cancer therapy using CNTs, 297–298; photodynamic therapy (PDT) using CNT, 299–300; photothermal therapy (PTT) using CNT for cancer, 298–299; properties, 288–292
carbon onions, 136, 147–148
carbonaceous NPs, 159
carbon-based nanomaterials, 136, 142–143
casein, 52
cationic gemini surfactants, 339–342
cationic gemini surfactants as genes and drug carriers: aggregated shape, 333–336; aggregation number, 333; delivery of nucleic acids, 338; effect of head region (head group and spacer), 338–345; effect of the hydrophobic tail, 345–347; introduction to, 329–330; physicochemical properties, 331–333
cationic polyethyleneimine (PEI) nanoparticles, 16
Caveolin-dependent endocytosis, 386–387
CD44 targeted alpha-tocopheryl succinate, 292
CD133 aptamer modified docetaxel liposomal formulation (CD133-DTX-LP), 246
cefquinome-loaded poly lactic-co-glycolic acid microspheres, electron microscopy, 242f

Index

cell death mechanisms, 390–391
cell membrane disruption, 387–389
cell organelles damage and mitochondrial shutdown, 389
cell-surface membranes, adhesion to, 112–113
cellular toxicity and CNTS, 290–292
cellular uptake, CNTs, 288–289
cellular uptake efficiency of RuPOP@ MWCNTs, 296f
cellulose, 52
cellulose nanocrystals (CNC), 126
central nervous system (CNS), 258
cerium oxide, 137
cetyltrimethylammonium bromide (CTAB), 313
chemical structure of gemini surfactants and their CMC values, 332t, 336–338
chemical structure of gemini surfactants used as gene delivery vectors, 340t
chemo-photothermal treatment, 220
chip-based nanoparticles, 3
chiral (SWCNTS), 143–144
chitosan (CS), 3, 219
chitosan glycosylated-cationic polyethyleneimine nanoparticles, 16, 52, 201
chondroitin sulfate-modified MWCNTs, 292
chromium, 141
CIP2A, 300
cisplatin, 300, 316
cisplatin niosomes, 247
Clathrin- and caveolin-independent endocytosis, 387
Clathrin-dependent endocytosis, 386
cleavage of acid-labile bonds, 162–164
cleavage of acid-labile linkages, 164f
clinical trials with intranasal approach for common brain diseases, 262–263t
clusters, 136
CNS disorders, 265
CNS pathological conditions, treatment of, 277–278
CNS pharmacological therapies, clinical failure of, 261
CNS systemic circulation, 56
CNT functionalization, outcomes of several, 291t
CNT in cancer vaccine, 300–303
CNTs, noncovalent surface modifications, 290
CNTs for oral cancer therapy, 300
CNTs in cancer therapy, 292–297
coacervation, 3
cobalt, 141
cognitive desire, 259
cold HPH method, 38–39
composite-based nanomaterials, 136
confinement methods (top-down techniques): high-pressure homogenization, 42–43; media milling, 43–44

contact lenses, nanomaterials-loaded, 15
controlled radical polymerization (CRP), 28–29
controlled radical polymerization (CRP), main types of, 28f
conventional drug delivery vehicles, limitations of, 16
coordination modulation, MOFS, 185–186
copper (Cu), 137
copper (Cu) oxide, 137
core-shell lipid-polymer hybrids SLNs, 100
core-shell MOF nanocomposites, 197
corneal diseases, 15
covalent functionalization, 289–290
covalent modification, MOFS, 185–186
cowpea chlorotic mottle virus (CCMV)-like particles, 95
CPt microspheres, 242
critical micellar concentration (CMC), 313, 320, 331–333
curcumin micellar nanoparticles (Cur-NPs), 244, 316
cyclodextrin, 3
cyclodextrin-based gemini surfactants, 337–338
cystic fibrosis, 14
cytoskeleton alteration, 389
cytotoxicity measurement of DCA-loaded Zr-fum composite via MTS assay against, 200f

D

DDS due to trio stimuli developed on CP5-capped UiO-66, schematic illustration, 193f
DDSs (thermally triggered decarboxylation mechanism), 201
dendrimer, basic structure of, 11f
dendrimer, defined, 248
dendrimer-mediated lung targeting, 248–249
dendrimers, 2, 11–12, 49, 52–53, 70, 71f, 136, 159, 248, 248f, 250, 258, 271, 271t, 273t, 276–278
dendrimers as multifunctional nanoplatforms, 248f
deposition, 356
detergent removal method, 8–9
DEX, 218
dextran, 3
diabetes mellitus, 14
diabetic retinopathy, 15
diagrammatic working of DOX/Fe(bbi)@ SiO2–FA, 188f
dialysis, 3, 32
dialysis method, PNPs formulation, 32f
diamond, 145
different mechanisms through which a drug can cross the BBB, overview of, 261f

different polymers used for the preparation of nanoparticles, advantages of, 4t
different types of single-walled nanotubes, 144f
dilution method, 9
dip transfer method, 97
direct pathways of nose to brain administration with the nasal cavity and drug delivery after an intranasal administration, 264f
DNA damage and mutagenicity, 389–390
DNA nanotubes, 13
docetaxel-lecithoid nanoparticles (DTX-LN), 243
docetaxel-lecithoid nanoparticles and docetaxel injection in rabbit, biodistribution of, 243f
dodecyltrimethylammonium bromide (DTAB), 367
double emulsification, 3–4, 36
DOX, 316, 318
Doxil (PEGylated liposome-encapsulated doxorubicin), 55, 312
doxorubicin, 233, 290, 294
doxorubicin-loaded poly (butyl cyanocrylate) nanoparticles, 233
doxorubicin-PAMAM dendrimer, 248
drop-on-drop deposition, 122–123
drop-on-solid deposition, 122
drug delivery *see* nanomaterials
drug delivery, application of nanomaterials, 13–14
drug delivery, nanoparticles in, 41f
drug delivery through stimuli based MOFS: ion-receptive MOFs, 189; magnetic MOFs, 188–189; multiple stimuli-responsive MOFs, 191–194; single-stimulus-based MOFs for drug delivery, 187–188; temperature-sensitive MOFs, 189–191
drug delivery to organs, 198–200
drug delivery to the brain, 15
drug nanocrystals (NCs), 24
drug NPs, 24
drug particle confinement by media milling, schematic representation, 43f
drug precipitation (bottom-up techniques), 42
drug release from enzyme-sensitive nanoparticles, 165f
drug release from photo-sensitive nanoparticles, mechanics of, 168f
drug release from pH-sensitive DDS, mechanism of: enzyme-sensitive nanoparticles, 164–165; ionic microenvironment-sensitive nanoparticles, 165–166; protonation or deprotonation, 162–164; redox-sensitive nanoparticles, 164
drug release from pH-sensitive nanoparticulate DDS composed of polyacidic polymer, mechanics, 163f

drug release from trigger-sensitive nanoparticles, modes of, 159
drug release from ultrasound-sensitive nanoparticles, mechanism, 170f
drug saturation concentration, increased, 112
dry powder inhalers (DPIs), 233–234
dual chemo-photothermal therapy, 220
dual/multi-responsive DDS, 170–171
dynamic light scattering (DLS), 374

E

EGF-EGFR (epidermal growth factor receptor) system, 14
egg white (ovalbumin), 100–101
Eituxan, 219
electric double layer (EDL), 360
electric field-sensitive nanoparticles, 169–170
electro responsive as exogenous stimuli, 90
electromagnetic radiation-based printing systems, 128–131
electrostatic stabilization of the nanoparticles, schematic representation, 360f
ELISA technique, 300
emulsion polymerization, 26–27
emulsion polymerization technique, description of, 26f
emulsion-diffusion, 3
emulsions, 50–52
emulsion-solvent evaporation, 3
encapsulation of ibuprofen within MOF-74-Fe(II) via oxidation, 190f
encephalitis, 275
endocytosis, 289
endogenous stimuli responsive (host environment-responsive), 88–90
endogenous triggers, 160
endogenous trigger-sensitive nanoparticles, 161
endosome, 289
endothelial cells, 244
engineered nanoparticles, 53–54
enhanced permeability and retention effect (EPR), 10, 220
entinan–MWCNTs, 301
enzymatic degradation, 289
enzyme responsiveness as endogenous stimuli, 89–90
enzyme-sensitive nanoparticles, 161, 164–165
epidermal growth factor (EGF), 300
epilepsy, 258–259
EPR effect (enhanced permeability and retention effect), 115, 56–157
erythrocyte membrane, 201
E-selectin, 242
estradiol-PEG MWCNTs, 294
eternal modification, MOFs, 187

Index

ethosomes, 52
ethylenediaminetetraacetic acid (EDTA), 367
ethylprednisolone-polyamidoamine dendrimer (PAMAM G4-OH) complex, 249
European Chemical Agency (ECHA), 383
European Food Safety Authority (EFSA), 383
European Medicines Agency (EMA), 383
exogenous stimuli responsive (external stimuli-responsive), 90–91
exogenous triggers, 161
exogenous trigger-sensitive nanoparticles: electric field-sensitive nanoparticles, 169–170; magnetic field nanoparticles, 167–168; photosensitive nanoparticles, 166–167; temperature-sensitive nanoparticles, 167; ultrasound-sensitive nanoparticles, 170
exopolysaccharides (EPSs), 95
extrinsic semiconductors, 142

F

fabrication methods for bioinspired nano-formulations: bottom-up approach, 98–99; top-down approach, 97–98
fatty acids, 4
FDA regulatory guidelines, 382
flexible MOF theranostic nanoplatform, 196–197
fluidized bed granulation, 116–117
fluorescent and magnetic MWCNTs, 297
folic acid (FA)-modified graphene oxide, 219
freeze casting technique, 98–99
freeze-drying, 116
fullerenes (C60), 2, 136, 142–143
functional nanomaterials (FNs), 148; carbon-based nanomaterials, 142–148; inorganic-based nanomaterials, 137–142; introduction to, 136
functionalization, CTNS, 289
functionalization of GO to prepare p-CCG, 216f
fundamental principles of inkjet techniques, 121f
fused deposition modeling (FDM), 123–126

G

gadolinium, 141
gallium arsenide, 142
Gd-MTX(methotrexate) nanoscale coordination polymeric material, 184
gelatin methacrylate (GELMA), 129
gemini surfactant, 329–347
gemini surfactant, effect of head region (head group and spacer): amino acids head groups, 344–345; glycosylated head groups, 343–344; heterocyclic head group, 342–343

gemini surfactant, effect of hydrophobic tail: *Aliphatic Tail Length*, 345–346; asymmetry, 346–347; degree of unsaturation and stereochemistry, 346
gemini surfactants and their aggregation number values, chemical structure of, 334t
gene delivery, nanotides for, 14
gene therapy, 14
general structure of gemini surfactants displaying the ionic head groups, hydrocarbon tails, and the spacer region, 330f
"Generally Recognized as Safe," silica as, 312
Geobacter, 358
germanium, 142
glaucoma, 15
glucose, 201
glucose dehydrogenase (GDH), 186
glutamic acid derivatives, 159
glycodendrimers, 12
GO sheets, 222
GO-based smart drugs, 219
gold (Au), 137
gold nanocages, 139–140
gold nanoparticle-based electrochemical sensor and biosensor, 139–140
gold nanoparticles, 5–8, 139–140
gold nanoparticles, attached with surface ligands, 8f
gold nanoshells, 139–140, 311
gold NPs, 159
GO-PEI, 219
GO-sheets, 219
graphene, 142–143, 145–146
graphene, history of, 209–210
graphene oxide, 146, 210, 221
graphene oxide, functionalization of, 214–216, 222
graphene oxide, preparation of, 211–213
graphene oxide, silanization of, 215f
graphene oxide (GO)-nanoflakes, 319
graphene oxide in nanomedicines, 216–218
graphene oxide preparation using Hummers–Offeman method, schematic presentation, 213f
graphene quantum dots (GQDs), 136, 142–143, 146–147
graphene sheets, rolled into CNTs, 287
graphene-based materials, toxicity of, 223
graphene-based materials in cancer diagnosis, 221
graphene-based materials in cancer therapy, 218–221
graphene-based nanomaterials: functionalization of graphene oxide, 214–216; graphene oxide in nanomedicines, 216–218; graphene-based materials in cancer diagnosis, 221; graphene-based

materials in cancer therapy, 218–221; introduction and history of graphene, 209–210; preparation of graphene oxide (GO), 211–213; structure and composition of graphene oxide, 210–211
graphite, 145
graphite or graphene oxide, different structures of, 212f
graphite oxide, 210
graphitic salt, 217
Guillain-Barré syndrome, 260

H

HAS transferrin and lactoferrin, 100–101
head region, gemini surfactants, (head group and spacer), 338–345
headache, 258
hemochromatosis, 14
hepatocellular carcinoma, 296
HER2-positive and basal-like subtypes, breast cancer, 314
heterogeneous nucleation, 27
high-energy methods, 37
high-pressure homogenization (HPH), 37–39, 42–43, 115
Hofmann model, 210
homogenization, SNLs preparation, 4
homo/hetero aggregation, 356
hot HPH method, 38
hot or cold homogenization, SLN, NLD creation, 38
hot-melt extrusion (HME), 126
HSA-based complexes, 101
HSA-coated complexes, 101
HSA-dye complexes, 100–101
human serum albumin (HAS), 100–101
Hummer's method, 210, 212, 217, 222
hyaluronic acid, 300
hybrid nanoparticles, 159
hybrid silica nanoparticles, 319–320
hydrocortisone acetate NCs, 116–117
hydroxypatite, bioinspired nano-formulations, 97

I

imidazolium, 342
impact of increasing the number of terminal lysine moieties on molecular packing parameter, 336f
incorporated carboxyl-functionalized diiodo-substituted BODIPY, 186
incorporating nanocrystals into FDM-printed solid dosage forms, methods of, 125
inflammation, 390
inhalation delivery of nanodrugs, 72–78

inhalational therapy, need for, 229–230
inhaled particles, 231–233
inkjet-based printing systems, 121–122
inorganic elements, 182
inorganic nanomaterials, 316
inorganic nanotubes, 13
inorganic-based nanomaterials, 136, 159; carbon nanodiamonds (CNDs), 147; carbon nanotubes (CNTs), 143–144; carbon onions, 147–148; carbon-based nanomaterials, 142–143; fullerenes (C60), 144–145; graphene, 145–146; graphene oxide, 146; magnetic nanoparticles, 141; metal and metal oxide-based nanomaterials, 137–139; nanogold, 139–140; nanoplatinum, 140–141; nanoporous-activated carbon, 148; nanosilver, 140; semiconductor quantum dots (QDs), 141–142
inorganic-based nanomaterials, classification of by composition, 138f
inorganic-based nanoparticles, 53
interfacial polymerization (IP), 27–28
internalization of SWNT constructs into DCs by confocal microscopy, 301f
intranasal drug administration see nose-to-brain drug delivery, nanostructured drug carriers for
intranasal oil-in-water nanoemulsion, 268
intrinsic semiconductors, 142
invasomes, 62t
ion responsiveness as endogenous stimuli, 89
ionic gelation, 3
ionic microenvironment-sensitive nanoparticles, 161, 165–166
ion-receptive MOFs, 189
IP interfaces, different kinds of, 27f
iron oxides, 137, 141
iron sulfate NPs, 159
isorecticular MOF (IRMOF-3), 196

K

Kaposi's sarcoma, 55

L

large unilamellar vesicles, 8–9
laser engraving, 98
LBL technique, 98
lectin-modified solid lipid nanoparticles, 50–52
Lerf-Klinovski's model, 210
ligand-mediated lung targeting, 249–250
lipid micelles, 35
lipid nanocapsules (LNC), 34
lipid nanomaterials, 316

Index

lipid nanoparticles *see* solid lipid nanoparticles (SLNs)
lipid nanosystems, 50–52
lipid-based bioinspired nano-formulations, 99–100
lipid-based nanocarriers, 24, 34–36, 278
lipid-based nanocarriers, schematic representation of, 35f
lipid-based NP, 272–276
Lipofectamine Plus (transfection agent), 341
liposome nanocarriers, active and passive targeting, 10f
liposome-mediated lung targeting, 245–247
liposomes, 8–11, 34, 49–52, 55f, 56, 63, 100, 159, 258, 271, 273t, 311
liposomes, advantages and disadvantages of, 9t
liposomes, defined, 245, 273
liposomes micro/nanoemulsions, 271t
liquid/monomer-in-liquid (Lm-in-L), 27–28
liquid/monomer-in-liquid/monomer (Lm-in-Lm), 27–28
liquid/monomer–solid (Lm–S), 27–28
low-energy methods, 36
lower critical solution temperature (LCST), polymers with, 89
lung targeting, advantages and disadvantages, 230f
lung targeting, advantages of, 234
lung targeting, challenges to, 235–236
lung targeting, micro- and nanocarrier-mediated lung targeting: dendrimer-mediated lung targeting, 248–249; ligand-mediated lung targeting, 249–250; liposome-mediated lung targeting, 245–247; microspheres-mediated lung targeting, 236–250; nanoparticle mediated lung targeting, 243–245; niosome-mediated lung targeting, 247–248
lung targeting, microspheres-mediated, 236–242
lung targeting, nanomaterials for: anatomy of lungs and drug deposition, 230–233; future perspective, 250; introduction to, 229–230; methods of drug delivery in lungs, 233–236; micro- and nanocarrier-mediated lung targeting, 236–250
lung targeting, nanoparticle-mediated, 243–245
lungs, anatomy of and drug deposition: fate of inhaled particles, 231–232; inhaled or targeting chemotherapies in lungs, history of, 233
lungs, method of drug delivery: challenges in lung targeting, 235–236; introduction to, 233–234; lung targeting, advantages of, 234
lungs, size-dependent deposition of particles, 231f

lyophilization-based stability, 371
lysosome, 289

M

macropinocytosis, 386
"magic bullet", 156
magnetic and/or metallic nanoparticles, 53
magnetic core-shell MOFs, 196–197
magnetic field, diagnostic usage, 169
magnetic field nanoparticles, 167–169
magnetic field-sensitive nanoparticles, mechanism of, 169f
magnetic MOFs, 188–189
magnetic nanoparticles, 3, 141
magneto-responsiveness as exogenous stimuli, 90–91
main carbon-based nanomaterials, schematic illustration, 143f
main uptake mechanism of NPs and the endocytic pathways, 385f
manganese, 141
Marangoni effect, 31
MCF-7 cells viability towards different doses of MIL-88B-on-UIO-66, 199f
MDR inhibitors, 316
mechanical dispersion method, 8–9
media milling, 43
melphalan, 337–338
melting solidification printing process (MESO-PP), 126–128
membrane nanotube, 13
meningitis, 275
mesoporous silica nanomaterials, 358–359
mesoporous silica nanoparticles (MSNs), 312, 358–359
mesoporous silicas NPs, 159
metal and metal oxide-based nanomaterials, 137–139
metal organic frameworks, characteristics of: particle size, 184; pore size of MOFS, 184; rationality and biodegradability of MOFS, 184–185; toxicological compatibility of MOFS, 185
metal organic frameworks (MOFs), 182
metal organic frameworks for drug delivery: challenges and future perspectives, 200–220; characteristics of metal organic frameworks, 183–185; drug delivery through stimuli based MOFS, 187–194; drug delivery to organs, 198–200; functionalization of MOFS, 185–187; introduction to, 182; multifunctional MOFS, 194–198; synthesis of MOFS and MOF nanoparticles (NMOFS), 183

metal oxide nanomaterials, 137
metallic nanoparticle, 271t
metallic nanoparticles, bioinspired nanoformulations, 96
"metalloid staircase", 142
metastatic melanoma, 337–338
methacrylic acid (MAA), 88
methyl viologen salts, 192f
methylene green (MG), 186
micelles, 34, 49, 53, 100, 159, 172f, 271, 271t, 278, 311, 313, 316
micro emulsion technique, SNLs preparation, 4
micro vesicles (MVs), 94–95
microbial origin bioinspired nanomaterials, 94–95
microemulsion and double emulsion methods, 36
microemulsions, 69t, 258
microfluidics, 29
microfluidics systems, scheme of, 30f
microfluidics-assisted nanofabrication, 29–30
microporous MOF, 183
microRNA (miRNA), 317
microspheres, defined, 236
microspheres-mediated lung targeting, 236–242
mitochondria, 289
mitochondria-targeted transportation of P-D-CS-CNTs in MB49 cell, illustration of, 298t
MM-398, 312
MOF-based drug administration, 201
MOFS, functionalization of: coordination modulation, 186; covalent modification, 185–186; external modification, 187; noncovalent modification, 186; surface modification, 185
MOFS and MOF nanoparticles (NMOFS), synthesis of, 183
molecular imaging techniques, 221
monomers, natural, 11
mononuclear phagocyte system (MOPS), 312
motor symptoms with Parkinson's, 259
mucoadhesive and surfactant agents for nose-to-brain formulations improvement, recent studies, 269f
mucoadhesive excipients, 278
multi-drug resistance (MDR), 316
multifunctional MOFS: core-shell MOF nanocomposites, 197; flexible MOF theranostic nanoplatform, 197–198; magnetic core-shell MOFs, 196–197; polymer mantle on MOFs, 194–196
multilamellar vesicles, liposomes as, 8–9
multiphase flow system (MFS), 29–30
multiple sclerosis, 258
multiple stimuli-responsive MOFs, 191–194
multistep template molding, 97

multi-walled CNTs (MWCNT), 143–144
mutual acceptance of data (MAD), 382
MWCNT, transferrin conjugated, 295f
MWCNTs in PTT, effectiveness of, 299
myeloperoxidase, 289

N

NaDDBs, 367
nanocapsules, 2, 25, 28, 34–35, 125, 172f
nanocarriers, 312
nanocoating, 136
nanocomposite film, 172f
nanocrystals, 2, 40–41, 53, 75t
nanocrystals, techniques for obtaining and post-processing, 113–116
nanodrug for lung delivery, ideal parameters for, 78f
nanoemulsion (NE), 34, 49, 59t, 60t, 73–74t, 100, 159, 258, 271, 273t, 278
nanoemulsion (NE), defined, 275
nanoemulsions by high-pressure homogenization and ultrasonication, schematic representation of, 38f
nanofibers, 136, 271t
nanofilms, 136
nanogel, 172f
nanogold, 139–140
nanoiron (nano Fe), 141
nanolipid carrier (NLC), 59t, 61t, 67t
nanoliposomes, 2, 10–11, 14
nanomaterial classification by dimension, 137f
nanomaterials, classification of by composition, 138f
nanomaterials, stability of: factors influencing the stability of nanoparticles, 359–360; introduction to, 355–356; lyophilization-based stability, 371; nanoparticles surface chemistry, 361–370; shelf-life (aging), 371–373; stability measurements of nanoparticles, 373–374; stability tests, 371; stabilization mechanisms, 360–361; transformation processes, 356–359
nanomaterials, types of, 3
nanomaterials used for the lung-mediated delivery, 237–239t
nanomedicine, 311–312
nanoparticle, 67t, 75t
nanoparticle aggregates (NPA), 172f
nanoparticle attached with opsonin, schematic representation, 363f
nanoparticle interactions at cellular and molecular levels, 383–387
nanoparticle physiochemical properties, 391f

nanoparticle-based drug delivery systems, 24
nanoparticle-mediated lung targeting, 243–245
nanoparticles, 2–3, 136–139, 172f
nanoparticles, attributes of, 50
nanoparticles, charge or electrostatic stabilization, 360
nanoparticles, different administration routes of: brain drug delivery and nanotechnology, challenges to, 56–63; brain drug delivery and role of nanoparticles, 56; brain drug delivery of nanoparticles, pharmacokinetics of, 63–64; challenges to oral delivery and nanotechnology, 64–70; inhalation delivery of nanodrugs, 72–78; oral delivery of the role of nanoparticles, 64; pharmacokinetics of oral nanoparticles, 70–72; pharmacokinetics of transdermal nanoparticles, 55–56; transdermal delivery and nanotechnology, challenges to, 54–55; transdermal delivery and role of nanoparticles, 54
nanoparticles, factors influencing the stability of: aggregation, 359; sedimentation, 359–360
nanoparticles, oral delivery of, 64–72
nanoparticles, preparation of through double emulsification method, 4f
nanoparticles, regulatory aspects of, 382–383
nanoparticles, stabilization mechanisms: charge of electrostatic stabilization, 360; steric stabilization, 361
nanoparticles and nanotoxicity, 382
nanoparticles for combination therapy: chemotherapy agents combined with multidrug resistance inhibitors, 316–317; combination based on photothermal/photodynamic therapy, 318–319; delivery of nucleic acids and chemotherapy agents combined with nucleic acid, 317–318; hybrid silica nanoparticles, 319–320; introduction to, 315–316
nanoparticles for nose-to-brain delivery with active pharmaceutical ingredient, examples of, 273t
nanoparticles surface chemistry: PEGylation, 362–366; stabilizers, 361–362; surfactants, 366–370
nanoparticles uptake in intestinal epithelium, schematic of, 71f
nanoplates, 136
nanoplatinum, 140–141
nanopore devices, 12–13
nanopores, 2, 12–13
nanoporous activated carbon, 136

nanoporous structures, 136
nanoporous-activated carbon, 148
nanoprecipitation, 3, 31–32
nanoprecipitation process, schematic representation, 31f
nanoribbons, 136
nano-robotics, 3
nanorods, 136, 145
nanosheets, 136
nanoshells, 2, 12–13
nanosilver, 140
nanosized amorphous particles, 53
nanospheres, 2–3, 25
nanostructured lipid carriers (NLC), 34–35, 50–52, 57t–59t, 258
nanosuspension, 73t
nanosystems, 60t, 76t
nanosystems, classifications of: dendrimers, 52–53; engineered nanoparticles, 53–54; lipid nanosystems, 50–52; micelles, 53; polymeric NPs (PNPs), 52
nanosystems, specific pharmacokinetic properties of, 51
nanotechnology, 1–2
nanotherapeutics-based pulmonary delivery, 72–78
nanotoxicity *see* nanotoxicology and regulatory aspects of nanomaterials and nanomedicines
nanotoxicological assessment methods, advantages and limitations of, 396t
nanotoxicology and regulatory aspects of nanomaterials and nanomedicines: nanoparticle interactions at cellular and molecular levels, 383–387; nanoparticles and nanotoxicity, 382; nanotoxicology assessment methods, 394–397; NP toxicity determinants, 391–394; proposed mechanisms of NPs toxicity, 387–391; regulatory aspects of nanoparticles, 382–383
nanotoxicology assessment methods: ex vivo methods, 395; in silico methods, 395; in vitro methods, 394–395; in vivo methods, 395
nanotubes, 2, 13, 136, 144f, 147, 216, 271t; *see also* carbon nanotubes (CNTs)
nanowires, 136–137
nasal cavity, anatomy of, 264–265
National Surgical Adjuvant Breast and Bowel Project Protocol B-27, 315
natural organic matter (NOM), 356–357
nebulizers, 233–234
neurodegenerative diseases, 259
neuro-infections, 258

neuroinflammation, 260
neutral lipids, 14
NGO-PEG, 219
NGO-PEG-DOX, 220
nickel, 141
niosome-mediated lung targeting, 247–248
niosomes, 52, 63–64, 271t, 273t
nitroxide mediated polymerization (NMP), 28–29
noncovalent modification, MOFs, 186
nonpolymer-based nanocomposites, 137
nose-to-brain drug delivery, nanostructured drug carriers for: common brain disease and the challenge of treatments, 258–263; introduction to, 257–258; nose-to-brain delivery, limitations from, 266–267; nose-to-brain route, 264–265; NPs for nose-to-brain targeting, 271–277; requirements of nose-to-brain formulations, 267–270
nose-to-brain drug delivery transport percentage (DTP), 272–273
nose-to-brain formulations, requirements for, 267–270
nose-to-brain limitations regarding nasal physiological conditions, drug physical-chemical characteristics, and formulation features, examples of, 267f
nose-to-brain route, 264–265
nose-to-brain route, limitations of, 266–267
Noyes-Whitney equation, 40, 112
nozzle-based printing methods, 123–124, 123f
NP cellular uptake and interactions with different mechanisms of cytotoxicity, 388f
NP properties and possible cytotoxic effects, 392t
NP toxicity determinants: dose-depend toxicity, 393–394; nanoparticle aggregation, 393; nanoparticle chemistry, 393; nanoparticle size and surface area, 391; nanoparticle surface chemistry, 391; NP repeated dose exposure, 393; NP shape, 392–393
NPs for brain delivery, and their advantages, 271t
NPs for nose-to-brain targeting: introduction to, 271–272; lipid-based NP, 272–276; polymer NP-based systems, 276–277
NPs in drug delivery, main advantages of, 41f
NPs toxicity, proposed mechanisms of, 387
nuclear magnetic resonance (NMR) spectroscopy, 373–374
nucleic acid nanoparticles (NANPs), 318
nucleic acids, delivery of, 338
nucleus, 289

O

ocular drug carriers, 14–15
olfactory mucosa pathway (direct pathway), 264
one-dimensional (1D) nanomaterials, 136
oral cancer therapy, 300
oral delivery and nanotechnology, challenges to, 64–70
oral delivery of the role of nanoparticles, 64
oral drug delivery, challenges of, 110–111
oral nanoparticles, pharmacokinetics of, 70–72
oral nanosystems, pharmacokinetics summary for, 73–77t
oral solid dosage forms, incorporation of nanocrystals, 116–117
oral solid dosage forms (OSDFs), 116
oral solid formulations of nanocrystals by 3DP, 121–128
organic poly-dentate ligands, 182
organic-based nanomaterials, 136, 159
Organisation for Economic Co-operation and Development (OECD), 382
organo-gels, 159
organophosphorus poisoning (OP), 273
osteosarcoma, 294
Ostwald ripening reaction, 42, 359
oxidative stress, 389
oxidoreductases, 165

P

paclitaxel (nab-paclitaxel), 316
pancreatic cancer, 298–299
"PAOM", 217–218
Parkinson's disease (PD), 259
partial glycerides, 4
partial wetting technique, 97
particle size of MOFS, 184
particle-assisted replication, 97
passive diffusion, 289
passive targeting, 156–157
pathways of transdermal drug delivery systems, 55f
PCL nanoparticles, 273t
PEG applications, 363–366
PEG-GO, 220
PEGylated graphene oxide, 220–221
PEGylated nanoliposomes with galactose, 14
PEGylated nanoparticles, 364–365
PEGylation of nanoparticles, 91, 201, 218, 362–366
PEGylation on the physiochemical properties of DNA NPs, effects of, 245f
PEI-modified GO, 219–220
peptide absorption, 70
peptide dendrimers, 12
pH, nanoparticles stability, 373
pH gelatin polymers, 201
phagocytosis (cell eating), 384–385
pharmaceutical nanocrystals, 111–113

Index

pharmacokinetics (PK), 78
pharmacokinetics summary for selected transdermal drug delivery nanosystems, 57–62t
phase inversion methods, 36–37
phase inversion methods, schematic representations, 37f
pH-controlled MOFs, 187–188
phospholipid bilayer (PLB), 201
photodynamic therapy (PDT), 220
photodynamic therapy (PDT) using CNT, 299–300
photosensitive nanoparticles, 166–167
photothermal therapy (PTT), 220
photothermal therapy (PTT) using CNT for cancer, 298–299
photothermal therapy/photoacoustic imaging, 27
photothermal/photodynamic therapy, combination based on, 318–319
pH-responsive diclofenac sodium@ZJU-101, 188
pH-responsiveness as endogenous stimuli, 88
pH-sensitive nanonparticulate DDS composed of polybasic polymer, mechanism of, 163f
pH-sensitive nanoparticles, 161–162
pinocytosis, 386
plants and animals, as bioinspired materials, 93
plasmid-incorporated lipoplexes, 14
platinum (Pt), 137
PLGA nanoparticles, 90–91
Pluronic, 367
PNPs, manufacturing techniques, 25f
polo-like kinase (PLK), 294
poly (D, L-lactide-co-glycolide) (PLGA), 3
poly (N-isopropylacrylamide) (PNIPAM), 89
poly (N-vinyl caprolactam), 89
poly (vinyl ether), 89
poly ethylene glycol (PEG), 3
poly lactic-co-glycolic acid (PLGA), 276
polyalkylcyanoacrylates, 52
polyamidoamine (PAMAM) dendrimers, 12, 277
polyanhydrides, 88
polybasic or cationic polymers, 162
polycaprolactone (PCL) NPs, 270
polycarbonates, 88
poly(dimethylsiloxane) (PDMS), 102
poly(e-caprolactone) (PCL), 88
polyethylene glycol, 367
polyethylene glycol diacrylate (PEGDA), 129
poly(ethylene glycol) (PEG), 358
polyethyleneimine (PEI), 219
polyethyleneimine (PEI) dendrimers, 12
polyglycolic acid (PGA), 52, 276
polyketals, 88
poly(lactic acid) (PLA), 52, 88, 276
polylactic-co-glycolic acid (PLGA), 52
polylactide (PLA), 3
poly(lactide-co-glycolide) (PLGA) microspheres, 240–241
polymer brush, 201
polymer mantle on MOFs, 194–196
polymer NP-based systems, 276–277
polymer-based bioinspired nano-formulations, 99
polymer-based nanocomposites, 137, 146
polymer-based NPs, 271
polymeric hydrogels, bioinspired nano-formulations, 96
polymeric micelles, 271, 271t
polymeric nanoparticles, 3, 52, 60t, 65t, 68t, 75t–76t, 258, 271t, 273, 278, 311
polymeric nanosystems, 358
polymeric NPs (PNPs), 24–25, 49
polymerization, 25
polymer-lipid nanoparticle (NP), 66t
polymers, 3
polymersomes, 159
poly(N-isopropyl acrylamide) (PNIPAM), 189
polypropilimine dendrimers, 12
polysilsesquioxane (PSilQ), 319
polyvinylpyrrolidone (PVP), 367
pore size of MOFS, 184
porous coordination polymers (PCPs), 182
porous organic-inorganic hybrid extended networks, 182
porous silica nanoparticles, 312
possible transformation phenomenon of synthesized nanoparticles, 357f
precipitation-chemical cross-linking, 3
preformed polymers, 30–31
pressure-assisted microsyringe (PAM), 123–124
pressurized metered-dose inhalers (PMDIs), 233–234
"printability" of polymeric solutions, evaluation of, 122
printing of nanocrystals using inkjet printing, 123f
printing of nanocrystals using selective laser sintering, 128f
protein aura, 17
protein-based bioinspired nano-formulations, 100–101
protonation or deprotonation, 162
Pseudomonas aeruginosa, 235
pyridinium, 342
pyrrolidinium, 342

Q

quantitative structure–activity relationships (QSAR), 395–397
quantum dots (QDs), 2–3, 13, 53, 136–137

R

rat serum (RSA), 100–101
rationality and biodegradability of MOFS, 184–185
reactive oxygen species (ROS), 318, 358
receptor-mediated transcytosis (RMT) mechanism, 63–64
redox responsiveness as endogenous stimuli, 88–89
redox-sensitive nanoparticles, 161, 164
Registration, Evaluation, Authorization, Characterization, and Hazard-restriction of chemicals (REACH), 383
RES accumulation, 362
Respimat Soft Mist Inhaler (SMI), 233–234
retention effect (EPR), 2
retina, accumulation of nanomaterials, 16
reverse microemulsion, 313–314
reversible addition-fragmentation chain transfer (RAFT) polymerization, 28–29
rifampicin (RIF)-based chitosan (CHT)-coated liposomes, 247

S

salting out, 3, 33
salting technique, PNPs manufacture, flow diagram, 33f
SC barrier, 54
schematic loading of PNIPAM on to UiO-66-PNIPAM, 191f
SDS, 367
SDS surfactants, 367
sedimentation, 356, 359–360
selective laser sintering (SLS), 128–129, 128f
SEM micrographs of freeze-dried (a) plain and (b) Enox-Alb MS of formula F8, 240f
semiconductor nanomaterials, 142
semiconductor quantum dots (QDs), 141–142
Shewanella, 358
sialic acid (SA)-modified microspheres (MS), 242
silica nanoparticles, 311
silica nanoparticles including solid, mesoporous, hollow mesoporous, and hybrid silica nanoparticles, 316f
silica-based nanomaterials, 312–313
silica-based nanoparticles, synthesis and functionalization of, 313–314
silicon, 142
silver (Ag), 137
silver nanoparticles (AgNPs), 358
single-phase continuous flow system (SFCFS), 29–30
single-stimulus-based MOFs for drug delivery, pH-controlled MOFs, 187–188

single-walled CNTs (SWCNT), 143–144
Sirolimus nanocrystals, 116
skin adhesives, 101–102
skin patches, 101
skin/epidermis, 54
small drug molecules, delivery of, 336–338
small interfering RNA (siRNA), 317
small unilamellar vesicles, 8–9
smart polyanionic or polycationic polymers, 161–162
synthetic polymeric materials as bioinspired materials, 93–94
solid lipid nanoparticles (SLNs), 3–5, 49–52, 57t–58t, 61t, 65t–68t, 77t, 258, 271, 271t, 273t
solid lipid NPs (SLN), 34
Solidification-Fusion 3D Printing Process (MESO-PP), 124
solumatrix fine particles, 53
solvent dispersion method, 8–9
solvent displacement, 3
solvent emulsification/evaporations, SNLs preparation, 4
solvent evaporation technique, method of SLNs preparation, 5f
solvent-based methods, 39–40
solvent-based methods, schematic representations, 40f
sophoridine-loaded poly(lactide-co-glycolide) microspheres, 241f
splenocytes, 312
spray drying, 3–4, 115
stability in relevant media, nanoparticles, 373
stability measurements of nanoparticles, 373–374
stabilization mechanisms, 367f
Staphylococcus aureus, 275
stereolithography, 129–131
stereolithography, obtaining solid dosage form with nanocrystals, 130f
steric stabilization, nanoparticles, 361
steroids, 4
stimuli-sensitive nanoparticles, recently designed, 172f
Stöber process, 313
Stoke's law, 357
stroke, 259
supercritical fluid methods, PNPs manufacture, 34f
supercritical fluids technology, 33–34
supercritical solution (RESS), 33–34
surface area, increase due to decreased particle size, 113f
surface ligands functional groups, schematic representation of, 362f
surface modification, MOFS, 185
surfactants used for nanostabilization, 368–370t

Index

SWCNT modified with chitosan, anticancer, 290
SWCNTs in glioma cells, distribution of, 293f
symbolic representation of binate-stimuli receptive DDS due to UMCM-1- NH2 NMOF locked via pillararenes, 192f
synthetic organic systems, 159

T

targeted drug delivery, 156
targeted drug delivery by nanoparticles, modes of, 157f; active targeting, 157–158; passive targeting, 156–157; trigger-sensitive targeting, 158
targeting, active and passive, 8–9
temozolomide (TMZ)-Lactoferrin (Lf) NPs, 52
temperature-sensitive MOFs, 189–191
temperature-sensitive nanoparticles, 167
theranostics, 100–101
thermosensitive hydrogels, 294
Tilmicosin-gelatine microspheres (TMS-GMS), 236–240
titanium dioxide, 53, 137
Tobramycin, 236
Toll-Like Receptor agonist, 301
top-down and bottom-up methods of nanocrystal formulation, 114f
top-down technique, NCs preparation, 41, 115
toxicity, 295f
toxicological compatibility of MOFS, 185
Tpp-modified curcumin, 242
transdermal delivery, bioinspired-based nano-formulations for, 101–102
transdermal delivery and nanotechnology, challenges to, 54–63
transdermal drug delivery (TDD), 54–57
transdermal nanoparticles, pharmacokinetics of, 55–56
transferosomes, 52, 56
transformation processes: biological transformations, 358–359; chemical transformation, 357–358; introduction to, 356–357; physical transformations, 357
transition-metal-catalyzed atom transfer radical polymerization (ATRP), 28–29
transmission electron microscopy (TEM), 374
transresveratrol (RES), 100
trigeminal nerve (direct pathway), 264
trigger classifications, 161
triggering stimuli, trigger-sensitive nanoparticles, 159–161
triggering stimuli, types of, 160f
trigger-sensitive nanoparticle for drug delivery: applications of trigger-sensitive nanoparticulate drug delivery system, 171; dual/multi-responsive DDS, 170–171; endogenous trigger-sensitive nanoparticles, 161–166; exogenous trigger-sensitive nanoparticles, 166–170; introduction to, 156; limitations of trigger-sensitive nanoparticular drug delivery system, 171–173; modes of drug release from trigger-sensitive nanoparticles, 159; modes of targeted drug delivery by nanoparticles, 156–158; trigger-sensitive nanoparticles for drug delivery, 158; types of triggering stimuli, 159–161
trigger-sensitive nanoparticle for drug delivery system, applications of, 171
trigger-sensitive nanoparticle for drug delivery system, limitations of, 171–173
trigger-sensitive nanoparticles for drug delivery, 158
trigger-sensitive nanoparticles, modes of drug release, 160f
trigger-sensitive targeting, 158
triglycerides, 4
triple-negative breast cancer (TNBC), 294, 314
Triton, 367
trypsin, 165
tumors, 13, 15, 184, 220f, 221, 233, 239t, 247, 249, 294, 295f–296f, 299–300, 312, 314–315
tumors on mice after treatments indicated, representative photos, 220
two-dimensional nanomaterials (2D), 136

U

ultrasonication or high-speed homogenization, 39
ultrasonication/high speed homogenization, 4
ultrasound-responsiveness as exogenous stimuli, 91
ultrasound-sensitive nanoparticles, 170
uncurable diseases, search for cure, 50
unilamellar vesicles, liposomes as, 8–9
uptake of MWCNTS by dendritic cells, 302f
uptake of siNEG AF456 in tumor xenografts, 295f
US Nanotechnology Characterization Laboratory (USNCL), 382

V

vaccines, cancer see cancer vaccines
viral CNS infections, 260
viral encephalitis, 260
viral origin, bioinspired materials, 95–96
virosomes, 95
virus-like particles (VLPs), 95
in vitro to in vivo (IVIV) nanotoxicology, 393–394

W

wax, 4
West Nile virus, 260
wet bead milling (WBM), 43, 115
Wilson's disorders, 14
Working Party on Manufactured Nanomaterials (WPMN), 382
World Health Organization, 258

X

x-ray diffraction, 373–374

Z

zero-dimension (0D) nanomaterials, 136
zero-valence iron (Fe0), 141
zigzag (SWCNTs), 143–144
zinc oxide, 53
zinc oxide (ZnO) nanoparticles, 70, 137
Zr-MTX (methotrexate) nanoscale coordination polymeric materials, 184